An Introduction to
Forensic Geoscience

An Introduction to Forensic Geoscience

Elisa Bergslien, PhD
Associate Professor
Buffalo State College
Buffalo, NY, USA

WILEY-BLACKWELL
A John Wiley & Sons, Ltd., Publication

Registered office: John Wiley & Sons, Ltd, The Atrium, Southern Gate, Chichester, West Sussex, PO19 8SQ, UK

Editorial offices: 9600 Garsington Road, Oxford, OX4 2DQ, UK
The Atrium, Southern Gate, Chichester, West Sussex, PO19 8SQ, UK
111 River Street, Hoboken, NJ 07030-5774, USA

For details of our global editorial offices, for customer services and for information about how to apply for permission to reuse the copyright material in this book please see our website at www.wiley.com/wiley-blackwell. The right of the author to be identified as the author of this work has been asserted in accordance with the UK Copyright, Designs and Patents Act 1988.

Library of Congress Cataloging-in-Publication Data

Bergslien, Elisa.
  An introduction to forensic geoscience / Elisa Bergslien. – 2nd ed.
    p. cm.
  Includes index.
    ISBN 978-1-118-22795-4 (hardback) – ISBN 978-1-4051-6054-4 (paper)
    1. Forensic geology. 2. Environmental forensics. I. Title.
  QE38.5.B47 2012
  363.25–dc23
                        2012002695

A catalogue record for this book is available from the British Library.

Set in 11.5/13 pt Plantin by Toppan Best-set Premedia Limited
Printed and bound in Malaysia by Vivar Printing Sdn Bhd

1 2012

Contents

COMPANION WEBSITE
This book has a companion website:
www.wiley.com/go/bergslien/forensicgeoscience
with Figures and Tables from the book

List of Tables and Figures

* indicates tables and figures that are located on the companion website.

Tables

Chapter Two

Chapter Three

Chapter Four

Chapter Five

Chapter Eleven

Figures

Chapter Three

Chapter Four

Chapter Five

Chapter Eleven

List of Color Plates

List of Cases

Note on cases: Several examples of real cases have been included throughout the text to help illustrate the potential forensic uses of geological materials. Some of the case titles in the text are followed by a citation, indicating a published authority that serves as the principal source of information for that case. Other cases are listed without an initial citation. These have been reconstructed by this author from a variety of primary and secondary source materials such as court transcripts, newspaper, magazine, and journal articles, books and other publically available resources. These reconstructions are the best the author can provide, given the information available, and the author apologizes for any mistakes or mischaracterizations that have inadvertently been included in this text.

Preface

The popularity of television shows such as *CSI* and *Cold Case*, not to mention all the cable documentary shows, created a recognizable increase in student interest in the sciences. Inspired by this increased interest in science and with the cooperation of the long-standing Forensic Chemistry program at Buffalo State College, which started in the 1970s, I developed a problem-based learning, forensic geoscience course that was offered for the first time in the spring of 2005. Based on my work in contaminant hydrogeology, I already had an appreciation for forensic geoscience from the civil law side and was very interested in becoming more conversant with criminal law applications as well.

In the United States, most forensic workers have strong backgrounds in chemistry and/or biology, but little or no training in the earth sciences. There are relatively few practicing forensic geoscientists in the United States and the discipline is currently outside the mainstream perception of forensic science. Indeed, it is not even mentioned in the National Academy of Sciences' 2009 report. My course was designed to give the students destined for careers in forensic laboratories a crash course in the basic principles of geology and a hands-on introduction to the many ways that geological materials could be forensically useful.

One of the major obstacles I encountered was a lack of materials to support such a course. The classic text *Forensic Geology* by Raymond C. Murray and John C. F. Tedrow was no longer readily available, Ray Murray's wonderful follow-up, *Evidence from the Earth*, is geared more toward laypeople, and most current forensic texts concentrate on forensic chemistry and the intricacies of DNA. As a result, I ended up writing a laboratory manual/textbook of my own for use in the class. I begged, borrowed, and bought literally dozens of books that addressed specialized topics, like pigment analysis, and procured hundreds of articles in an attempt to synthesize reference materials describing the disparate ways in which geoscience can be applied forensically. I also had the blessing of an immeasurable quantity of help and support from a variety of experts in the field.

At the time, I did not have much thought beyond giving my students the materials they needed to be successful. Then, at the 2006 Geological Society of America's annual meeting, I presented some of the materials I had developed and got a strongly favorable response. Afterward, I was flooded with requests for background case information, suggestions on how to develop forensic exercises, and, most commonly, pointers to materials. Eventually, several people told me that I should make my materials into a book, because there was nothing else like it available. I set about transforming the course materials I had developed into a textbook that is meant to be a detailed introductory exploration of those areas of geology most likely to be of use to a forensic science student. I sincerely hope that this book fulfills the wishes of the many colleagues who have offered support and that it provides both a useful introductory text for undergraduate courses on forensic geoscience and a practical basic reference for the forensic science community.

In the intervening time since I started my course and the completion of this book, there have been other forensic geoscience books published, all of which make significant contributions to the field. However, they generally assume that the reader already has some level of background knowledge or are compilations of papers. This book takes a completely different tack: it was written with the assumption that the reader does not necessarily have anything other than a general background understanding of the natural sciences.

Each geoscience topic is introduced from basic concepts, which are developed with increasing complexity, in order to give a taste of the wide range of possible forensic applications. The chapters contain lists of further reading to appropriate textbooks and journal articles that readers can use as starting points for more detailed study of the geoscience topics presented. A variety of reference tables have been compiled for the text so that this book can serve as a basic reference in a laboratory setting. Applicable case studies are also presented in each chapter, many of which have references to additional information, others of which were developed from primary sources. The goal of this book is to give readers a familiarity with the amazing range of ways in which geosciences principles and geological materials can be used forensically, thus the subject matter presented typically goes into more depth than would a traditional introductory geology textbook. However, it is very important to note that this text is by no means an exhaustive study.

Companion Website: Many of the chapters have additional information available on the companion website at www.wiley.com/go/bergslien/forensicgeoscience.

Cautionary Note: This book is an introductory textbook that focuses on providing training information for students. It does *not* presume to dictate forensic laboratory methodology or prescribe definitive protocols or definitions. Any procedures described were developed primarily for use by students and may not reflect the actual procedures employed by a particular forensic laboratory. This book will hopefully serve as a useful starting point for geologists, attorneys, members of law enforcement agencies, forensic scientists, and, of course, students, but as the myriad disciplines that fall under the general heading of geoscience testify, there are a wide range of possible approaches to questions and a number of different analytical methods that can be used in any given situation. The materials here are meant to help individuals understand what geological materials may potentially prove useful, and help place forensic geoscience into context in the wider field of forensic science. For actual cases situations, experts in appropriate sub-disciplines should be consulted.

Acknowledgments

I would like to express my profound gratitude to Raymond Murray, without whose encouragement this book would never have been written. My thanks to Jack Crelling, whose kindness in offering help as I started teaching forensic geology was of incalculable value. I would also like to thank Alastair Ruffell, Bill Schneck, Erich Junger, and Laurance Donnelly. Other people who provided help are Maureen Bottrell, Nelson Eby, Marianne Stam, and Kevin Williams. Thank you to Darrel Kassahn for his help turning some of my sketches into usable graphics. I am grateful to the forensic geology community, and to anyone I forgot, my apologies and thanks.

I am grateful to Ian Francis, Delia Sandford, and Kelvin Matthews for their patience and to Tim Bettsworth for his clarity. Thanks to my anonymous reviewers for their comments. I hope that in the future I will have the opportunity to add several new cases and the additional chapters suggested.

Finally, I have to thank my family for putting up with this project. I dedicate this book to my children, Nathan and Leta, with all my love.

Chapter 1
A Brief History of Forensic Science and Crime Scene Basics

The word *forensic* in this context means "the application of scientific methods and techniques to the investigation of crime" and encompasses a wide range of endeavors, from gathering and analyzing evidence to offering expert testimony in a court of law (Forensic, 2001). Forensic specialists exist in an almost endless variety of scientific disciplines, including anthropology, biology, entomology, chemistry, serology, psychology, and, of course, geology. An associated term is *criminalistics*, which is the "application of scientific techniques in collecting and analyzing physical evidence in criminal cases" (Criminalistics, 2011).

There is no consensus as to exactly when science first entered the realm of law enforcement, though it was certainly in use in some areas of the world long before it became a recognized field of study. In the Western world, there were initially large social barriers between the shady world of the Bow Street Runners (London's first professional police force, founded around 1749) and the rarified ivory towers of the gentleman scientist. Most early scientists were independently wealthy and often of high social rank. The early history of law enforcement, on the other hand, was unfortunately rife with tales of corruption, incompetence, and even murder. It is no wonder that the scientists of the time would have considered it well beneath them to even speak to a "copper," much less work with or, goodness forbid, for them.

The birth of the modern science of geology is usually linked to the 1785 presentation of a paper entitled *Theory of the Earth* by Scotsman **James Hutton** (1726–1797) to the Royal Society of Edinburgh. Based on years of observations of geological processes in action and the layering of exposed rock (stratigraphy), Hutton explained that the Earth must be much older than previously thought.

An Introduction to Forensic Geoscience, First Edition. Elisa Bergslien.
© 2012 Elisa Bergslien. Published 2012 by Blackwell Publishing Ltd.

He hypothesized that there had to be several cycles of deposition, uplift, deformation and erosion, in order to form the sequence of rock layers exposed, and each of these cycles must have occurred slowly, as demonstrated by the geological processes in action today; therefore, the Earth must have an extremely long history. In 1795, Hutton published his two-volume *Theory of the Earth*, expanding on his previous work and presenting one of the fundamental principles of geology: *uniformitarianism*. Uniformitarianism is the concept that the geological processes at work today, shaping the Earth, are the same processes that have been active throughout geologic history. This concept is often referred to using the phrase "the present is the key to the past." Thus, the surface of the Earth has not been shaped by random, unknowable events but, for the most part, by processes that we can see in action right now. This concept also establishes a link between geologists and the world of forensics, where scientists in both are using clues from the past to work out a sequence of events to determine what happened.

One of the first criminal cases to mention geologic evidence occurred in 1786 in Kirkcudbright, Scotland. A couple returned home to find their adult daughter lying dead on the floor, her throat slashed. It was established that she did not commit suicide and that her attacker was left-handed. The only other clues to the identity of the murderer were footprints found in boggy ground near the family's cottage. Plaster casts were made of the footprints, and the boots of all the men who attended the young woman's funeral were examined. None matched. Eventually, the authorities did find a boot that matched the plaster casts. It belonged to a laborer named Richardson. He was left-handed and had several scratches on his face.

Initially, Richardson appeared to have an alibi, since he had been at work with two other men that day. Additional questioning, however, revealed that Richardson left the other men to go to the blacksmith's and that he had been gone for much longer than expected. When he returned, Richardson had scratches on his cheek and muddy feet. During a search of his cottage, investigators found stockings that were bloodstained and covered in mud that was identical to the mud near the farm cottage of the victim. Apparently, the mud contained a significant amount of sand and was unlike the soils found elsewhere in the area. It turned out that the young woman was pregnant and that Richardson was her lover. He was found guilty of her murder and confessed before execution. While this case is more famous for its precedent-setting use of plaster casts, forensic geology also played an important supporting role.

In 1810, **Eugene Francois Vidocq** (1775–1857) was appointed the head of the new French *Brigade de la Sûreté*, or Sûreté for short, the world's first plain-clothes investigative police agency (Figure 1.1). Vidocq, a former criminal himself, is credited with creating the first police files (a card-index system where the physical appearance of apprehended criminals was recorded), being a pioneer in the field now known as *criminology* (the study of criminals and crime as a social phenomenon) and with introducing the science of ballistics into police work. He is also credited with being the first to make plaster casts of foot and shoe impressions, though that is apparently not quite true, and he was a master of disguise and surveillance. Vidocq recognized that sometimes an expert from outside law enforcement might be of some assistance in solving crimes. Vidocq served as the inspiration for some of the famous detectives in literature, including Edgar Allan Poe's French sleuth Auguste Dupin.

Figure 1.1 Portrait of Eugène François Vidocq by Achille Devéria.

An important development for both geology and forensic science occurred in 1828, when **William Nicol** (1770–1851), a Scottish physicist/mineralogist and lecturer at the University of Edinburgh, invented the polarizing light microscope, one of the key tools for identification of geological materials (Figure 1.2). He developed the *Nicol prism*, using a rhombohedron of "Iceland spar," which is a very clear, well-formed calcite crystal, first by bisecting the crystal in a plane passing through its obtuse angles (those > 90°) (see ABCD on Figure 1.3a) and then cementing the halves back together again with Canada balsam, a transparent resin. When light (S) enters the resultant structure, it is split into two polarized rays (light waves in which the vibrations occur in a single plane). One of these rays (O ray) undergoes total internal reflection at the balsam interface and is reflected to the side of the prism. The other ray (E ray) is not reflected at the interface and leaves through the second half of the prism as plane polarized light (Figure 1.3b). Nicol also developed a method for preparing geologic samples by cementing the sample to a glass slide and then grinding the rock down until it was thin enough to see through so that the inner structures of geological materials could be examined. While Nicol was not involved in any criminal cases, the descendants of his microscope are found in crime laboratories around the world.

Dr Mathieu Joseph Bonaventure Orfila (1787–1853) is considered the father of forensic toxicology. A Spanish-born scientist, he became a professor of medical jurisprudence (1819) and of chemistry (1823), in France, and in 1813 he published the first scientific book on everything known about poisons at the time, *Traité des poisons tirés des règnes mineral, végétal et animal, ou toxicologie générale: Considérée sous les rapports de la physiologie, de la pathologie et de la médedine légale* (often called *A Treatise of General Toxicology*). It included information about methods of detecting

Figure 1.2 Swift and Son Petrological microscope with Nicols prism attachment.
Source: 1882 copy of *Practical Micropscopy* by George Davis.

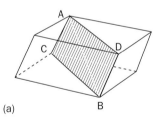

(a)

Figure 1.3a Plane of bisection through a crystal of Iceland spar.
Source: 1882 copy of *Practical Micropscopy* by George Davis.

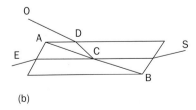

(b)

Figure 1.3b Path of light rays through a Nicol prism.
Source: 1882 copy of *Practical Micropscopy* by George Davis.

poison, but he found the methods in use at the time to be unreliable. Much of Orfila's fame came from his work to refine detection methods in order to achieve greater accuracy. However, in the case of arsenic, another scientist got there first.

James Marsh (1794–1846), an Englishman, invented a process that could detect the presence of the gas arsine, which is produced when arsenic is heated. The method of arsenic detection that existed at the time, called the *Rose method*, after its inventor, was complex and somewhat unreliable. Marsh decided to improve the method and to make it more demonstrative so that a jury would be able to understand the results. He placed the suspected samples into a closed flask, dissolving them in a solution of arsenic-free zinc and weak sulfuric acid. If arsenic is present, arsine gas forms and is led through a long drying tube to a glass tube in which the gas is heated (Figure 1.4). Heated arsine gas decomposes into arsenic and hydrogen gas, which is ignited at the end of the tube. Arsenic is deposited as a "mirror" just beyond the heated area, on any cold surface held in the burning gas emanating from the jet. A black mirror of arsenic is formed when a glazed porcelain dish is held in the flame (air deprivation). A white mirror of arsenic is formed on a black plate held over the flame (excess air). A more refined form of the Marsh test is still used today. This method is so sensitive that it can be used to detect minute amounts of arsenic in foods or stomach contents. The case that made the Marsh test, and Dr Orfila, known to the public was that of Marie Lafarge, in 1840.

Marie Lafarge was a pretty, young widow of 24, whose husband, Charles Lafarge, had apparently died of cholera (Figure 1.5). However, Marie was known to be unhappy with her arranged marriage, and Dr Lespinasse, the attending physician, told Charles' mother that he suspected her son had been poisoned. The doctor interviewed the servants and was told that Marie had been observed

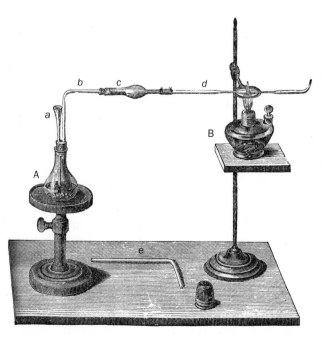

Figure 1.4 Apparatus for the application of Marsh's arsenic test.
Source: 1867 copy of *Micro-chemistry of Poisons* by Theodore G. Wormeley, MD.

Figure 1.5 Image of Marie Lafarge(?).
Source: Frontispiece of the 1841 translation of
the *Memoirs of Madame Lafarge*, published by
Carey & Hart.

sprinkling a white powder over Charles' meals, including a cup of eggnog she had
given him just hours before his death. Marie had purchased large amounts of
arsenic in the months preceding her husband's death, ostensibly for killing rats, a
common enough practice at the time. Dr Lespinasse directed the police to seize the
glass Charles had been drinking from and to transport his body for autopsy. A local
pharmacist tested the eggnog and found arsenic. There was clearly a large body of
circumstantial evidence against Marie; however, the prosecution would also need to
prove that there was arsenic inside of Charles.

Marie Lafarge's trial was reported worldwide by sensationalist newspaper articles.
By all accounts, Charles had been a disagreeable character who owned a rat-
infested forge on the brink of bankruptcy, and Marie had been forced unwillingly
into the marriage. Because the experts brought in by the prosecution were unable
to demonstrate the presence of arsenic in Charles' stomach contents, the public
was divided over her guilt or innocence. Midway through the trial, Mathieu Orfila
was brought in by the prosecution and asked to test for arsenic using Marsh's
method. Orfila examined the work done by the previous experts and found that
they had performed the tests incorrectly. Not only did he establish the presence of
arsenic in Charles Lafarge's body, he also proved that it had not originated from
the soil surrounding Lafarge's coffin. Arsenic, like formaldehyde, was commonly

used to preserve bodies, and many old cemeteries have high levels of arsenic in the soil even now. Based on Orfila's results, Marie was found guilty of murder and spent the rest of her life in prison.

What might be the first use of a microscope in a murder case occurred during the Praslin Affair in France in 1847. This story was so sensationalized at the time that it is hard to separate fact from invention, and the storyline was further distorted by the fictitious version presented in Rachel Field's 1938 book *All This, And Heaven Too*, which was made into a movie starring Bette Davis in 1940. An outline of the tale is roughly as follows.

Charles Laure Hugues Théobald, the Duc de Choiseul-Praslin, married Altarice Rosalba Fanny Sébastiani, in 1824, when he was 19 and she only 16. Fanny was the independently wealthy daughter of one of Napoleon's generals, and she bore Charles nine children. Initially, the marriage was reported to be a happy one, though as the years passed there was greater and greater discord. One point of conflict included a governess named Henriette Deluzy-Desportes, who was eventually dismissed without a reference at the duchess's insistence. In alternate versions of the story, the duchess was upset either because Henriette was her husband's mistress or because Henriette came between the duchess and her children. Whatever the case, Charles continued to call on Henriette even after she had been dismissed and the whole situation was the subject of much gossip in Parisian society.

In some versions of the story, the duchess declared, in June 1847, that she was going to seek a divorce, a scandalous and expensive endeavor that could separate the duke from his children. Alternately, the duke was attempting to force his wife to write a letter of recommendation for Henriette so that she could secure a post at a nearby school. Whatever the reason, things shortly came to boiling point.

On the morning of August 18, servants heard screams and the bell to the duchess's bedroom beginning to ring. The sounds of crashing and more screams were heard through the door, which was locked or wedged shut. Some of the servants ran outside and looked up at the windows to the bedroom. At one point, the shutter opened and they could see a man that looked like the duke. Believing that the duke was fighting off intruders, the servants rushed back to the bedroom to find the door open and the duchess dead. Her throat had been cut, her face battered and she might have been stabbed, though she was not hacked to death or chopped up with a saber.

At this point, the duke walked into the room and screamed. He said that he had only just been awoken by the noise. The police were summoned and an investigation headed by M. Allard, Vidocq's successor, began. The duchess's jewelry was untouched, which indicated that burglary could not have been the motive. Under a sofa in the room, they found a pistol, which turned out to belong to the duke, covered in blood. At this point, the duke's story apparently changed, and he claimed that he heard his wife's calls for help, rushed to her room and found her covered in blood. He dropped the pistol when he picked his wife up and, seeing she was dead, he went back to his room to wash. The inspectors followed a trail of blood to the duke's room, finding within a bloodstained handkerchief, a bloodstained dagger hilt, and a piece of bloodstained bell-pull rope. The severed end of the bell-pull was found under the duchess's body and the duke was arrested for murder.

There were several lines of evidence gathered against the duke, but the one of interest here began with the question of whether you could tell that the pistol had been dropped in a pool of blood or if it had been used to beat the duchess. The esteemed pathologist **Auguste Ambroise Tardieu** (1818–1879) examined the pistol under a microscope and discovered chestnut hair and fragments of skin tissue both near the butt of the pistol and near the trigger. His microscopic investigation confirmed that the duchess had been battered with the pistol. Based on this and other evidence, the duke was clearly linked to his wife's murder. Before being taken to prison, the duke swallowed poison and died three days later, still proclaiming his innocence. Several newspapers of the day, though, suggested that the duke's death was a masquerade and that his connections to the royal family allowed him to be spirited safely away.

While the Marie Lafarge trial brought the public its first experience with forensic chemistry, a fictional character is most responsible for popularizing the idea of using science to solve crime. Sherlock Holmes, a character created by **Sir Arthur Conan Doyle** (1859–1930) was introduced in *Beeton's Christmas Annual* in 1887 with *A Study in Scarlet*. The literary detective was a composite of Conan Doyle's medial school professors and based chiefly on a Dr Joe Bell. The exploits of Eugene Vidocq probably also played a role as well, though this was apparently denied by Conan Doyle. Published from 1887 into the 1920s, the coldly logical Holmes and his companion Dr Watson solved seemingly inexplicable crimes using observation, deductive reasoning, and scientific experimentation.

> "For example, observation shows me that you have been to the Wigmore Street Post-Office this morning, but deduction lets me know that when there you dispatched a telegram."
>
> "Right!" said I. "Right on both points! But I confess that I don't see how you arrived at it. It was a sudden impulse upon my part, and I have mentioned it to no one."
>
> "It is simplicity itself," he remarked, chuckling at my surprise – "so absurdly simple that an explanation is superfluous; and yet it may serve to define the limits of observation and of deduction. Observation tells me that you have a little reddish mould adhering to your instep. Just opposite the Wigmore Street Office they have taken up the pavement and thrown up some earth, which lies in such a way that it is difficult to avoid treading in it in entering. The earth is of this peculiar reddish tint which is found, as far as I know, nowhere else in the neighbourhood. So much is observation. The rest is deduction."
>
> "How, then, did you deduce the telegram?"
>
> "Why, of course I knew that you had not written a letter, since I sat opposite to you all morning. I see also in your open desk there that you have a sheet of stamps and a thick bundle of postcards. What could you go into the post-office for, then, but to send a wire? Eliminate all other factors, and the one which remains must be the truth."
>
> (From *The Sign of Four* by Arthur Conan Doyle, published 1890)

Sir Arthur Conan Doyle actually became involved in a few real criminal cases, and though he achieved some positive results he was profoundly angered by the corruption and racism he found in the judicial system. Conan Doyle was a firm believer in overturning false convictions, and his work was partially responsible for the establishment of the Court of Criminal Appeal. In one case in 1906, he applied forensic geology, along with many other lines of evidence, to argue that an English solicitor, George Edalji, accused and convicted of mutilating farm animals, was not guilty. The soil on the shoes that the solicitor wore on the day of the crime was a black mud, quite dissimilar to the yellow sandy clays of the crime scene.

Despite Doyle's evidence, Edalji ended up with only a partial clearing of his name. He was found innocent of cattle mutilation, but still considered guilty of writing anonymous letters about the crimes. The commissioners appointed to consider the case decided that because Edalji might have "brought his troubles on himself" he should be awarded no compensation for the three years he served in prison. Conan Doyle called the whole affair a blot on the record of English justice. "What confronts you," he wrote, "is a determination to admit nothing which inculpates another official, as to the idea of punishing officials for offences which have caused misery to helpless victims, it never comes within their horizons" (Womack and Hines, 2001).

Johann (Hans) Baptist Gustav Gross (1847–1915) is widely regarded as one of the founders of modern criminalistics and even credited with coining the term *criminalistics*. A criminalist, according to Gross, was someone who studies crime, criminals, and the scientific methods of their identification, apprehension, and prosecution. He was a public prosecutor and judge in Graz, Austria and in 1893 published the first treatise describing the application of scientific disciplines to the field of criminal investigation, *Handbuch für Untersuchungsrichter als System der Kriminalistik* (published in English in 1907 under the title *Criminal Investigation*) a groundbreaking text that was used all over the world. It detailed the types of assistance that could be expected from the fields of microscopy, chemistry, forensic medicine, toxicology, mineralogy, botany, serology, ballistics, anthropometry, and fingerprinting. Gross wrote in his *Handbook* that "dirt on shoes can often tell us more about where the wearer of those shoes had last been than toilsome inquiries" (Murray and Tedrow, 1992: 3).

Georg Popp (1867–1928), a forensic scientist who ran a laboratory in Frankfurt, Germany, may be one of the first scientists to use geologic evidence in a criminal case. In October 1904, a seamstress named Eva Disch was found murdered in a bean field. At the crime scene, a dirty handkerchief was found that contained bits of coal, snuff, and grains of the mineral hornblende. This information was used to locate a suspect who worked part-time at both a coal-burning gasworks and in a quarry where the mineral hornblende was abundant. He also used snuff. In his pants' cuffs were two layers of dirt. The lower layer matched the soil at the crime scene, while the upper layer, which contained crushed grains of the mineral mica, matched the soil found on the path between the crime scene and the suspect's home. Confronted with this and other evidence, Karl Laubach confessed.

In 1908, Popp again highlighted the utility of forensic geology with the Margarethe Filbert case. The decapitated body of a woman, eventually identified as Margarethe Filbert, was found in a state forest near Rockenhausen, Bavaria. The body had clearly been dragged to the location where it was found, and investigators were initially at a loss for a motive or suspects. It was unclear whether the attack was sexual or a robbery. The victim's skirt and petticoat were thrown back over her head and her left glove had been removed. Eventually, witnesses placed the owner of an adjacent field near the site on the day of the crime. The suspect claimed to have been inspecting his other fields that day, far away from the crime scene, and his wife supported him. The situation reached an impasse. The local district attorney, Sohn, aware of the Eva Disch case sought out Popp's assistance.

Georg Popp studied the available evidence, including soil encrusted on the dress shoes of the suspect, a farmer named Andreas Schlicher. It had been established that Schlicher's wife had cleaned his dress shoes the night before the murder and that he had only worn them once, on the day of the murder. Schlicher claimed that he was nowhere near where the crime took place, so Popp collected soil samples from various locations around the crime scene and the suspect's land for study. Assisted by a geologist named Fisher, Popp found that the ground around the suspect's home was covered with green goose droppings, and the suspect's fields had a soil that contained fragments of porphyry, milky quartz, and mica, while the soil around the crime scene contained decomposed red sandstone, angular quartz, and iron-rich red clay. He also took samples at a castle where evidence linked to the crime had been found. There the soil contained coal, lots of brick dust, and broken pieces of cement from the crumbling castle walls.

The area in front of the heel on the suspect's dress shoes was thickly caked with mud, and Popp applied a variant of the geologic *law of superposition* (though he probably didn't think of it that way) by reasoning that, since the shoes were only worn for one day, the layers of soil in the mud would correspond in sequence to the places that Schlicher visited on the day of the murder. By carefully removing the layers of soil, Popp uncovered the following stratigraphic layers: first a layer of goose dropping directly in contact with the shoe, followed by a layer containing fragments of red sandstone. Next came a layer that contained coal, brick dust and cement fragments. Clearly, the suspect had not been walking in his own fields, as no porphyry, or milky quartz, was found. Just as clearly, the suspect had been at the scene of the crime, and then hidden evidence at the castle. Helping to clinch matters, reddish-brown wool fibers, like those from the dress of the victim, were found in the soil layer with the decomposed sandstone.

Schlicher was found guilty and apparently confessed after sentencing. Based on her clothing, Schlicher had assumed that Margarethe was rich and had decided to rob her. In reality, she was almost penniless and, in a rage, Schlicher cut off her head and hid it. It was finally recovered after his confession. Following this success, Georg Popp went on to make other contributions to the field of forensic science, including expanding the use of botanical identification in forensic investigation.

Edmond Locard (1877–1966) also put the ideas of Hans Gross and Sherlock Holmes into practice. A great fan of both, Locard stated, "I hold that a police expert, or an examining magistrate, would not find it a waste of time to read Doyle's novels . . . If, in the police laboratory at Lyons, we are interested in any unusual way in this problem of dust, it is because of having absorbed ideas found in Gross and Conan Doyle" (Locard, 1929: 277). Born in 1877, Dr Locard studied medicine and law at Lyons, becoming the assistant of the pioneer criminologist Alexandre Lacassagne, professor of Forensic Medicine at the University at Lyons. In 1910, Locard established the first police crime laboratory with the Sûreté (Figure 1.6). Armed with only two rooms in an attic, two assistants, a microscope, and a rudimentary spectrometer, Locard soon became renowned to forensic scientists and criminal investigators throughout the world for his pioneering work. In 1920, Locard published *L'enquête Criminelle et les Méthodes Scientifiques*, a seven-volume work in which appears a passage that may have given rise to the forensic precept that "every contact leaves a trace," commonly referred to as *Locard's Exchange Principle*:

> Whenever two objects come into contact, there is always a transfer of material. The methods of detection may not be sensitive enough to demonstrate this, or the decay rate may be so rapid that all evidence of transfer has vanished after a given time. Nonetheless, the transfer has taken place. (Murray and Tedrow, 1992: 7)

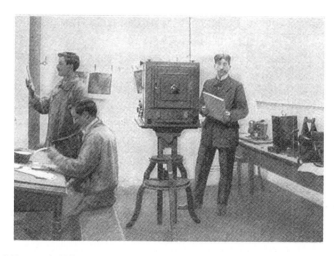

Figure 1.6 Edmond Locard, Director of the Laboratory of Police Technique at Lyon, France, next to an enlarging camera used to detect forged signatures.

One example of his work is the Emile Gourbin case. One morning in 1912, Marie Latelle was found strangled to death in the parlor of her parents' villa. Emile Gourbin, a bank clerk in Lyon, France, and Marie's fiancé, came immediately under suspicion of the murder. He was known to be jealous of Marie's flirting. Gourbin was arrested but produced what appeared to be a very strong alibi. According to the doctors, Marie died around midnight, but at that time Gourbin was at a friend's house playing cards. According to his friends, who lived miles away from the scene of the crime, they shared dinner with Gourbin, consumed a great deal of wine, and played cards into the night. They all went to bed around 1 o'clock in the morning.

Locard was called in and, after examining the scratches on the victim's neck, he took scrapings from underneath Gourbin's fingernails. Examining the scrapings under a microscope, Locard noticed a pink dust in the scrapings that turned out to be a powder consisting of rice starch, bismuth, magnesium stearate, zinc oxide, and an iron oxide used as a red pigment, commonly called Venetian red. Essentially, the skin cells collected from under Gourbin's fingernails were covered in pink face powder. Locard asked the police to search Marie's room for face powder. The box they recovered had powder of the same composition as that from the scrapings, and it turned out that the face powder had been custom-made for Marie by a Lyons druggist. This evidence led to Gourbin's confession. He had tricked his friends into thinking they went to bed at 1 a.m. when in reality Gourbin had gotten them drunk and turned their clock ahead. According to his confession, Gourbin flew into a rage after Marie refused to marry him. The jury found Gourbin guilty of premeditated murder, mostly because he had moved the clock ahead in order to sneak out to meet Marie.

An important American pioneer was **August Vollmer** (1876–1955) who, in 1905, was elected town marshal of Berkeley, California and, in 1909, was appointed as the City's first Chief of Police. It was the start of a highly distinguished career. Vollmer should probably be credited as the man who contributed the most to the professionalization of the American police force and for promoting the application of scientific principles to police work in the United States. In 1907, he established a police school within his department, which included instruction from university professors on subjects such as evidence procedures and applications of the law. It was the first school of its kind in the world. Based on the positive results from the departmental police school, in 1916 Chief Vollmer was instrumental in establishing the first School of Criminology at the University of California at Berkeley.

The same year, **Albert Schneider**, a professor in the College of Pharmacy at the University of California and part-time criminologist for the Berkeley police department, was apparently the first to use a vacuum apparatus to collect trace evidence. He also published articles on police use of the microscope. Through a series of steps, Vollmer and Schneider, among others, created the first police crime laboratory in the United States.

Edward Oscar Heinrich (1881–1953) is the most famous of the faculty associated with Vollmer's Cop College. Known as the Wizard of Berkeley, Heinrich is a remarkable figure in the history of the American justice system and credited with solving more than two thousand crimes. Also called the American Sherlock Holmes, an accolade he apparently disliked, Heinrich did remarkable work with trace evidence, including geological materials such as sand, soil, and paint pigments. Heinrich graduated from UC Berkeley with a degree in chemistry in 1908 and held a series of different jobs, including serving as the Chief of Police for the city of Alameda, California, before being appointed to the faculty at the University of California at Berkeley and opening a private laboratory in 1919.

One of his more famous cases was known as the Flapjack Murder. On the night of August 2, 1921, a stranger, dressed in a heavy overcoat with the collar turned up and wearing driving goggles, drove up to the Holy Angels Catholic Church. He rang the doorbell of the parish house and asked to speak with Father Patrick Heslin. The stranger explained that his friend (or relative) was dying and had requested the last rites. Father Heslin grabbed his religious paraphernalia and went into the night with the stranger.

When Father Heslin failed to return by the next morning, his housekeeper contacted the archdiocese. Shortly thereafter, the archdiocese received a ransom note demanding $6,500 for the safe release of the priest. The note said, "You will get instructions [for the delivery of the ransom] . . . about nine o'clock, perhaps tonight. Had-to-Hitt Him four Times And He is unconscious from pressure on Brain So Better Hurry and no fooling. Tonight at 9 o'clock" and warned against contacting the police. The second letter failed to arrive and the archdiocese contacted the police on August 5. A vast manhunt ensued, with no sign of Father Heslin. Edward Heinrich examined the ransom note and announced that the flowery style of writing meant that the letter's author was a baker and decorator of cakes. Apparently, this announcement was taken with a pinch of salt.

A week passed and Archbishop Hanna offered a reward for information leading to the missing priest. One of the people who showed up to claim the reward was William Hightower, a former baker, who claimed to have information about the location of the body. He told a somewhat confused story about how he found out about the body. Then he led a party of police and journalists toward a sign for Albers Flapjack Flour on Salada Beach. Hightower took the party directly to the body, which was buried only a couple of feet down in the sand. The priest had been shot twice and his skull crushed. In and around the impromptu grave, investigators also recovered some white cord, boards from a tent floor, a tent peg, and .45-caliber cartridges.

There were several lines of evidence linking Hightower to the murder of Father Heslin. For example, Heinrich examined Hightower's jackknife and found threads that appeared to match the white cord found with the body. He also recovered sand from several of Hightower's possessions that was similar to the sand where the body had been buried. Based on this and other evidence, Hightower was convicted and sentenced to life imprisonment in San Quentin. He was released in 1965 at age 86.

Edward O. Heinrich's most spectacular triumph was probably the D'Autremont attempted train robbery. On October 11, 1923, a Southern Pacific express train going southbound through the Siskiyou tunnel near the border between Oregon and California was stopped by three armed men. They ordered the engineer to halt the train at the far end of the tunnel, leaving the passenger cars in darkness. Shortly afterward, there was a large explosion as the bandits attempted to open the mail car. They used too much explosive, causing the entire car to burst into flames, killing the mail clerk inside. Upon hearing the explosion, the train's brakeman ran forward, only to be gunned down. At some point in these bungled proceedings, the bandits also shot the engineer and the fireman to death. The fire prevented the robbers from getting any loot and they fled the scene empty-handed. The sound of the explosion quickly brought rescue crews to the scene. After learning what happened, the rescue crews promptly turned into a posse that combed the area.

After a protracted search failed to turn up any sign of the bandits, the evidence collected from the scene was passed on to Heinrich. He examined a pair of overalls that had been recovered and announced that "the man who wore these overalls is a left-handed, brown-haired lumberjack not more than 25 years old, about 5 feet 8 inches tall, thickset, clean-shaven; he has recently worked in lumber camps in northwest Oregon or Washington." It was a seemingly amazing announcement, but Heinrich proceeded to explain the source of each of these pieces of information. For example, he found a couple of hairs on the overalls and, using a microscope, was able to discern both the hair color and approximate age of the suspect. Because the pockets on the overalls were more worn on the left side, he knew the suspect was left-handed. By examining spots of sap, wood chipping, and needles from the pockets, Heinrich knew the suspect was a lumberjack who had recently been cutting fir trees.

In addition to constructing a description of one of the suspects, Heinrich found one other clue that previous investigators had missed. Buried deep in one of the pockets was a small piece of paper that turned out to be a registered-mail receipt. He was able to tease the receipt number from the paper, and the Post Office traced it to one Roy D'Autremont of Eugene, Oregon. Roy fit the description Heinrich provided and had been missing since October 11, the day of the holdup. His two brothers were also missing and a massive manhunt ensued. It took almost four years for the D'Autremont brothers to be located and brought to justice.

During the 1930s, several important forensic laboratories were established and the application of scientific methods to investigate crime was becoming commonplace. In 1932, under the direction of J. Edgar Hoover, the **Federal Bureau of Investigation** organized a national laboratory intended to offer forensic services to all law enforcement agencies in the country (Figure 1.7). Its official birthday was set as November 24, 1932; the date was arbitrarily decided because the founding of the lab took place over several months. Forensic geology was in use at the laboratory as early as 1935 and by early 1939 heavy mineral separations and mineral identifications were standard practices for cases involving soils. There is an informative article on the history of the FBI laboratory on its website (http://www.fbi.gov/about-us/lab/forensic-science-communications/fsc/oct2007/research/2007_10_research01_test4.htm). On April 10, 1935, the **Metropolitan**

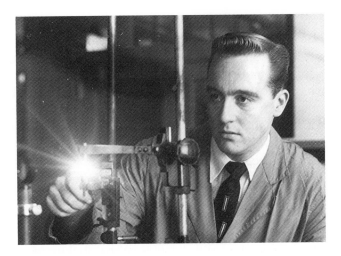

Figure 1.7 Federal Bureau of Investigation laboratory scientist.
Source: Photograph courtesy of the FBI.

Police Forensic Laboratory opened at the Police College, Hendon, United Kingdom.

There have been several modern pioneers in forensic science; too many to include here. However, there are a few people worth a special mention. **Walter McCrone** (1916–2002) was a world-famous chemical microscopist, applying microscopy to a wide range of analytical problems, including forensic science. He published over 600 articles and was editor and publisher of the professional journal *The Microscope*. He is perhaps most famous for his work on the Shroud of Turin and the Vinland Map. McCrone taught microscopy and materials science beginning in 1944 at what is now the Illinois Institute of Technology. In 1956, he left to set up his own research company, McCrone Associates, in Chicago and developed the multi-volume *McCrone Particle Atlas*. His most lasting legacy is found in the skills of the literally thousands of forensic scientists throughout the world whom he trained.

In 1975, **Ray Murray** and **John Tedrow** published *Forensic Geology: Earth sciences and criminal investigation*, the first textbook in the field. The text outlines principles for the collection, examination, and evaluation of evidence, and discusses the conclusions that can be drawn for presentation in a court of law. The approach presented in this text still serves today and the book is found on the reference shelves of a number of crime laboratories.

Now that you have an understanding of the history of forensic geology, we can look at the procedures used at a modern...

Scene of the Crime

Arriving at the scene of a horrendous crime, the crime scene investigator surveys his surroundings. A brutal murder occurred here just hours before and, as everyone knows, the first 24 hours following a crime are critical. His steely gaze sweeps the room and he spots what may be a vital piece of evidence. It is most likely the one piece of evidence that will blow the case wide open and identify the murderer. He marches across the room, bends over and–

Unlike on the television shows, the last thing in the world a real crime scene investigator is going to do at this point is go tromping across the crime scene to pick up evidence. There are strict rules that must be followed in order to preserve the integrity of the evidence collected and to ensure that as much information as possible is collected at the scene of a crime. No one interested in solving a crime would risk compromising the scene the way they do on television. Classic unsolved crimes are often so because of improper crime scene procedures, leading to doubts about the quality or even legitimacy of the evidence collected. Was that bloody glove really found at the crime scene or was it planted? Who really dropped that cigarette? Once the integrity of a crime scene is lost, there is no getting it back.

Remember Locard's Exchange Principle, which can be paraphrased as "every contact results in an exchange of material"? Every person passing through a crime scene will leave something behind and will pick something up. This applies not only to the bad guys but also to everyone else. Each person passing through the scene following the crime will obscure or confuse matters to some degree. Thus, the number of people entering and leaving a crime scene must be strictly controlled, as must the conduct of each person at the scene.

The first officers to arrive are responsible for securing the scene. This means that they must do their best to ensure that everything stays exactly the way it was when they arrived. No one should be allowed to wander around the scene, no toilets should be flushed, no phones used, and no towels moved. The only exception to this is when life-saving measures must be taken. Not only does the evidence at the scene need to be preserved, nothing should be inadvertently added either. Random cigarette butts have led to no end of problems and it would be impossible to find the original footprints of a perpetrator if another dozen people have just wandered around the same ground.

Police agencies have detailed policies describing the duties of the first officer at the scene. The following is a general description of what will happen at the crime scene and is by no means authoritative, nor does it include all of the responsibilities of the police, such as ensuring that suspects are not still present. Instead, this description concentrates on the procedures necessary for the collection of trace evidence and does not necessarily even apply to the collection of biological evidence, which often requires special handling.

Technically, a crime scene includes not just the obvious location of criminal activity, a murder scene for example, but any areas where evidence may be collected. Sometimes "a" crime scene may actually comprise two or more discrete locations miles apart. For example, if a body is found dumped in one location, and the murder scene is somewhere else, both of those areas are part of the crime scene. If the car that was used to transport the body is also located, it too becomes part of the scene of the crime.

Each crime scene is different, which means that the initial responding officer must approach it carefully and be ready to adapt to the situation at hand. This officer is placed in the unfortunate position of having to accomplish several things at once. One of their initial duties is to secure the scene. This could mean something as simple as locking a door or stringing up rope or crime scene tape, or it could mean setting up barricades and chasing away sightseers. Not only should the visible crime scene be protected, but likely zones of the perpetrator's entry and exit should also be preserved. The routes leading to the crime scene may yield evidence just as important as what is found at the scene itself. If possible, there should be a buffer zone and enough area to create entry and exit pathways for investigators. Everyone working the scene would generally enter and leave along the same route to minimize their impact on the scene.

Another of the first responder's responsibilities is to take detailed notes, recording their precise time of arrival, information about the condition of the scene and creating a chronological record of when additional personnel arrived and what was done. It can sometimes be vital to know which lights were turned on or off, what doors were open, if any telephones were off the hook or if there were any distinct odors. Obviously, if injured people are at the scene, the administration of first aid takes priority over protection of the crime scene, but if at all possible the exact location of the victims should be noted, photographed, and/or sketched, and emergency medical responders should be directed into and out of the scene so as to minimize their impact. If there is a body at the scene, once death has been established, it should remain untouched until the arrival of the coroner or medical examiner. The original positions of anything that the medical team moves, such as furniture, need to be recorded. Otherwise, nothing at the scene should be moved unless absolutely necessary and the telephones at the scene should not be used.

Once the scene is under control, witnesses and possible suspects should be located and separated as soon as possible. Everyone needs to be kept from discussing what they have seen and thereby inadvertently, or intentionally, tampering with each other's memories. Interestingly, the tendency to tidy up can be a stress reaction, unrelated to guilt or innocence, and any witnesses at the scene must be kept from indulging in any housekeeping, such as sweeping up the dirt on the kitchen floor or wiping up spilled liquids.

One of the next people to arrive at the scene would typically be the lead investigator(s), the person(s) who would be ultimately responsible for oversight of the entire case. It is usually the lead investigator's job to determine what other personnel are needed at the scene and to perform tasks such as securing search warrants, if necessary. It would usually be the lead investigator's decision to call in others, such as crime scene technicians, to assist with the investigation.

Processing the Crime Scene

Processing the crime scene is a lengthy, exacting process that involves detailed documentation of the conditions at the scene and collection of all pertinent physical evidence. In larger cities, a team of investigators who specialize in the identification, collection, and preservation of physical evidence would process the crime scene. Many different names are used for these teams, such as crime scene team (CST), crime scene investigators (CSI), or evidence technicians. In areas with smaller forces, specially trained police officers might collect the evidence. In whichever case, the lead investigator would not be the person doing this work. Nor would the CST be involved in the interrogation of suspects or interviewing witnesses. Furthermore, the people who collect the evidence very rarely are the same people to perform any analysis.

The CST would usually be called in after the scene has been secured. This team could consist of just one or two crime scene technicians or it could be a whole group of people with a range of different types of expertise. If it has not already been done, the first thing the CST would do is to set up a command post outside of the crime scene with a sign-in sheet for everyone entering and leaving the scene. This sign-in also serves as part of the chain of custody for any items or samples collected.

Next, the CST would conduct a survey, or preliminary walk-through, to make observations and create plans for processing the scene. Nothing would be touched yet; everyone at this point is conducting a visual examination. The goal here is to take note of as much as possible about the scene and gather information about such things as points of entry and exit, items out of place, and items that are missing. Anything could potentially be evidence. The only exception to the "no touching" rule is with transient evidence, evidence that by its very nature, or due to the conditions at the scene, will disappear or lose its evidentiary value if not preserved or protected. Evidence such as wet shoe prints, bloodstains in the rain or small quantities of spilled liquid must be processed as quickly as practicable. In this case, the crime scene technicians have to work swiftly to preserve and record what they can.

Following the walk-through, the CST will come up with a search plan. Since each scene is different, the resulting plans will be based on the situation at hand. In all cases, searches must be performed systematically and thoroughly. Members of the CST work their way methodically through the area following a specific pattern, for example working an outdoor scene using a line search pattern (Figure 1.8a) or spiraling around inside a room working from bottom to top (Figure 1.8b). Each item of evidence would be tagged with a numbered freestanding marker or a flag, again without touching anything. In complex crime scenes, once an investigator

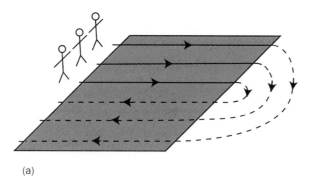

(a)

Figure 1.8a Example of an outdoor line search.
Source: With thanks to Darrel Kassahn

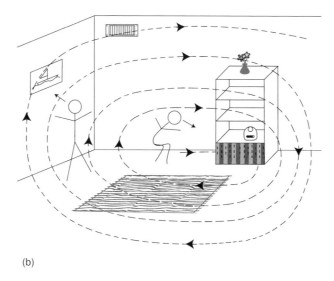

(b)

Figure 1.8b Example of an interior spiral search.
Source: With thanks to Darrel Kassahn

finishes searching an area, they often trade places with someone else who would then search the same area over again, often using a reverse pattern, i.e. working from top to bottom rather than bottom to top, to recheck the area for anything that was missed the first time.

Still without moving anything, the scene would be photographed, sketched, measured, and possibly videotaped. As one crime scene analyst from Virginia put it, "Shoot your way in, and shoot your way out." Everything should be photographed, from far away and from close-up. The CST photographer would start by taking pictures of the overall crime scene from a variety of angles and vantage points. Each item of evidence would be photographed several times, initially from a distance to orient the evidence within the overall scene, then from close-up. All close-up photographs should include a scale, a direction up indicator (if not horizontal) and be taken at a 90° angle. Film is cheap, as is media for digital storage of myriad image files, so everything should be well photographed, even if its relevance is not immediately apparent. A photograph of dusty cobwebs on a staircase could turn out to be important if a suspect claims that the victim rolled down the stairs. There is no harm in having extra photographs, while it can be potentially be devastating to miss a critical image.

In addition to the photographs, an investigator would also sketch the scene to record the exact locations of each item of evidence (Figure 1.9). The sketch map

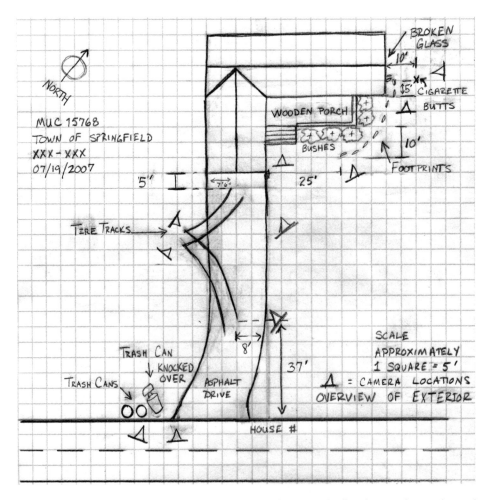

Figure 1.9 Example of a crime scene sketch with north arrow indication and notations showing the location of each camera.

needs to indicate north, which should be determined at the scene using a compass. Investigators must then make careful location measurements for each item of evidence from at least two fixed reference points using tape measures or measuring wheels. In large outdoor scenes, a global positioning system can be used to determine locations. There needs to be a scale on every sketch as well. It is useful to make sketches showing the locations from which photographs of the scene have been taken.

Once the scene has been sketched and photographed, and the locations of all of the marked pieces of evidence have been thoroughly documented, investigators can begin to invasively search the scene. At this point investigators can start to disturb the scene somewhat in order to locate hidden evidence. Close-up photographs must be taken of anything discovered during this phase of the search.

Only after the scene has been thoroughly documented can the collection of evidence begin. The order in which evidence is collected is prioritized: fragile materials are collected first, then moving from the most accessible items to the least accessible ones. Processing of the scene would also progress from less invasive techniques to more invasive techniques. Invasive techniques include using powder to look for latent fingerprints, or using chemical indicators to search for blood or drugs. If possible, trace evidence should be collected prior to the use of fingerprint powder or chemicals since it might inadvertently be moved or contaminated by the technician performing such activities. The goal here is to ensure that no evidence is compromised by premature removal or treatment.

Once an investigator is ready to collect an item, they must start a *chain-of-custody* form or tag that gives a complete description of the item, the time, date and location of recovery, who collected it, a case number and the tag number associated with the item in the photographs (Figure 1.10). If the item is large enough, the investigator will also mark it for identification. The chain-of-custody form started here must accompany the evidence from the point of collection, to the laboratory or storage locker, all the way to the courtroom and must list every person who ever had possession. For anything collected at the crime scene to be admissible as evidence, it must have been legally obtained, its origins identifiable by a crime scene technician and there must be an intact chain of custody.

Different types of evidence have different handling and storage requirements. For example, liquid bloodstains should not be stored in plastic, since they would immediately start to grow mold and decompose. Sand and soil samples should not be stored in glass. Many of the mineral components of such a sample are harder than glass and could break loose minute pieces of the sample bottle. This might make it difficult or impossible to separate any glass native to the sample from the glass introduced from the sample bottle. Items must be packaged securely and individually in order to prevent cross-contamination. If items of clothing are being collected, each item should be packaged separately, including the shoes (i.e. left shoe packed separately from right shoe). All evidence containers must be sealed at the scene. Once an item has been picked up, the spot where it was located should once again be photographed.

Once evidence collection is complete, items will be arranged for transport to the appropriate laboratory for analysis or to a police storage locker. The lead investigator would usually be responsible for authorizing the removal of evidence from the scene. The sign-in sheet forms another part of the chain of custody because the identity of every person transporting evidence is recorded as they sign out.

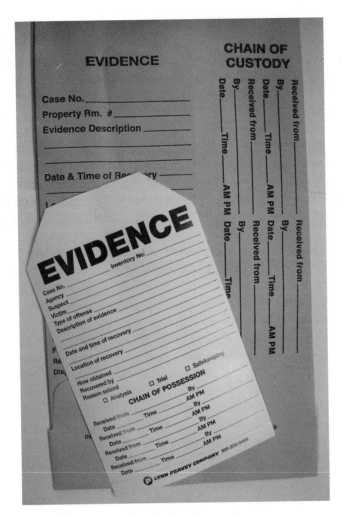

Figure 1.10 Evidence bag and tag with chain-of-custody forms.

It is important to collect all pertinent evidence prior to the release of the scene. Once the lead investigator releases the scene, it would most likely take a search warrant to return, and any evidence collected later could be viewed with suspicion. For a much more authoritative introduction to this topic, see Barry Fisher's (2005) *Techniques of Crime Scene Investigation* or a similar text.

Types of Evidence

Pretty much anything can be evidence. Common types of physical evidence include fingerprints, glass, documents, and fibers, plus biological and geological materials. The items of evidence collected for a case and submitted for analysis are often referred to as *questioned samples*. The first stage of analysis for most trace evidence is *identification*. If the evidence provided is a white powder, what is it? Is it heroin, ground aspirin, or talcum powder? In some cases, for example in a drugs bust, all that might be necessary is an identification. There are also times where an identification is all that is possible.

The next step after identification is *comparison*: a determination of whether two (or more) samples could have a common origin, like with DNA. For geological materials, it is usually much more fruitful to approach this step by applying the principle of *exclusion*; working through a series of steps to see if you can demonstrate that two (or more) samples could or could not have the same point of origin. For example, if you have soil samples taken from a crime scene and from a suspect's boots, it is usually not possible to prove that they are from the same source (you would need some significantly atypical properties for that). Instead, you determine whether they are *excluded* from potentially having the same source (i.e. could not have) or are *not excluded* from having the same source (i.e. could have come from the same place). This might seem like a strange semantic quibble, but it is actually a vital step in understanding the appropriate mindset for analysis.

In this example, the soil samples collected from the crime scene would be called *control samples*, or samples of known origin. The samples taken from the boots would be the questioned samples. In addition, for this type of investigation, *alibi samples* might be collected from alternative locations a suspect reports visiting. This way the analyst can determine whether the soil on the suspect's boots has more in common with the soil from a public park the suspect said they visited than with the soil from the crime scene.

Forensic comparison sometimes starts with comparing the questioned, control, and alibi samples with *reference samples* of known origin that have been extensively examined and are well documented. This type of evaluation is used to determine the key characteristics of each evidentiary sample. The next step is to see how many key characteristics each of the evidentiary samples share in order to determine which of them, if any, are related and to what degree. *Substrate samples* (background blanks) are also sometimes needed to determine what might be unique about questioned samples or if there are components of the sample that might somehow interfere with the analysis technique. For example, in an arson case or when investigating a suspected toxic chemical dump, it is important to distinguish what is in the background from what might have been added in the questioned samples.

There are two basic levels of comparison. The first is *class comparison*, or determining that the samples being compared come from the same group. For example, if you determine that a bloodstain is of type A blood, then you know that it comes from someone in the class of people who have type A blood, which encompasses around 26% of the population. This result also excludes anyone who does not have type A blood, or the remaining 74% of the population. Most of the time, this level of comparison is the best one can do with geological materials. This means that you can identify a soil as a silty clay loam, but you cannot tell exactly which silty clay loam. For geologic evidence, you would typically report that a sample was either *excluded from* or *not excluded from* belonging to a particular source/class.

The second level of comparison is *individualization*, where it is possible to determine to a high degree of probability that the two samples being compared have the same origin or source. This is how DNA examination works, but is rarely possible with geological materials. For DNA, we have a finite set of possible sources (the total population of humans), a finite number of characteristics to compare, and an ever-growing database of samples, making it possible to statistically calculate the probability that two samples came from the same person. There are no such known parameters for geological materials (i.e. exactly how many different types and combinations of geological materials exist and where exactly they all are), nor is there a forensic database for comparison, which makes it impossible to statistically calculate the probability of a common source.

However, you do sometimes run into mixtures of minerals that are very rare, have unusual elemental signatures or materials that might have some characteristics formed under very specific weathering conditions, which make those samples distinct from more common geological materials. When you have a distinctive sample, you can look at the ranges of common materials in the same general area, and if there are no other comparable materials, you can consider your sample to be normatively individualized.

Most often, trace evidence units handle geological materials. In practice, this includes not only obviously geological materials, such as rocks, fossils, and soil, but also building materials, glass, potting soils, kitty litter, and the mineral components of pigments, inks, cosmetics, and other household products. According to a contact in the FBI Mineralogy Division Trace Evidence Unit, the work performed there is: identification ~12% of the time; class characteristics ~81% of the time; *provenance* (determining the source or origin of something) ~5% of the time; *fractography* (studying patterns of breakage, usually to match them) ~2% of the time; and body recovery ~0.1% of the time. The real point here is that forensic geologists have to be very flexible and able to deal with a wide range of materials.

Further Reading

This was just a brief introduction to the history of forensic science, specifically to the application of geology in law enforcement, and to crime scene investigation. It must be emphasized that crime scene investigation is a very complex and specialized topic, and the information here is by no means authoritative. Every police agency and forensic laboratory has their own set of standard operating procedures (SOPs) and quality assurance/quality control (QA/QC) programs. For an introduction on how to create and maintain an appropriate forensic laboratory notebook, please go to the companion website (www.wiley.com/go/bergslien/forensicgeoscience). Parties interested in more background on forensic geology are also directed to:

Murray, R. C. (2004) *Evidence from the Earth*. Mountain Press Publishing Co., Missoula, MA. [A nice overview of the topic.]

Murray, R. C. and Tedrow, J. C. F. (1992) *Forensic Geology*, 2nd edn. Prentice Hall Inc., Englewood Cliffs, NJ.

Ruffell, A. and McKinley, J. (2008) *Geoforensics*. John Wiley & Sons Ltd, Chichester, UK. [A more advanced book, very informative.]

Wilson, C. and Wilson, D. (2003) *Written in Blood: A history of forensic detection*. Carroll and Graf Publishers, New York.

For more information on crime scene processing, the following books give a good introduction to the topic (there are lots of other well-written books out there).

Federal Bureau of Investigation (2008) *FBI Handbook of Crime Scene Forensics*. Skyhorse Publishing, Inc., New York. [Inexpensive and useful.]

Gardner, R. M. (2005) *Practical Crime Scene Processing and Investigation*. CRC Press, Boca Raton, FL. [A concise "how to" approach.]

Ragle, L. (1995) *Crime Scene*. Avon Books, New York. [A very informative mass-market book on the topic.]

References

Criminalistics (2011) Merriam-Webster Online. Encyclopedia Britannica, http://www.merriam-webster.com/dictionary/criminalistics, accessed 10/11/11.

Doyle, A. C. (1887) A Study in Scarlet. In: *Beeton's Christmas Annual.* Ward Lock & Co., London.

Doyle, A. C. (1890) Sign of the Four. In: *Lippincott's Monthly Magazine.* Lippincott's, Philadelphia and London.

Field, R. (1938) *All This, And Heaven Too.* Macmillan, New York.

Fisher, B. J. (2005) *Techniques of Crime Scene Investigation.* Taylor & Francis, Boca Raton, FL.

Forensic (2001) *The Oxford Essential Dictionary of Difficult Words.* American Edition. Berkley Books, New York.

Gross, H. (1893) *Handbuch für Untersuchungsrichter als System der Kriminalistik.* 2 volumes. *J. Schweitzer Verlag, Munich.*

Locard, E. (1929) L'analyse des poussieres en criminalistique. *Revue Internationale de Criminalistique* **September** (4–5): 176–249. Which was translated into English in Locard, E. (1930) The analysis of dust traces. Part I. *American Journal of Police Science* **1**(3): 276–98.

Murray, R. C. and Tedrow, J. C. F. (1975) *Forensic Geology: Earth sciences and criminal investigation.* Rutgers University Press, New Brunswick, NJ.

Womack, S. and Hines, S. (2001) *The True Crime Files of Sir Arthur Conan Doyle.* Berkley Prime Crime, New York.

Chapter 2
Minerals: The Basic Building Blocks of Geology

Minerals play a role in both forensic science and in our everyday lives in many different ways. Many of the products or materials you use on a daily basis are geological in origin. This includes obvious items like gemstones and gold and less obvious things such as windows, many of the required components in cell phones, and the drywall (or wallboard) in your home. All of these things depend on one or more different types of mineral deposits. It is important both economically and in terms of potential criminal activity to understand how minerals form and how they are distributed on the surface of the Earth.

Mineralogical Fraud

In this chapter, you will find a few examples of how the understanding of a little basic geoscience would have saved several people a great deal of time, pain, and money. For the first example, we go back to the 1800s.

In January 1848, gold was discovered in California at a place called Sutter's Mill, a sawmill on the South Fork of the American River. As news of this discovery leaked out, it traveled first to San Francisco, then back east and across the sea, setting off the great Gold Rush in which tens of thousands of people, mostly men, immigrated to try their luck. In January 1848, there were approximately 14,000 non-Indians in California, mostly of Mexican descent, with only a few thousand US citizens in the area. Less than five years later, the population had increased to more than 223,000. In an era when the weekly wage was around seven dollars, the chance to make $25 or $35 or more in a single day was an enormous attraction.

Following the discovery of gold in California came more discoveries of gold, silver, and other precious metals, both small and large. The Pike's Peak Gold Rush

An Introduction to Forensic Geoscience, First Edition. Elisa Bergslien.
© 2012 Elisa Bergslien. Published 2012 by Blackwell Publishing Ltd.

to Colorado started in 1858. A year later came the discovery of the Comstock Lode of gold and silver in Nevada. For a while it seemed like new mines were being opened everywhere in this unexplored land. Even into the 1860s, the territory west of the Mississippi was only roughly understood geographically and geologically. This meant that the western United States was seen as a potentially infinite source of fabulous natural resources, and that the nation was basically composed of two disconnected developed coasts.

On the other side of the country, politically and geographically a world away from California, Abraham Lincoln was elected president in 1860, and in 1861 the Civil War started. The primary tool of exploration had been military surveys modeled on the Lewis and Clark Expedition (1804–806), which had been commissioned by President Thomas Jefferson. Veterans Meriwether Lewis and William Clark had both scientific and commercial goals as they headed the first US transcontinental expedition to the Pacific Coast. But government-sponsored exploration of the west ground to a halt as all available resources were diverted to the war. The Civil War ended in 1865, the nation was weary, and the national infrastructure devastated. The westward expansion intensified after the war was over, as unemployed former soldiers sought ways to re-establish themselves. The United States of America was still basically a coastal country with a large, undeveloped void in the center. Even in the midst of war, President Lincoln had understood the importance of uniting the two coasts of the country, and in 1862 he signed the Pacific Railroad Act, authorizing the Central Pacific and Union Pacific Railroads to build a transcontinental railroad. It was a project that he did not live to see to its completion.

As part of the exploration effort, and as a means of employing some of the ex-soldiers, Congress, on March 2, 1867, authorized four geological survey teams to cover strips of territory from east to west. For example, Major John Wesley Powell, a geologist who lost his right forearm at the Battle of Shiloh, led the team that explored the Green and Colorado Rivers, including the Grand Canyon. Clarence Rivers King led the survey team that followed the Fortieth Parallel from Wyoming to California, along the route of the transcontinental railroad. Also of note, diamonds were discovered in South Africa in 1867, and for most Americans it seemed just a matter of time before they were discovered in the United States as well.

The Great Diamond Hoax

Now, with the background and mood of the nation understood, comes the hoax. There are many different accounts describing the following sequence of events. Several modern versions of the story appear to be based on a fictitious, and overly melodramatic, version of events written in 1937. The narrative presented here was pieced together from newspaper stories of the time, a few more recent articles, and a couple of books that have been written about the hoax. All of these sources are contradictory and confusing, but the general outline is as follows.

In 1871, two prospectors, Philip Arnold and John Slack, approached a businessman named George D. Roberts to show him a small leather bag containing uncut diamonds and other precious stones and to seek his financial support. The pair claimed to have discovered the gemstones in "Indian territory" and hinted at Arizona. The oft-repeated story of Arnold and Slack showing up at the Bank of California appears to be apocryphal. They had been involved in prospecting and mining ventures long prior to 1871 and had several contacts within the industry. Based on the sack of stones, Roberts telegraphed Asbury Harpending, who was in London seeking investors for a different mining scheme, and asked him to return to San Francisco at once and become involved in a new venture. William Ralston, the president of the San Francisco branch of the Bank of California and key player in the Comstock Lode, also became involved at this point, through either Roberts or Harpending. It should be noted that Harpending's memoirs sometimes recount events quite differently from the newspaper accounts of the time.

At this point, possibly financed by Roberts and one of the others, it appears that Arnold and Slack made a second trip to their "diamond fields," returning with another large sack of diamonds and rubies. With this display, Arnold convinced Roberts that a large investment would be necessary to stake out the diamond fields and start a large-scale recovery of gemstones. For whatever reason, Slack agreed at this point to sell his quarter interest in the venture for $100,000. Roberts approached two of his friends, William Lent and General George Dodge, to raise the money to buy out Slack.

Slack ended up with $50,000 up front, and Arnold was still in possession of some $25,000 from a previous mining venture, the Pyramid Range Silver Mountain Company through which he almost undoubtedly knew Harpending. Using this money, the pair left for England, where records show that in July 1871 they purchased over $19,000 worth of uncut diamonds and rubies from Leopold Keller and Company. There were almost certainly other stops to purchase more stones as well.

Upon their return to San Francisco, Arnold and Slack volunteered to make another trip to their diamond fields to bring back "a couple of million dollars' worth of stones" (Harpending, 1915: 147). The stones would be held by the investors as a guarantee for their investment. After supposedly visiting their diamond fields, the pair telegraphed from Reno, Nevada to have someone meet their train at Lanthrop, California. Harpending was sent to go meet them and reports in his memoirs that Arnold claimed they had collected the promised amount of stones in two packs, one for each man. On their way back, however, they found "the water in a river they had to cross extremely high" and stated that they lost one of the packs off their raft during the crossing. Anyone familiar with the American southwest would find this report suspicious, since the area is mostly desert and there would not be any swollen rivers to cross during the late summer in any of the territories in which the diamond field could be located.

Harpending reports giving Arnold and Slack a receipt for the pack of gemstones and taking it back to his home to show the other investors. He (or Roberts) dumped the sack out onto his billiard table, where it made "a dazzling, many-colored cataract of light" (Harpending, 1915: 149). A portion of the stones was taken to New York City for appraisal and to drum up more investors. Someone arranged for Charles Lewis Tiffany, founder of Tiffany and Co. (yes, that Tiffany) to appraise some of the stones. After examining the stones for a couple of days, Tiffany reported that they were worth $150,000. Since the portion taken to New York was only one-tenth of the stones Arnold and Slack had collected, Harpending set the value of the whole pack at $1,500,000. It turns out that

Tiffany was a silversmith by training, and neither he nor the lapidary he apparently involved had any training in assessing uncut gemstones.

The results of the assessment generated a new wave of investment in Arnold and Slack's diamond field, though at some point the major investors decided not to include too many people from outside of San Francisco. The investors decided to form a mining company to survey and secure the land. A long series of transactions started, during which money was exchanged and shares were divvied up. Concerns about the status of mining claims on Federal or Indian lands led the men to put money toward securing the necessary legislation to solidify their land claims. Congressman General Ben Butler, who ended up with 1000 shares in the new mining company, helped smooth the passage of the General Mining Act of 1872, which was ostensibly to promote the development of the mining Resources of the United States.

After some persuasion, Arnold and Slack now agreed to have their find inspected, as long as their visitors wore blindfolds to and from the site (the blindfolds might also be apocryphal). It appears that at some point Arnold went on another buying spree abroad in order to prepare the diamond field for the inspection. Henry Janin, a well-respected mining engineer, was engaged by the investors to examine the field. The composition of the inspection team varies from report to report, but at least a few of the investors, including Harpending, also went along. The team took a train to Rawlins, Wyoming and spent four days on horseback, while Arnold deliberately led them in circles to disguise the location of the diamond field. It turned out to be a mesa in northwestern Colorado, just south of Wyoming and just east of the Utah boarder.

Immediately upon arriving, the team started looking for gemstones, which they found in abundance. They were only in the field for seven days, with Arnold and Slack doing the majority of the gemstone collection. The inspectors returned with stories of precious stones, not just diamonds, but rubies, sapphires, and emeralds, lying all over the place. Some basic geology would have come in handy at this point. The gemstones that the investors reported finding do not usually occur together like this, and certainly not scattered randomly about the surface of the landscape. It was later disclosed that Arnold selected all of the sites that were explored and that Janin was kept busy surveying the claim area, erecting posts, securing water and timber rights, and was thus prevented from doing any digging on his own. Based on his inspection, Janin placed a value of $400,000,00 on the diamond field.

After Janin's findings were reported, the San Francisco and New York Mining and Commercial Company was incorporated with a capital of $10,000,000 and stock in the company was offered to 25 of the most influential men in San Francisco, on July 11, 1872. Sometime during these proceedings, Slack disappeared, possibly not even returning with the inspection team. Arnold collected another $150,000, either for signing a quitclaim or for his original share in the previously collected stones, and he sold his stock in the company to Harpending (or through Harpending) for another $300,000. He then promptly took off for Kentucky, his total take from the scheme something in the neighborhood of $8 million in today's money.

Things began to fall apart when Clarence King, the man who headed the geologic exploration of the Fortieth Parallel, became involved (Figure 2.1). There are several different versions of how King first heard about the diamond fields.

However he did, he was determined to find out more. The purported diamond fields reportedly lay in the area that had been explored by his survey team. In fact, the mesa described by the diamond hunters sounded exactly like one where King himself had placed a survey marker. It would be a tremendous blow to the survey teams if they had somehow missed a mineral deposit of this magnitude during their five years of exploration. In King's words, "It was a matter of self-defense."

After gathering intelligence, King and some of his men went immediately to visit the diamond fields, arriving in late October 1872. They set up camp once they found the claim notice posted by Janin and began to explore. King soon noticed that wherever he found a diamond he also found about a dozen rubies, an unnaturally uniform distribution, and that the gemstones were only found in disturbed ground or in anthills with footprints around them. Gemstones are denser than quartz, so they should settle more deeply into the sediment layers, rather lying on the surface. However, when King dug trenches into the sediment in areas that were rich on the surface, he did not find any gemstones. After four or five days of exploration, King set out for San Francisco, arriving on November 10.

He first approached Janin with his findings, and then met the directors of the mining company at Ralston's office at the Bank of California. There King presented a report on his findings and declared that "the diamond fields upon which are based such a large investment and brilliant hope are utterly valueless, and yourselves and your engineer, Mr Henry Janin, the victims of an unparalleled fraud." King had arrived in time to stop the sale of another 100,000 shares of stock at $100 a share. Another inspection party, this time headed by King, set out for the diamond fields. Following the new inspection, King's report was published and the San Francisco and New York Mining and Commercial Company dissolved, with all the accompanying fallout from distraught investors.

John Slack had long since disappeared and was never heard from again. Philip Arnold purchased a house and 500 acres, in his wife's name, in Elizabethtown, Kentucky and lived with his family raising horses, sheep, and pigs. A San Francisco grand jury indicted both men for fraud, and eventually Arnold settled out of court with William Lent, the man who lost the most money, for $150,000. Clarence King was hailed in the newspapers as a hero and eventually became the first director of the United States Geological Survey. Asbury Harpending was suspected of complicity with Arnold and Slack, since he had worked previously with Arnold on other mining ventures, and because he was in London at the same time that Arnold was on one of his gemstone-buying sprees. Harpending ended up dissolving all of his real-estate holdings and leaving San Francisco, eventually moving to Kentucky. William Ralston repaid the 25 invited investors out of his own pocket and accepted the losses, which amounted to something over $500,000. In August 1875, following a series of poor investments and a run on the bank, the Bank of California failed. Ralston was refused admittance to a board of directors meeting held on August 27, and following what was apparently his normal routine went to the North Beach for a swim. He was found floating face down in the bay, possibly the victim of a stroke or as a result of suicide.

There have been several articles about the Great Diamond Hoax, most of which are wildly inaccurate. The account present here, which was the best this author could do to make sense of the whole mess, was based on the newspaper reports of the time (some of which are also quite ridiculous), Harpending's account (which was written years later, is extremely self-serving, and incorrectly reports some details), and other primary and secondary source materials. As with many good cons, the mood of the day played an enormous role in taking what should really have been a fairly transparent scheme and elevating it into a phenomenon. King was able to expose the fraud because he had a basic understanding of how, and where, minerals form. Thus, we will start our tour of forensic geoscience with just that information.

Minerals

Minerals are the basic building blocks of the vast majority of the materials on Earth. Mineral formation and distribution is related to specific geological processes, which means that *some* places on Earth have something akin to a mineral fingerprint: a unique mineral or mixture of minerals that occurs in that condition only in that place. This wide variety, and occasionally unique distribution or assemblage, makes minerals an important forensic tool. In order to understand rocks, sand, and soil, it is important first to have an understanding of minerals, and in order to understand minerals we must have some basic understanding of atoms. The atomic structure of a mineral is reflected in its physical and chemical characteristics.

Atoms are the smallest divisible components of matter that still have the identifiable properties of a specific element (e.g. size, mass) and they are the fundamental units of each mineral's crystal structure. An atom is composed of three different basic particles: *protons*, positively charged matter; *electrons*, negatively charged matter; and *neutrons*, neutral (without charge) matter. An *element* is a

Figure 2.1 United States Geological Survey photograph showing the members of the Fortieth Parallel Survey. From left to right: James T. Gardner, Richard D. Cotter, William H. Brewer, and Clarence King.

Source: Courtesy of the United States Geological Survey.

collection of the same kind of atoms. You can identify an atom as being of a specific element, say an atom of carbon, but the parts of an atom are interchangeable, so you cannot tell if a proton is from an atom of carbon or from an atom of uranium. While this model is not really correct, the most easily grasped concept of an atom is that of Niels Bohr, who developed the "planetary" model (Figure 2.2), where each atom consists of a cloud of electrons orbiting a central nucleus that contains both protons and neutrons (except the most common form of hydrogen, which has a single proton in its nucleus).

On a periodic table, the elements are identified by an atomic symbol, such as H for hydrogen, and arranged in order by the number of protons (Figure 2.3). The number of protons in an element equals the *atomic number*, which is usually listed in top of the cell on a periodic table. If you know the atomic number, you know the element. For example, any atom with six protons is carbon, and any atom with 92 protons is uranium. The number of protons plus the number of neutrons equals the *atomic weight*. The mass of an electron is negligible by comparison, so it is not included in calculating atomic mass. Atomic mass can vary since the same element can include differing numbers of neutrons. *Isotopes* are atoms of the same element with differing numbers of neutrons. Most isotopes are actually stable, which means that they do not break apart. Unstable isotopes, which do break down or alter, are called *radioactive isotopes*.

Magnesium (atomic symbol Mg) has the atomic number 12; therefore, every atom of Mg has 12 protons. There are three stable, naturally occurring isotopes of magnesium, one with 12 neutrons, one with 13 neutrons, and one with 14 neutrons. Thus, the atomic mass of Mg can be 24, 25, or 26. There are also several unstable or radioactive isotopes of Mg, with masses both lower (Mg-20, Mg-21) and higher (Mg-27 through Mg-37) than their stable counterparts (Figure 2.4).

In a neutral atom, the number of electrons is equal to the number of protons. Technically, electrons do not really orbit the nucleus (i.e. move in fixed little ellipses or circles), instead they are most likely to be located at various specific distances from the nucleus, called *orbitals*, which can be visualized as thin, three-dimensional shapes centered around the nucleus. The shapes range from simple hollow spheres to complex forms that look like multiple hourglasses stuck together to things that cannot really be visualized easily at all (Figure 2.5). Each orbital is associated with a specific energy level. *Shells* are clusters of orbitals in various energy regions, and historically are labeled starting from the innermost shell K, L,

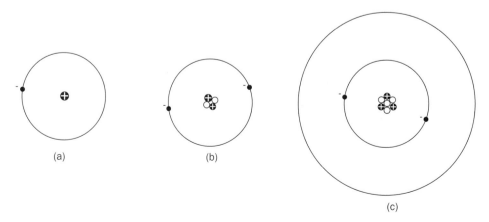

(a)   (b)   (c)

Figure 2.2 Basic models of atoms of (a) hydrogen, (b) helium, and (c) lithium.

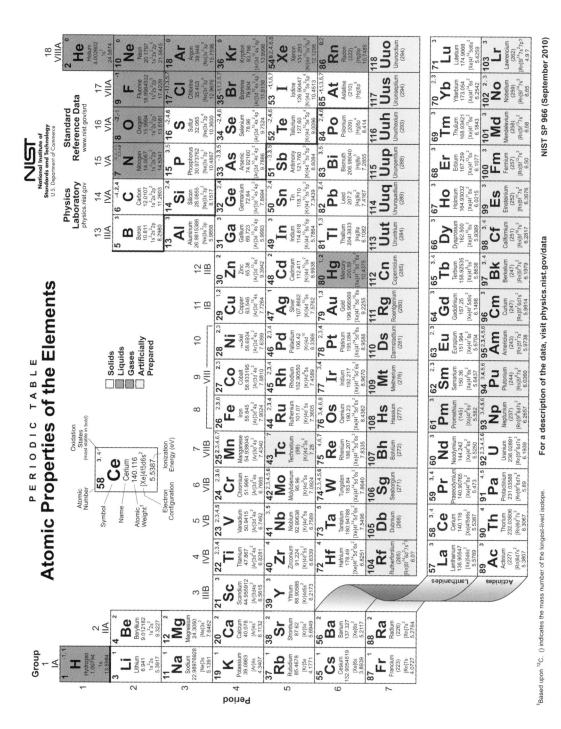

Figure 2.3 Periodic table of the elements. Please see Color Plate section.

Source: Modified from Dragoset *et al.*, (2010), Periodic Table: Atomic Properties of the Elements. National Institute of Standards and Technology.

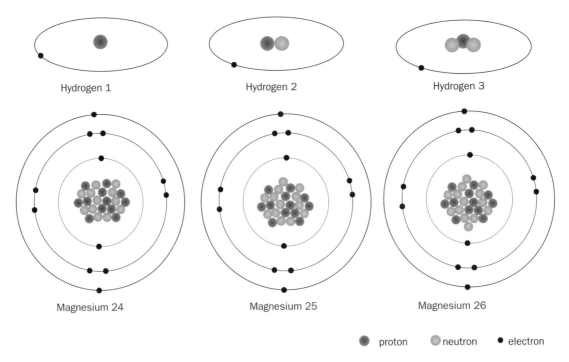

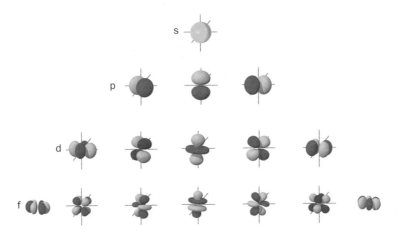

Figure 2.4 Isotopes of hydrogen and magnesium.

Figure 2.5 Examples of electron orbitals.

M, N, O, P, and Q. Within each shell are a series of subshells (or shapes) labeled s, p, d, f, and g. Subshells are filled by electrons in a specific order, and each shell can hold only a certain number of electrons. The K-shell can hold 2 electrons, the L-shell 8, the M-shell 18, the N-shell 32, etc. (Figure 2.6).

If you have ever wondered why the periodic table looks the way it does, count the number of elements in each row (also called a *period*) (Figure 2.3). You will see that the first row has 2 elements (H and He), the second and third rows have 8, the next two rows have 18, and the final two rows hold 32 elements, when you squeeze the displaced lanthanide and actinide sequences back in place. The elements were arranged in rows so that the elements in each column (group) have similar properties, which are the result of the similar status of their outer shells.

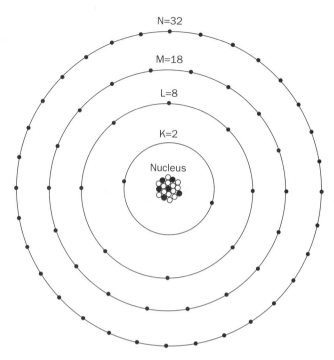

N=32
M=18
L=8
K=2
Nucleus

Figure 2.6 Sequence of number of electrons in the shells of an atom.

Though the rows of the periodic table do not quite line up with the historically designated K thru Q shells, because the order in which the subshells fill is rather more complicated, the periodic table is a visual representation of the atomic structures of the elements. An understanding of the periodic table is important for geoscientists because it helps them to predict and interpret an element's behavior in the environment.

You can think of these shells as amusement park rides, each with a certain number of seats available on the ride. For example, the K ride has 2 seats, the L ride has 8 seats, and so on. A stable electronic configuration for an atom is one with a completely filled outermost collection of orbitals (which is not quite synonymous with shell). When all of the seats on the ride are full, everything is in balance. If one or more seats are empty, things are a bit off kilter. Thus, atoms often lose, gain, or share electrons to obtain a stable configuration. Noble gases, in the far right column in the periodic table (helium, neon, argon, krypton, xenon, and radon), have completely filled outer orbitals, so they are stable and under normal circumstances non-reactive (i.e. inert). Elements in the leftmost column, like sodium (Na) and potassium (K), only have one electron in their outermost shell. The outermost shell(s) gets increasingly full as you move across the periodic table toward the column of noble gasses. Elements toward the left of the periodic table tend to lose electrons, while elements toward the right tend to gain electrons. Elements near the middle of the table can do either, depending on the circumstances. Thus, the columns in the periodic table group elements that behave in a similar way chemically.

The periodic table can also be broken into three broad groups of elements based on certain physical and chemical qualities. All of the elements on the left-hand side of the periodic table, excluding hydrogen, are called *metals*, thus are malleable and good conductors of heat and electricity. The elements boron, silicon, germanium, arsenic, antimony, tellurium, polonium, and astatine are *metalloids*, with properties between those of metals and nonmetals. This group of elements is located around that strange bold line that runs through the right-hand side of most periodic tables, which is there to help you locate the boundary between the metals (left-side) and the nonmetals (right-side) quickly. Hydrogen and all of the elements to the right of the metalloids are called *nonmetals*. Nonmetals are brittle and poor conductors.

When elements lose or gain electrons, the charge balance on the atom becomes unequal. Charged atoms are called *ions*. Positively charged atoms (+), which have lost electrons, are called *cations*, while negatively charged ions, which have gained electrons, are called *anions*. Elements from the left-hand side of the periodic table, such as sodium and potassium, lose electrons to become cations, while elements from near the right-hand side of the periodic table, such as fluorine, chlorine, and oxygen, gain electrons to become negatively (−) charged, or anions.

The drive to attain a stable electronic configuration in the outermost shell, along with the fact that this sometimes produces oppositely charged ions, causes atoms to bind together. When atoms become attached to one another, we say that they are *bonded* together. Typically, only the outermost electrons, or valence electrons, are involved in bonding.

Types of Bonding

Ionic bonding occurs when one or more electrons are donated by one atom and received by another, resulting in charge imbalances in each of the atoms, turning them into ions. The atoms are bonded together by the force of attraction between ions of opposite charge. Each ion is surrounded by ions of opposite charge. Ionic bonds are strong bonds, though not the strongest, and are the most important bond type in the majority of minerals.

The best mineral example of an ionic bond is found in halite (NaCl), more commonly known as table salt (Figure 2.7). Halite is formed of sodium (Na) and chlorine (Cl) atoms. Looking at a periodic table, you will see that sodium is located in the far left column on the periodic table, which means that it has only one electron in its outermost shell. Chlorine, on the other hand, is located one column in from the right-hand side of the periodic table, so it is only missing one

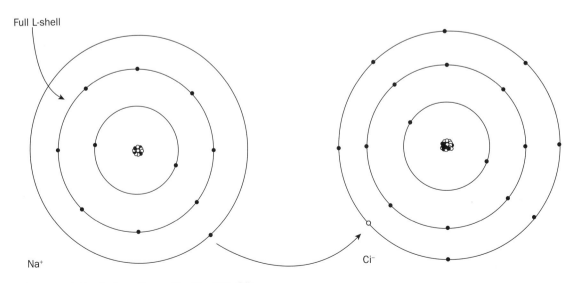

Figure 2.7 Ionic bonding of halite (NaCl).

electron. Each sodium atom donates its electron to a chlorine atom, turning into a cation. Each chlorine atom becomes anionic when it accepts the donated electron. The oppositely charged atoms attract each other to form sodium chloride (NaCl). This type of attraction is quite strong, as demonstrated by the 801 °C melting temperature of halite. However, ionic bonds are susceptible to breakdown in polar liquids (liquids that are composed of molecules that have a charge imbalance to their structure). If you dump salt into water, a highly polar liquid, it dissolves easily.

Covalent bonding occurs when electrons are shared between two or more atoms. In some cases, it would be structurally unsatisfactory to lose electrons. Instead, they are shared so that each atom has a stable electronic configuration (completely filled outermost shell) part of the time. Since all of the atoms need to hold on to all of the electrons to remain stable, these bonds are the strongest chemical bonds. Pure covalent bonds are rarely found in minerals, the best example being that of diamond (which is still not purely covalent). In the structure of a diamond, each carbon atom is surrounded by four other carbon atoms each of which share one electron with the central carbon atom, which in turn shares one of its electrons with each of the surrounding carbon atoms. The resultant structure is very strong, though somewhat brittle. Compounds formed with covalent bonds may also have a slight charge to their structure by virtue of the placement of their electrons, called *polarity*.

For example, a water molecule is composed of two hydrogen atoms and one oxygen atom (H_2O) (Figure 2.8). Hydrogen only has one electron, so it would be structurally unsound to donate that electron to another element, because that would leave only a naked proton. To be stable, hydrogen needs two electrons, which would fill the K-shell. Oxygen has six electrons in its outer shell but, because it is an L-shell, needs eight to be stable. So, each hydrogen atom bonds to the oxygen atom by sharing its one electron and in turn gets a part share in one of the oxygen atom's electrons. This results in a stable configuration for each atom. It also results in a highly polar molecule. The communal electrons basically become trapped between the atoms they are being shared by, so the hydrogen sections of the

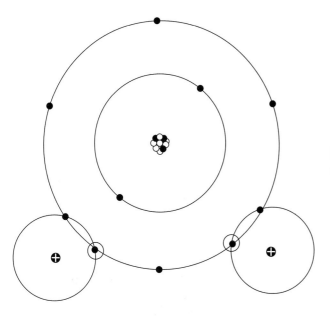

Figure 2.8 Covalent bonding of a water molecule (H_2O).

molecule are something like unshielded protons, giving those portions of the molecule a slight positive charge. The oxygen portion of the molecule has just acquired a part share in two more electrons, resulting in a slight negative charge.

In nature, most mineral bonds are actually of mixed character: somewhat ionic and somewhat covalent. The more prevalent the covalent nature of the bond, the harder and more brittle the resulting mineral. Together, ionic and covalent bonds are called *strong bonds*. There are also two common *weak bonds* that are important to understand: metallic bonds and van der Waals bonds.

Metallic bonding is the most common type within native metals like copper and gold. In this case, all of the electrons are constantly being shared by all of the atomic nuclei, and can move freely from one nucleus to another. Because the electrons are not fixed in a certain position, minerals that are constructed from nearly pure metallic bonds are very soft. Picture yourself bending a piece of copper wire. Copper is very malleable, which means that it can be bent or shaped easily without breaking. Pure gold is also extremely soft, which is why other materials are mixed in to create gold jewelry. Otherwise, you would have to get your jewelry fixed every time you bumped into a table. Materials bonded with metallic bonds are excellent conductors of electricity because the electrons can move freely through the material.

Van der Waals bonding does not involve sharing or the transfer of electrons. These bonds are the result of electrostatic attraction between charged structures, such as polar molecules like water or the unsatisfied bonds found on the surfaces of crystalline structures. Molecules with a charge imbalance created by their structure, such as water, are called permanent dipoles. Polarity can also be created temporarily by the movement of electrons. Electrons normally move around freely in their shells and at any given point they could all be located on one side or the other of an atom. This movement results in temporary dipoles, which last for as long as the electrons are unevenly distributed. The final form of charge imbalance is an induced dipole. This happens when the electrons in an atom or molecule are all attracted to a positively charged section of a neighboring atom or molecule. Though van der Waals bonds are weak, they make up for this to some degree by sheer numbers and are very important in a number of environmental processes. They form the basis of bonding in many soft minerals and are very important in processes that involve water, a strongly polar molecule.

Since the oxygen portion of the molecule has a slight negative charge and the hydrogen portions have a slight positive charge, water molecules are attracted to each other (Figure 2.9). This is what gives water so many of its unique and important physical and chemical properties. The same slight charges that occur in water molecules occur in mineral structures, giving the surface of the mineral a slight charge.

Van der Waals bonding usually results in a zone of weakness along which a mineral breaks very easily. Mica and graphite, like in your pencil, are both good examples of this. Each of these minerals is formed of sheets that are van der Waals bonded together, thus each mineral is easily pulled apart into thin layers (Figure 2.10). Several different bond types are usually present in a mineral, determining many of its physical properties.

For something to be considered a *mineral* it must:

- be a *solid*, which means having a fixed size and shape. Solids can be deformable, which means that the size or shape can be changed by the application of sufficient force (think of making something out of Play-Doh);

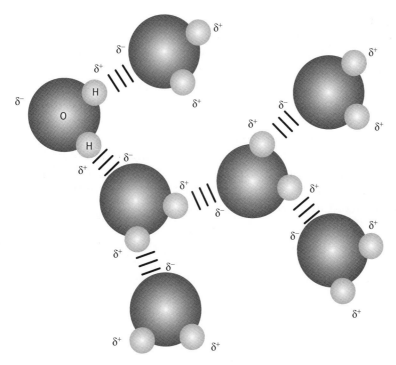

Figure 2.9 Hydrogen bonding of water molecules.

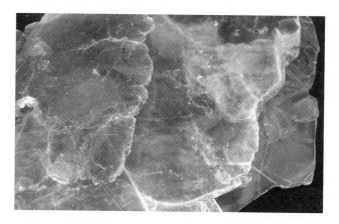

Figure 2.10 Van der Waals bonding found in sheets of mica.

- be *naturally occurring*, i.e. forms naturally in the environment without the aid of humans. Anything manufactured or created by humans is synthetic and does not count as a mineral;
- be *inorganic*, i.e. not the product of biological activity or processes (i.e. shells, bone, tree sap, and the like do not count as minerals);
- have a *definite chemical composition*. This means that every time we see the same mineral it has the same basic elemental composition that can be expressed by a chemical formula. Some variation does occur (called *substitution*), usually not too much, and only of a specific sort. Ionic substitution, also called solid solution, occurs because some elements (ions) have the same approximate size and charge, and can thus substitute for one another in a crystal structure. The mineral olivine ranges from Fe_2SiO_4 to Mg_2SiO_4, with magnesium (Mg^{+2}) freely substituting for iron (Fe^{+2}). The ions have the same charge (+2) and are about the same size, thus

they can substitute for one another in the crystal assembly without fundamentally changing the structure. This means that olivine can have a range of compositions expressed as the formula $(Mg,Fe)_2SiO_4$, with any amount of iron or magnesium from 0 to 100%. This kind of substitution occurs in many minerals, such as plagioclase feldspars, which range from $NaAlSi_3O_8$ (called albite) to $CaAl_2Si_2O_8$ (anorthite) with the $NaSi^{+5}$ group substituting for $CaAl^{+5}$ (called a complex solid solution);

- have a characteristic *crystalline structure*, which means that the atoms are arranged within the mineral in a specific ordered manner with an atomic arrangement that repeats over and over. Such an orderly arrangement needs to fill space efficiently and keep a charge balance. Since the size of atoms depends largely on the number of electrons, atoms of different elements have different sizes. Halite (NaCl) is arranged in a cubic fashion (Figure 2.11). Each sodium ion is surrounded by six chlorine ions and each chlorine ion is surrounded by six sodium ions. The charge on each chlorine is −1 and the charge on each sodium is +1 to give a charge-balanced crystal. This atomic arrangement is reflected in the shape of the mineral crystal. The small cubes that you see if you spill some salt on the table are not the result of a complicated manufacturing process but the result of the packing arrangements occurring at the atomic level.

It should be noted that the last two requirements, for a definite chemical composition and characteristic crystalline structure, might sound the same, but it is possible for the same atomic building blocks to create multiple, different crystalline structures. Minerals with the same chemical composition but with different crystal structures are called *polymorphs*. The most commonly known examples are the minerals diamond and graphite (Figure 2.12). Both are chemically composed of the element carbon (C), but in a diamond the carbon atoms are all covalently bonded, creating a very strong structure, while in graphite there is a mixture of strong bonds, which form little plates, with weak bonds between each plate. This means that when you use a pencil to draw a line, you are breaking the weak bonds between the plates.

For a material to be considered a mineral, it must meet all five of the requirements listed above. For example, *glass* can be naturally formed (for example the volcanic glass called obsidian) and is a solid, but its chemical composition is not always the same, and it does not have a crystalline structure. In fact, that is part of the definition of glass, that is it *amorphous*, or lacking an orderly crystalline structure. Thus, glass is not a mineral. Ice is also naturally formed, is solid, does have a definite chemical composition that can be expressed by the formula H_2O,

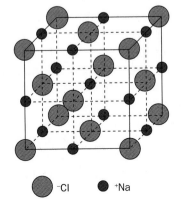

Figure 2.11 Cubic crystalline lattice of halite (NaCl).

●⁻Cl   ●⁺Na

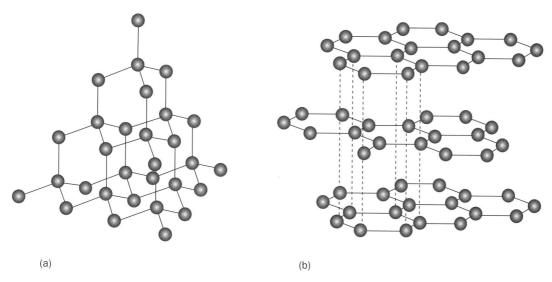

Figure 2.12 Polymorphs of carbon: (a) diamond, (b) graphite.

and does have a definite crystalline structure when solid. Thus, ice is a mineral, but liquid water is not (since it is not solid). *Amber* (fossilized tree resin) is a solid that occurs naturally in the environment, but it lacks a crystalline structure, and because it is the product of biological activity it is not a mineral.

Mineral Fraud Now: The Story of Bre-X

Now that you are armed with a basic understanding of what a mineral is, let's look at another extraordinary case of mineral fraud. Just in case you had been thinking that something as melodramatic as the Great Diamond Hoax could not happen today, behold the story of Bre-X. In 1993, the junior mining company Bre-X Minerals Ltd, based in Alberta, Canada, purchased the rights to part of an area called Busang in Borneo, Indonesia and began exploring for gold. Initial drilling results were encouraging, and in January 1994, Bre-X announced that the company had found some two million ounces of recoverable gold. It was the first of a long line of ever-more-exciting reports about what was at one point thought to be the richest gold deposit in the world.

Bre-X started out as a penny stock at 30 cents a share introduced in 1989 on the Alberta Stock Exchange in a *blind pool*. As such, there was no detailed plan of how invested funds were to be used. Instead, it was essentially a venture-capital deal in which the public would be blindly betting that the company they invested in would somehow strike it rich. Because Bre-X stocks were initially placed in this fashion before ending up on other stock markets, many of the normal disclosure laws, including issuance of formal prospectuses, did not apply. Following the acquisition of rights to Busang, Bre-X used the Internet to promote its stock, posting company financial statements, press releases, and an anonymous running commentary on the status of Busang that hinted at future increases in stock value. This commentary would have been illegal if it had occurred in any of Bre-X's published materials, but as it was posted unofficially on the Internet, it fell into gray territory.

Based on the potential of the Busang gold field, Bre-X obtained financial support from the investment firm LOM Securities, Inc. and raised enough capital for more drilling. On September 20, 1994, Bre-X reported a potential resource of 3–6 million ounces of gold. In November, the *Financial Post*, a leading Canadian business newspaper, published its first article on Bre-X. For the next two years, the gold reserve estimates kept climbing and the stock price continued to rise. On April 23, 1996, Bre-X was listed on the Toronto Stock Exchange, opening at $186 per share and rising with every new ecstatic press release.

One warning note was sounded by the *Financial Post's* mining reporter Peter Kennedy in a March 23, 1996, article. Quoting a variety of sources, Kennedy pointed out that the cyanide assay method being used by Bre-X was destructive and samples could not be retested. He discussed the validity of the cyanide-leaching method versus the more commonly applied fire-assay technique and, most troublingly, revealed that there had been no independent corroboration of the Bre-X assay results. David Walsh, the chief executive officer (CEO) of Bre-X, and John Felderhof, chief geologist and Bre-X's exploration manager in Indonesia, immediately went into damage control mode and issued a flood of press releases and statements. As a result, no other news outlet picked up Kennedy's story.

The stock price recovered and with each new announcement Bre-X stock rose in value. In May 1996, the price hit over $200 per share and stockholders approved a 10-for-1 split. A 10-for-1 split means every share purchased before the split was now worth 10, a process that allows more shares to be sold. In June 1996, Bre-X announced that Busang contained 39 million ounces of gold, and then one month later upped that estimate to 47 million ounces. In August 1996, Bre-X was listed on the American NASDAQ Stock Exchange.

Through late 1996 and early 1997, due to Indonesia's interesting business laws and idiosyncratic government, and Bre-X's own questionable dealings, the company went through a shaky period where its mining rights and partnership status were in dispute. A protracted series of legal and political maneuvers followed while Bre-X tried to sort out its interests. The company had to make deals to satisfy damaged partners, and the Indonesian government demanded a say in choosing Bre-X's development partner. To vastly oversimplify, since Bre-X was a junior company, the traditional pattern would be for a larger mining company, with the resources and infrastructure for development, to buy the junior company out. Because of the Indonesian government's demands, Bre-X would be at a severe disadvantage in any deals they negotiated.

In an attempt to protect its position and obtain clear mining rights, in late 1996, Bre-X signed an agreement with Indonesian President Suharto's son Sigit Harjojudanto, promising him a $40-million-dollar consulting fee and 10% of future contracts and profits. Meanwhile, Suharto's daughter Siti Hardiyanti Tutut Rukmana had struck her own deal with Toronto-based Barrick Gold Corporation, in which that company promised her all of the construction contracts for the mine if she ensured that Barrick was made Bre-X's mining partner. Former American President George Bush even sent President Suharto a letter gently lobbying for Barrick. Under Suharto's daughter's influence, the Indonesian government told Bre-X that it must partner with Barrick and give them a 75% share. This upset both Bre-X and Suharto's son, setting off a power struggle within the Indonesian government. Meanwhile, by the end of 1996, Bre-X insiders had made millions selling some of their shares. Felderhof reportedly made $46 million, and Jeannette Walsh, David's wife, made $25.2 million. All told, insiders netted slightly less than $100 million dollars.

Through a series of maneuvers inside and outside the Indonesian government, the Barrick deal fell apart in January 1997. On January 23, a fire swept through the Bre-X Busang encampment destroying, among other things, company records, the Busang seismic survey results, and the Bre-X geologist's field notes. On February 17, 1997, the American mining company Freeport McMoRan Copper & Gold Inc. was made the official partner of Bre-X. In the new deal, Freeport got 15%, Bre-X got 45%, and Indonesian interests got the remaining 45%. This meant that Bre-X's share in the gold deposit had been cut in half (dropping from 90% to 45%), causing the stock to drop. As a damage-control measure, Bre-X held a conference call with mining analysts on February 19, and during the call, Felderhof commented, "If you ask me what is the total potential, I would feel very comfortable with 200 million ounces," suddenly doubling the supposed size of the gold deposit.

Right after the partnership deal was announced, Freeport sent geologists to Busang to begin a due diligence study of the gold supply. Thus began the first independent study of the site as Freeport conducted its own drilling. They started on March 1 and in their first three holes found no gold at all. While it is true that drilling results vary, Freeport used a process called *twinning* and drilled their holes just 1.5 meters away from the holes that Bre-X reported as being rich in gold. There should have been some comparable amount of gold in the cores, but there was not. The Freeport geologists scheduled a meeting with Bre-X geologist Mike de Guzman to go over the core sample results. At the time, de Guzman, Walsh, and Felderhof were all at the Prospectors and Developers Convention in Toronto, Canada, where Felderhof received the Prospector of the Year Award.

It took several days for de Guzman to return to Indonesia. On March 19, 1997, on his way to the meeting with Freeport's chief geologist, 17 minutes into the flight, de Guzman jumped or fell from a helicopter over the Indonesian jungle. Initially, journalists described it as an accident. Bre-X stock was briefly halted on the Exchange, pending an announcement. Later, David Walsh announced that de Guzman had committed suicide and read excerpts from a suicide note declaring that de Guzman killed himself because he had been diagnosed with Hepatitis B. It was eventually disclosed that de Guzman's note had some suspicious elements, for example his wife's name was misspelled. A body presumed to be that of de Guzman was located five days later, though its advanced state of decay made a positive identification difficult. Apparently, the identification was based on a single fingerprint, and though dental records were made available, they were never used. These circumstances led to speculation that de Guzman either faked his death and escaped or that he was murdered. To further complicate matters, it was soon discovered that de Guzman had four wives, with children, who knew nothing about each other.

Shortly after de Guzman's death, the Indonesian newspaper *Harian Ekonomi Neraca* leaked some of the preliminary results of Freeport's findings and suggested that Busang might not even be worth mining. The article got picked up by an international news service and from there got posted on the Internet. When the stock market opened in Toronto on March 21, Bre-X shares fell by $2.25 a share, or by a total of approximately $500 million. David Walsh responded by issuing a denial and threatening to sue. Then, six days after the newspaper report, on March 26, 1997, Bre-X announced that independent mining consultants Strathcona Minerals Services Ltd had concluded that the size of the Busang deposit might have been overstated. Later the same day, Freeport McMoRan formally issued a drilling report that stated, "To date analyses of these cores, which remain incomplete, indicate insignificant amounts of gold."

Before trading began on the Toronto Stock Exchange on March 26, Bre-X's stock was halted pending the press releases. When trading resumed the next day, Bre-X stock fell $13 to $2.50 a share. In a frenzied selloff, over ten million shares changed hands and the Toronto Stock Exchange's computers crashed from all the activity. In approximately 30 minutes, $3 billion of the company's value was wiped out, as were many paper millionaires who held the stock. The stock was halted again on March 31, pending new information, but the company released nothing and the stock began to trade again an hour late on April 1. The stock traded for 22 minutes and volume was so heavy that once again the Toronto Stock Exchange's computers crashed. The computers came up again an hour later, but trading on Bre-X did not resume. A similar pattern was repeated for the next two days. Bre-X stock continued to bob around the $2 to $3 range while rumors flew about and everyone awaited the formal results from the investigation by Strathcona.

Strathcona Minerals Services Ltd was brought in, by Bre-X, as a neutral third party to conduct its own assessment of the Busang gold field. Strathcona's preliminary investigation led to the March 26 press releases, and they became the second company to independently drill in Busang. Their report, officially released on May 5, 1997, stated, "We believe there to be virtually no possibility of an economic gold deposit" and went on to conclude that the Bre-X gold samples had been "salted." Literally, this means that someone would add gold dust to the sample, as if shaking it out of a hypothetical saltshaker, before it was sent out for analysis.

It turns out that there had been a long list of red flags and discrepancies since the beginning of the Busang saga, but virtually no one wanted to pay attention to them. *Northern Mining* carried an article that revealed a July 1996 report by the Australian metallurgical consultants Normet, which stated that in the Bre-X samples they examined approximately 90% of the gold particles were "mostly rounded with beaded outlines." This implies alluvial (riverbed) gold, or gold that has been exposed on the surface of the Earth to transport by water. The Busang area is supposed to be a volcanic hard-rock area where the gold should be virtually all "finely distributed with sharp, pointed edges." That both types of gold were showing up in the Bre-X supplied samples is highly unlikely and should have raised suspicions.

Bre-X also did not follow several normal industry practices, such as splitting cores in half. Traditionally, one half of the core would be crushed to assay the amount of gold it contained while the other half would be kept in reserve for retesting if necessary. Bre-X should also have been keeping a library core or a small section of about 10 centimeters out of every 2 meters drilled. Finally, it was revealed that rather than shipping the cores directly from the field to an assay laboratory, the cores were marked up, possibly crushed, and shipped to a warehouse in Samarinda, where they sometimes sat for several weeks before being shipped out. According to one report, the cores that went through Samarinda ended up having gold in them, while the cores that did not had no gold in them.

Following the release of Strathcona's report, Bre-X stock plummeted to 80 cents a share. On May 8, Bre-X filed for bankruptcy protection and Felderhof was fired. Shortly after that, the stock was delisted. Bre-X faced a number of lawsuits by angry investors who had lost billions, including the Ontario Municipal Employees Retirement Board (loss of $45 million), the Quebec Public Sector Pension fund ($70 million), and the Ontario Teachers Pension Plan ($100 million).

David Walsh, the CEO of Bre-X, moved to the Bahamas. He died in June 1998, of a brain aneurysm. In May 1999, the Ontario Securities Commission filed eight counts of illegal insider trading against Mr Felderhof, alleging that he pocketed $84 million by using non-disclosed information. One day later, the Royal Canadian Mounted Police (RCMP) announced it was ending its investigation into Bre-X without laying criminal charges against anyone. An RCMP spokesperson was quoted as saying, "There was no doubt that fraud took place, but we have no evidence." Two major obstacles to the investigation were that key witnesses were dead and the fact that witnesses in Indonesia and the Philippines were difficult or impossible to locate and could not be compelled to testify. Despite the dropping of criminal charges, civil class action suits continued. Bre-X finally went bankrupt in 2002.

John Felderhof was the only Bre-X official ever charged. His trial began in October 2000, and lasted on and off for several years, with Mr Felderhof *in absentia*. He had been living in the Cayman Islands, which does not have an extradition treaty with Canada, since the collapse of Bre-X, though some reports placed him in other countries. On July 31, 2007, Mr Felderhof was found not guilty of insider trading.

A recent television version of this story tries to pin all of the blame for these events on de Guzman, but its account contains several factual errors. For better information on the events, the Canadian Broadcast Company has a nice website of archived news clips that outline the Bre-X story (http://archives.cbc.ca/IDD-1-73-1211/politics_economy/bre-X/).

Mineral Groups

Now we go back to the story of how minerals form, which in turn explains why you find different minerals in different geologic settings. The varieties of minerals we see distributed around the world are all based on the types of chemical elements available during their formation. There are over 4000 currently accepted minerals, but because of the limited abundance of elements present in the Earth's crust, only about 20 to 30 of these minerals are common. The exact distribution of minerals at a specific location is related to the local bedrock, local depositional environment, and the local climate. There can be significant differences over relatively short distances, making minerals a very useful forensic tool. Table 2.1 lists the most abundant elements in the Earth's crust (these numbers will vary a bit depending on which reference you consult). Note that carbon, one of the most abundant elements in the biosphere, is not among the top eight.

The most common minerals are those based on oxygen and silicon, the two most abundant elements in the Earth's crust. These minerals are called the *silicates*. Silicate minerals form approximately 95% of the continental crust; the oceanic crust is composed almost entirely of silicate minerals. All of the minerals in this group are based on a structure called the silicon–oxygen tetrahedron, composed of four oxygen atoms covalently bonded to one silicon atom $(SiO_4)^{4-}$ to form a little

Table 2.1 Approximate elemental composition of the continental crust.

Element	Weight %
Oxygen (O)	47.2
Silicon (Si)	28.8
Aluminum (Al)	7.96
Iron (Fe)	4.32
Calcium (Ca)	3.85
Sodium (Na)	2.36
Magnesium (Mg)	2.20
Potassium (K)	2.14
All others	1.17

Source: Data compiled from Wedepohl (1995).

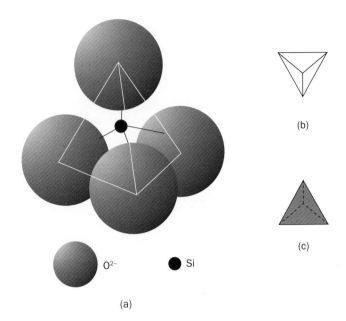

Figure 2.13 (a) Ball and stick model of a silicon–oxygen tetrahedron, (b) geometric symbol representing a silicon–oxygen tetrahedron with the point facing up, (c) geometric symbol representing a silicon–oxygen tetrahedron with the point facing into the page.

Figure 2.14 Neosilicates (single-tetrahedron) structure.

Figure 2.15 Single chain inosilicate structure.

three-sided pyramid (Figure 2.13). This anionic structure has a charge (negative 4), so it is not complete and is capable of bonding to itself and to cations in many different ways to form a variety of different minerals.

The silicates are divided into seven groups based on the manner in which the silicon–oxygen tetrahedra are linked together. In the *neosilicates* (single-tetrahedron), the tetrahedra are isolated and are only linked together by ionic bonds with metal cations (Figure 2.14). Olivine and garnet are examples of neosilicates. Slightly more complex are the *single chain inosilicates*, where the tetrahedra form a chain by sharing two oxygen atoms each (Figure 2.15). The minerals of the pyroxene group have this structure. In *double chain inosilicates* the tetrahedra form a double chain by sharing either two (on the outside of the chain) or three (on the inside) oxygen atoms (Figure 2.16). The minerals of the amphibole group, which includes hornblende, are an example of this structure.

The *phyllosilicates*, or sheet silicates, form two-dimensional sheets because all of the tetrahedra share all three basal oxygen atoms (Figure 2.17). Mica and most clay minerals have this structure. In the *tectosilicates* (framework silicates), the tetrahedra all share all four oxygen atoms to create a three-dimensional network,

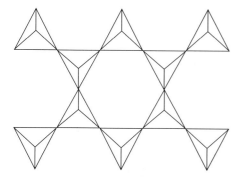

Figure 2.16 Double chain inosilicate structure.

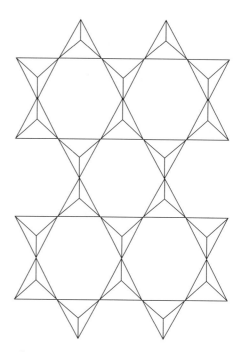

Figure 2.17 Phyllosilicate or sheet silicate structure.

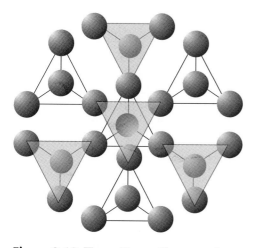

Figure 2.18 Tectosilicate (framework silicates) structure.

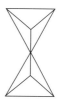

Figure 2.19 Sorosilicate structure.

forming minerals like feldspar and quartz (Figure 2.18). The *sorosilicates* share one oxygen between two tetrahedra, creating isolated double-tetrahedra structures that are linked together by metal cations (Figure 2.19). Though there are many members of this group, most are quite rare, epidote being the most common. Finally, *cyclosilicates* (ring silicates) contain rings of three, four, or six linked SiO_4 tetrahedra. Examples of this group include beryl and tourmaline (Figure 2.20).

Silicate minerals are also divided roughly into two groups based on their chemical composition. Minerals that contain significant amounts of iron and magnesium are called *ferromagnesian* (or mafic) minerals, such as olivine, hornblende, biotite, and the pyroxenes. Minerals that do not contain iron and magnesium, such as quartz and the feldspars, are called *non-ferromagnesian* (or felsic) minerals.

While the silicates clearly dominate, there are many other important mineral groups. About 20 elements can be found in the environment in their natural state and are referred to as *native elements*. The most important examples of this group are the native metals: gold, silver, copper, platinum, and iron. Also in this group are the elements arsenic, sulfur, diamond, and graphite.

The *sulfide* mineral class, in which metal cations are bonded to sulfur, includes the majority of the important metal ore minerals. Most minerals in this group are

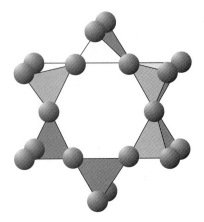

Figure 2.20 Cyclosilicate (ring silicate) structure.

opaque, fairly soft, heavy, and metallic-looking. The most commonly known members of this group are pyrite (fool's gold) and galena.

Oxides are composed of oxygen anions bonded to one or more different metal cations. Most members of this group are hard, heavy, and refractory, a term that means that they are hard to work with. Therefore, even though they might contain valuable elements like iron, they are not considered ores because it is almost impossible to extract the elements of economic interest. Some important oxides are spinel, magnetite, and hematite.

Halides are minerals that have elements from Group 17 of the periodic table (F, Cl, Br, and I) as the dominant anion. Each of these elements forms large ions with a charge of negative one, thus they tend to bond almost purely ionically with cations that have a charge of positive one or two. The resulting structures are highly symmetrical. Examples include halite (NaCl), sylvite (a common salt substitute), and fluorite.

The *carbonates* are built around the anionic $(CO_3)^{2-}$ complex and, unlike the silicon–oxygen tetrahedra, never share oxygen atoms. The bonds forming this complex are strong, but not as strong as covalent bonds, and are susceptible to breakdown when exposed to hydrogen ions (i.e. acid). This is the basis of the common effervescence (fizz) test that is used to identify carbonate minerals. Examples include calcite, dolomite, and aragonite.

There are many ways to analyze minerals, from using polarized light microscopes to very sophisticated instruments that determine elemental composition and atomic arrangement. Each of these methods is useful, but the best way to initially get acquainted with the most common minerals, and learn to understand their properties, is to work with hand samples. The only tools you need for this are a hand lens (like a jeweler's loupe) or magnifying glass, a piece of copper (US pennies are now zinc with a thin coat of copper so you are better off buying a sheet of copper from a hobby store and cutting it up, unless you have a US penny that predates 1982), a piece of glass, a streak plate (a tile of unglazed porcelain), a steel file or nail, a small magnet, and a small dropper bottle of dilute hydrochloric acid (2–3% HCl). More sophisticated approaches to mineral identification will be introduced in later chapters.

Properties of Minerals

Since there is a fairly limited set of common elements available for mineral formation and strict rules govern how those elements can be arranged, many minerals can be identified simply based on some common physical properties that

Figure 2.21 Examples of mineral luster, from left to right: metallic luster (two pieces of galena and one cube of pyrite), resinous luster (middle rear copal, middle front orpiment), and vitreous (two pieces of quartz). Please see Color Plate section.

Figure 2.22 Examples of the range of possible colors of the mineral fluorite. Please see Color Plate section.

are the byproduct of atomic arrangement and bonding. These physical properties allow us to distinguish between minerals and thus to identify many of them without having to resort to complex optical or analytical methods. Among the common properties used are:

- **Luster** is the way that light reflects off the fresh surface of a mineral. The two major classes of luster are *metallic* and *nonmetallic* (Figure 2.21). Minerals that look like metals have a *metallic* luster. Nonmetallic minerals can have a range of different appearances, the most common of which are *vitreous* (like freshly broken glass), *resinous* (like resin or dried honey), *pearly* (iridescent or like a pearl), *greasy* (as if covered with a thin layer of oil), *silky* (silk-like), *dull/earthy* (lacking reflection/like dry soil), or *adamantine* (brilliant, like a diamond) luster. This is usually the first property assessed when identifying a mineral. You need to look at a fresh, unweathered surface to determine a mineral's true luster.
- **Color** is ascertained by looking at a fresh, clean surface in reflected light. It is usually a mineral's most apparent characteristic, but this one is a bit dangerous, since most minerals occur in more than one color (Figure 2.22). Slight variations in chemistry or impurities can cause a mineral to have a range of different colors.

Each of the colors of a mineral is given a *variety* name. For example, quartz appears in white (milky quartz), pink (rose quartz), purple (amethyst quartz), grey (smoky quartz), clear (rock crystal), and orange (citrine), just for starters. There are some minerals where color is diagnostic, such as sulfur, which is always an opaque yellowish-green, and gold, which is always gold, but in many cases color is not particularly helpful. Color can also vary over a small area, so care should be taken to ensure that any color comparisons made are looking at a truly representative sample of the reference materials.

- **Transparency** (technically *diaphaneity*) is the ability of a mineral to transmit light. This property is not diagnostic in itself but is an important aspect of the concept of color. If light is transmitted through a sample with little disturbance, the mineral is considered *transparent* (like clear glass). If light passes through a mineral but is disturbed or distorted, the sample is *translucent* (like foggy or sandblasted glass). If no light passes thorough, the sample is *opaque*. This property should be recorded along with the color information (i.e. transparent purple is quite different from opaque purple)

- **Hardness (H)** is a measure of the resistance of a mineral to scratching or abrasion. The stronger the atomic bonds are, the stronger the resulting mineral is and the harder it is to scratch. In 1822, Austrian mineralogist Friedrich Mohs created a scale of relative hardness using a set of ten minerals, where higher-numbered minerals will scratch lower-numbered minerals. Each mineral can scratch those with a lower number on the scale but cannot scratch those with a higher number on the scale. Two minerals of the same hardness can scratch each other. Mohs' scale of hardness is shown in Table 2.2.

To test a mineral's hardness, you try to scratch an unweathered surface of the unknown mineral with an object of known hardness, such as your fingernail or a piece of glass. You also have to check that you have really made a scratch, and have not just left a trail of powder behind. If the unknown mineral is scratched by your fingernail, then you know it has a hardness of less than 2.5. If the mineral scratches glass, then you know that it has a hardness of greater than 5.5.

Table 2.2 Mohs' scale of hardness.

Mohs Number	Mineral	Hardness of reference items
1	Talc	
2	Gypsum	2.5 Fingernail
3	Calcite	3.5 Copper
4	Fluorite	
5	Apatite	5.5 Glass 5.5 Steel knife blade
6	Orthoclase	6.5 Ceramic streak plate
7	Quartz	
8	Topaz	
9	Corundum	
10	Diamond	

In addition to the common objects listed above, you can also use commercial *hardness kits* or a set of metal scribes inset with materials of known hardness. This is most useful if you are working with minerals harder than glass.

- **Streak** is the color of a mineral after it has been ground into a powder. This property is a bit more reliable than color, since the streak of a mineral is usually the same, regardless of the variety. All quartz has a white streak, regardless of the color of the crystal, and all hematite has a reddish-brown streak. The easiest way to determine a mineral's streak is to use a square of unglazed porcelain called a *streak plate*. They typically come in both white and black, to allow you to ascertain the streak of a range of different minerals. One limitation of this method is that streak plates have a hardness of about 6.5, so anything harder than the streak plate will not leave a streak and instead will scratch the porcelain. To determine the streak of minerals harder than around 6, you need to crush a small piece of the mineral with a hammer.

- **Crystal habit** is the external shape of a complete crystal of a mineral, which only occurs when a mineral is unrestricted in its growth. In nature, perfect crystals are rare. The faces that develop on a crystal depend on the space available for the crystals to grow. If crystals grow into one another, or in a confined environment, it is possible that no well-formed crystal faces will be developed. Crystal habit is a description of the shapes and aggregates that a certain mineral is likely to form, the external shape of the crystal, with all of its bonds intact. This is different from the more complex classification scheme called *crystal form*, which describes how a set of crystal faces are related to each other by symmetry. For now, we will stick to using purely descriptive terms that can be applied to well-formed individual crystals and to distinctive aggregates (groups) of crystals (Figure 2.23 and Table 2.3).

(a)   (b)   (c)   (d)   (e)   (f)

(g)   (h)   (i)   (j)

Figure 2.23 Some common crystal habits: (a) acicular, (b) bladed, (c) cubic, (d) dodecahedral, (e) octahedral, (f) tabular, (g) dendritic, (h) fibrous, (i) globular, (j) rosette. Please see Color Plate section.

Table 2.3 Common crystal habits.

Individual crystals	
Term	**Description**
Acicular	Long, slender, needle-like crystals
Bladed	Flattened, elongated crystals, like knife blades
Blocky	Shaped like a block (like the wooden blocks children play with)
Cubic	Cube-shaped
Dodecahedral	Shaped like a dodecahedron (with 12 faces)
Equant	Any shape approximately of equal height, width and depth, ideally a cube or sphere, but this term is used to describe minerals that have all of their boundaries of approximately equal length
Octahedral	Shaped like octahedrons (with eight faces)
Platy	Thin, flat crystals like plates, wider than bladed and thinner than tabular
Prismatic	Columnar or "pencil-like," elongated crystals that are thicker than needles. One of the most common of crystal habits
Tabular	Rectangles, like a table or like a book, thicker and shorter than bladed crystals

Aggregates, or clusters, of crystals	
Term	**Description**
Botryoidal	Smooth bulbous or globular form that looks like a bunch of grapes
Dendritic	Tree-like growths, branching growth like the pattern of frost that forms on windows in winter
Drusy	Small crystals that cover a surface, like the inside of a geode
Fibrous	Aggregate of slender, parallel crystals that look like fibers
Globular	Radiating individual crystals that form spherical groups
Lamellar	Layered like a stack of playing cards
Massive	Mass of crystals with no distinct shape
Reniform	Round, kidney-shaped masses
Reticulated	Lattice-like groups of slender crystals
Rosette	Petal-like crystals arranged in a flattened radial habit around a central point
Radiated	Radiating groups of crystals
Stellated	Radiating individuals that form a star-like shape

- **Cleavage** is the tendency of a mineral to break or split along planes of weakness in a crystal. The planes that crystals are most prone to split along are planes where atomic bonding is weakest. This is different from crystal form, which requires all bonds to be intact. Crystal cleavage is a smooth break producing what appears to be a flat crystal face. Cleavage is reproducible, meaning that a crystal can be broken along the same parallel plane over and over again, and the same mineral will always have the same cleavage. All cleavage planes of a mineral must match that mineral's symmetry. The tendency for minerals to cleave or not and in which directions is very characteristic and therefore helpful for their identification. For example, mica breaks easily into thin sheets (Figure 2.10), while halite (salt) breaks into smaller and smaller cubes (Figure 2.24).

  Cleavage is described in terms of how easily the cleavage is produced. From easiest to hardest to produce the terms are *perfect*, *imperfect*, *good*, *distinct*, *indistinct*, and *poor*. The angle between cleavages is also important to note and may be diagnostic. Minerals with just one cleavage plane, like mica, are said to have *basal cleavage*. The pyroxene and amphibole groups of minerals are distinguished primarily by angle of cleavage. Pyroxenes have two cleavage planes at approximately 90°, while amphiboles have two cleavage planes that are not at 90°, meeting at 56° and 124° instead. Three cleavage planes that meet at 90° angles result in *cubic cleavage* (Figure 2.24), and three cleavage planes that don't meet at 90° result in *rhombohedral cleavage* (Figure 2.25a, b). Minerals with four cleavage planes are said to have *octahedral cleavage*, and minerals with six cleavage planes have *dodecahedral cleavage*.

  It is possible to identify cleavage without damaging a crystal, by looking carefully inside a crystal. Narrow cracks, called cleavage traces, that run parallel to planes of symmetry and corresponding to incipient cleavage zones can often be seen (Figure 2.25a and Figure 2.26).

  If a mineral breaks irregularly, like glass shattering, this is referred to as *fracture*. The most common fracture type is *conchoidal* (Figure 2.27), a smoothly curved

Figure 2.24 Halite has perfect cleavage in three planes that intersect at right angles. Hitting a piece of halite with a hammer results in smaller and smaller, roughly cubic, pieces.

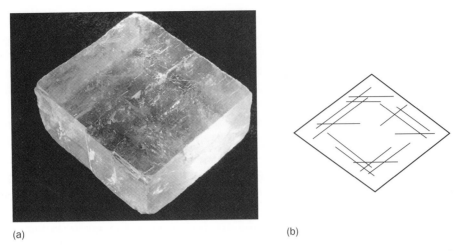

(a)   (b)

Figure 2.25 (a) Cleavage planes visible in a crystal of calcite, (b) calcite crystals have three planes of cleavage that do not meet at right angles. Perfect crystals of calcite are rhombohedrons.

Figure 2.26 Photograph of transparent oligoclase, a plagioclase that displays 90° cleavage (arrow 1) as well as conchoidal fracture (arrow 2).

Figure 2.27 Close-up of conchoidal fracture.

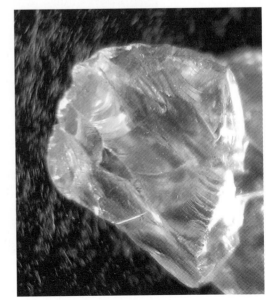

fracture familiar to people who have examined broken glass. Quartz has this fracture type, and almost all specimens that have been broken demonstrate this fracture type very well. While quartz has no cleavage, it is possible for a mineral to display both cleavage and fracture (Figure 2.26).

- **Specific gravity (SG)** is the ratio of the weight of a substance to the weight of an equal volume of water at 4 °C, at which point water is defined as weighing 1 gram per cubic centimeter (g/cm^3). For example, if a ruby weighs 1 gram (5 carats) and an equal volume of water weighs 0.250 grams, the SG of the ruby is 4 (calculated as $1/0.250 = 4$). Determination of SG is fairly easy and, most importantly for the gem trade, nondestructive. In crystalline materials, SG is determined both by the kinds of atoms present (atomic weight) and how the atoms are packed and bonded together. In general, atoms with a higher hardness also have a higher SG.

The simplest way of estimating SG is by performing a *hydrostatic weighing* test. This test takes advantage of the fact that imporous solids placed in water are buoyed up and therefore weigh less than in air. The loss in weight is equal to the weight of the water displaced. To determine SG, the gemstone is first weighed dry in air and then again while submerged in water. Technically, this procedure is best preformed with an analytical balance equipped with a hydrostatic weighing apparatus, but you can get a good estimate of SG by doing the following.

First, place your sample on an accurate balance and record its dry weight (s_{dry}). Next, partially fill a graduated cylinder with water, deep enough that you can fully submerge your sample without it touching the bottom but not so full that you will spill any water. Put the graduated cylinder onto a scale and weigh the whole thing (w_{cw}). Now tie your sample using a thin piece of fishing line or wire. Carefully lower the mineral into the graduated cylinder so that your sample is completely covered by water but does not touch the side or bottom of the container. You might also want to gently jiggle your sample to ensure that no air bubbles are trapped underneath. Once the mineral is settled in the water, tie it to a support. You can take the measurement while simply holding the mineral, but it will not be as accurate since your hand will be moving slightly. Record the weight of the graduated cylinder with the water and submerged mineral (w_{cwm}). Subtract the weight of the graduated cylinder plus water (w_{cw}) from the weight of the graduated cylinder plus water with the mineral submerged (w_{cwm}) to obtain the weight of the water displaced by the mineral (w_{water}). The SG of the sample is equal to the weight of the dry mineral (s_{dry}) divided by the weight of the displaced water (w_{water}).

$$SG = \frac{s_{dry}}{w_{cwm} - w_{cw}} = \frac{s_{dry}}{w_{water}}$$

Your results should always be greater than one and unitless. While this method will generally suffice for most common minerals, there are several more sophisticated methods of determining SG, which must be employed when working with small samples.

A hydrostatic weighing apparatus is a more accurate way of determining SG. The first step is once again to weigh the sample on the balance in air (a). Next, take a beaker that is two-thirds full of water and suspend it above the balance on a bridge or from a ring stand. The beaker should not be in contact with the balance. Now, immerse the suspension cage in the water and hang it from the stand that sits on the pan of the balance. Place your sample in the submerged cage so that it dangles in the water without touching the side or bottom of the beaker, ensuring that you do not spill any of the water. Record the weight of the sample suspended in water (w). SG is determined as follows:

$$SG = \frac{a_{\text{weight of stone in air}}}{a_{\text{weight of stone in air}} - w_{\text{weight of stone in water}}}$$

For example, if you examine a stone sold as a topaz and find that the gem weighs 1.950 grams in air and 1.214 grams when immersed in water, you find that your sample has the following SG:

$$SG = \frac{1.950}{1.950 - 1.214} = 2.65$$

Consulting your reference tables, you can see that this stone is clearly not topaz.

Other Special Properties

- **Magnetism**: usually magnetite will be the only mineral that will be attracted by a magnet. For a more sensitive analysis, a magnetic separator can be employed.
- **Reaction to acid**: minerals that contain carbonate ions (CO_3) will react to dilute hydrochloric acid by effervescing, or fizzing. This test is used to identify calcite (vigorous fizzing), dolomite (much less vigorous fizzing), and azurite (a blue mineral that will fizz).
- **Taste**: halite, the mineral that is used as table salt, has a very distinctive taste, but you should not use a taste test on laboratory specimens.
- **Odor**: minerals that contain sulfur have a distinctive rotten-egg smell.
- **Feel**: very soft minerals, like talc, often have distinctive soapy or greasy feel.

By using basic physical properties and the identification table included here (Table 2.4; note that an extended version of Table 2.4 is available on the companion website: www.wiley.com/go/bergslien/forensicgeoscience), or better yet those found in a textbook on mineralogy, and the associated flow charts (Figure 2.28), it is possible to quickly differentiate most common, and many rare, minerals. This information is important, because minerals are the basic components of most types of forensic geologic evidence, such as rocks, sand, and, most commonly, soil. By examining hand samples using the simple tools described here, many minerals will quickly become familiar, making it easier to identify them under less ideal circumstances.

Gems and gemstones, which are just minerals that people have taken a special fancy to, are introduced in a later chapter on optical methods for two reasons. First, gems and gemstones are not commonly found in forensic samples, and instead are themselves objects of crime. Second, the methods that are often used to identify common minerals, such as scratch tests or hydrochloric acid, are damaging and anathema to the identification of potentially invaluable gem materials.

This procedure is meant for hand samples, so named because they would fit neatly into your hand, and only applies to common minerals. Also, note that some properties are subjective, like light-colored versus dark-colored. You need to keep an open mind while examining minerals.

1. Determine whether the specimen is a metallic or non-metallic. If you are uncertain about the luster, it is probably non-metallic.

2. For **Metallic and Submetallic Minerals**

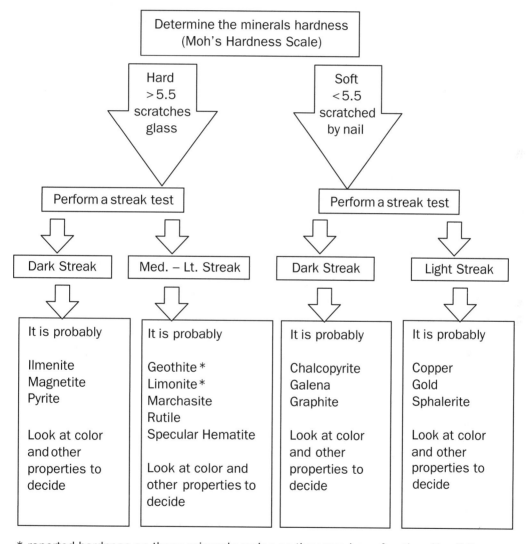

* reported hardness on these minerals varies so they may be softer than H = 5.5.

Figure 2.28 Mineral identification procedures.

2. For **Non-metallic Minerals**
Section I. Light Colored

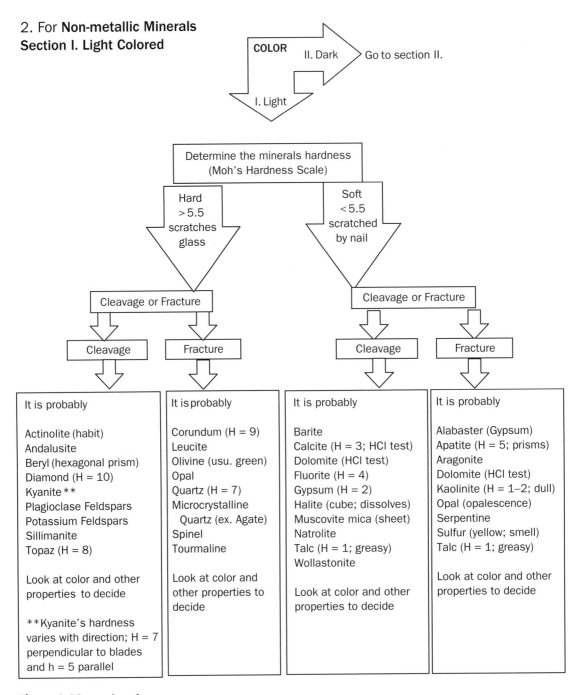

Figure 2.28 *continued*

2. (cont.) For **Non-metallic Minerals**
Section II. Dark Colored

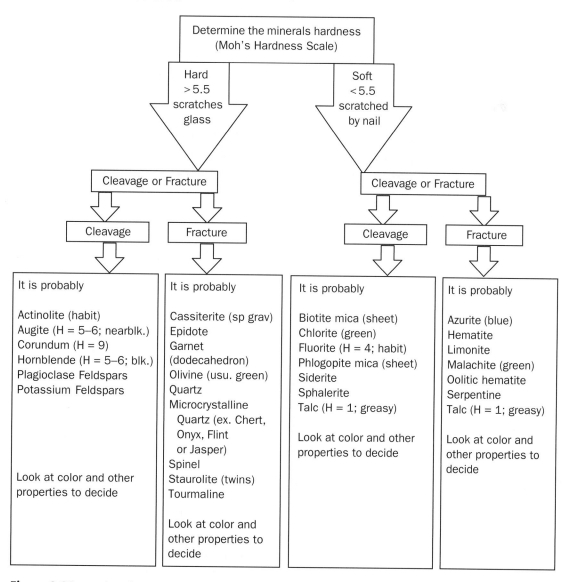

Determine the minerals hardness
(Moh's Hardness Scale)

Hard
>5.5
scratches
glass

Soft
<5.5
scratched
by nail

Cleavage or Fracture

Cleavage or Fracture

Cleavage

Fracture

Cleavage

Fracture

It is probably

Actinolite (habit)
Augite (H = 5–6; nearblk.)
Corundum (H = 9)
Hornblende (H = 5–6; blk.)
Plagioclase Feldspars
Potassium Feldspars

Look at color and other
properties to decide

It is probably

Cassiterite (sp grav)
Epidote
Garnet
(dodecahedron)
Olivine (usu. green)
Quartz
Microcrystalline
   Quartz (ex. Chert,
   Onyx, Flint
   or Jasper)
Spinel
Staurolite (twins)
Tourmaline

Look at color and
other properties to
decide

It is probably

Biotite mica (sheet)
Chlorite (green)
Fluorite (H = 4; habit)
Phlogopite mica (sheet)
Siderite
Sphalerite
Talc (H = 1; greasy)

Look at color and other
properties to decide

It is probably

Azurite (blue)
Hematite
Limonite
Malachite (green)
Oolitic hematite
Serpentine
Talc (H = 1; greasy)

Look at color and
other properties to
decide

Figure 2.28 *continued*

Table 2.4 Listing of the physical characteristics of common or important minerals (short version).

Mineral	Color/ Diaphaneity	Streak	Luster	H	SG	Cleavage	Habit
Actinolite (Amphibole)	Green Transparent to translucent	White	Vitreous	5–6	3.1–3.3	Perfect at 56° and 124°	Prismatic, radiating, massiv
Albite – Anorthite (Plagioclase)	White, gray or colorless Translucent to opaque	White	Vitreous to dull	6–6.5	2.6–2.75	Perfect in one, good in another at ~90°	Blocky, tabular, platy
Anhydrite	White, gray, colorless, blue Transparent to translucent	White	Vitreous	3.5	3.0	In three directions	Tabular, prismat
Apatite	Variable, green, yellow, blue, reddish-brown Transparent to translucent	White	Vitreous, greasy, resinous	5.0	3.1–3.2	Indistinct basal, conchoidal fracture	Prismatic, massi
Aragonite	White, colorless Transparent to translucent	White	Vitreous to dull	3.5–4	2.9	Distinct in one, poor in another	Prismatic, blade
Augite (Pyroxene)	Dark green to black Transparent to opaque	White to gray	Vitreous	5.5–6	3.2–3.6	Two good to perfect at ~90°	Prismatic, massi
Azurite	Deep blue, pale blue Translucent to opaque	Blue	Vitreous to dull	3.5–4	3.7	One good, one fair, Conchoidal fracture	Bladed, botryoidal, nodular, massive
Beryl	Green, blue, yellow, red, colorless, pink Transparent to translucent	White	Vitreous	7.5–8	2.6–2.9	Imperfect basal	Hexagonal prism
Biotite Mica	Black, dark brown Transparent to translucent	White to gray	Vitreous to pearly	2.5–3	2.7–3.4	Perfect basal	Sheets (Lamellar
Calcite	Variable, colorless, white, gray Transparent to translucent	White to gray	Vitreous to earthy	3.0	2.7	Perfect Rhombohedrons	Variable
Chalcedony (Quartz): Chert, Flint, Agate, Onyx	Gray, black, white Translucent to opaque	White	Dull to porcelaneous, pearly or subvitreous	7	2.5–2.6	None, conchoidal fracture	Cryptocrystalline mineraloid

ineral	Color/ Diaphaneity	Streak	Luster	H	SG	Cleavage	Habit
halcopyrite	Brassy-yellow Opaque	Greenish-black	Metallic or semi-metallic	3.5–4	4.2–4.3	Poor in one direction	Disphenoid or massive
hlorite	Green, white, yellow Transparent to translucent	Pale green	Vitreous, dull, pearly	2–3	2.6–3.4	Perfect basal	Tabular, massive
orundum	Variable Transparent to translucent	White	Adamantine to vitreous	9	3.9–4.1	None (parting), conchoidal fracture	Prismatic, bladed
olomite	Colorless, white, gray Transparent to translucent	White	Vitreous to pearly	3.5–4	2.9–3.0	Rhombohedral	Variable
nstatite (Pyroxene)	White, colorless, gray, light brown Translucent	White	Vitreous to pearly	5–6	3.2	Two perfect at ~90°	Prismatic, massive, lamellar
luorite	Varied Transparent to translucent	White	Vitreous	4.0	3.1–3.3	Perfect octahedral	Cubic, Equant
alena	Silver gray to lead Opaque	Lead gray	Metallic	2.5–2.8	7.2–7.6	Perfect cubic	Cubic, octahedral
arnet	Dark red to reddish brown Transparent to translucent	White or pink	Vitreous to resinous	7–7.5	3.4–4.2	None, conchoidal fracture	Dodecahedron, massive
raphite	Dark gray to black Opaque	Black	Metallic to dull	1	2.1–2.2	Perfect basal	Massive, laminar
ypsum	Colorless to white Transparent to translucent	White	Vitreous to pearly to silky	2.0	2.3	One good, two distinct	Variable
Ialite	Colorless to white Transparent to translucent	White	Vitreous	2.5	2.1–2.2	Perfect cubic	Equant, cubic
Iematite oolitic or cher	Rust red, brown, black Opaque	Dark rich red	Earthy	1–6.5	5.2–5.3	None	Tabular, botryoidal, massive
Iematite specular	Silver to gray Opaque	Dark rich red	Metallic	5.5–6.5	5.2–5.3	None	Tabular, massive

continued

Table 2.4 *continued*

Mineral	Color/ Diaphaneity	Streak	Luster	H	SG	Cleavage	Habit
Hornblende	Black, dark brown, dark green Opaque/ Translucent	Gray or pale green	Vitreous	5.0–6	3.0–3.5	Two perfect at 56° and 124°	Prismatic
Hypersthene (Pyroxene)	Brown, gray, green Translucent	White	Vitreous to pearly	5–6	3.4–3.9	Two perfect at ~90°	Prismatic, massive, lamellar
Limonite (Goethite)	Yellowish-brown to brown, orange Opaque	Pale brown	Varies	5.0–5.5	4–4.3	None	Massive, botryoidal
Magnetite	Black Opaque	Black	Metallic	5.5–6.5	5.2	Fracture	Equant
Malachite	Banded light and dark green, green Opaque to translucent	Green	Dull, silky	3.5–4	3.9	One good, hard to see	Botryoidal, globular
Microcline (Potassium Feldspar)	Off-white, yellowish, pink, brown, green Translucent to opaque	White	Vitreous, pearly, dull	6–6.5	2.5	Perfect in one, good in another at ~90°	Blocky, tabular
Muscovite Mica	Silver, white, yellow Transparent to translucent	White	Vitreous, silky, pearly	2.5–3.5	2.8–2.9	Perfect basal	Laminar, tabular
Oligoclase/ Labradorite/ Anorthite (Plagioclase Feldspars)	Off-white, gray, pale green, yellow, brown/gray to black/white, colorless Transparent to opaque	White	Vitreous to dull	6–6.5	2.6–2.8	Perfect in one, good in another at ~90°	Blocky, tabular
Olivine	Dark to pale green to yellowish or black Transparent to translucent	White or gray	Vitreous	6.5–7	3.2–4.4	Two poor at 90°, conchoidal fracture	Granular massive, tabular,

Table 2.4 *continued*

Mineral	Color/ Diaphaneity	Streak	Luster	H	SG	Cleavage	Habit
Orpiment/ Realgar	Orangey-yellow to yellow/orange to red Transparent to translucent	Yellow/ orange	Earthy to resinous	1.5–2	3.5–3.6	One perfect, one good	Massive, tabular, prismatic
Orthoclase (Potassium Feldspar)	Off-white, yellow, red, brown, orange Opaque to translucent	White	Vitreous to dull	6.0	2.5–2.6	Two good at 90°, conchoidal fracture	Blocky, tabular
Phlogopite Mica	Pale brown to brown Transparent to translucent	White	Vitreous to pearly	2.5–3	2.9	Perfect basal	Tabular, prismatic, lamellar
Plagioclase Feldspars	Off-white, colorless, gray, reddish, black Transparent to opaque	White	Vitreous to dull	6–6.5	2.6–2.8	Perfect in one, good in another at ~90°	Blocky, tabular, platy, bladed
Potassium Feldspars/ Alkali Feldspars	Pink, orange, white, brown, green Transparent to opaque	White	Vitreous to dull	6–6.5	2.5–2.6	Two good to excellent at 90°	Blocky, tabular
Pyrite	Brassy yellow (fool's gold) Opaque	Greenish-black	Metallic	6–6.5	5.0–5.2	Poor	Cubic, pyritohedral
Quartz	Any Transparent to translucent	White	Vitreous	7.0	2.65	None, conchoidal fracture	Prismatic, variable
Rutile	Black, reddish brown, yellow Opaque to translucent	Brown	Adamantine, submetallic	6–6.5	4.3	One distinct	Prismatic, acicular
Spinel	Red, but varies Transparent to opaque	White	Vitreous	8	3.5–4.1	None, conchoidal fracture	Octahedrons
Sulfur	Yellow Transparent to opaque	Yellow	Earth, dull vitreous	1.5–2.5	2.1	Two very poor	Massive, blocky

continued

Table 2.4 *continued*

Mineral	Color/ Diaphaneity	Streak	Luster	H	SG	Cleavage	Habit
Talc	White, silver, pale green, gray Translucence to opaque	White	Pearly to greasy or dull	1.0	2.6–2.8	Perfect basal, hard to see	Massive or lamellar
Tremolite (Amphibole)	White, gray, greenish, colorless Transparent to translucent	White	Vitreous, silky or dull	5–6	2.9–3.1	Perfect at 56° and 124°	Prismatic, fibrou

Summary

Readers should now be familiar with the basic properties of minerals and the principles behind mineral identification. They should also now have a feel for the wide variety of minerals found on the surface of the Earth and their potential as evidence. Minerals are the basic building blocks of most types of geologic evidence, so this information will come into play repeatedly in the following chapters.

Further Reading

Francis, D. (1997) *Bre-X: The inside story*. Key Porter Books Limited, Toronto, Ontario. [One of the mass-market books available, it is a decent version of the story, but the CBC online material, and archived newspaper articles, are better resources.]

Klein, C., Hurlbut Jr., C. S., and Dana, J. W. (1993) *Manual of Mineralogy: After James D. Dana*. John Wiley & Sons Ltd, New York. [A classic textbook used in many a college's mineralogy classes.]

Mottana, A., Crespi, R., and Liborio, G. (1978) *Simon and Schuster's Guide to Rocks and Minerals*. Simon and Schuster, New York. [One of the better guidebooks, currently available in its sixth edition.]

Perkins, D. 2001. *Mineralogy*, 2nd edn. Prentice Hall, Upper Saddle River, NJ. [A common introductory text used in mineralogy classes.]

Woodard, B. (1967) *Diamonds in the Salt*. Pruett Press, Boulder, CO. [A not entirely factual, but not completely ridiculous, account of the Great Diamond Hoax.]

References

Dragoset, R. A. Musgrove, A., Clark, C. W., and Martin, W. C. (2010), *Periodic Table: Atomic Properties of the Elements* (Version 4), NIST SP 966. [Online] Available: http://physics.nist.gov/PT [accessed 01/10/2012]. National Institute of Standards and Technology, Gaithersburg, MD.

Harpending, A. (1915) *The Great Diamond Hoax and Other Stirring Incidents in the Life of Asbury Harpending*. Edited by Wilkins, J. H. The James H. Barry Co., San Francisco.

Wedepohl, K. H. (1995) The composition of the continental crust. *Geochimica et Cosmochimica Acta* **59**(7): 1217–1232.

Chapter 3
Rocks: Storybooks of the Earth

With a basic understanding of mineralogy under your belt, it is now time to tackle the topic of rocks. Unlike with minerals, there is no straightforward technical definition for a rock. Rocks can be composed of consolidated aggregates of minerals, naturally cemented rock fragments, chemical precipitates, and/or naturally occurring solid accumulations of organic matter. A simple rule of thumb is that anything solid and not synthetic can be called a rock. There are three major classes of rock: *igneous*, *sedimentary*, and *metamorphic*, each typical of different geologic environments. Each class of rock has distinctive characteristics that result from its composition and processes of formation. Rocks have played a variety of roles in crimes. For example, they have been used as murder weapons and to weigh down bodies thrown into bodies of water. When examining trace evidence, such as sand or soil, its geologic components can be both pure minerals and fragments of rock, thus it is important to understand and be able to identify both types of materials.

This chapter goes into considerably more detail on rock classification than an average introductory geology book because the level at which this topic is usually introduced is insufficient for forensic differentiation. Despite the fact that there are only three major classes of rocks, there is an almost infinite amount of variation in kind within those classes. Rocks tell the geologic story of their formation and can potentially be used for *geosourcing*, or tracking down their points of origin. A stray rock found in the trunk of a car can be used to point back to a crime scene, or aid in locating a body. Plus, rock is the parent material for sediments and soils. It is important to get a true feel for the great diversity of rock types found on the surface of the Earth for a better understanding of their potential as evidence.

An Introduction to Forensic Geoscience, First Edition. Elisa Bergslien.
© 2012 Elisa Bergslien. Published 2012 by Blackwell Publishing Ltd.

The Cindy Rogers/Cheryl Renee Wright Case (Rapp, 1987)

In Lake County, California, just before 10 p.m., August 11, 1980, a woman named Cindy Rogers, her father Frank, and a family friend moved their lawn chairs out from a patio to be nearer the swimming pool where Cindy's son was playing. Shortly afterward, Frank got up to turn on the pool area lights. As he sat back down, a shot rang out and, to his horror, he watched his daughter die. Cindy had been killed by a bullet through the heart fired from a high-powered rifle with a telescopic sight. Witnesses nearby saw a man, wearing a jacket and carrying a gun, run past their house into the darkness. Soon afterward, they saw a car start to move. The headlights came on, the engine started, the car turned left onto Highway 20 eastbound, and accelerated away.

Around 10:50 p.m., a car traveling east approached a police roadblock, pulled to the side of the road briefly, then executed a U-turn, and fled westbound, back the way it came. Police officers pursued the car, following it along the highway and then down a private driveway. At the end of the trail, the officers found a silver Camaro abandoned behind a residence. It had skidded to a stop because a log blocked the road. A search of the car produced a loaded rifle, an empty rifle case, binoculars, a corduroy jacket, a knit cap, a ski mask, and some gravel on the floorboards of the front seat, among other items.

The previous day, while traveling north on Interstate Highway 5, 19-year-old Cheryl Renee Wright got a flat tire. Around 9 p.m., two service station employees saw her in the company of a man driving a silver Camaro. The man discussed getting the tires on her Vega fixed with the service station employees, then claimed to have a friend about six miles away who owned a Vega. He said that he could borrow tires from this friend to fix Cheryl Wright's car. She left the service station with the man, who drove toward the northbound I-5 ramp, which leads to Highway 20. The tow service operator at the station, who had lived in the area for more than 20 years and knew almost everyone, didn't know anyone in the area with a Vega. On August 17, 1980, Cheryl's badly decomposed body was found in Colusa County buried under loose gravel at an abandoned oil well off Highway 20. She had been killed by a shot to the head with a .25-caliber bullet fired from a semiautomatic pistol.

Finding Cindy Rogers' murderer turned out to be pretty straightforward. On July 20, 1980, Cindy had filed three felony charges against her new husband, Gerald Frank "Jerry" Stanley. They had married on July 7, 1980, four days after they first met. Nine days later came actions that led Cindy to file felony charges against her new husband. Two days after that, Stanley burned down her house. Two days later, he set fire to her car. The silver Camaro police found abandoned on the night of the murder had been rented on August 4 by Mrs Stanley, Gerald's mother. Gerald Stanley's fingerprints were found inside the car.

Meanwhile, in neighboring Colusa County, police were looking for Cheryl Wright's murderer. After witnesses reported seeing Cheryl with a man who matched Gerald Stanley's description traveling in a light-colored Camaro, a Colusa County deputy sheriff inspected the impounded vehicle. He noticed the small rock fragments on the floor of the vehicle and realized that they resembled the gravel that had been used to cover Cheryl's body. He took samples of the rocks in the car and collected four bags of gravel from the site where the body had been found. These samples were taken to the California Department of Justice Crime Laboratory, which in turn sent them to the California Division of Mines and Geology for identification.

These rocks turned out to be of great importance in linking Stanley with the murder of Cheryl Wright. The gravel from both the car and oil well site contained angular and sub-angular clasts of granite, granodiorite, quartzite, quartz-mica schist and rhyolite, consistent with a processed crushed rock called "3/8-inch pea gravel." These materials were distinctly different from the local rock, which consisted of andesite, dacite, serpentine, and quartz, as well as chert and siltstone. Based on the composition of the gravel, the Sheriff's department was able to determine that the gravel at the oil well site was from a crushed-stone company located in Bakersfield, some 320 miles away. According to company records, no other loads had been delivered to any locations within 50 miles of the crime scene.

In 1984, Gerald Stanley was convicted of the first-degree murder of Cindy Rogers Stanley, arson of an inhabited dwelling, and burglary of an inhabited trailer coach. The jury also determined that the case fit special circumstances that would make him eligible for the death penalty. In capital cases, where the death penalty is an option, the trial has two phases. During the guilt phase (where the jury makes a finding of guilty or not guilty), past convictions and unrelated criminal activity are not admissible as evidence. This is to ensure that the verdict is based solely on the merits of the case in hand.

Following a guilty verdict, there is a separate penalty phase in which evidence of a defendant's past criminal record and information about other criminal activities that have not gone to trial can be presented. Thus, the jury was initially unaware that Gerald Stanley had been convicted in 1975 of the murder of his second wife and that his third wife disappeared, her fate still unknown. It was in this second phase of Gerald Stanley's trial that the evidence regarding the murder of Cheryl Renee Wright was presented, in order to support the prosecution's contention that the death penalty was justified. There was never a separate trial for her. Based on an examination of several special circumstances that would make him eligible for the death penalty, the jury returned a verdict of death. A series of appeals and motions were filed following his 1984 conviction and Gerald Frank Stanley is currently on death row at San Quentin, still alive as of this writing. The gravel found in the Camaro was a key piece of evidence linking Cheryl Wright's murder to Gerald Stanley.

The Rock Cycle

The first step in rock classification is determining which of the three major classes it belongs to: igneous, sedimentary, or metamorphic. Rocks are the product of their environment of formation, and, more importantly, they can be altered by their environments into new forms. This concept is called the *rock cycle* (Figure 3.1). Planet Earth can be considered the ultimate recycler, acting on a global scale, as material is moved and altered by surface and subsurface processes. To visualize this, imagine a pool of molten rock, deep inside of the Earth. Now imagine there is a

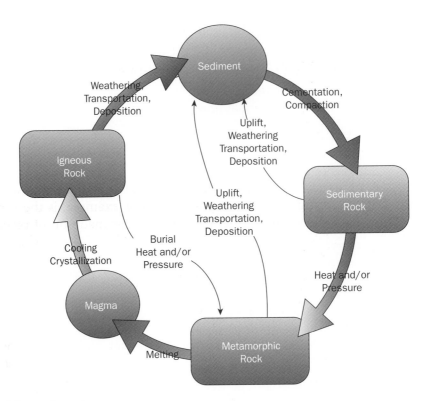

Figure 3.1 The rock cycle.

violent eruption that carries the molten rock to the surface of the Earth, where it cools to form *igneous rock*.

Our igneous rock experiences prolonged exposure to wind and rain, and significant changes in temperature. Through time, it breaks down into smaller and smaller pieces. These pieces are then carried by water, wind, or ice to a new location, where they are deposited. As the sediment piles up, the layers on the bottom are subjected to increasing pressure and chemical changes. Those sediments eventually become welded together to form *sedimentary rock*, and over time our sedimentary rock gets buried deeper and deeper by the addition of new layers of material.

Now imagine that our sedimentary rock is located on the margin of a continent. Over the course of geologic time, that continental margin collides with the margin of another continent. In extremely slow motion, the margins crumple and huge mountains rise. Deep in the heart of those mountains, our sedimentary rock is being subjected to enormous heat and pressure, causing it to alter into *metamorphic rock*. If our metamorphic rock is buried deeply enough, or heated up enough, it will eventually melt, starting the process all over again.

This generalized story is just the basic outline of the rock cycle. There are also shortcuts through the cycle. Imagine that, rather than melting, our metamorphic rock is eventually exposed at the surface again, as the overlying layers of rock are broken down and transported away. Our metamorphic rock will also eventually be disaggregated into sediments that could become components of a sedimentary rock. Or imagine if the molten rock we started with never erupted to the surface and instead stayed located deep inside the Earth. It could have cooled in place and eventually become part of a collision zone, transforming directly from igneous rock into metamorphic rock. As you can imagine, there are many different ways that rocks can move through this cycle, altering into new forms, but also often carrying some components of information about their previous state into their new incarnation.

Properties of Rocks

The key characteristics of rocks are somewhat different from those of minerals' and so some new terminology applies. The first thing you examine when classifying a rock will typically be its *texture*, or the general appearance and arrangement of the grains or crystals composing the rock (Figure 3.2). For example, is the rock crystalline (composed of inter-grown mineral crystals) or made up of pieces of sediment? Crystalline rocks can be further differentiated as *foliated*, where the minerals are aligned into layers or patterns, or *non-foliated*, where the minerals are randomly oriented. If the components of the rock, crystals, or particles are small (usually less than 1 mm), the rock is referred to as *fine-grained*, while rocks with components that are easily identifiable by the unaided eye (usually larger than 1 mm) are called *coarse-grained*. In contrast to minerals, *color* is often a diagnostic property for rocks, while Mohs' hardness scale is not very useful, since rocks are usually heterogeneous mixtures of multiple materials. Table 3.1 can be used to help determine the rock class of a sample of unknown origin. Once you know which

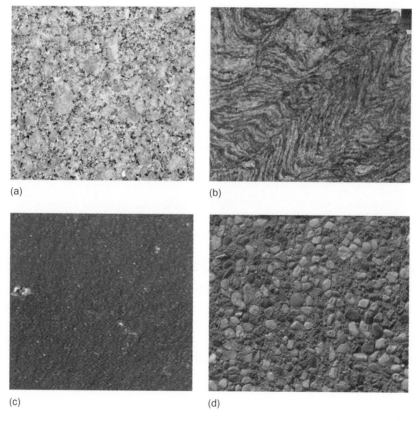

(a)

(b)

(c)

(d)

Figure 3.2 Examples of rock textures: (a) coarse-grained, crystalline, non-foliated rock (granite), (b) coarse-grained, foliated rock (gneiss), (c) fine-grained, crystalline, non-foliated rock, (d) rock formed of pieces of sediment (conglomerate).

Table 3.1 Classification of hand samples by rock type.

Texture	Characteristics		The rock is most likely
Rock is layered	Layers are flat; material is soft or crumbles easily		Sedimentary
	Layers have visible particles of gravel or sand		Sedimentary
	Layers are made of aligned, interlocking mineral crystals that are all about the same size		Metamorphic
	Crystals in the different layers are of different sizes and/or orientations		Sedimentary
	Layers are folded or deformed		Metamorphic
No layers, no visible grains	Looks like mud or dry clay		Sedimentary
	Looks like glass	Heavy, shiny, like window-glass (ex. obsidian)	Igneous
		Light, duller opaque black (ex. anthracite coal)	Metamorphic
	Does not look like glass	Very soft rock, easily scratched by fingernail	Sedimentary
		Very hard	Igneous
		Reacts to dilute hydrochloric acid	Sedimentary
Visible grains present	Contains fossils	Whole or pieces of shells, bones, plant material	Sedimentary
		Fossils are deformed or distorted	Metamorphic
	Grains are gravel, sand, and/or silt		Sedimentary
	Grains are rounded, smooth		Sedimentary
	Grains are obviously fragments of other rocks		Sedimentary
	Grains are mineral crystals	Crystals are randomly oriented in hard rock	Igneous
		Large crystals in a fine-grained matrix	Igneous
		Crystals are soft, easily scratched or broken	Sedimentary
		Flat crystals are foliated (lay parallel to each other)	Metamorphic
		Crystals react to dilute hydrochloric acid — Crystals all same size	Metamorphic
		Crystals react to dilute hydrochloric acid — Crystals different sizes	Sedimentary
		Crystals are halite or gypsum	Sedimentary
		Crystals are olivine, feldspar, pyroxene (silicates)	Igneous
		Crystals are garnet, serpentine, galena, sphalerite	Metamorphic
Rock looks metallic and scaly or smooth			Metamorphic
Rock contains bubble-shaped cavities (vesicles) or looks like meringue			Igneous
Rock is hard and looks ropy or blobby			Igneous

class your rock sample belongs to, you can further narrow your classification by referring to the detailed tables in the appropriate section.

Igneous Rocks

Igneous rocks, as previously described, form by the cooling and crystallization of molten rock (called *magma* if it is inside of the Earth and *lava* if it has erupted to the surface). The composition of igneous rock varies, based on the chemistry of the original magma source and its subsequent path toward cooling into its final state as a rock. The primary source of magma is deep inside of the Earth, thus we must start with understanding the Earth's basic structure.

The Earth is formed of a series of layers of different chemical compositions and/or physical behaviors (Figure 3.3). The outermost layer is called the *lithosphere* (brittle sphere) and is formed of strong, solid rock with an average thickness of 100 kilometers. The layer directly beneath the lithosphere is called the *asthenosphere* (weak sphere), which is also solid, but hot enough that it is able to bend and flow. Think of the juxtaposition of the hard candy layer (lithosphere) over the soft chocolate (asthenosphere) in an M & M candy (or Smarties in the United Kingdom).

The lithosphere is broken into a series of plates that move on top of the flowing asthenosphere. Places where the plates run into each other are called *convergent boundaries* (Figure 3.4a), and are the setting for volcanoes to form. Places where the plates are moving away from each other are called *divergent boundaries* (Figure 3.4b), and are places where magma from deep inside the Earth can rise to the surface in seams called *mid-ocean ridges or rift valleys*. Finally, the places where the

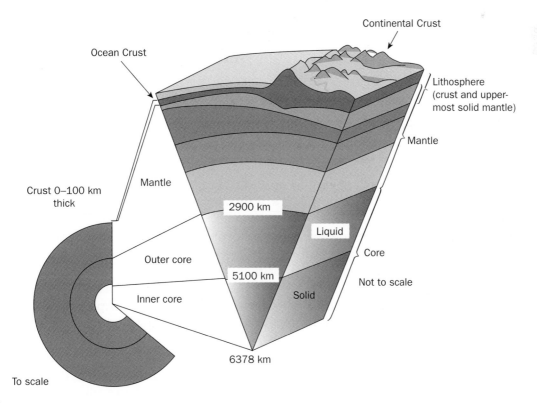

Figure 3.3 Geophysical layers of the Earth.

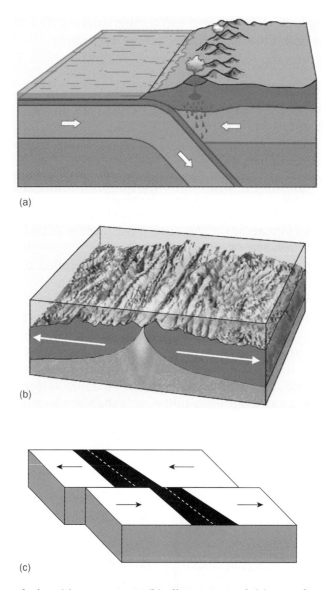

(a)

(b)

(c)

Figure 3.4 Plate boundaries: (a) convergent, (b) divergent, and (c) transform.

plates rub along each other moving in opposite directions are called *transform boundaries* (Figure 3.4c). Here no new rock forms (no volcanoes), but you get many earthquakes.

Magma typically forms inside the lower lithosphere or in the asthenosphere, and is generated or located at convergent and divergent plate boundaries. The other significant source of magma is *hotspots*, or places that are not associated with plate boundaries, where magma from deep inside the *mantle* (a layer defined by chemical composition, that includes the lowermost lithosphere, the asthenosphere, and a layer called the mesosphere) rises up to the surface. The Hawaiian Islands are an example of this last phenomenon. Molten rock is less dense than solid rock; therefore, it rises toward the surface of the Earth. If the magma cools inside of the Earth, it forms something called a *pluton*, or *intrusive* rock formation. If the molten rock rises up and erupts to the surface, thus becoming lava, the resulting solid rocks are called *volcanic*, or *extrusive*, *rocks*. Depending on the source of the magma, and the changes it goes through as it rises to the surface, different types of rocks are formed.

Classification of Igneous Rocks

The classification of igneous rocks is based on both the texture (Figure 3.5) of the rock and the mineral composition. Remember that, for crystalline rock, texture

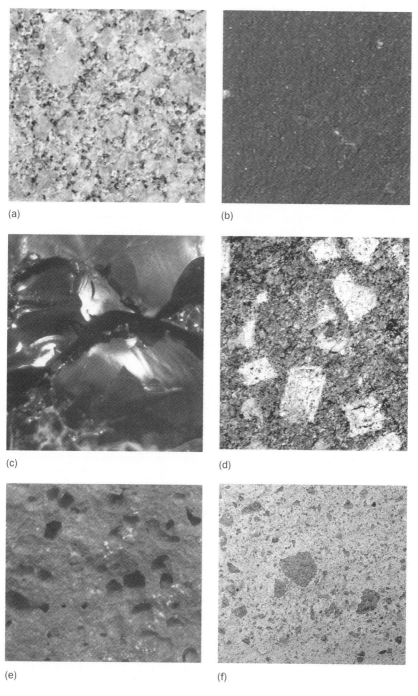

Figure 3.5 Igneous rock textures: (a) phaneritic (granite), (b) aphanitic (basalt), (c) glassy (obsidian), (d) porphyritic (andesite), (e) vesicular (basalt), (f) pyroclastic (volcanic tuff).

means the size, shape, and arrangement of the mineral grains. Crystal grains start to grow as the molten rock begins to cool. The longer it takes the magma to cool, the larger the crystals can grow.

Textures

- **Phaneritic**: coarse-grained rock formed of individual mineral grains that can be seen with the unaided eye. This texture results from a slow cooling of molten rock.
- **Aphanitic**: fine-grained rock formed of individual mineral grains too small to be seen with the unaided eye or even under the microscope. This texture results from moderately fast cooling of molten rock.
- **Glassy**: an amorphous texture that lacks a regular internal structure, the molten rock cools so fast that crystals do not have time to form.
- **Porphyritic**: results when a slowly cooling magma is suddenly erupted or forced near to the surface, quickly cooling the remaining molten material. This results in a rock with some large mineral crystals called *phenocrysts* in a matrix of very small crystals.
- **Pegmatitic**: results when magma cools extremely slowly, creating a rock with huge mineral grains.
- **Vesicular**: results from rapid cooling of magma or lava with a high gas content. The escaping gas forms bubbles in the cooling rock, leaving behind cavities called *vesicles*, which can be anywhere from <1 mm to several centimeters in diameter. Highly vesicular rocks, like pumice, can be so full of void space that they float.
- **Pyroclastic**: results from explosive volcanic eruptions when particles are expelled into the air from a volcanic vent. The resulting rock is composed of fragments and glassy particles.

Most rocks fall into the first two texture categories. As the molten mass solidifies, the type of rock that forms depends on the chemical composition of the magma. Different minerals will form, depending on the original chemistry of the magma, any changes that occur during crystallization, and the cooling stage of the material. This also means that some minerals will not occur together naturally in a rock. The complex succession of minerals that crystallize from molten rock is called *Bowen's reaction series*, after the scientist who first worked this process out (Figure 3.6).

In an idealized sequence, igneous rocks form in the following way: magma that starts out deep inside of the Earth is chemically similar to the middle layers of the Earth, and is rich in iron and magnesium as well as silicon and oxygen. If you solidified this magma, it would form the coarse-grained rock *peridotite*, the fine-grained rock *komatiite* (though, strictly speaking, the chemistry of a komatiite is different from that found in the modern mantle), and the porphyritic rock *kimberlite*, each of which are rich in the mineral olivine. Collectively, these rocks are called *ultramafic rocks* because they are high in iron and magnesium. Most of the mantle is composed of the rock peridotite, which is composed of the minerals olivine and pyroxene. Komatiites, or volcanic ultramafic rocks, are extremely rare because the conditions under which they could form only existed very early in Earth's history. They are high in magnesium, low in titanium, and are generally restricted to the oldest parts of the Earth's crust – places over two billion years old. The conditions under which komatiites formed no longer exist on the Earth. The

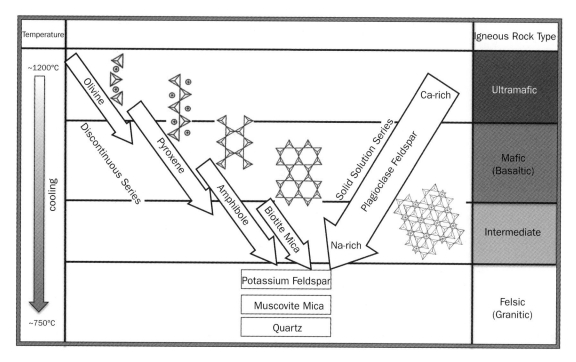

Figure 3.6 Bowen's reaction series.

process by which kimberlites form is complex and poorly understood, but they are of great significance, due to their relationship with diamonds.

As ultramafic magma starts to cool and travels toward the surface, it differentiates (separates), meaning that certain minerals will begin to crystallize out of the melt first. Iron and magnesium get used first, initially forming the mineral olivine, next creating pyroxene minerals, and then evolving to form amphibole minerals. From the same melt, calcium-rich plagioclase feldspar starts forming shortly after olivine does. In a continuous series, plagioclase feldspar keeps forming, but as the calcium is used up, more and more sodium starts to be incorporated. Magma at this stage solidifies to produce *mafic rocks*. The plutonic form is called *gabbro* (Figure 3.7a) and the volcanic form *basalt* (Figure 3.7b). To be considered a gabbro, a rock must contain pyroxene, calcium-rich plagioclase feldspar, some amphibole, and some olivine. Basalt has the same mineral composition, but the crystal grains are much too small to see. Basalt is the most common volcanic rock on the surface of the Earth and forms the vast majority of the ocean floor.

At the next stage in the magma's evolution, much of the iron and magnesium has been used, so olivine no longer forms. Instead, pyroxenes and amphiboles are the most common ferromagnesian minerals. Much of the calcium from the original melt has also been used, so the plagioclase that forms now incorporates a mixture of calcium and sodium. Rocks formed at this stage, known as the *intermediate rocks*, are the plutonic *diorite* (Figure 3.7c) and the volcanic *andesite* (Figure 3.7d). To be called a diorite, a rock must contain amphibole, plagioclase feldspar that contains a mixture of calcium and sodium, some pyroxene, and usually some biotite. Andesite is the fine-grained volcanic version with the same chemical composition.

Finally, at the lowest temperatures, almost all of the original iron, magnesium, and calcium have been used up. Pyroxene no longer forms, and the plagioclase that forms is based on sodium. Potassium feldspar now dominates, and since the melt is

Figure 3.7 Common igneous rocks: (a) gabbro, (b) basalt, (c) diorite, (d) andesite, (e) granite, (f) rhyolite. Please see Color Plate section.

still rich in silicon and oxygen, the mineral quartz forms. Also common are the minerals biotite and muscovite. Molten rock at this stage would solidify to form the plutonic rock *granite* (Figure 3.7e) and the volcanic rock *rhyolite* (Figure 3.7f). These rocks are referred to as *felsic*. Granite is the most common igneous plutonic rock on Earth and forms the bulk of the continents. Granite contains essential quartz, sodium-rich plagioclase, and potassium feldspar, usually with some minor amounts of hornblende, biotite, and/or muscovite. Rhyolite has the same mineral composition, but because it is fine-grained it usually looks pink and fluffy.

Due to historical precedent and philosophical differences, the technical classification of igneous rocks can be one of most confusing aspects of geology. Because the boundaries between igneous rock types are arbitrarily set and can be based on different characteristics, there are actually several different schemes in use around the world and over 1500 names. Since 1973, the International Union of Geological Sciences (IUGS) Subcommittee on the Systematics of Igneous Rock has been constructing a standard classification scheme. The following is a *simplified* version of the IUGS system. The key to any work done with igneous rocks is to work consistently with the same classification system, and to recognize that alternative systems exist.

In the IUGS system, rocks are classified based on the percentage of five minerals or minerals groups: quartz (Q), plagioclase feldspars (P), alkali feldspars (A), feldspathoids (F), and ferromagnesian minerals. This system works in the field for coarse-grained rocks as follows: the first major division is based on the percentage of ferromagnesian minerals (minerals rich in iron and/or magnesium) into (1) *feldspathic rocks*, in which feldspars and quartz dominate and there is less than 90% ferromagnesian minerals, and (2) the *ferromagnesian rocks*, which are dominated by minerals like olivine, pyroxenes, and amphiboles and composed of 90% or more ferromagnesian minerals.

For feldspathic rocks, the next major division is based on whether the rock contains quartz or feldspathoids (also called *foids*). The feldspathoid minerals – such as nepheline, magnesium-rich olivine (foresterite), nosean, perovskite, melilite, leucite, sodalite, haüyne, and melanite – do not occur with quartz. Rocks of this composition are also usually quite rare. (For reference, see Tables 3.2–3.4 on the companion website: www.wiley.com/go/bergslien/forensicgeoscience. Table 3.2 can be used for the identification of feldspathic phaneritic rocks with quartz. Table 3.3 is used for rocks containing feldspathoids, and Table 3.4 is used for ferromagnesian rocks.) For volcanic rocks, determining mineralogy is impossible without a chemical analysis, which means that the IUGS method, partly shown in Table 3.5 on the companion website, is much less commonly used.

Now that you are probably thoroughly confused, you should be relieved to know that for most practical purposes igneous rocks can be sufficiently identified using a simple field classification scheme (Table 3.6). It would only be under very special conditions that a detailed chemical examination and technical classification would be necessary. The vast majority of the time, sufficient classification can be performed using Table 3.7.

A Volcanic Crime

A man's body was discovered concealed in a forest in Yokohama, Japan. The soil where the body was found contained distinctive volcanic material ejected from Mt Fuji: black, strongly magnetic, porous basalt that contained phenocrysts of bytownite (a calcium-rich plagioclase). Upon investigation, authorities also found fragments of this distinctive rock on the floor mat and in the trunk of the victim's wife's car. The area where the couple lived was covered with loamy soil developed from thick layers of volcanic ash that did not contain basalt fragments. Using the unique rock fragments, investigators were able to link the woman to the crime. Apparently, she killed her husband and, with the help of her daughter, transported the body 100 kilometers to the west of their home, abandoning it in the forest where it was found.

Table 3.6 Simplified field identification scheme for common igneous rocks.

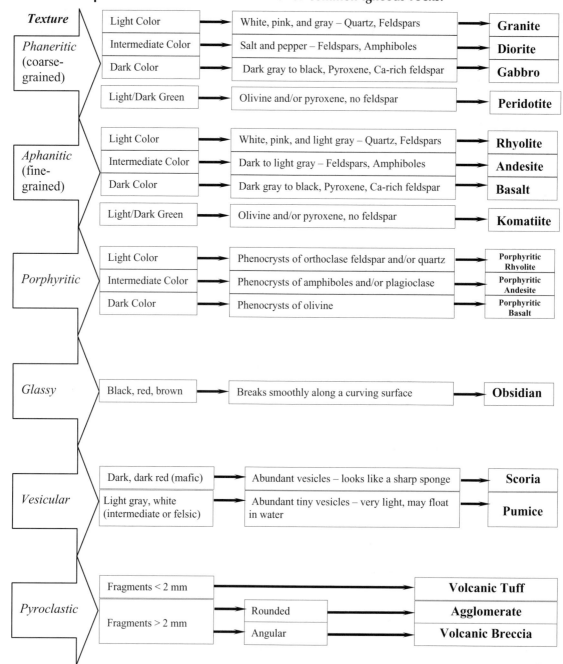

Table 3.7 Field classification of igneous rocks.

	Felsic 86–100% Silica (K-spar > Plagioclase)	Intermediate 46–85% Silica (K-spar < Plagioclase)	Mafic 16–45% Silica (<50% Olivine)	Ultramafic <15 % Silica (>50% Olivine)
Texture	**Rock Name**			
Pegmatitic	Pegmatitic Granite	Pegmatitic Diorite	Pegmatitic Gabbro	Pegmatitic Peridotite
Phaneritic	GRANITE (if no quartz SYENITE)	DIORITE	GABBRO	PERIDOTITE
Porphyritic	Porphyritic Granite	Porphyritic Diorite	Porphyritic Gabbro	Porphyritic Peridotite
Aphanitic	RHYOLITE (if no quartz TRACHYTE)	ANDESITE	BASALT	KOMATIITE (very rare)
Glassy	Obsidian			
Vesicular	Pumice (lightweight, fluffy rock)		Scoria (many vesicles) or Vesicular Basalt (few vesicles)	
Pyroclastic	Agglomerate (dominantly > 64 mm and rounded) Pyroclastic breccia (dominantly > 64 mm and angular) Lapilli tuff (pyroclasts 2–64 mm) Tuff or ash tuff (<2 mm)			

A Question of Source (Ruffell and McKinley, 2008)

One day, as a thick slab of rough-cut granite was being moved at the Belfast docks it fell from the forklift and broke, revealing a series of powder-filled drill-holes. The powder turned out to be very pure cocaine and local police were left with a complicated problem. Where, between the quarry and its arrival in Belfast, were the holes drilled and cocaine secreted inside of the granite? The first step was to determine to point of origin for the granite, which turned out to be a commercial variety of Vigo granite.

Vigo granite is an evenly grained, relatively homogeneous material with few cracks that is commonly used as ornamental stone. For example, it is often polished for use as high-quality flooring or as cladding on the exterior of buildings. Vigo granite is quarried out of the mountains of northwest Spain for export. This information gave investigators a starting point, but cut-stone slabs are often stored, sometimes for months, in various places along their routes. They may stay in the original quarry for a time before being moved into storage, where they may sit for a while longer before being moved to the docks, pausing again before final shipment. Once the slabs arrive at a new port, they once again may remain in storage for a time prior to use. When you have pinned down the point of origin, additional evidence must be examined in order to work out the full shipping route. It turns out that ornamental stone has a long history of use in smuggling, so this type of investigation is rather common.

Sedimentary Rocks

Sedimentary rocks are formed at the Earth's surface and can be composed of weathered pieces of pre-existing rocks (sediments), organic material, or created by chemical processes. While igneous rock forms the cores of the continents, sedimentary rocks make up some 75% of the exposed surface of the Earth. Different types of sedimentary rocks are created in the different environments on Earth. For example, rocks formed along a river (an alluvial setting) are different from the rocks formed by glacial processes. This means that there are more common types of sedimentary rocks than igneous rocks, but this also means that the different types of sedimentary rocks are generally more distinctive and easier to recognize.

The most common sedimentary rocks are composed of pieces of other rocks that have been altered, or *weathered*, by exposure in a particular climate and physical environment. Weathering not only breaks rocks and minerals down physically into smaller pieces but also alters minerals chemically, dissolving them or changing them into different minerals. In general, minerals are most stable at the same physical and chemical conditions under which they form. Thus, the minerals that form deeper inside the Earth, such as olivine and calcium-rich feldspar, are unstable at the surface of the Earth, and break down easily. Minerals that form at near-surface conditions, such as quartz, or on the surface of the Earth, such as kaolinite, are stable and hard to alter.

As indicated, there are two primary forms of weathering: chemical weathering and physical weathering. With *chemical weathering*, the rocks and minerals are altered via chemical reaction, creating new minerals that are different from their source. When a piece of granite is chemically weathered, the quartz pretty much stays the same, but feldspar breaks down to form clay minerals such as kaolinite. Heat and humidity are the most important factors in chemical weathering. Hot, wet climates are areas of intense chemical weathering, while little to no chemical weathering occurs in very arid climates, or in areas that are below freezing most of the time, where there is little or no liquid water.

Material undergoing *physical* (or mechanical) *weathering* is cracked, crushed, or scraped apart, but all of the resulting pieces have the same composition as their source. For example, imagine breaking a piece of granite apart with a hammer. You end up with smaller and smaller pieces of the minerals found in the original granite. The rock fragments resulting from weathering are technically called *clasts* or *detritus*. As with chemical weathering, some climates promote physical weathering, while in others physical weathering is of less importance. The relative intensity of physical and chemical weathering that occurs in different climate zones creates different collections of sediments with distinctive characteristics.

After weathering creates the clasts, *erosion* is the process by which the clasts are transported through the action of an "agent," such as wind, flowing glacial ice, or running water. The mode and duration of transport leaves its mark on the clasts, usually rounding and/or sorting them. Generally, the further a particle is transported, the smoother the surface will become (this is called *rounding*), the smaller the particle becomes, and the better the particles will be *sorted* by size (this will be discussed in more detail in Chapter 5).

Next, the clasts are *deposited* in an environment: as mud on a lake bottom, or as sand on the ocean shore, or blown into dunes in a desert. The environment of deposition will also generate distinctive characteristics, which can be preserved as the sediment enters the final step of *lithification*, the process by which unconsolidated sediment hardens (lithifies) and becomes sedimentary rock. The term *diagenesis* refers to all of the physical, chemical, and biological changes that occur during lithification, such as compaction, cementation, recrystallization, and replacement. All of these events leave their mark on the final product.

Thus, sedimentary rocks contain information about the conditions under which the sediments were deposited, how the clasts forming the rock were transported, and even information about the original source(s) of the clasts. This potentially gives the rocks very distinctive characteristics. One of the most noticeable features of clastic sedimentary rocks is *stratification*, or the layers (strata) that form as sediments are deposited through time. Strata thicker than 1 cm are called *beds* (Figure 3.8a), while thinner layers are called *laminations* or *laminae* (Figure 3.8b). Strata are usually visible as a result of differences in color, grain size, or mineralogy. The upper and lower surfaces of these layers are called *bedding planes*. Sedimentary rocks are also where you find fossils. Most of Earth's history has been ascertained through interpretation of the layers of sedimentary rock.

 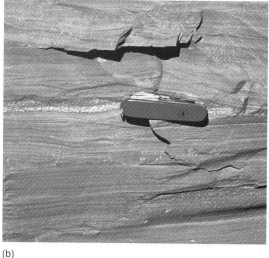

(a)   (b)

Figure 3.8 (a) Photographic example of sedimentary bedding in the Wadi Degla, northern Egypt with USGS scientists for scale, (b) photographic example of laminae with pocketknife for scale.
Source: Courtesy of the United States Geological Survey.

(a)   (b)   (c)   (d)

(e)   (f)

Figure 3.9 Examples of common sedimentary rocks: clastic rock (a) conglomerate, (b) arkose, (c) quartz sandstone, (d) shale; chemical sedimentary rock, (e) chert; biological sedimentary rock, (f) coquina. Please see Color Plate section.

Sedimentary rocks can also be formed via a purely chemical process, like halite forming on the shores of an evaporating lake, or be generated from organic material, such as limestone forming from the microscopic calcite shells of organisms. The method of formation clearly differs from that of sedimentary rocks created out of pieces of other materials, but the resultant rocks are still distinctive products of their environments.

Classification of Sedimentary Rocks

As with igneous rocks, there are actually several schemes by which sedimentary rocks can be classified. To simplify, there are three basic categories of sedimentary rock (Figure 3.9):

- **clastic** or **detrital** (formed of rock fragments; also called terrigenous)
- **chemical** (the result of precipitation or chemical reaction)
- **biological** or **organic** (composed primarily of organic material; also called biochemical).

Clastic sedimentary rocks generally are composed of three things:

- **clasts**: the large particles forming the rock such as gravel, sand, or silt
- **matrix**: the fine-grained material surrounding clasts and/or
- **cement**: usually silica, calcite, or iron oxide: material that holds the rock together.

Table 3.8 Modified Udden–Wentworth scale for sediments and associated rock types.

Measurement range (mm)	Grain Size	Clast	Rock
> 256	Gravel	Boulder	Conglomerate (or breccia)
64–256		Cobble	
4–64		Pebble	
2–4		Granule	
1–2	Very coarse sand	Sand	Sandstone
0.5–1	Coarse sand		
0.25–0.5	Medium sand		
0.125–0.25	Fine sand		
0.0625–0.125	Very fine sand		
0.0039–0.0625	Mud	Silt	Siltstones
< 0.0039		Clay	Shale (or claystone)

Clastic sedimentary rocks are classified by the size of the component pieces first (the clasts) and then by composition. The scale usually used to classify the grain size of the clasts is the Udden–Wentworth scale, which can be simplified, as shown in Table 3.8.

Particles larger than about 0.25 mm can be distinguished by the unaided eye. For example, grains of table salt are usually between 0.25 and 0.5 mm. Grains smaller than 0.25 mm can be distinguished with a hand-lens or jeweler's loupe. Silt and clay can only be distinguished using a microscope or with some form of sedimentation testing, such as a pipette analysis. The sedimentary rocks associated with the different ranges of grain size are also listed in Table 3.8.

The shapes of the individual particles in clastic rocks are also important. *Rounded* particles (sediments with smoothed surfaces) generally have been transported far from their points of origin, while *angular* particles have not traveled far from the place where they originally weathered free (Figure 3.10). *Sorting* refers to the range of particle size in a sediment deposit or a sedimentary rock. When all of the clasts are approximately the same size, the rock or deposit is referred to as *well sorted*, while rocks or deposits that contain a wide range of clast sizes are referred to as *poorly sorted* (Figure 3.11). In general, though not always, the better the sorting, the further the clasts have been transported.

Thus, the fragments inside of a sedimentary rock can tell you a great deal about the source of the sediment, through composition, how far the particles traveled, through roundness and sorting, and even about the mode of travel, through sorting and surface texture. Particles carried by the wind tend to be well sorted and to have a *frosted* surface created by all of the high-velocity impacts that occur between the particles (Figure 3.12). Particles carried by glacial ice tend to be very poorly sorted because the ice acts more or less like a bulldozer sweeping up everything in its path. The clasts can also get distinctive scrape marks (called *striations*) created as they grind against each other while trapped in the ice (Figure 3.13). If you have ever played with pebbles that you found in a stream, you already have a good idea

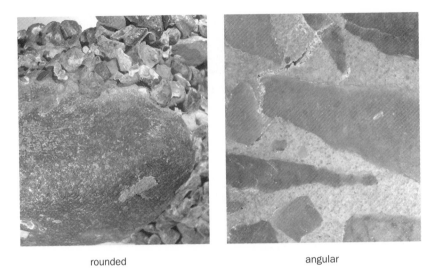

rounded   angular

Figure 3.10 Rounded (conglomerate) and angular (breccia) clasts.

(a)   (b)   (c)

Figure 3.11 (a) Well-sorted sand versus (b) poorly sorted sand versus (c) an extremely poorly sorted till. Please see Color Plate section.

of the features created by this form of transport. The clasts get well sorted and well rounded, but because the water cushions impacts the particles do not have scrape marks or a frosted surface. Instead, they tend to be very smooth.

Other important features common to clastic sedimentary rocks are formed during or after deposition, including *mudcracks, raindrop impressions, ripple marks, included fossils, animal burrows*, and *cross-bedding* (Figure 3.14). These features can be preserved as sediments lithify, giving geologists more information about the environment of formation, and giving forensic workers additional characteristics to help differentiate and source a particular rock. A sedimentary rock displaying one or more of these structures can help to more uniquely characterize it. Some highly

Figure 3.12 Frosted grains.

Figure 3.13 Striations on a glaciated ledge.
Source: Courtesy of the United States Geological Survey.

generalized relationships between sedimentary rock types and their environments of formation can be found in Table 3.9. It is important to remember that while sedimentary rocks reflect the conditions under which they formed these are not necessarily the same conditions that you find in that same area now.

Chemical sedimentary rocks (Figure 3.15) are formed when the ions released into a solution by chemical weathering precipitate out to form crystalline solids or when minerals that have already been deposited are altered via chemical reaction. The most common minerals formed via precipitation are calcium carbonate (calcite; $CaCO_3$), iron oxides (hematite; Fe_2O_3, and other forms), and cryptocrystalline silica (chalcedony, chert; SiO_2, and opal; $SiO_2 \cdot nH_2O$). Because these rocks tend to form in marine environments, chemical sedimentary rocks often may contain fossils.

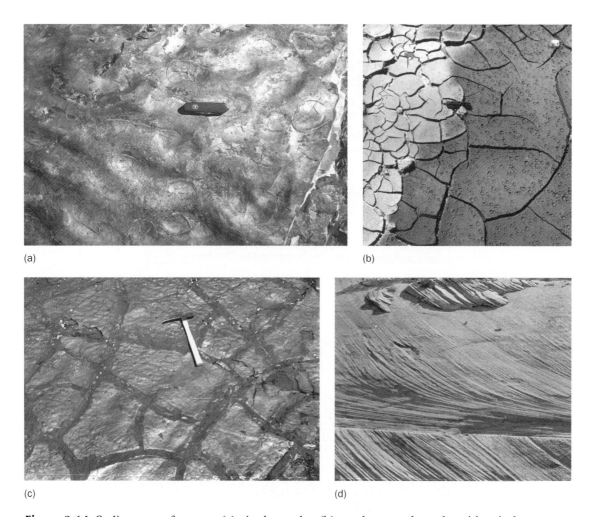

(a)

(b)

(c)

(d)

Figure 3.14 Sedimentary features: (a) ripple marks, (b) modern mud cracks with raindrop impressions, (c) ancient mud cracks, (d) cross-bedding in the Navajo sandstone.
Source: Courtesy of the United States Geological Survey.

The most common chemical sedimentary rocks are *limestone* ($CaCO_3$), formed by the precipitation of the mineral calcite, and *dolostone* [$Mg,Ca(CO_3)_2$], formed from limestone altered chemically such that magnesium replaces some of the original calcium. These rocks are part of a group of rocks called *carbonates*, all of which are composed of minerals containing CO_3. The easiest way to identify a carbonate is to use dilute hydrochloric acid (HCl), which will cause calcium carbonate to effervesce (fizz). Limestone will effervesce quite vigorously, while dolostone will only react weakly, often only if it is powdered. Other common chemical rocks are *chert* (flint) formed by precipitation of silica (silica ooze), *halite* (salt) formed by the precipitation of halite (the mineral and rock name are the same), and *gypsum*. Also of interest are *travertine limestone*, a limestone that is formed by precipitation in caves, and *oolitic limestone*, composed of tiny spheres of concentrically layered calcite.

Biological sedimentary rocks (Figure 3.16) are formed from the skeletons or shells of organisms, or out of other types of biological materials. Most limestone is

Table 3.9 Simplified scheme to relate sedimentary rock types with environments.

	Environment	Sedimentary Rock Type	Grain Size, Shape, Sorting
Continental	Alluvial fans	Breccia, Arkose, Conglomerate, Sandstone	clay to gravel, angular, poorly sorted
	Caves, Geothermal	Travertine Limestone	crystalline white/gray
	Desert	Quartz Sandstone (Quartz Arenite)	sand, rounded, well sorted, frosted grains
	Floodplains	Shale, Siltstone	clay to silt, well sorted
	Glacial	Tillite	all sizes, rounded to angular grains, poorly sorted
	Lake (Lacustrine)	Shale, Siltstone, Limestone, Evaporites (gypsum, halite)	clay to fine sand (coarsening upward), variable sorting
	Rivers (Alluvial/Fluvial)	Sandstone, Conglomerate, Siltstone, Shale	clay to gravel (fining upward), mostly rounded to some angular, moderate sorting
	Swamps (Paludal)	Peat, Lignite, Coal, Black Shale	clay to silt, variable sorting
Transitional	Beach, Tidal Zone	Quartz Sandstone, Coquina, Conglomerates	sand (marine shells), rounded to angular, moderately well sorted
	Deltaic	Sandstone, Siltstone, Shale (in sequence)	clay to sand (coarsening upward), poor sorting
	Lagoon	Siltstone, Shale, Limestone, Oolitic Limestone	clay to silt (marine shells), poorly sorted
	Tidal Flat	Siltstone, Shale, Limestone, Dolostone	clay to silt (marine shells), variable sorting
Marine	Abyssal Plain	Shale, Chert, Micrite, Chalk, Diatomite	clay or crystalline (microscopic marine shells), well sorted
	Continental Slope and Rise	Lithic Sandstone, Arkose, Siltstone, Shale, Limestone	clay to sand (marine shells), poorly sorted
	Continental Shelf	Sandstone, Shale, Siltstone, Fossiliferous Limestone, Oolitic Limestone	clay to sand (marine shells), poor to moderate sorting
	Reef	Fossiliferous Limestone	variable to crystalline, few to no grains (coral and marine shells)

(a)

(b)

(c)

(d)

Figure 3.15 Chemical sedimentary rocks: (a) limestone, (b) travertine limestone, (c) chert, (d) oolitic limestone. Please see Color Plate section.

actually biochemical in nature, or a combination of microscopic carbonate shell and chemically precipitated carbonate cement. Other common biological sedimentary rocks are *coal*, formed from plant fragments; *coquina*, formed from visible fragments of shells; and *amber*, which is fossilized tree resin.

This is just the briefest introduction to a complex topic. Most geology students take at least one full semester course on sedimentary rocks just to get a feel for the variety of rocks formed on the surface of the Earth. As with igneous rocks, there are several classification schemes employed by sedimentologists, but in most cases, because the features of sedimentary rocks are so distinctive, a simplified classification scheme is sufficient (Table 3.10).

(a)   (b)   (c)

(d)

Figure 3.16 Biological sedimentary rocks: (a) fossiliferous limestone, (b) coquina, (c) coal, (d) two different examples of amber. Please see Color Plate section.

Another Case of a Rocky Substitution (Murray and Tedrow, 1992)

A shipment of expensive Scotch whiskey arrived in Canada from overseas. When the shipment was unpacked, it contained blocks of limestone rather than expensive single malt. The rocks had been cut into approximately the same size and weight as a bottle, and placed neatly into each compartment of the shipping container. Limestone is an extremely common sedimentary rock, and there are deposits found in many locations around both the United Kingdom and Canada. However, careful examination of the fabric of the limestone revealed details of composition such as microfossils (dinoflagellates and acritarchs) that allowed researchers to link the rocks to a quarry in central England. They also tracked down a suspect who had access to that quarry, worked for the liquor distributor, and had been spotted taking home many samples of the rock.

Table 3.10 Field classification of sedimentary rocks.

Composition	Properties			Rock Type	
Clastic sediment (Rock fragments and/or mud)	>2 mm	Angular fragments		**Breccia**	
	>2 mm	Rounded fragments		**Conglomerate**	
	Mostly sand (1/16–2 mm)	Often white, tan, pink, generally light	Mostly quartz, few or no rock fragments (can scratch glass)	**Quartz Sandstone**	Sandstone
		Often red, pink, or buff	Feldspar fragments	**Arkose**	
		Often gray, dark colors	Mostly rock fragments	**Graywacke**	
	<1/16 mm (no visible grain to unaided eye)	Color varies Mostly silt, slightly gritty	Often scratches fingernail	**Siltstone**	Siltstone
		Very fine grained silt, yellowish	Softer than fingernail but some particles scratch glass	**Loess**	
		Color varies Mostly clay, smooth, can be scratched by fingernail	Breaks easily into sheets	**Shale** (Most common sedimentary rock on Earth)	Mudstone
			Massive, crumbles, very smooth	**Claystone**	
Biological or biochemical sediment	Effervesces with dilute HCl	Visible pieces of shell (<90%)		**Coquina**	Limestone
		Some shells/fossils in limestone matrix		**Fossiliferous limestone**	
		White to off-white, powders easily, may have fossils		**Chalk**	
		Dark, very fine-grained, usually has conchoidal fracture		**Micrite**	
	Does not react to HCl	Soft, crumbles easily but can scratch glass, usually gray		**Diatomite**	
	Plant fragments	Brown, visible plant material		**Peat**	
		Black, will rub off		**Bituminous Coal**	
		Black, does not rub off		**Anthracite Coal** (technically metamorphic)	

Table 3.10 *continued*

Composition	Properties		Rock Type	
Chemical sedimentary rocks	Crystalline, often gray, white, or clear	Same hardness as fingernail, salty taste	**Halite**	
		Can be scratched by fingernail	**Gypsum**	
	Has conchoidal fracture, scratches glass		**Chert**	
	Effervesces with dilute HCl	Strong effervesces, hardness greater than fingernail, will not scratch glass	**Limestone**	Limestone
		Spherical grains effervesce readily, very light in color	**Oolitic Limestone**	
		Microcrystalline masses, visible layering, cave formation	**Travertine Limestone**	
		Effervesces only when powdered, hardness greater than fingernail, will not scratch glass	**Dolostone**	

The Lady in the Lake: A case of justice delivered or justice denied?

The following is a brief recounting of a complex and controversial case in which a rock served as a key piece of evidence and a key part of the controversy. This is not so much an example of how important geologic evidence is as an example of the complexities of investigation and trial. It also supports the contention that there needs to be more forensic investigators skilled at interpreting geologic evidence and a better understanding of the Earth Sciences by the population in general.

Be aware that there is a significant amount of sensationalism and misinformation in newspaper reports about this crime. A rough outline of part of the story, based mostly on the Cater Walsh Reporting Ltd transcript of the summing-up conducted on January 26, 2005, is as follows. On July 17, 1976, it was the beginning of school holidays and Carol Ann Park (a schoolteacher), her husband Gordon, and their three children planned to go on a trip to Blackpool for the day. Shortly before they were to leave, Carol said she was not feeling well and decided not to go. Gordon took the children to Blackpool by himself, where they visited a funfair and an exhibition on the television show *Dr Who*. When they returned home that evening, Carol was gone.

According to Gordon, there was no note, no sign of a struggle or robbery, and strangely Carol had left her rings, including her wedding ring, on a small dressing table in the bedroom. Carol and Gordon had a stormy marriage, with at least one period of formal separation during which Gordon had been awarded custody of their children. Carol had also left Gordon at least twice previously, also during school holidays, so he did not initially report her missing. He also did not call

Carol's family, stating, "They were not in the habit of keeping others informed of their spats" (Cater Walsh, 1995: 51). It was not until Carol did not return at the beginning of the next school term, several weeks later, that Gordon reported her missing. A missing person's inquiry was launched, but to no avail. At the time, the police informed Gordon that he would be their main suspect should a body ever be found. Time passed, the children grew up, and Gordon eventually remarried.

Then, 21 years later, in August 1997, two divers recovered Carol's body from Coniston Waters (a lake). She had been severely beaten in the face with a heavy instrument (there is disagreement as to whether it was sharp or blunt), and her body, dressed in a nightgown, was tied into a fetal position, wrapped in plastic garbage bags, and then put into a canvas bag pulled shut with a drawstring. A piece of lead pipe was tied to the outer bag as a weight and her body had then been dumped into the lake, where it had landed on a ledge.

According to the examining pathologist, her body had to have been tied up within 2–3 hours of death, prior to the onset of *rigor mortis* (stiffening and contraction of the muscles causing the body to become rigid), though alternatively she could have been tied up 24–48 hours later, after rigor mortis had passed. The pathologist thought the latter unlikely, because the body would have begun to decompose at that point, which would have prevented the advanced development of adipocere seen on the body. *Adipocere* is a white, water-insoluble material created as bacteria break down the neutral fats in the body. It tends to form best when bodies decompose in the absence of oxygen under cold, humid conditions, such as in the relatively watertight wrappings that concealed Carol's body while it lay underwater. Examination also revealed that Carol's hands showed signs of defensive wounds (fractures) and that her eyes had been covered with an adhesive medical dressing.

Police initially arrested Gordon, charging him with murder, but the charges were dropped in January 1998 due to insufficient evidence. Six years later, in January 2004, Gordon was arrested again and this time he was brought to trial. Understandably, after such a long lapse in time, witness statements were a bit of a mess. One neighbor reported seeing a blue or gray Volkswagen Beetle going up the driveway to the house on a day when she knew that the Park family had gone out for the day, most likely the day of the Blackpool trip. She knew that Gordon was not behind the wheel of the car and she remembered thinking it odd that the car had stayed for some 20 minutes before leaving. Other witnesses reporting seeing Carol at various points that day, but a clear timeline could not be reconstructed from their statements. In sum, the majority of the statements appear to support the contention that Carol was seen alive after Gordon and the children had left the house, but doubt remained.

Police had searched the Park's home, called Bluestones, at the time of the missing person's report in 1976 and found no trace of Carol. Given the lapse in time, there was little now that could be used to link anyone to the crime. The trial was long, complex, and based almost solely on circumstantial evidence. The major physical evidence brought out at the trial basically consisted of the complex knots on the ropes used to tie the body (Gordon owned a boat) and on a rock. We will, obviously, concern ourselves with the latter.

After the initial recovery of the body, the police made additional dives in order to search for evidence. One of these dives occurred on September 30, 1997, when a significant number of items, including clothing and cosmetics, were recovered. Strangely though, the diver involved did not remember picking up a rock. The diver stated that, "once he started to grab for clothing, the silt came up and he had to feel about to recover the other items . . . he would not have consciously gathered a stone. He had no recollection of doing so. He speculated that it must have been wrapped up with the other items . . . He said he had no recollection of it at all when he looked at it. Indeed, he said that if he had noticed a rock at all, he probably would have discarded it . . . He agreed that there was nothing on the dive log or on the exhibit label which referred to a rock" (Cater Walsh, 1995: 84). This testimony is one of the primary reasons for the controversy surrounding this case. The rock was not recorded separately as evidence until October 4, five days later, when the contents of the bag filled by the diver were unpacked, photographed, and labeled by a different officer. Also disquieting is the fact that neither the rock nor any of the other associated items were ever actually directly associated with the body or with Carol.

The rock, labeled PDB 5/19, was presented in the trial as exhibit six. Geologists for both the Crown and the defense examined PDB 5/19 and two rocks that the police had taken from the garden wall at Bluestones, the former family home. Each of the geologists also visited Bluestones and Coniston, where the lake is located, to collect a few additional samples for comparison. The evidentiary rock was described as a grayish siltstone or fine-grained sandstone with some orange discoloration at one end caused by chemical weathering. The topic of when such chemical weathering (oxidation) could have occurred was unaddressed. The comparison rocks were generally identified as a "fine-grained sandstone or siltstone dominated by quartz, along with a substance which is described as muscovite vita [sic]" (Cater Walsh, 1995: 93).

The geologist working for the Crown initially identified PDB 5/19 as containing a diagenetic (i.e. the product of chemical change in parent material and not clastic) form of the mineral monazite. Monazite is actually a class of yellowish-brown rare-earth phosphate minerals, the most common of which is (Ce, La, Nd, Th)PO_4. He also identified monazite in four other samples collected from Bluestones, but not in any of the samples collected in Coniston. However, after seeing the report from the geologist working for the defense, he re-examined all of the samples. Using newer equipment (probably an upgraded scanning electron microscope), he now found that PDB 5/19 did not contain monazite after all. The mineral in question lacked phosphorous, a required component, and instead contained calcium. The geologist then tentatively identified the mineral as synchysite.

Upon re-examination, he also found that two of the rocks from Coniston actually did contain true monazite. However, now he reported that the evidentiary rock and the rocks from Bluestones contained the newly identified calcium rare-earth mineral, while the Coniston rocks did not. He also described textural similarities (remember that means the pattern of the mineral grains) between rock PDB 5/19 and the Bluestones samples.

The geologist for the defense also performed examinations, using bulk mineralogy, chemical analysis, and color comparisons. According to this geologist, there did not appear to be any truly significant differences between the Bluestone samples, the Coniston samples, and the evidentiary rock. They were all typical examples of the Windermere supergroup (a stratigraphic association). According to his bulk chemical analysis, there was a high degree of similarity between PDB 5/19, three of the samples from Coniston, and two of the stones from Bluestones. He agreed that the calcium rare-earth mineral identified by the geologist for the Crown was a difference between the evidentiary rock and the other Coniston samples examined; however, he opined that this was simply because they simply had not found it yet. There were no published data for the area describing the degree of occurrence of monazite or any calcium rare-earth minerals, so its rarity or commonality could not be determined. Both scientists were working from a very limited dataset. The defense geologist was fairly sure that, if they kept looking, the textural features and diagenetic calcium rare-earth mineral identified by the Crown's geologist would be found in samples from Coniston. In addition, Coniston Waters lies in a glacial valley carved out during the last ice age. Glaciers commonly re-distribute material, and there was no way of demonstrating that rock PDB 5/19 wasn't carried to Coniston Waters by the ice long ago.

One final point of interest raised by the geologist for the defense concerns the lack of diatoms (microscopic algae that form distinctive silica shells) on the evidentiary rock. Rock PDB 5/19 showed no signs of the moss or algae commonly found on stones from a garden wall, nor did it show any evidence of prolonged submersion. When examined for diatoms, the rock was found to have just a single broken fragment of diatom. Usually, rocks that have been submerged in a lake would show a number of diatoms. The geologist offered three possible explanations for this: that the rock had been protected somehow while submerged, that the rock was actually from the shore and not the water, or that the rock was actually from somewhere else entirely. There appears to have been considerable discussion of this issue, but no report on whether the body or its wrappings contained diatoms. Because the exact origin of the rock was unclear, it was also impossible to determine whether it had originally been wrapped in clothing or not.

This is just a very brief summary of the evidence presented by both of the geologists involved with the case. It took three days for the geologic evidence to be presented at court. At the summation of the case, jurors were instructed to assess how the expert evidence "helps you to decide whether or not . . . it is a strange coincidence that rocks with this association were found at Bluestones and apparently in the one recovered from the lake, but not in the Coniston samples" (Cater Walsh: 109).

Gordon Park, who was in his 60s at the time, was found guilty and sentenced to life with a minimum 15 years. Attorneys filed an appeal in 2008, partially on the grounds that a new forensic report by different geologists concluded that rock PDB 5/19 was in fact indistinguishable from other rocks found widely in the Coniston Waters area, and was therefore meaningless as evidence. The appeal was dismissed in November 2008, with the statement that the rock was "only one element of a strong circumstantial case against Park." There are actually several other strange circumstances associated with the Lady in the Lake case and interested parties should read the judge's summing-up (available at www.freegordon.com).

Metamorphic Rocks

Metamorphic rocks are created through the transformation of existing rocks by intense heat (>200 °C) and/or pressure (>300 MPa), or via interaction with hot ion-rich fluids. These are solid-state transformations, i.e. no melting, that result in the alteration of existing minerals and the creation of new minerals. As with the other major categories of rock, in different geologic settings there are different types of metamorphism, which in turn result in different types of rocks.

Mountain-building events (orogenies), and the extreme pressures and increased heat of burial associated with crustal collisions at convergent plate boundaries, cause large-scale *regional metamorphism* (also called *dynamothermal metamorphism*). As you can probably guess, this type of metamorphism is usually confined to areas along the edges of continental landmasses or to areas that are currently undergoing some type of collision (like where India is colliding with Asia, creating the Himalayas). Continental collisions occur under what are called *differential stress* conditions, which means that the stress (force being applied to the rocks) varies with direction. In a highly simplified way, the rocks are tightly squeezed in the left–right direction (along the axis of the collision) but are under much less pressure in the up–down direction (perpendicular to the axis of the collision). This creates a distinctive texture to the rocks, called foliation, as the minerals reorganize and orient to the stress field.

Another major component of regional metamorphism is *burial pressure*. Pressure increases with depth due to the weight of the overlying material. For example, one cubic meter of granite weighs around 2691 kg (one cubic foot equals 76.2 kg or 167.9 lb), which means that the pressure at the bottom of a block of granite approximately the height of an average person (1.67 m or 5.5 ft) would be around 4.5 metric tons. Imagine how that pressure translates in the heart of a mountain with kilometers of overlying material. Even outside of areas undergoing convergence, burial pressures can be enormous. Changes caused solely by increased pressure, without associated increases in temperature, are called *static metamorphism*.

As rock is forced deeper into the Earth, it is also subjected to higher and higher temperatures. Under normal circumstances, the temperature of the Earth increases at a rate of ~20–30 °C per kilometer deeper into the crust. This is called the *geothermal gradient*. At convergent boundaries, there are some places where slabs of crust are being forced back into the Earth quickly. This material is subject to extreme pressure, but because it takes rock a long time to heat up it is at a lower temperature than its surroundings. On the other hand, there are also some zones of high heat, often associated with volcanism, where temperatures are greater than the geothermal gradient would suggest. The range of heat and pressure conditions found in zones of regional metamorphism create diagnostic sets of metamorphic rocks, which can be used to reconstruct the tectonic history of an area. This also means that metamorphic rocks tend to be very distinctive of particular areas, making them useful for geosourcing.

Another important type of metamorphism is *contact metamorphism*, where contact with molten rock (lava or magma) in essence bakes the surrounding rocks. An *aureole*, or halo, of metamorphosed rock is created along the contact zone, and because this occurs under conditions of uniform stress the rocks that are formed lack the distinctive texture created by regional metamorphism (called *non-foliated*). There is no hard-and-fast distinction between regional metamorphism and contact metamorphism, as they often grade into each other. In general, contact metamorphism occurs at a much smaller scale than regional metamorphism does, may also be caused by small igneous intrusions, and occurs in hotspot zones (not associated with plate boundaries).

The majority of metamorphic rocks are associated with the conditions described above; however, there are also some more localized events that can create them. For example, the conditions that occur specifically along fault zones are referred to as *dynamic* (or cataclastic) *metamorphism*. The rocks formed there, which are subject to crushing and ductile flow, are called *mylonites*. A significantly different form of metamorphism, called *hydrothermal alteration* (or metasomatism), occurs when rocks are infiltrated by high-temperature, ion-enriched fluid. Rich ore deposits are often the result of this type of metamorphism. *Impact* (or shock) *metamorphism* is the result of a violent impact, as with an extraterrestrial object or during an extremely violent volcanic eruption, creating conditions of extremely high pressure.

All forms of metamorphism cause the original texture and features of rock to be altered and eventually destroyed, as are any fossils located within the rock. Metamorphic rocks that have been intensely altered are referred to as *high-grade metamorphic rocks*, while those that have only been slightly altered are called *low-grade metamorphic rocks*. When affected by intense directionalized pressure, such as that associated with the squeezing convergence of regional metamorphism, a new texture called *foliation* can be created (Figure 3.17). Foliation is the planer arrangement and segregation of elongated minerals like mica and hornblende (i.e. these minerals sort by type and become oriented so that their long axis is perpendicular to the plane of highest stress). The resulting rocks typically have a layered appearance, with alternating bands of light- and dark-colored minerals. (Gneiss, pronounced "nice," is a good example.)

Contact metamorphism, and other forms of metamorphism that occur under conditions with approximately equal application of pressure in all directions, does not create foliation. These rocks are referred to as *non-foliated*. Rocks formed of minerals that are equidimensional, i.e. roughly the same shape/length in all orientations, such as quartz or calcite, will also form non-foliated metamorphic rocks, such as quartzite and marble, respectively. These minerals cannot be oriented

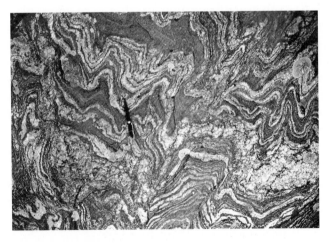

Figure 3.17 Intensely foliated gneiss from Black Canyon of the Gunnison National Park, Colorado.
Source: Courtesy of the United States Geological Survey.

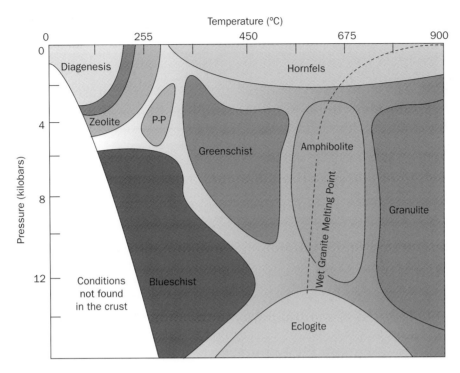

Figure 3.18 Metamorphic facies.

by pressure since there is no long axis to rotate. The presence or absence of foliation is usually the first property examined when classifying a metamorphic rock.

In addition to changing the texture of rock, metamorphism also changes the mineral content. Some minerals, like quartz, are stable, but many other minerals are not. There are also minerals, called *index minerals*, that are only formed under specific metamorphic conditions. In regional metamorphic terrains, the temperature and pressure regime creates a distinctive distribution of index minerals across a large area, which can be broken down into distinct *facies* (Figure 3.18). In geology, facies are the characteristics of a body of rock that reflect its origin and can be used to distinguish it from surrounding facies. For example, assemblages of minerals or fossils can be used to identify facies.

Starting at the exterior of a zone of regional metamorphism and working inward works something like this. In areas of low-grade metamorphism, with temperatures ranging between 200 and 500 °C, chlorite, muscovite, and biotite are important index minerals (Table 3.11). Chlorite is typically light to dark green and gives the metamorphic rock greenschist its distinctive color. Muscovite and biotite, however, are familiar silicate minerals that are also part of Bowen's reaction series and are therefore commonly found in igneous and sedimentary rocks. The difference here is that the heat and pressure of metamorphism cause these platy micas to start recrystallizing into larger and larger grains, which create foliated textures in rocks and give slates and phyllites their distinctive shiny look. Intermediate-grade metamorphic rocks, which form at temperatures between 500 and 550 °C, are typified by the minerals garnet and staurolite. These index minerals, like chlorite, are only formed under the appropriate metamorphic conditions. High-grade metamorphic rocks, such as gneiss, which have been subjected to temperatures above 550 °C, contain the distinctive minerals kyanite and sillimanite.

Table 3.11 Index minerals in the Barrovian zones with pelitic protoliths. Index minerals are in bold.

Rock Types	Metamorphic Zone	Temperature (°C)	Starting from mudrocks/shale, a typical mineral assemblage includes
Slates and Phyllites	Chlorite Zone	200–425	**Chlorite**, Albite, Muscovite, and Quartz
	Biotite Zone	425–500	**Biotite** starts to replace Chlorite, Albite, Muscovite, and Quartz
Phyllites and Schists	Garnet Zone	500–530	**Garnet**, Biotite, Albite, Muscovite, and Quartz Chlorite disappears
Schists	Staurolite Zone	530–550	**Staurolite**, Garnet, Biotite, Muscovite, Oligoclase, Quartz Albite disappears
	Kyanite Zone	550–700	**Kyanite**, Garnet, Biotite, Muscovite, Oligoclase, and Quartz Staurolite phases out
Schists and Gneisses	Sillimanite Zone	>700 (melting starts around 900)	Biotite, Garnet, **Sillimanite**, Oligoclase, and Quartz Kyanite phases out.

Another important factor is the parent rock, or *protolith*, that is being altered. *Metamorphic zones* are named for groups of minerals that would form at certain temperatures from specific types of parent rocks. The Barrovian zones listed in Table 3.11 only apply if the rocks being metamorphosed are mudrocks or shale (called *pelitic rocks*). If the parent rock is basalt (mafic) or rich in calcite (calcareous), the products of metamorphism are somewhat different. A *skarn*, for example, is a high-grade metamorphic rock formed by the contact metamorphism of a calcareous parent rock (limestone) and a silica-rich magma. The resulting rock contains minerals such as garnet, epidote, and wollastonite. A *metamorphic facies* is the set of all minerals that may be found together even when the parent rocks have different chemical compositions. A condensed version of the mineral assemblages associated with common metamorphic facies is listed in Table 3.12.

Classification of Metamorphic Rocks

The classification of metamorphic rocks is typically based on texture (foliated versus non-foliated) and composition. A field classification guide for the most common metamorphic rocks can be found in Table 3.13. The first step is to decide whether the rock is foliated. If so, to what degree? If the rock has a strong tendency

Table 3.12 Metamorphic facies and associated mineral assemblages.

Metamorphic Facies		Temperature (°C)	Pelitic Protolith	Mafic Protolith	Calcareous Protolith
Zeolite	**Do not always occur**	100–200	Interlayered Smectite/ Chlorite	Interlayered Smectite/ Chlorite	Calcite
Prehnite-Pumpellyite		150–300	Albite, Chlorite, Prehnite, Pumpellyite	Albite, Chlorite, Prehnite, Pumpellyite	Calcite
Greenschist		300–450	Albite, Biotite, Chlorite, Garnet, Muscovite, Quartz	Actinolite, Albite, Chlorite, Epidote, Sphene	Calcite, Dolomite, Epidote, Quartz, Tremolite
Epidote Amphibolite		450–550	Albite, Biotite, Garnet, Muscovite, Quartz	Albite, Epidote, Hornblende	Calcite, Diopside, Epidote, Quartz, Tremolite
Amphibolite		500–700	Biotite, Garnet, Muscovite, Plagioclase, Quartz, Staurolite, Kyanite, or Sillimanite	Biotite, Garnet, Hornblende, Plagioclase, Sphene	Calcite, Diopside, Quartz, Wollastonite
Granulite		700–900	Garnet, Plagioclase, Quartz, Kyanite, or Sillimanite, Hypersthene, Potassium Feldspar	Augite, Garnet, Hornblende, Hypersthene, Olivine, Plagioclase	Calcite, Diopside, Hypersthene, Plagioclase, Quartz,
Blueschist		150–350 P > 5–8 Kb	Albite, Aragonite, Jadeite, Quartz, Lawsonite, Paragonite	Albite, Garnet, Glaucophane, Lawsonite, Sphene	Aragonite, Muscovite
Eclogite		350–750 P > 8–10 Kb	Coesite, Plagioclase, Potassium Feldspar, Sillimanite	Garnet, Pyroxene (Omphacite)	Aragonite, Diopside, Hypersthene, Plagioclase, Quartz

Table 3.13 Field classification of metamorphic rocks.

Texture	Characteristics		Rock Type
Foliated	Fine-grained or no visible grains	Dull shine, breaks into thin sheets (slaty), color varies – usually dark	Slate
		Shiny, mica minerals often dominate, wavy or crenulated	Phyllite
	Medium- to coarse-grained	Visible sparkles, breaks into scaly sheets, chlorite, mica, and garnet common. Will scratch glass	Schist
		Visibly banded, does not break into sheets. Will scratch glass	Gneiss
		Even-grained, lacks hydrous minerals	Granulite
Non-foliated or foliated	Medium- to coarse-grained	Visible crystals of amphibole	Amphibolite
Non-foliated	Fine-grained or no visible grains	Glassy texture, black, breaks unevenly or conchoidal fractures	Anthracite Coal
		Very hard, dull, dark color – opaque. Will scratch glass	Hornfels
		Color usually shade of green or lime, contains serpentine, harder than fingernail	Serpentinite
		Can be scratched with your fingernail, slippery feel, contains mostly talc	Soapstone
	Medium- to coarse-grained	Light colors, glassy shine, fused quartz grains that will break mid-grain. Will scratch glass	Quartzite
		Should react to HCl – might need to be powdered, resembles sugar cube	Marble
		Distorted pebbles, conglomeratic texture but breaks across grains	Meta-conglomerate

to split into thin plates and the minerals are to fine-grained to be seen, it is *slate*. If the rock not only splits into thin plates but also has barely visible crystals of muscovite, chlorite, graphite, or talc that give the surface a sheen, or the whole rock is slightly rippled, it is *phyllite*. If the rock is strongly foliated, breaks parallel to its foliation, and is coarse-grained such that individual crystals are visible, it is *schist*. Finally, if the rock has clear segregation of minerals into distinct bands, but shows no tendency to split into sheets, it is *gneiss*.

A more detailed explanation of the different types of *foliated textures* created by differential stress is as follows:

- **Slaty texture** (Figure 3.19a) results from low-grade metamorphism. These rocks are very fine-grained, with individual mineral layers that are not visible to the naked eye, for example shale. Foliation is present on a microscopic scale and causes the rocks to split into thin layers. Rocks with the ability to break into sheets are called *fissile*, a property which is similar but distinct from mineral *cleavage*. *Cleavage* is related to atomic structure, while *fissility* is due to mineral orientation in a rock.

- **Phyllitic texture** (Figure 3.19b) is also fine-grained and due to low- to intermediate-grade metamorphism. The foliation is generally just barely visible to the naked eye. Phyllites typically have a glossy sheen and a wavy or crenulated surface.

- **Schistose texture** (Figure 3.19c) results from intermediate- to high-grade metamorphism and results in medium to coarse grains, as like-type minerals are welded together. The foliation is generally visible to the naked eye and these rocks tend to split into sheets.

- **Gneissic texture** (Figure 3.19d) is typical of high-grade metamorphism. Minerals are present in layers, segregated by ion migration. These rocks do not split into thin sheets and the foliation is more a visible property than a physical

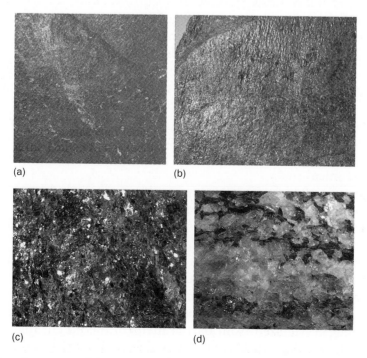

(a)   (b)

(c)   (d)

Figure 3.19 Foliated metamorphic textures: slaty, phyllitic, schistose, gneissic. Please see Color Plate section.

property. Gneisses are usually identified by their distinctive banded appearance (Figure 3.17).

High-grade metamorphism can also result in a medium- to coarse-grained rock that is lacking in water-bearing minerals and is known as a *granulite*. Most of the minerals that survive or are produced under these conditions are *equidimensional* (i.e. of equal dimensions). This produces a rock that does not have obvious foliation, instead appearing similar to a phaneritic igneous rock, and often contains minerals that are only produced via metamorphic reactions.

Non-foliated metamorphic rocks are those composed of equidimensional minerals with no preferred orientation, or are formed under uniform stress conditions, and are said to have a *granoblastic texture*. The result is a metamorphic rock where the grains form a mosaic. Common examples are marble, which forms from limestone or dolostone (calcite and dolomite are not elongated); quartzite, which forms from quartz sandstone or chert; anthracite, a very high-grade coal formed from bituminous (sedimentary) coal; and soapstone, which is composed primarily of talc. Not all quartzites, marbles, and soapstones are pure. Some contain impurities like clays that were originally interlayered with or mixed inside of the original quartz sand or lime mud. These clay impurities metamorphose into layers of micas or other minerals, which may give marble (in particular) a banded, gneissic appearance, or which may give a slight foliation to some quartzites and soapstones.

Other important non-foliated, or weakly foliated, metamorphic rocks include serpentinite, a dark-green rock formed by hydrothermal alteration of an ultramafic parent rock; amphibolite, a medium- to coarse-grained rock, usually black, formed of hornblende and plagioclase that is sometimes found foliated (i.e. amphibolite schist); greenstone (or metabasalt), a green rock formed from metamorphosed basalt that contains chlorite and epidote; and hornfels, a hard, fine-grained massive rock that is usually dark in color and may superficially resemble basalt. Hornfels are formed by contact metamorphism and can be rather nondescript. They contain less magnetite than basalts do and are less dense.

Some metamorphic rocks have a texture in which large crystals called *porphyroblasts* are surrounded by smaller crystals, similar to the igneous porphyritic texture (Figure 3.20). Porphyroblasts can occur in both foliated and non-foliated metamorphic rocks. Some examples include eclogite, which is medium- to coarse-grained and consists mostly of large garnet crystals in green clinopyroxene (called omphacite), and garnet mica schist, a schist with large porphyroblasts of garnet and obvious sheets of mica.

A few other potentially important metamorphic rocks include mylonite, a fine-grained layer rock that is the result of dynamic metamorphism and can have any mineralogy; metaconglomerate, a metamorphosed version of a sedimentary conglomerate in which the clasts are still recognizable but often have been stretched; and migmatite, a metamorphic rock that has been subjected to such high temperatures and pressures that it has partially melted, resulting in a deformed, swirling appearance similar to that of a highly deformed gneiss, but in this case while the dark bands have undergone metamorphism the minerals in the lighter bands have actually crystallized from partial melts of the parent material.

This sounds rather complicated, but at a basic first level of characterization, it is actually rather straightforward. First, decide whether the rock is foliated. Then look at the grain-size and the basic character of the rock. This will generally lead to a type name.

Figure 3.20 Porphyroblastic garnets in a garnet mica schist.

An Investigation of Bosnian War Crimes (Brown, 2006)

In 1997, the United Nations International Criminal Tribunal for the Former Yugoslavia (UN ICTY) began excavations of mass gravesites located in northeastern Bosnia. The graves were associated with a massacre of civilians in and around Srebrenica that occurred in July 1995. Intelligence indicated that three months after the initial executions occurred the original mass graves were exhumed and, over a very short period, the bodies transported to an unknown number of secondary gravesites. Forensic workers in a variety of fields were brought in to investigate both the primary and secondary gravesites in order to bring additional evidence as part of war crimes indictments and to help document the scale and organization of the original crime and subsequent attempts at concealment. The rapid creation of multiple secondary gravesites was believed to be part of an attempt at creating the defense that the deaths were the result of small-scale skirmishes that occurred in the normal course of war and that there was not an organized mass genocide.

Northeastern Bosnia lies in the central Balkan Mountains, which themselves are part of the Alps–Himalayas convergence orogenic belt, which stretches across most of Europe and Asia. The area is geologically complex, with NW–SE trending units of igneous, sedimentary, and metamorphic rock ranging in age from Paleozoic to Cenozoic. Much of the territory shows evidence of regional metamorphism and is dominated by folded and variably metamorphosed clastic and carbonate rocks, meta-sandstones, meta-conglomerates, and meta-limestones, grading into schist. There are also small outcrops of epidote, amphibolite, and serpentinite. Srebrenica, however, lies in an area of extrusive volcanics, dominated by andesite, dacite, and pyroclastic material. This complexity greatly improves the possibility of linking transported geological material to its source.

Part of the investigation involved taking a variety of soil samples from all 5 primary sites and from 19 secondary sites. A number of mineralogical and pollen analyses were conducted on the samples with the goal of linking secondary sites to primary sites. One example of the success of this tactic comes from the secondary grave called Hodzici Road (HZ3), where investigators found a striated clast of serpentinite. One primary gravesite, Lazete I, was located in an area with this type of serpentinite exposed, allowing researchers to link the two sites. Overall, the mineralogy and pollen analysis of samples from the secondary gravesites indicated that in each case the composition of the grave soil was foreign to the site and could not have originated there. This evidence, linked with forensic investigation of clothing, documents, and ballistic evidence proved successful at trial.

Summary

This chapter introduced readers to the major classes of rocks found on the Earth and the conditions of their creation. Rocks tell the stories of their formation and can therefore be useful for geosourcing. Mineral content is but one component used for identification. The rocks themselves can have several textural or physical features that can be used for forensic comparison. Often, this wealth of information is ignored, which is a shame.

Further Reading

Boggs Jr., S. (2006) *Principles of Sedimentology and Stratigraphy*, 4th edn. Pearson-Prentice Hall: Upper Saddle River, NJ. [A useful textbook.]

Mottana, A., Crespi, R., and Liborio, G. (1978) *Simon and Schuster's Guide to Rocks and Minerals*. Simon and Schuster, New York. [One of the better guidebooks, currently available in its sixth edition.]

Raymond, L. A. (2002) *Petrology: The study of igneous, sedimentary, and metamorphic rocks*, 2nd edn. McGraw-Hill: New York. [An advanced textbook.]

References

Brown, A. G. (2006) The use of forensic botany and geology in war crime investigation in NE Bosnia. *Forensic Science International* **163**: 204–10.

Cater Walsh (2005) *At the Crown Court Order No. T2004/7403, Courts of Justice, Crown Square, Manchester. Wednesday, 26th January 2005*. Cater Walsh Reporting Ltd. Kidderminster, Birmingham.

Murray, R. C. and Tedrow, J. C. F. (1992) *Forensic Geology*, 2nd edn. Prentice Hall Inc., Englewood Cliffs, NJ.

Rapp, J. S. (1987) Forensic geology and a Colusa County murder. *California Geology* **40**: 147–53.

Ruffell, A. and McKinley, J. (2008) *Geoforensics*. John Wiley & Sons Ltd, Chichester.

Chapter 4
Maps: Getting a Sense of Place

There are a variety of ways in which maps and an understanding of geographic location can play a vital role in criminal investigation. Many studies have demonstrated that offenders usually commit crimes in areas near their current, or former, homes or in other places with which they have a high degree of familiarity. This applies to crimes from burglary and arson to rape and murder. Many so-called serial killers had a geographic element to their crimes, such as where they picked victims or where they disposed of bodies. Developing a sense of place and learning to interpret spatial information, such as regional topography or local line-of-sight, are key skills for a forensic investigator. For a forensic geoscientist, maps provide a wealth of information about where a particular sample may, or may not, have originated.

The Importance of Local Knowledge

On September 11, 2001, the United States of America suffered the worst terrorist attack ever on American soil. Based on a range of types of intelligence information, the following month US and British forces began retaliatory strikes against the Taliban regime in Afghanistan. In a videotaped message, broadcast by al Jazeera on October 7, 2001, Osama bin Laden praised the September 11 attack and condemned the hour's-old aerial bombing of Kabul. When this footage was broadcast in the United States, Jack Shroder, a professor of geography and geology at the University of Nebraska-Omaha, and former director of the National Atlas of Afghanistan, called out to his wife, "I know where he is."

An Introduction to Forensic Geoscience, First Edition. Elisa Bergslien.
© 2012 Elisa Bergslien. Published 2012 by Blackwell Publishing Ltd.

Dr Shroder first visited Afghanistan in 1973, where he worked developing detailed maps of the country and served as the head of Kabul University's seismic station. He was briefly jailed and then expelled following the 1978 Afghan Communist coup. When Dr Shroder saw the videotape, he recognized the sheered and faulted rock formations behind bin Laden. Though the image quality was poor, the rocks showed quartz veins and had been weathered into distinctive spheroidal shapes. This was enough information for Dr Shroder to tell a *San Francisco Chronicle* reporter that the rocks indicated bin Laden was probably in a ravine located in Pashtun tribal territory in a southwestern province of Afghanistan. He was able to give much more detailed information to the US government, pinpointing the location where bin Laden was videotaped within approximately 20 miles. Though bin Laden had long since moved, this example highlights the continued importance of field geology. In an age of satellite photos, global positioning units and Google Earth, nothing replaces the intimate knowledge of a field worker.

Global Location Systems

One of the most important field skills that any forensic worker or geologist can have is the ability to read maps. The first step in that process is gaining an understanding of the different ways that information about relative position is shared. Using latitude and longitude is the most common way of communicating the location of any place on the Earth's surface. Imagine a grid system with east–west and north–south lines that encircle the Earth. The east–west lines are called lines of *latitude* (or parallels), which run parallel to each other and to the equator, which is a special line that circles the Earth exactly midway between the North Pole and the South Pole (Figure 4.1). The location of a given parallel is designated as the angular distance between that parallel and the equator as measured from the center of the Earth. This angular distance ranges from 0° (zero degrees) at the equator to 90° (ninety degrees) north at the North Pole and 90° south at the South Pole. Hence, latitude indicates how far a place is located to the north or south of the equator. Because the Earth bulges out around the equator rather than being a perfect sphere, the length of a degree of latitude varies from 110.574 km (68.708 statute miles) near the equator to 111.694 km (69.403 statute miles) near the poles.

Lines of *longitude* (or meridians) run north and south, perpendicular to the equator (and all other parallels) meeting at both poles. This means that the distance between meridians varies with latitude. The *prime meridian* is a line of longitude running through the location of the former Royal Observatory in Greenwich, United Kingdom, which was arbitrarily defined as 0° longitude in 1884. The location of a given meridian (or line of longitude) is the angular distance between that meridian and the prime meridian. Angular distance ranges from 0° at the prime meridian to 180° on the opposite side of the globe. Thus, longitude indicates how far a place is located to the east or west of the prime meridian. The length of a degree of longitude varies from 111.319 km (69.171 statute miles) at the equator to 0 km (0 statute miles) at the poles.

The prime meridian is also the basis of global timekeeping, called Universal Time (UT), and formerly known as Greenwich Mean Time (GMT). The Earth rotates at approximately 15° per hour (360°/24 hr = 15°/hr), therefore when it is 12

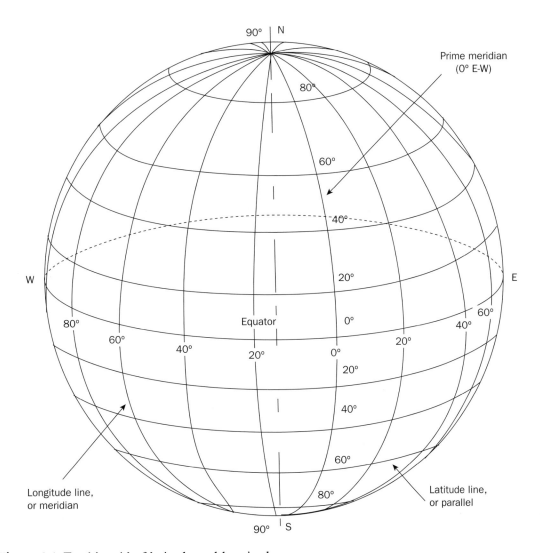

Figure 4.1 Earth's grid of latitude and longitude.

noon at the prime meridian, it is 11 a.m. at 15°E (into Europe) and 1 p.m. at 15°W (toward Iceland). Not all countries follow this practice, but most do. The International Date Line, at which the day and date change, is located around longitude 180°, though the actual boundary zigzags around countries.

The lines of latitude and longitude create a reference grid system that can be used to locate any spot on the globe. Measurements for the geographic coordinate system are usually expressed in degrees, minutes, and seconds. There are 60 seconds (″) in a minute, 60 minutes (′) in a degree, and 360 degrees (°) in a circle. When reading the location of a position using the geographic coordinate system, latitude is always expressed first, then longitude. For example, Baghdad, Iraq is located at 33°20′ north latitude, 44°26′ east longitude. Hemisphere must be specified, and either spelled out as in the example shown previously or using signs. In latitude, the northern hemisphere is specified as positive (+) and the southern hemisphere is specified as negative (−). With longitude, the western hemisphere is negative (−) and the eastern hemisphere is positive (+). For example, the summit of Mount St Helens in the United States is at + 46°12′01″ and −122°11′20″.

Confusingly, coordinates can also be expressed as decimals, or as partial decimals. For example, 56°34′21″ can be converted into a decimal as follows: first divide the seconds by 60 (21/60 = 0.35) to convert them into minutes. Next, add the decimal you just calculated to the number of minutes in the coordinate (0.35 + 34 = 34.35). Finally, divide the minutes you just calculated by 60 to convert them into degrees (34.35/60 = 0.5725). Therefore, 56.5725° is the same as 56°34′21″. You can convert a decimal back into degrees, minutes, and seconds by following this procedure in reverse: multiply 0.5725 by 60 to get 34.35 minutes, and multiply 0.35 by 60 to get 21 seconds. There are also maps that will express latitude and longitude in partial decimals as degrees and minutes, such as 56°34.35′. Make sure that you understand the system that your map or directions are using if you do not want to get lost.

In addition to latitude–longitude, there are several other positioning systems currently in use. In the late 1700s, the United States created a grid system called the US Public Land Survey System (PLSS) for the purposes of locating property lines and for use on legal documents (Figure 4.2). This system is *not* used in the states formed from the original 13 colonies (Virginia, New Jersey, Massachusetts, New Hampshire, Pennsylvania, New York, Maryland, Connecticut, Rhode Island, Delaware, North Carolina, South Carolina, and Georgia), or in Maine, Vermont, Tennessee, Kentucky, Texas, West Virginia, Hawaii, and parts of Ohio. Those areas used a variety of earlier referencing systems, such as *metes and bounds*, which are usually based on arbitrarily chosen local reference points, such as rivers, large rocks, or even trees (that have long since disappeared). PLSS is used in the remainder of the states, plus some parts of Canada. More information for this system can be found at the National Atlas website (http://nationalatlas.gov/ mapmaker).

PLSS is based on grids that are unique to specific regions and are located relative to a state's (or territory's) *principal meridian* (N–S line) and *base line* (E–W line perpendicular to the principal meridian) (Figure 4.2). There are 37 named principal meridians that form the basis of this system. *Range lines*, which run parallel to the principal meridian, are spaced out every six miles and *township lines*, which run parallel to the base line, and are also positioned every six miles (Figure 4.3). A *township* is an approximately six-mile by six-mile square area of land bounded by township lines and range lines. Ranges are numbered R1E, R2E, and so on to the east of the principal meridian, and R1W, R2W, and so on to the west of the principal meridian, while township lines are numbered T1N, T2N, and so on to the north of the base line, and T1S, T2S, and so on to the south of the base line. Thus, each township can be identified by its position relative to the principal meridian and base line as follows: T3S R2W would be the square located 12–18 miles south and 6–12 miles west of the intersection of the principal meridian and base line.

Each township is further divided into 36 *sections*, which are one-mile square areas of land, or 640 acres, which are numbered from 1 to 36 starting in the northeastern corner and incrementing east to west and then west to east on alternating rows (6–1, 7–12, 18–13, etc.) (Figure 4.3). A full map reference to a location includes the state, principal meridian name, township and range designations, and the section number: Nebraska, Sixth Principal Meridian T7N, R2W, Sec. 5. The sections can be further divided into a hierarchy of *quarters*, which are written in order of increasing size and direction (Figure 4.3). For example, NW¼ SE¼ would be the northwest quarter of the southeast quarter of the section. Thus, the location of a particular 10-acre parcel of land could be given as NW¼

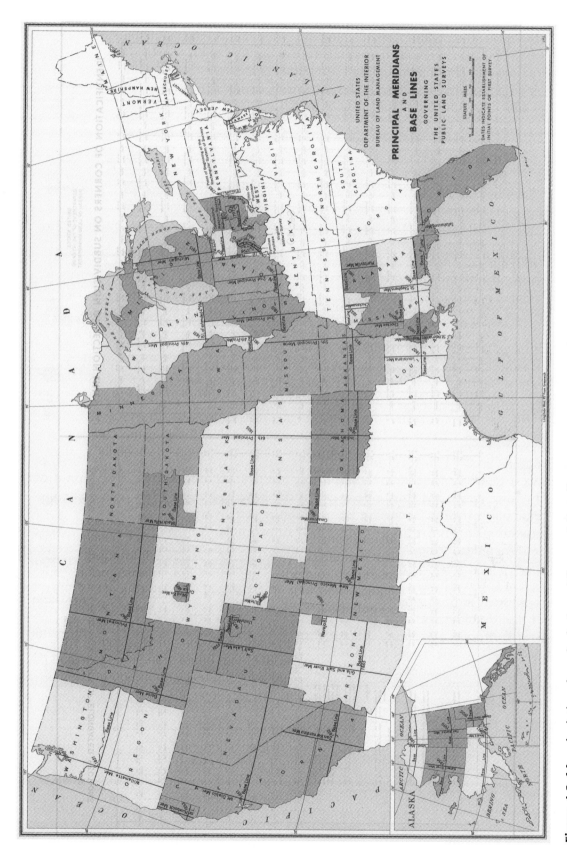

Figure 4.2 Map depicting the principal meridians and base lines used for surveying states (colored) using the Public Land Survey System (PLSS), also called the Township and Range System.
Source: Courtesy of the Bureau of Land Management.

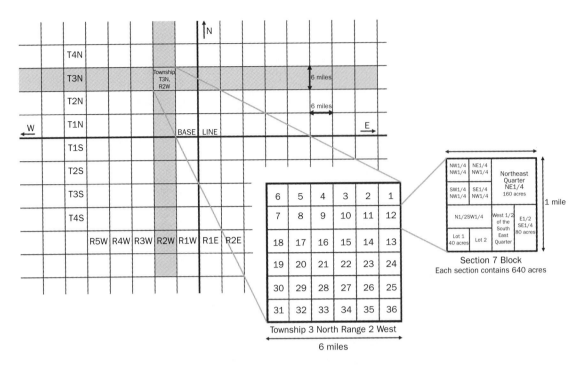

Figure 4.3 Township and range gird. The starting point of the grid is the principal meridian and base line. Each 6-square-mile section is called a *township*. The townships are further broken into 36 square-mile numbered *sections*. Sections can be further subdivided into *blocks*.

SW¼ SE¼ Sec. 4 T3S R4E. In words, this means that the parcel is located in the northwest quarter of the southwest quarter of the southeast quarter of section 4 of the township that is the third one south of the base line and the forth one east of the principal meridian. Township information is usually included in red on topographic maps in the United States where available.

The difficulties of the PLSS system are fairly obvious, and using straight longitude and latitude coordinates can be awkward, especially if you are attempting to locate a specific house. These complexities were felt on a much grander scale when different nations tried to communicate locations. Thus came, following World War II, the development of a global military navigation grid, the Universal Transverse Mercator System (UTM) (Figure 4.4a). UTM divides the Earth into 60 sectors, each 6° wide, running from 84° North to 80° South, to form a series of rectangular grids measured in meters [a separate grid system, the Universal Polar Stereographic (UPS), is applied at the poles].

The sectors are numbered, beginning at the International Date Line (180° longitude) from Zone 01 in the west (from 180° to 174° west longitude) to Zone 60 in the east (from 174° to 180° east longitude). The continental United States lies between Zone 10 in the west to Zone 19 in the east (Figure 4.4b), while the United Kingdom lies in Zones 29–31 (Figure 4.4a). Each Zone is subdivided into an eastern and western half by drawing a north–south line perpendicular to the equator, called the *central meridian*. This is the only line in the Zone that will run from pole to pole and be perpendicular to the equator. The remaining grid lines are drawn parallel to the central meridian line to form a rectangular grid that will be increasingly skewed toward the poles from geographic north to south.

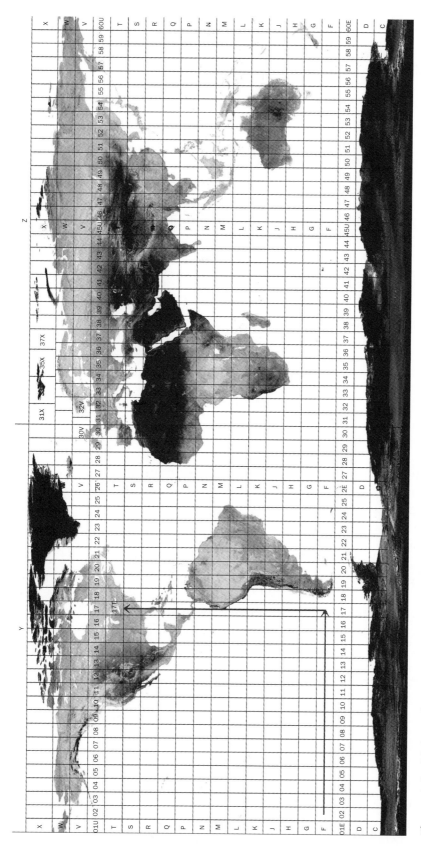

Figure 4.4a Universal Transverse Mercator System (UTM).
Source: Courtesy of the United States Geological Survey.

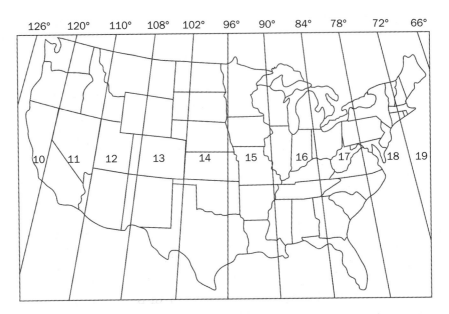

Figure 4.4b UTM overlain on map of the United States.
Source: Courtesy of the United States Geological Survey.

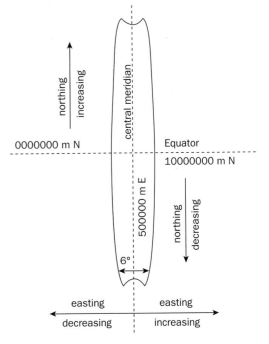

Figure 4.5 Numbering system for UTM coordinates within a zone.
Source: Courtesy of the United States Geological Survey.

The location of each point in the grid is defined by its *easting* coordinate, or distance within the zone measured in meters from west to east, and its *northing* coordinate, or distance up or down from the equator measured in meters (Figure 4.5). To eliminate the need for negative numbers, the equator is defined as having northing of 10,000,000 meters, and the central meridian for each Zone is assigned an easting value of 500,000 meters. Thus, any easting values less than 500,000 lie to the west of a Zone's central meridian and any values greater than 500,000 lie to the east. UTM coordinates are given with the Zone first, then the easting, and ending with the northing. For example, Mount Rushmore National Monument (the one with the carvings of four US presidents) is located at approximately 13N E623740 N4859587; or in the northern hemisphere, Zone 13, 623,740 meters east of the western edge of the zone and 485,9587 meters north of the equator.

Full-line black UTM grids appear on USGS topographic maps produced between 1978 and 1992, and have reappeared on more recent maps. On the remainder of maps produced from the mid-1950s on, UTM grids are indicated by blue tick marks on the map's perimeter. One northing and one easting label is written out in full on each topographic map in the form $^{47}08^{000m}$N (along the left-hand side of the map) and $^{7}15^{000m}$E (along the bottom edge of the map). The remaining values are written in UTM shorthand and do not end in ^{000m} as $^{7}15$ or $^{47}08$. So, for example, if you were looking for a point at 17N E706212 N4699150, it would be located 212 meters east of the tick mark labeled $^{7}06$ and 150 meters to the north of the tick mark labeled $^{46}99$ in Zone 17 of the northern hemisphere, or roughly in the middle of the town of Machias, New York, United States of America (Figure 4.6).

Since the UTM system predates the existence of satellites, it was originally applied using regional ground-based surveys to determine the location of grid boundaries. Each of these surveys is identified on the basis of its location and year, such as the North American Datum of 1983 (NAD83). The most current datum, and the one most global positioning systems (GPS) use, is the World Geodetic System of 1984 (WGS84). Most USGS maps are referenced to the North American Datum of 1927 (NAD27), which has some significant differences from WGS84. Maps that use NAD83 can be treated as virtually identical to WGS84.

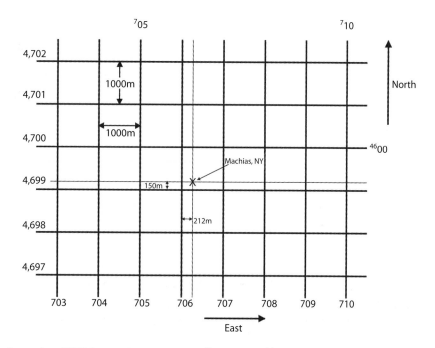

Figure 4.6 Locating UTM coordinates 17N E^{7}06 212 N^{46}99 150.

Check the lower left-hand corner of the map to determine which datum was used. These systems vary not because one is wrong and another is right but because they are different ways to try to describe the Earth, which is not a perfect sphere.

Technically, the grid system described here is the *civilian UTM grid and coordinate system*. The US Department of Defense actually uses the Military Grid Reference System (MGRS), which divides the UTM Zones into sectors, 8° high north–south, that are identified by *designator letters*, starting with the letter C (80° to 72° south latitude), to letter X (72° to 80° north latitude), skipping the letters I and O. This method makes it easier to pin down locations faster and reduces the likelihood of a transcription or reporting error.

Where Am I? Where Are You?

When Hurricane Katrina made landfall on the morning of August 29, 2005, the protective levees broke, causing widespread flooding in New Orleans. The usual landmarks for orientation, such as street signs, house numbers, and even whole sections of the city were wiped out. As rescue workers attempted to locate people calling for help, they ran into a significant problem: successfully communicating location. Locals were used to using street maps and regional knowledge, but volunteers from out-of-town needed other tools to help them pinpoint locations. The US military used the MGRS, and other rescue workers ended up using PLSS maps, while many civilian volunteers used longitude and latitude on their GPS receivers. The different systems in use, and the different formats for presenting location data, resulted in significant confusion and delays.

Technically, this should not have happened, because four years earlier the Federal Emergency Management Agency (FEMA) had officially endorsed a plan for a unified mapping system called the United States National Grid (USNG). The USNG is based on the MGRS, with each section of the country identified by an alphanumeric Grid Zone Designation (GZD) that is composed of the UTM Zone, followed by the alphabetic MGRS latitude band (e.g. 18S). In this case, it is very important to note that "S" does not mean south: it is the code for the MGRS latitude band. Each GZD section is divided up into 100,000-meter by 100,000-meter squares that are identified by a two-letter code (e.g. UJ) (Figure 4.7).

Precise locations are described by grid coordinates, which are reported as an even number, four to ten digits long. For example, the Jefferson Memorial in Washington, DC is located at 18S UJ 23350546. To read this, the coordinate numbers are split in half (2335 0546) to give the *easting* and *northing* position, thus the monument is located 23,350 meters to the east and 5460 meters to the north of the origin of square UJ (Figure 4.8). The first two digits are always read in thousands of meters (i.e. 23.350 thousand meters). Maps that use USNG are overlain with a labeled 1000-meter grid. While use of this system would greatly enhance emergency response, it has not yet seen wide-scale adoption in the United States, partially due to the lack of easily available maps. More information on the system can be found at the Federal Geographic Data Committee (FGDC) website (http://www.fgdc.gov/usng). NOAA has a useful page with interactive tools for converting between USNG, UTM, and latitude–longitude coordinate systems (http://www.ngs.noaa.gov/TOOLS/usng.shtml).

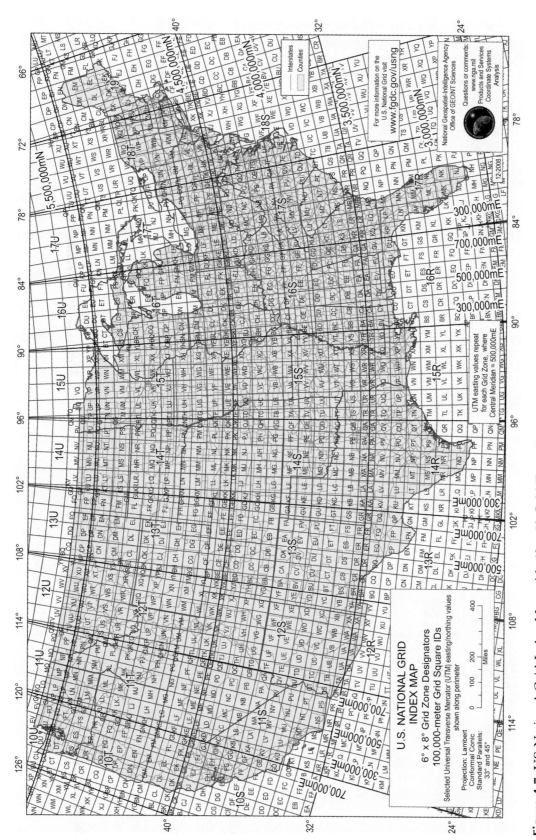

Figure 4.7 US National Grid Index Map with 6° × 8° Grid Zone Designators and 100 000-meter Grid Square IDs.
Source: Courtesy of the National Geospatial-Intelligence Agency.

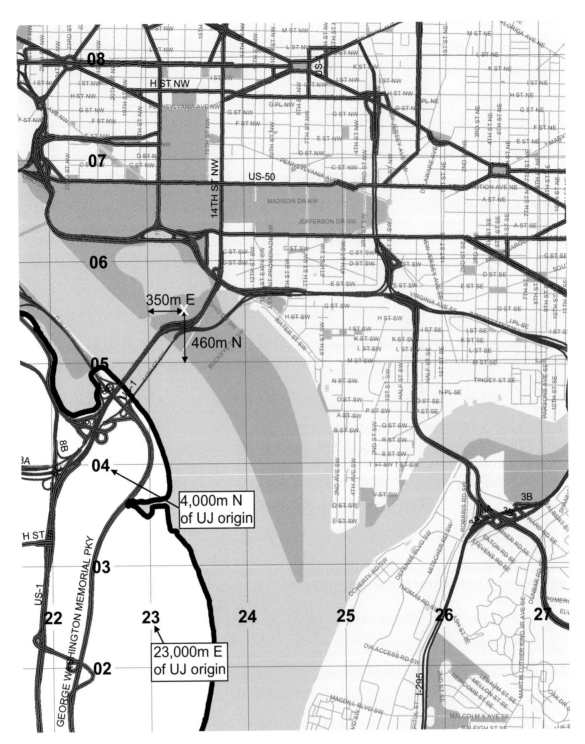

Figure 4.8 Location of the Jefferson Memorial using USNG coordinates.
Source: Section of Washington, DC map from the Office of the Chief Technology Officer (OCTO), USNG, Digital Globe.

Maps in the United Kingdom

UK Ordnance Survey maps also use a National Grid System that divides the country into a series of 100-kilometer by 100-kilometer squares, which are identified by a pair of letters (e.g. TQ). Once again, a grid overlies each square. The lines that run north–south are called *eastings*, because they are numbered in a west to east direction, while the perpendicular lines running east–west are called *northings* (because they are numbered from south to north). Each square on the map is subdivided into smaller grids by grid lines spaced 1 km apart, numbered from 00 in the southwest corner to 99, moving to the east and north. Grid references are written with the letters of the square first, then the numbers of the eastings, followed by a space and the numbers of the northings. For example, TQ 22 79 means 22 kilometers east and 79 kilometers north in grid square TQ.

More precise locations are provided by subdividing the one-kilometer grid squares into 100-meter intervals to create six-figure map references. For example, the Science Museum in London is located at TQ 266 793, or 26.6 kilometers east and 79.3 kilometers north in grid square TQ. As you can see, this system works the same way that the USNG system does, but there are significantly better educational materials available at the British Ordnance Survey website (http://www.ordnancesurvey.co.uk/oswebsite/education) and maps can be found at the Ordnance Survey's Get-a-map website (http://www.ordnancesurvey.co.uk/oswebsite/getamap).

The Global Positioning System

A Global Positioning System (GPS) is a global, satellite-based radionavigation system that provides three-dimensional location information, as well as precision velocity and timing services. The Navigation Signal Timing and Ranging Global Positioning System (NAVSTAR GPS) was the first, becoming fully operational in April 1995. It is currently composed of a network of 31 communication satellites placed into medium Earth orbit for the US Department of Defense and a series of ground-based tracking stations monitored by the US Air Force and the National Geospatial-Intelligence Agency. Originally, the system was intended solely for military use, but in the 1980s the government made the system available free for civilian use.

Orbiting the Earth twice a day, each satellite communicates simultaneously with the ground-based tracking stations and the other satellites, in order to determine its own location. Each satellite also continuously transmits a microwave signal reporting the current time, as determined by an on-board atomic clock, parameters to calculate the location of the satellite, and a report on the general health of the system. The satellite transmissions are picked up by using a GPS receiver. Most currently available GPS units are multi-channel microwave receivers with data processors/computers and an accurate clock. By determining the delay between the reported location of each satellite and the reception of the satellite's signal, and applying some corrections, a GPS receiver can calculate the distance to each satellite.

To determine an exact position (a *fix*), a receiver needs to *acquire* (pick up) information from a minimum of four satellites to triangulate its location relative to the satellites. Civilian units are typically accurate to within approximately 10 meters (around 32.81 ft) horizontally and within 20 meters (65.62 ft) for elevation. Most GPS receivers can report position in latitude/longitude and in UTM coordinates utilizing a variety of different datums. The GPS system is standardized on the World Geodetic System 1984 (WGS84) UTM datum, which can be significantly different from the older NAD27 datum, which is referenced on many USGS maps. Most modern units allow users to select the reporting datum so that the information displayed matches the reference system of whatever maps you are using. Some mapping GPS units are also able to report locations based on PLSS maps or other specialized coordinates.

Because GPS receivers are based on microwave transmissions, you need to use them in as open an area as possible. They will not work inside of buildings, in caves, or even in heavily forested areas. It is also important to realize that GPS receivers do not replace the use of a map and compass, and are instead an augmentation to traditional navigation methods. This fact is well understood by anyone who has had the batteries die, dropped their receiver, or become lost in an area with poor/no reception.

Currently, there are two fully operational systems, the United States NAVSTAR GPS and the Russian *Glo*bal'naya *Na*vigatsionnnaya *S*putnikovaya *S*istema (GLONASS), which re-attained global coverage in October 2011. GLONASS had fallen into disrepair in the mid-1990s after the collapse of the USSR. The Galileo system, the first one intended primarily for civilian use, is being developed by the European Union and the European Space Agency. It is in the initial deployment stage and the first two in-orbit verification (IOV) satellites were launched on October 21, 2011. Intended to be fully operational by 2012/13, funding issues have now pushed this out to 2019.

China is deploying a system called the BeiDou-2 ("Compass") Navigation Satellite System. It had nine functional satellites in orbit in 2011, will provide regional coverage by sometime in 2012, and is expected to ring the globe by 2020. In addition, the Indian government is developing the Indian Regional Navigational Satellite System (IRNSS). The first of the satellites is expected to launch in 2012 and the system is expected to achieve regional coverage by 2014.

Where Were You the Night of...?

The morning of October 18, 1999, in Spokane, Washington, William Bradley Jackson called 911 to report that his 9-year-old daughter, Valiree, was missing. The last time Jackson had seen her was 8:15 a.m. and her backpack was still sitting on the front porch. Police, canine units, and local volunteers immediately began a search of the neighborhood, but she was nowhere to be found. However, a detective searching the Jackson home noticed bloodstains on Valiree's pillow.

The police soon informed Jackson that they believed he had something to do with his daughter's disappearance. Five days later, the police obtained a warrant to search the Jackson's home and vehicles. Shortly thereafter, they obtained warrants that allowed them to install GPS tracking devices on Jackson's vehicles while they were still impounded. The vehicles were then returned to Jackson but he was not informed about the GPS units.

Data from one of the GPS units showed that on November 6 Jackson drove his truck to a storage unit and then to a remote location on a logging road where the truck sat motionless for about 45 minutes. On November 10, Jackson made another trip to a remote location where he remained for about 16 minutes. He then traveled back to the site on the logging road, where he stopped for about 30 minutes. After that, Jackson stopped several other places, including the storage unit. When the police investigated, they discovered Valiree's body in a shallow grave at the logging road site and they found two plastic bags with duct tape containing hair and blood at the other remote site. Based partially on the GPS records, Jackson was convicted of murder and sentenced to 56 years.

GPS units found similar use in a more infamous murder case, the 2002 disappearance of 8-month-pregnant Laci Peterson. Her husband, Scott, reported that he had left home early Christmas Eve morning to go fishing and came home to find her gone. For a variety of reasons, police were suspicious of Scott and obtained court permission to covertly attach GPS tracking devices to the undersides of the vehicles that he was known to drive, eventually including several rentals. The GPS devices showed that in January 2003 Scott made at least five trips, using a variety of vehicles, including three different rentals, to a marina not far from the location where the bodies of his wife and unborn son eventually washed ashore. The prosecution argued that these visits showed that Scott wanted to monitor the activities of the police and the possible discovery of his wife's body. The evidence provided from the GPS devices was partially responsible for Scott Peterson's eventual murder conviction.

Maps

A *map* is simply a graphic representation of the natural and/or artificial features of a part or the whole of a planet's surface or of a section of space. Maps are frequently used in forensic work, to help record the locations of evidence at outdoor crime scenes, for example, or to help organize searches. The most common type of map is a *planimetric map*, which depicts the locations of major geographic and cultural features such as rivers, roads, towns, and parks. Highway maps and road atlases are forms of planimetric maps, so you should already be familiar with their basic features, but several other types of maps can be valuable in forensic work. One liability of planimetric maps is that they do not provide any information about differences in elevation, only horizontal separation, which makes them very useful in an urban setting and much less helpful in the wilds.

For elevation information, you need a *topographic map* (Figure 4.9). Topographic maps are flat representations of the three-dimensional shape of the Earth's surface and show the relative position, shape, and size of physical features such as valleys, hills, and steep slopes. This information is conveyed using *contour lines*, imaginary lines along which all points are at the same elevation above (or below) a particular datum, usually sea level. Topographic maps are usually produced by government agencies and named for local landmarks such as towns or geographic features. *Bathymetric maps* are similar to topographic maps except that they are used to depict the relief of the bottom of a body of water using contour lines called *isobaths* (lines of equal depth below water).

Topographic maps can be quite useful for forensic investigators, because they depict the relief of the terrain. Topographic maps most often are set up to depict sections of the Earth's surface called *quadrangles*. A quadrangle is bounded on the

Figure 4.9 USGS topographic map. Portion of the Corfu, NY NW4 Attica 15' Quadrangle 1950 edition. Please see Color Plate section.
Source: Courtesy of the United States Geological Survey.

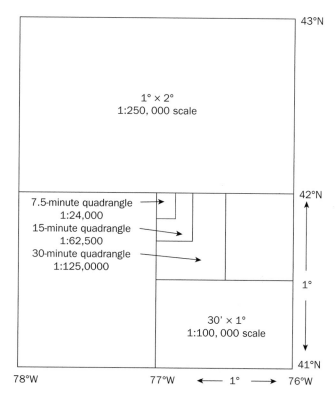

Figure 4.10 USGS quadrangle sections and their associated scales.

top and bottom by lines of latitude and on the left and right by lines of longitude. The two most common quadrangle maps are 15-minute quadrangles, which depict an area that measures 15 minutes of latitude by 15 minutes of longitude, and 7½-minute quadrangles, which depict an area 7½ minutes of latitude by 7½ minutes of longitude (Figure 4.10). Each 15-minute map can be divided into four 7.5-minute maps. Topographic maps in the United States are typically drawn at a 1:24,000 scale, while maps in the United Kingdom are drawn at a 1:25,000 scale.

How to Read a Topographic Map

Contour lines, which are usually brown or orange, connect points of equal elevation above sea level. The lines will always separate points of higher elevation from points of lower elevation. The lines are spaced out evenly at a fixed *contour interval,* such as every 10 feet or 5 meters; the exact interval will be indicated in the key to the map. Every fifth line, usually, will be printed slightly wider, indicating that it is an *index line* and will have elevation values printed out along the line in several places. The elevation of the thinner, intermediary, lines is determined by locating nearby index lines and counting up or down using the appropriate contour interval.

For steep terrain, contour lines will be close together, while for relatively flat areas they will be widely spaced. Contour lines never cross each other, except in the rare case that there is an overhanging cliff. In this instance, dashed lines will be used to represent the hidden contours. Merged contour lines are used to indicate vertical cliff walls. Concentric circles represent hills, while circles with *hachure marks* (short hatch lines perpendicular to the contour lines), which are made on the downhill side of the line, are used to indicate closed depressions. Contour lines will form a V pattern when crossing streams, and the apex of the V will always point in the upstream (uphill) direction.

Topographic maps are available in the United States from the United States Geological Survey (http://topomaps.usgs.gov/) and can be viewed online at the National Map Viewer website (http://nationalmap.gov). Information on Canadian topographic maps can be found at the Natural Resources Canada website (http://maps.nrcan.gc.ca/topo_metadata/). Maps of the United Kingdom are available online at the Ordnance Survey's Get-a-map website (http://www.ordnancesurvey.co.uk/oswebsite/getamap).

Geologic maps are used to show the distribution of geological materials (Figure 4.11). Different colors and patterns are used to symbolize different rock, sediment, or soil types. Geologic maps are usually printed on top of lightly inked planimetric maps (called *base maps*) that show the locations of roads, rivers, and major cultural features, so that it is easier to locate yourself on the map. *Geologic cross-sections* show what the topography and layers of underlying rock would look like if you were able to make a vertical cut through the ground. The closest easily visualized equivalent is a road cut or a cliff where layers of rock are visible. A cross-section is the same idea; it just shows the layers that exist in unexposed rock beneath your feet.

In the United States, a variety of maps are published by the United States Geological Survey and are available at the USGS National Geologic Map Database at (http://nationalmap.gov/ustopo/index.html). Maps of Canada are available on-line from Natural Resources Canada at (http://ess.nrcan.gc.ca/prodser_e.php#map). Maps of the United Kingdom can be found at the British Geological Survey website at (http://www.bgs.ac.uk/geoindex/index.htm).

All good maps will include the following information: the *location name* and *map type* (e.g. Geologic Map of New York State or Topographic Map of Bloomington, Indiana 7.5-minute quadrangle), a *north arrow* or other indicator of orientation, a *scale* used to convert map features to actual size, and a *legend* (or key) explaining the symbols and colors used on the map.

For orientation, at minimum all maps should include a *north arrow*, which indicates the direction of geographic north, called *true north*. This can be roughly defined as the point in the northern hemisphere where the Earth's axis of rotation meets the Earth's surface. It is important to know that true north is typically not where your compass is going to point. The actual direction a magnetized needle will point is called *magnetic north*. This is usually depicted on maps in the form of a partial compass-rose like symbol showing the difference (called *declination*) between true north (usually indicated with a star) and compass north (MN) (Figure 4.12).

The swirling motion of the molten iron in the Earth's outer core generates the Earth's magnetic field, which changes through time, called *secular variation*, and varies in a complex manner over the Earth's surface. Magnetic north is currently located in the Canadian Arctic, but it is slowing drifting northwest, so technically any declination listed on a map would only be exact for the year the map was printed. Depending on your location, changes in declination over time can be quite large (~1° every three years) or quite small (almost 0°). There are also variations due to changes in magnetic activity. For example, *magnetic storms* can cause rapid and erratic changes in declination. For most people on Earth, magnetic storms cause variations much too small to be of concern, but these changes become of greater importance the closer you get to the magnetic poles. If you happen to be working on the northern side of Victoria Island in Canada, for example, magnetic storms can cause compass errors of greater than 2° more than 75% of the time.

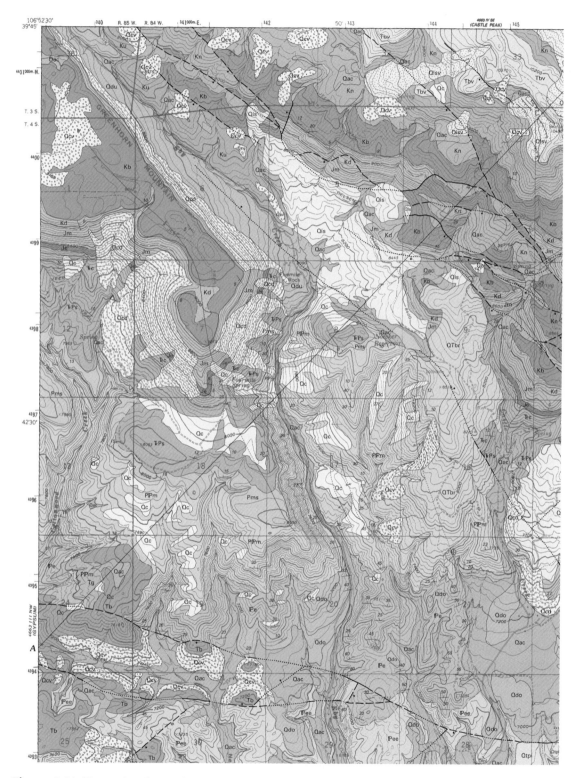

Figure 4.11 Example of a geologic map from a section of the Geologic Map of the Eagle Quadrangle, Eagle County, Colorado by David J. Lidke, 2002. Please see Color Plate section.
Source: Courtesy of the United States Geological Survey.

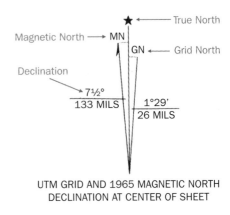

Figure 4.12 Example of an orientation compass rose.

UTM GRID AND 1965 MAGNETIC NORTH
DECLINATION AT CENTER OF SHEET

If the map you are working with is more than a decade or so old (or less depending on how close you are to the poles), it is advisable to get an updated declination. The easiest way to correct for changes in magnetic declination is to use an online declination calculator from an appropriate agency to find out current declination and set your compass appropriately. Note that if magnetic north is east of true north, the local declination will be reported as a positive number, while if magnetic north is west of true north the local declination will be reported as a negative number. NOAA has a website you can use to determine the estimated value of magnetic declination using latitude and longitude (or by entering a zip code for places in the United States) (http://www.ngdc.noaa.gov/geomagmodels/Declination.jsp). Alternatively, you can look up the most recent declination map at the Geological Survey of Canada's Geomagnetism webpage (http://gsc.nrcan.gc.ca/geomag) or use their declination calculator for cities in Canada (http://geomag.nrcan.gc.ca/apps/mdcal-eng.php).

If you do not know current declination, many maps will have a statement of magnetic declination, such as "11° 48′ approximate mean declination 1978 for center of map, annual change increasing 1.0′" that can be used to make a reasonable estimate. Since the annual change is listed as increasing, it is added to the listed declination (a decreasing declination would be subtracted). So current declination could be estimated by multiplying the annual change by the number of years since the map was printed (so 1.0′ × 32 = 32.0′), indicating that declination in 2010 would be approximately 11°48′ + 32′ = 12°20′. However, magnetic secular variation is not constant with time, therefore the older the map, the more likely this estimate will be significantly in error. For the example given above, the actual declination in 2011 was listed as being around 11°41′W, therefore this method should really only be used if the map is relatively recent (i.e. only a year or two old really).

Finally, if are working from a known position, you can make your own measurement of declination. First, you need to take a magnetic bearing to a distinct landmark, like a tower or mountain peak, that is both represented on your map and visible in your field of view. Your measurement will be more accurate the greater the distance to the landmark. To take a bearing, orient the compass so that the direction-of-travel arrow or sight is pointed at the landmark and determine the magnetic *azimuth* (from 0° to 360°) by turning the compass housing so that it is aligned with the compass needle to get a measurement in degrees (i.e. read the number from the housing where it meets the base of the direction-of-travel arrow or sight). Next, draw a light pencil line on your map that goes from your position, through the position of the landmark you are using, and off through the edge of the map. Place your compass on the map with its origin (the pivot point of the needle) centered where your pencil line intersects the edge of the map and with the 0° marking on the compass aligned with geographic north. Ignore the compass needle and measure the bearing on the map using your compass like a protractor (i.e. read the numbers at the point where your pencil line crosses the graduated circle of the compass on the side opposite the map) (Figure 4.13). The difference between your bearing and the map measurement is the declination. If you can, take measurements for multiple landmarks and use the average of your results as your declination.

If you have absolutely no idea what local declination is, set your compass to *zero declination* and simply record all of your measurements using the local magnetic

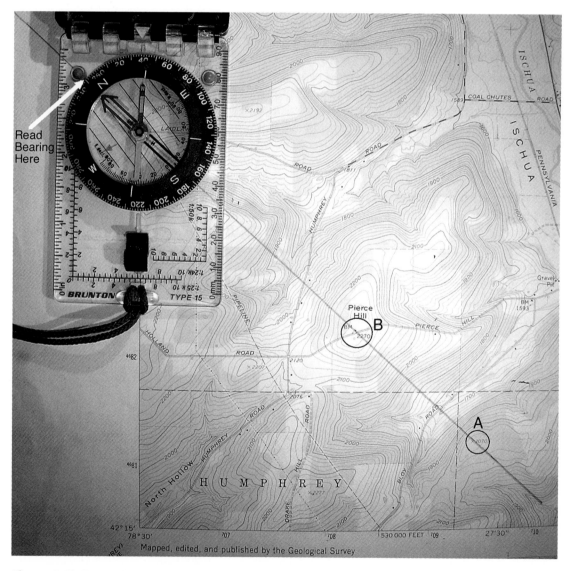

Figure 4.13 Reading a map bearing with a compass.

north, making sure that you clearly note this in your records. When you return to the laboratory, you can obtain the declination of the field site using the resources listed previously.

Some maps also indicate *grid north* (GN) on their compass rose (Figure 4.12) or use a set of north–south and east–west grid lines that intersect at right angles (these are especially common on topographic maps). Because the Earth is a sphere, the grid lines printed on a flat map cannot really align exactly with either true north or magnetic north and still be at 90° on a map, thus the need for a grid north designation. The declinations indicated on such maps are not true declinations (D) but are more accurately described as *grid declinations* (G). Everything described previously still holds, but you must apply a correction using the angle between grid north and true north, called the *convergence angle* (C), to find the true declination. If the grid north and magnetic north lines are both on the same side of true north, then you can obtain true declination simply by adding grid declination and the convergence angle (D = G + C). If they are on opposite sides of true north, you subtract the convergence angle from the grid declination (D = G – C) to determine true declination.

Maps must also include a *scale* that can be used to convert map features to actual size (Figure 4.14). Scale can be expressed in many different ways. Some maps use a simple *statement* or *verbal scale* such as "One inch equals one mile" or "1 cm = 1 km." Other maps, such as topographic maps, use a *representative fraction* or *fractional scale* in which the map is designed to be at a particular fraction (or ratio) of reality. For example, many topographic maps are 1:24,000, which means that one unit of distance on the map is the equivalent to 24,000 of the same units of distance in reality. Features that are 2 cm apart on this kind of map would be 48,000 cm (or 480 meters) apart on the ground. More commonly, a *scale bar* or *graphical scale* is created by drawing a line or bar of a specific length and visually presenting actual distance as a block of space on the map. The great advantage of this type of scale is that it remains accurate even if the map is reduced or enlarged. That is not true of the verbal or fractional scales. Even maps "not drawn to scale" generally use a relative scale so that all the objects in the picture are roughly the right size relative to each other.

Finally, all maps need to contain a *legend* (or key) explaining the symbols and colors used on the map. The symbols and patterns used on maps vary by map type. On topographic maps, artificial features are usually shown in black or red, water features are blue, green is used for vegetation, and brown is used for the contour lines (lines of equal elevation). Other topographic map symbols used by the USGS are shown in Figure 4.15 and can be found at (http://erg.usgs.gov/isb/pubs/booklets/symbols).

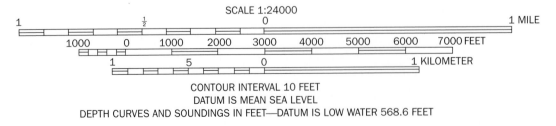

Figure 4.14 Example of a map scale.

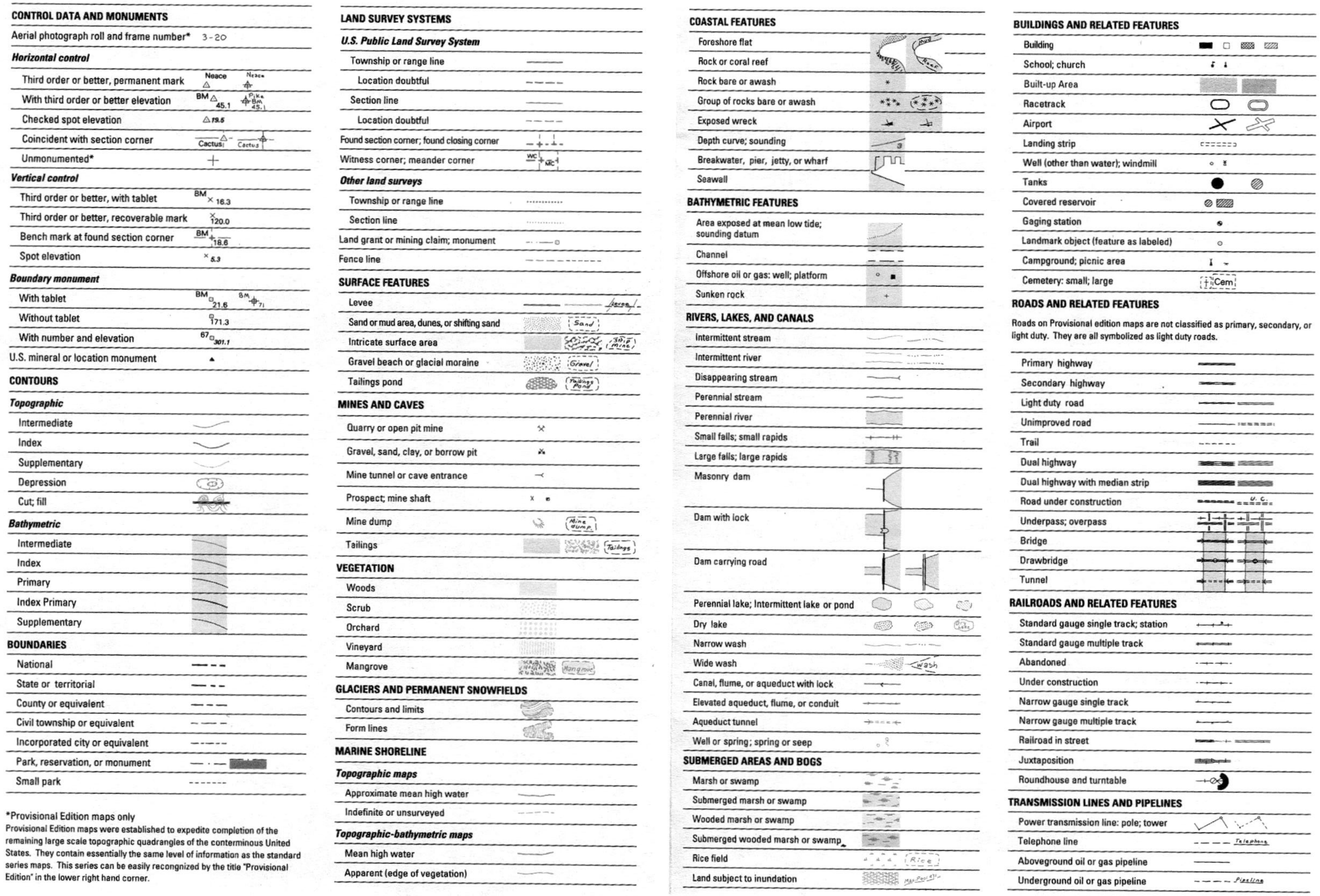

Figure 4.15 Symbols used on topographic maps produced by the USGS. Variations will be found on older maps. Please see Color Plate section.

Source: Courtesy of the United States Geological Survey.

In contrast to topographic maps, geologic maps are strikingly colorful. The colors and patterns are used to represent different *geologic units* (a distribution of a certain kind of rock of a given age range) (Figure 4.16). Each geologic unit is also assigned a set of letters to symbolize it on the map. Usually, the symbol is formed with an initial capital letter followed by one or more small letters. The capital letter represents the age of the geologic unit. Geologists have divided the history of the Earth into *eons* (the largest formal division), *eras*, *periods*, *epochs*, and *ages*. (There currently is a proposal to create two informal supereons: the Pregeozoic and the Geozoic.) The most common division of time used in letter symbols on geologic maps is the period (Table 4.1). Occasionally, the age of a rock unit will span more than one period, in which case both capital letters are used. For example, SD would be used to indicate that the rock unit began to form in Silurian time and was completed in Devonian time. The small letters indicate either the name of the unit, if it has one, or the type of rock, if the unit has no name.

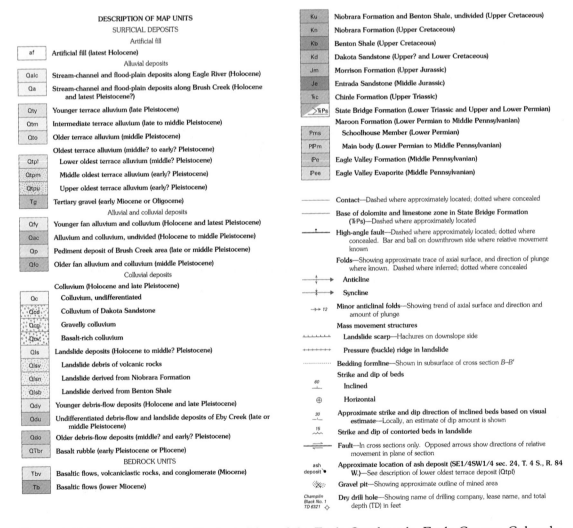

Figure 4.16 Simplified key for Geologic Map of the Eagle Quadrangle, Eagle County, Colorado by David J. Lidke, 2002. Please see Color Plate section.
Source: Courtesy of the United States Geological Survey.

Geology with Intent to Harm (Murray, 2004; Murray and Tedrow, 1992)

Authorities had a problem. New cars from Detroit were arriving in New Jersey dented and with smashed windows. In a case of repeating vandalism, someone was throwing rocks at the cars as they were being transported east by rail. Since it was clearly impossible to monitor the whole rail line, police asked the Pennsylvania Geological Survey to look at the rocks found in some of the cars to see if they could narrow down the possible sites where the vandalism was occurring. Complicating matters, there were actually two railroad routes used to transport the cars from Detroit to New Jersey, both of which passed through exposures of an incredible variety of rock types. Some of the areas had relatively unique rocks, but the tracks also passed through multiple areas where similar types of rock would be exposed.

Luckily, the rocks submitted for examination turned out to be uncommon. They were coarse-grained (pegmatitic) gneiss containing feldspar, quartz, biotite mica, chlorite, and thin crystals of what was thought to be apatite. An examination of geologic maps allowed researchers to narrow the possible source areas down to two places, the geologic province known as the Reading Prong in southeastern New York and a section of the geologic province called the Piedmont in eastern Pennsylvania. These were the two areas where the railroad tracks intersected exposures of this type of metamorphic rock. More detailed research suggested that the rocks from the Piedmont usually had little biotite and rarely had apatite, but the rocks from the Reading Prong were rich in biotite and commonly contained apatite. Researchers were able to narrow down the most likely point of origin to a stretch of tracks in the vicinity of West Point, north of New York City. This enabled police to concentrate on the West Point section of track, where they spotted several culprits in the act of vandalism and were able to make multiple arrests.

In a similar case of assault and vandalism, hurled rocks injured several people and damaged property. The rocks recovered at the scene were angular blocks of an unusual igneous type that exists in relatively small areas of the southern United States. Using geologic maps, investigators found that the nearest outcrop of this type of rock was 20 miles from the crime scene. Searching the area, police located a suspect who drove a pickup truck with similar rocks in the bed.

Remote Sensing and Other Resources

In addition to traditional paper maps, there are many other imaging tools available online that can be of use in forensic investigations. Basic road maps are available from sites like MapQuest (www.mapquest.com), Maps On Us (www.mapsonus.com), Streetmap (www.streetmap.co.uk), and Google Maps (http://maps.google.com). The National Geographic Society has an online Map Machine (http://ngm.nationalgeographic.com/map) that can overlay population density, natural disaster, seismic hazard, and other data onto world maps. The Perry-Castañeda Library Map Collection (http://www.lib.utexas.edu/maps) is one of the most comprehensive resources for digitized maps, including scanned versions of

Table 4.1 Abbreviations for geologic periods.

Neogene	N	Quaternary	Q
Paleogene	Pε	Tertiary	T
Cretaceous	K		
Jurassic	J		
Triassic	℞		
Permian	P		
Pennsylvanian	ℙ	Carboniferous	C
Mississippian	M		
Devonian	D		
Silurian	S		
Ordovician	O		
Cambrian	Є		
Precambrian	pЄ		

CIA maps, and is searchable by region, country, state, and city. There is also a directory of map resources for the United Kingdom on the British Cartographic Society website (http://www.cartography.org.uk/default.asp?contentID=705).

What might potentially be of more interest are the various forms of remotely sensed data. *Remote sensing* is defined as the non-contact recording of electromagnetic data, which includes visual as well as ultraviolet (UV), infrared (IR), thermal, and microwave wavelengths. These types of data can be used in a number of sophisticated ways, such as for tracking environmental pollution and looking for heat signatures, but the most commonly available data types include aerial photographs and satellite images.

Aerial photographs, or pictures of the Earth taken from the air, have been produced since before the invention of the airplane. There is a range of types of aerial photos available, taken from a variety of heights. They can be black and white, color, or false-color images, depending on the source. *False-color* means that the colors in the image have been assigned to wavelengths that our eyes cannot see (like ultraviolet or infrared) in order to make things that would normally be invisible visible. False-color is also used to enhance subtle differences. The National Aerial Photography Program (NAPP), which began in 1987, actively generates cloud-free photographs taken from 20,000 feet of the conterminous US (http://edc.usgs.gov/guides/napp.html). The photos are taken over five- to seven-year cycles and are centered over one-quarter sections of 7.5-minute USGS quadrangles, covering areas of approximately 5.5 square miles.

A related program, the National High Altitude Photography (NHAP) program, ran from 1980 to 1989 (http://edc.usgs.gov/guides/nhap.html). The NHAP photographs were taken from an altitude of 40,000 feet. For some areas, Digital Orthophoto Quadrangles (DOQs), or aerial photographs with geo-referencing (i.e. map coordinates), are also available (http://eros.usgs.gov/products/aerial/doq.html).

LIDAR (light detection and ranging), which can measure three-dimensional structures, might also prove useful (http://gisdata.usgs.gov/Lidar). There are banks of historical aerial photographs for the United States available as well (http:// eros.usgs.gov/products/aerial/historical.html).

The first crude satellite photographs of Earth were taken in 1959, by the US *Explorer 6* satellite. The first television images of Earth taken from space were broadcast by *TIROS-1*, the world's first weather satellite, which was launched in April 1960. Since then, dozens of *Earth-observing satellites*, with a wide range of differing capabilities, have been launched into orbit by several different countries and private companies. There are currently at least 68 Earth observation satellite missions in operation and many different types of satellite images available.

The National Aeronautics and Space Administration (NASA) operates a few Earth-observation missions, the most pertinent of which are the Landsat mission (Figure 4.17), the Earth Observing System (EOS) program, with the *Terra* satellite, and the *EO-1* mission. *Landsat-5* and *Landsat-7* are currently operating, and most images from previous *Landsat* satellites are freely available (http:// landsat.gsfc.nasa.gov). Data from many NASA missions can be centrally accessed at the USGS's National Satellite Land Remote Sensing Data Archive (http:// edc.usgs.gov/archive/nslrsda). Information about the European Space Agency's (ESA) Earth Observation Satellite images is also available online (http:// earth.esa.int/images), and you can tour Canada by satellite at the Canada Centre for Remote Sensing (CCRS) website (http://www.ccrs.nrcan.gc.ca).

In addition to the missions run by government agencies, there are now several civilian companies that operate satellites. DigitalGlobe, based in the United States, is a privately held company that operates the high-resolution *QuickBird* satellite, launched October 2001, and the *Worldview-1* satellite, launched September 2007, which is billed as producing the world's highest-resolution commercial satellite

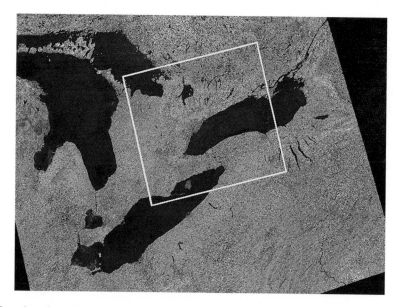

Figure 4.17 Landsat image of the Great Lakes Region of North America ID:LE70170302010125EDC00, taken 5.5.2010. Please see Color Plate section.
Source: Courtesy of the United States Geological Survey's Global Visualization Viewer.

imagery (www.digitalglobe.com). GeoEye, also based in the United States, operates the *IKONOS* and *OrbView-3* satellites, both of which have a resolution of less than one meter (www.geoeye.com). Spot Image, based in France, is the commercial operator of the *SPOT* satellites (www.spot.com).

Most of the data from the sites listed above is only available for a fee, though some sites offer free sample images and the NASA *Landsat* data is being released to the public on a rolling schedule. However, there are several free resources that can prove useful. Google Earth is a free, limited-capability program that superimposes satellite, aerial, and other imagery over a virtual globe (http://earth.google.com/). Terraserver-USA can be used to view Microsoft's collection of aerial photography of the United States (http://terraserver-usa.com), which is different from the commercial site called Terraserver, which sells imaging data (www.terraserver.com). Flash Earth (www.flashearth.com) and ACME Mapper (http://mapper.acme.com) are also potentially useful tools. Flash Earth allows you to compare the map results from a variety of different online sources (i.e. Google versus Yahoo versus Microsoft, etc.) to find the version that gives the best results. ACME Mapper is based on Google Maps, but lets you overlay topographic maps and digital orthophoto quads onto the images.

This list is just the tip of the iceberg as far as digital imagery is concerned. Many other websites host a range of different types of remote-imaging data, as well as different tools to manipulate the data. To truly do this topic justice, you should invest in a book on remote sensing and get another book that gives you an introduction to geographic information systems (GIS).

Summary

In this chapter, several different location systems and map types were briefly introduced. All of this location information may appear overwhelming at first, but becomes second nature with practice. The best thing to do is to spend time performing activities like orienteering, map-based scavenger hunts, and geocaching (GPS-based scavenger hunts). Several low-cost GPS units and applications for smart phones will help by automatically converting between coordinate systems and providing map information. Everyone, however, should get practice working with paper maps. There will be occasions when cellphone service is unavailable, satellite coverage blocked, or the digitally provided information turns out to be inaccurate or inadequate.

Further Reading

Most of the information in the chapter references online materials that are freely available. The links provided are your best resource for obtaining regionally appropriate materials. Some books of interest include:

Compton, R. R. (1985) *Geology in the Field*. John Wiley & Sons Ltd, New York.

El-Rabbany, A. (2006) *Introduction to GPS: The global positioning system*, 2nd edn. Artech House Publishers, Norwood, MA.

Maltman, A. (1998) *Geological Maps: An introduction*, 2nd edn. John Wiley & Sons Ltd, New York.

Slocum, T. A., McMaster, R. B., Kessler, F. C., and Howard, H. H. (2008) *Thematic Cartography and Geovisualization*, 3rd edn. Prentice Hall, Englewood Cliffs, NJ.

Spencer, E. W. (2006) *Geologic Maps: A practical guide to the preparation and interpretation of geologic maps*. Waveland Pr Inc., Long Grove, IL.

References
Murray, R. C. (2004) *Evidence from the Earth*. Mountain Press Publishing Co., Missoula, MT.

Murray, R. C. and Tedrow, J. C. F. (1992) *Forensic Geology*, 2nd edn. Prentice Hall Inc., Englewood Cliffs, NJ.

Chapter 5
Sand: To See the World in a Grain of Sand

Most of the time, full-sized rocks do not figure largely in criminal cases. Most forensic geological material falls into the category of trace evidence, which is, by its very nature *trace* (i.e. very, very small). In previous chapters, we generally examined minerals and rocks at the hand-sample scale, but now we will start looking at the issues associated with the analysis of geological material at a smaller scale, starting with sand. As mentioned previously, the term *sand* technically refers to a specific size range of clasts created under certain conditions of physical and chemical weathering. However, the word sand is also used colloquially to refer to any mixture of sand-sized materials such as rock fragments, minerals, shells, glass, and other debris. The natural sand that you are probably most familiar with is that found along the beach, though it is generated in other areas. Oddly enough, beaches are also common sites for criminal activity, probably due to their status as tourist destinations with large transient populations.

The Sands of War

You might not know this, but during World War II six lives were lost on the American mainland as a direct result of enemy action. In one of the lesser-known tales of World War II, forensic geoscience, and the identification and sourcing of sand, play an important role.

The idea of using the jet stream to carry balloon bombs, or *fusen bakudan*, apparently dates back to a 1933 Japanese military scientific laboratory program to develop new war weapons. The Japanese had been studying the powerful air currents in the upper atmosphere, especially the ones we now called *jet streams*,

An Introduction to Forensic Geoscience, First Edition. Elisa Bergslien.
© 2012 Elisa Bergslien. Published 2012 by Blackwell Publishing Ltd.

since the 1920s. The *fusen bakudan* were only one of several ideas under development and the program was formally stopped in 1935. Then, in 1942, a joint Army–Navy research project began development of balloon bombs that could be launched from submarines in order to retaliate for the Doolittle Raid. Two submarines were adapted for this purpose, but the program was discontinued in August 1943, when the submarines were suddenly redeployed to Guadalcanal. Thus arose the necessity to develop a balloon that could launch from the Japanese homeland. The Imperial Japanese Army and the Imperial Japanese Navy each began separate projects to design balloons capable of surviving the trip to the American mainland and deploying an explosive payload.

The Central Meteorological Observatory in Tokyo worked from 1942 to 1944 collecting detailed observations and some 200 of the balloons created for the submarine program were launched during the winter of 1943/44 to collected additional meteorological data and test project feasibility. These balloons were never intended to reach the United States, but based on the data they collected the meteorologists estimated that the jet stream would be capable of carrying a balloon across the Pacific from Japan to North America in an average time of 60 hours, roughly three days. The polar-front jet stream that flows from Asia, across Japan, to North America is thousands of kilometers long, a few hundred kilometers wide, and is typically located somewhere between about 9 and 15 km above the Earth's surface. It usually moves between 250 and 500 km per hour, though speeds of over 700 km per hour have been recorded.

The balloon attack formally began on November 3, 1944, the birthday of former ruler Emperor Meiji. The jet stream is strongest from November to March, which would help to ensure that the balloons reached the States, but a winter launch missed the fire season of the western United States and limited the possible scope of damage that the balloons could inflict. From November on until the cancellation of the project, balloons were launched on every clear day. The total number of balloons launched is unclear, but lies between 6000 to over 9000. They expected at least 10% to reach North America.

Though some were spotted earlier, the first published report of a balloon bomb reaching land was from Kalispell, Montana in early December 1944. It was described as "a huge paper balloon bearing Japanese ideographs and armed with an incendiary bomb capable of starting a major conflagration" (Christian Science Monitor, 1944). According to the FBI, the balloon had "characters indicating that it was completed on Oct. 31, 1944 [and] was painted red and yellow with a rising sun emblem" (Los Angeles Times, 1944). Other discoveries quickly followed in Oregon and Washington. There also had been earlier news reports of a "phantom plane," a parachute in the air and several explosions near Thermopolis, Wyoming, which was in reality almost certainly a balloon bomb.

There were actually several short newspaper stories published about the balloons in 1944, and in some areas their existence was common knowledge, though there is no evidence of an associated widespread panic. The major concerns appeared to be determining where the balloons came from and if they carried any passengers. "That they were made in Japan there was no doubt but it was found highly improbable the balloons started their flights in Japan because of their limited gas capacity" (Washington Post, 1945). *Newsweek* even suggested, "the best conclusion, therefore, is that any saboteurs [carried by the balloons] probably would be Germans" (Newsweek, 1945).

Sometime in January 1945, the US Government Office of Censorship requested a news blackout on all stories about the balloon bombs to keep the Japanese from getting any information about the fate of their balloons, and possibly to keep the public from wild speculation about what the balloons might be carrying. Meanwhile, the materials recovered from the balloons that landed were examined for clues as to the balloons' origin. The Japanese had done their best to ensure that there were no markings or stamps on any of the parts or materials used to construct the balloons that could be used to trace their places of manufacture, thus denying the Americans military targets (though a few items with location information did make their way into some of the balloons). Initially American Military Intelligence did not believe that the balloons could be coming over directly from Japan. Instead, they thought that the balloons could possibly be coming from North American beaches launched by landing parties from enemy submarines, from German prisoner-of-war camps, from Japanese–American internment camps, or possibly even from some small Pacific Islands.

This is where geology enters the story. The Military Geology Unit (MGU) of the United States Geological Survey (USGS) had been formally established in June 1942 with the intention of providing the US military with geologic information that could help invading armies, such as detailed terrain descriptions, as well as the locations of water supplies, important mineral reserves, and sources of construction materials. When sand from some of the ballast bags was recovered from the balloon bombs, it was delivered to the USGS and the men and women working with the MGU. There the geologists used stereomicroscopes and polarized light microscopes to perform a detailed examination of the few handfuls of sand that they had been provided with.

Ballast sand samples from the balloons that had landed in Holy Cross, Alaska, and Glendo, Wyoming were delivered to Clarence S. Ross, a mineralogist and petrologist with the USGS. He could tell almost immediately that the sand was beach sand and that it was not from North America. He found nothing in the sand, such as mica, to indicate a granitic (i.e. continental landmass) source. Instead, the sand contained around 52% hypersthene, and lesser percentages of augite, hornblende, garnet, high-titanium magnetite, and high-temperature quartz. Both the hypersthene and the augite were of volcanic origin, while the garnet and the hornblende were of metamorphic origin. Hypersthene (technically referred to as *ferroan enstatite* these days) is an uncommon mineral, so to find such a high percentage of it concentrated in the sand was quite unusual. Based on the mineral composition, the geologists knew that they were looking for a beach that had both volcanic and metamorphic rocks located in-land but no in-land granite. This composition also eliminated the Pacific Islands, which are mostly the result of basaltic volcanism, so their beach sands would contain neither the high temperature quartz nor the metamorphic minerals.

Julia Gardner, a paleontologist, examined the sand for coral and mollusks. She found that there was, in fact, no coral at all. In modern oceans, most coral is confined to water that is greater than 18 °C, which tends to keep it located in tropical and subtropical waters between 30° north and 30° south latitudes. Most coral also requires highly saline water that is clear enough to permit high light penetration. This meant that the ballast sand was from a beach located in a cold-water region, which eliminated almost all of the beaches south of Tokyo Bay as potential sources.

Kenneth E. Lohman, an MGU specialist in diatoms (a form of microscopic photosynthetic algae with silicate shells), recovered more than a hundred different species of recent and fossil diatoms. The fossil diatoms were all Pliocene in age, which helped to limit the age of the parent rock. Kathryn Lohman searched for foraminifera, a microscopic form of single-celled life that produces calcium carbonate shells. These organisms are almost entirely marine in origin and have a wide variety of forms, which can be used to help determine their source. The diatoms and foraminifera that the Lohmans found all pointed to Japan. In fact, the pair located an 1889 paper that described diatoms and foraminifera that matched the species found in the ballast samples and that came from an area around Sendai, on the northeast coast of Honshu. This narrowed the source area down to the eastern coastline of Honshu, somewhere north of Tokyo. More specifically, they identified Shiogama beach, eight miles northeast of Sendai, and another possible source zone called the Ninety-nine League Beach at Ichinomiya, forty miles southeast of Tokyo.

Given this intelligence, orders were dispatched for aerial reconnaissance of selected areas of the eastern coastal areas of Honshu. Some of the photos showed what might have been partly inflated balloons and possible launch areas. According to one newspaper, "an analysis of the sand ballast of the balloons revealed their launching points and these, too, were bombed" (Seattle Times, 1945), which is an oversimplification of events. The balloon attacks had already ceased as a result of the press blackout and redirection of the Japanese war effort.

Due to the self-censorship of the American press, the Japanese only learned of one landing. The Chinese newspaper *Ta Kung Pao* had picked up a story from American sources that a balloon reached Thermopolis, Wyoming and reported it in late December 1945. Strangely enough, most of the American papers did not have information about that incident because it had mistakenly been reported as a "phantom plane," instead listing the Kalispell, Montana landing as the first incident. According to later newspaper reports, the Japanese also thought that the Thermopolis balloon bomb had not exploded.

The news of six deaths in Oregon never reached them either. The only casualties of the balloon attacks occurred on May 5, 1945, near Bly, Oregon. The Reverend Archie Mitchell and his pregnant wife had taken five children from their Sunday school on a fishing trip/picnic. The minister was at the car, or walking some distance behind the group, when he heard his wife call out that they had found something. Moments later, there was an explosion, apparently caused by someone trying to move the balloon. Mrs Elsie Mitchell, 26, and all five of the children, ranging in age from 11 to 14, were killed in the explosion.

The press blackout was officially ended with a single press release on May 22, 1945, when the Army and Navy disclosed information about the attacks "so that the public may be aware of the possible danger and to reassure the nation that these attacks are so scattered and aimless that they constitute no military threat" (Chicago Daily Tribune, 1945). Basically, spring had started, the school year was ending, and the military wanted to prevent any additional accidents like what happened in Oregon.

In all, over 345 balloon fragments were discovered during the war, but due to the news blackout and low-key response, the Japanese had no idea where their balloons were going or what effect they were having. In April 1945, the project was declared a failure and the attacks were abandoned. Plus, American B-29s destroyed two of the three helium plants supplying the balloon project, as well as several of the factories supplying parts, which contributed to the decision to abandon the project. Thus, the balloon bombs were halted before the fire season started again. According to Kiyoshi Tanaka, one of the engineers leading the project, if the Japanese General Staff had learned that even one death had resulted from the balloon attack, 10,000 more would have been launched.

While the intelligence from the MGU did not directly result in the destruction of the Japanese balloon bomb program, it was still vital to the American war effort. For one thing, it kept American war efforts from being diverted to look for a fictitious Pacific Island launch site or for saboteurs lurking on American soil. And the MGU had successfully identified one of the launch sites. If the program had not already been abandoned as fruitless, this information probably would have resulted in bombings of the suggested launch sites.

As it was, there were several near misses by the balloons. A few landed near the Hanford atomic bomb plant, but failed to explode, and one struck a power line running from the Bonneville Dam to the Hanford plant, temporarily halting work. A balloon with one bomb still attached landed on a runway at an army airfield in Paine. Another of the balloons came within a few miles of Detroit, near the Chrysler tank arsenal, a Ford plant, and other large war plants. Apparently, a boy in Washington State carried around a live anti-personnel bomb for a few days before authorities managed to take it away, averting what could have been a very nasty accident. Balloon shrouds and other fragments were found as far east as Michigan (three incidents), as far south as Mexico (three incidents), and as far north as the Northwest Territories, Canada (four incidents). The majority of the balloons were found in British Columbia, Canada (57), Oregon (45), Alaska (37), and Montana (32), all numbers not including post-war discoveries.

One final note of caution, while over 6000 balloons and possibly as many as 9000 were launched, to date only 361 or so have been found. This means that there might be several more hiding out there in the less accessible areas of the western United States. The most recent reports of balloon fragments being found were in 1987 in North Dakota and in Oregon in 1992. The bulk of this story was based on newspaper and magazine articles from the 1940s and on John McPhee's "The Gravel Page" article from the January 29, 1996 issue of *The New Yorker*. A somewhat more detailed version of this story can be found on the companion website (www.wiley.com/go/bergslien/forensicgeoscience) and additional resources can be found in the Further Reading section at the end of this chapter.

An Introduction to Sand

All of the geological material on the surface of the Earth, as well as everything else, is subject to a process called *weathering*, or physical and chemical breakdown. *Physical* (or mechanical) *weathering* means that rock is literally broken into smaller and smaller pieces, without changing mineral composition. Thus, the fragments you find will have the same mineralogy as the parent rocks from which they came. *Chemical weathering* means that some or all of the components of the original rock are changed into new minerals, with different chemical compositions than the original rock. A familiar example of this process is the creation of rust on your car. The iron in the steel is chemically altered by exposure to air and water, creating a new substance commonly known as rust. The bright-red colors you see in many desert rocks are the result of the same process. Some minerals, like halite, might even be dissolved away entirely. There are specific chemical reactions involved in these transformation processes, so it is still possible to determine the nature of the original rock.

Physical and chemical weathering, to a greater or lesser extent, is occurring everywhere on the surface of the Earth, breaking rock into pieces called *sediments* that are categorized by size range (Table 5.1). Thus, the term *sand* technically refers to a specific size range of geologic particles.

Sand, a term used for any sediment that is composed of ≥ 50% sand-sized particles, is found virtually everywhere on the surface of the Earth. It is most commonly found along rivers, lakes, and coastlines, wherever wind and water are plentiful. In the high energy of these environments, the larger rocks and sediments get broken down into smaller pieces, while the wind and water winnow away the fines, resulting in high concentrations of sand-sized fragments. Because local rocks are the source for most sand, the sands formed in different geologic environments around the globe can have very different characteristics, with varying compositions and distinctive microtextures to the individual grains, which can allow identification of specific source areas. This discussion tends to focus on beach sands because beaches are common recreation areas that are typically highly accessible. The same techniques and terminology can be applied to any type of sand, and, indeed, with few exceptions, to any type of sediment.

The composition of a particular sand depends primarily on what native materials are available (Basu, 1976; Basu and Molinaroli, 1989). For example, in Hawaii, there are no quartz sand beaches because there is no local source for quartz. The Hawaiian Islands are formed of basalt erupted from a source deep inside of the

Table 5.1 The size range of sedimentary particles.

Grain size class	Diameter
Gravel	Any sediment >2 mm at any point
Sand	Any sediment 2–0.0625 mm
Silt	Any sediment 0.0625–0.0039 mm*
Clay	Any sediment <0.0039 mm*

*Visually, you cannot distinguish between silt and clay. Silt is gritty, while clay is smooth or soapy, like chalk dust. You can feel this texture difference if you rub a sample on your teeth, but I do not recommend it. For simplicity, everything smaller than 0.0625 mm can be called *fines* or *mud*.

Earth. There you will find carbonate sand beaches composed of broken pieces of organisms (coral, etc.), black sand beaches formed of volcanic basalt fragments rich in ferromagnesian minerals like magnetite and hornblende, and there are even some green-sand beaches composed almost purely of the mineral olivine. Most common beaches in North America, and indeed on most of the continents, are composed of quartz because it is the most resilient common mineral and is a major component of granite, the igneous rock that forms the bulk of the continental landmass.

As you can already probably guess, composition in large part determines the color of sand. For example, white-sand beaches are usually composed primarily of quartz. Due to the strength of its mostly covalent chemical bonds, quartz is a very durable mineral that is resistant to physical weathering and is highly resistant to chemical weathering. Therefore, quartz is often the most common mineral found in a high-energy beach environment. "White"-sand beaches also usually contain small quantities of various *accessory minerals* (minerals that are rarer and only form a small fraction of the sample) such as garnet, magnetite, and limonite, which can give the sand a colored tint and can make it possible to distinguish potential source areas. Sands formed in other environments will also potentially have distinctive coloration that is determined by the local materials available.

In western New York, near the Great Lakes and Niagara Falls, there is no local source for the replenishment of large amounts of quartz. The rocks there are mostly limestone, dolostone, and shale. A vast amount of quartz sand was derived from glacial deposits and collected on the shores of post-glacial lakes like Lake Warren and Lake Iroquois, two of the precursors to the modern Great Lakes. However, this sand moves readily between the lakes and is commonly transported away via the current. To stabilize many of the beaches, sand has to be imported. This is actually true in several places, so if you are playing locally on a nice, wide, clear quartz-sand beach, but the regional bedrock does not contain quartz, there is a good chance that at least some of the sand was carried in by trucks and placed there artificially.

Black-sand beaches, such as those in Hawaii, are created when molten mafic lava hits seawater, causing such rapid cooling (quenching) that the solidifying lava shatters, forming sand-sized fragments that are washed onto shore. Basalt is very susceptible to weathering, so these beaches are short-lived, geologically speaking. You need to have an active volcano located nearby to maintain a black-sand beach.

Another common beach sand is formed of broken pieces of the calcium-carbonate shells of sea organisms like oysters, clams, snails, calcareous algae, foraminifera, and coral. These types of beaches are great for shell collecting, but can often be very difficult to walk on barefoot, as the shells can be very sharp. If you look closely, these beaches may also include fossilized foraminifera, shark teeth, or the bones of whales and fish. In general, fossils are typically heavier than their modern counterparts, and may be different in color because of mineral staining that occurs with prolonged burial. The mixture of modern organisms washed in from the ocean and fossils that are usually weathered out from inland rocks will give the sand a special character. By identifying the organisms that make up the sand, it can be possible to locate the source beach.

Additional coloration of coastal sediments may occur as a result of the influence of color-producing minerals such as garnet (pink/red), hematite (rust-red/orange), limonite (yellow), magnetite (black), and olivine (green). Large concentrations of rare minerals give some sands very distinctive properties. Weathering causes the breakdown of geologically unstable minerals, such as feldspars and ferromagnesian minerals, while more stable minerals, such as quartz and zircons, become enriched. Clay minerals, which are the product of the chemical weathering of unstable minerals, are also very stable in the environment and can become concentrated. Beyond examining mineralogy and texture, there are also tools that look at elemental composition, or other factors, making it possible to distinguish sands of similar appearance.

Characterizing Sand

For sediments, the term *texture* refers to the size, shape, and surface condition of the clasts (pieces) composing the sample. These factors are primarily determined by how and for how long the individual grains were transported. In general, the larger the particle, the less it has traveled, so gravel is usually quite close to its source, while clay could have travel halfway around the world before being deposited.

Grain Size

The grain size scale used by sedimentologists is the Udden–Wentworth scale (Table 5.2), which brackets together particular size ranges into sediment *classes*. Grain size can be determined in a variety of ways. The largest clasts, such as pebbles and coarse gravel, can easily be measured directly with a caliper. For most large clasts, their shapes can be approximated by triaxial ellipsoids, and grain size is reported in terms of the lengths of the longest (L), shortest (S), and intermediate (I) major axes (Figure 5.1) (or their maximum, minimum, and intermediate grain diameters). By common practice, these axes should be orthogonal to each other, though they do not necessarily need to intersect at a common origin. In standard practice, L is assigned to the longest dimension of the clast as measured using a caliper, I is assigned to the longest dimension perpendicular to L, and S is assigned to the shortest dimension that is perpendicular to both L and I.

With forensic sand samples, which usually have a limited amount of material, it is often possible to manually measure individual particles under a microscope that is fitted with an *ocular* (eyepiece) *scale*. In this case, it is not possible to determine values for all three axes, so grain size is reported in terms of the longest dimension, the average dimension, or the maximum and minimum visible dimensions. For a more detailed discussion of this approach, see McCrone (1982). Obviously, the smaller the clast, the more difficult it becomes to measure it directly.

For bulk samples of sand or finer particles, when there are too many clasts to measure individually or they are too small to measure directly, particle sizes are assessed using an indirect method. The most common approach for sand and soil samples is to use a nested series of wire-mesh sieves (for a detailed description of this technique, see Chapter 7). For sand, a minimum of 30 to 50 g of material is generally necessary to perform an analysis using this method. Particles that fall through are finer than that mesh size, while particles that are retained in a specific sieve are larger than that mesh size, and finer than the mesh size of the sieve above. The result divides the sample into *bins* of particular size ranges (i.e. a particle size distribution). Dry, or wet, sieving is usually the most appropriate approach for grain size analysis of samples composed primarily of sand and gravel.

Table 5.2 Udden–Wentworth grain size scale.

Wentworth Size Class		Millimeters	Phi (⊠) units	Sieve No.
Gravel	Boulder	4096	-12	
		1024	-10	
		256	-8	
	Cobble	64	-6	
	Pebble	16	-4	
		4	-2	5
	Granule	2.00	-1	10
Sand	Very coarse sand	1.00	0	18
	Coarse sand	0.50	1	35
	Medium sand	0.25	2	60
	Fine sand	0.125	3	120
	Very fine sand	0.0625	4	230
Silt	Coarse silt	0.0310	5	
	Medium silt	0.0156	6	
	Fine silt	0.0078	7	
	Very fine silt	0.0039	8	
Clay	Clay	0.0020	9	
		0.00049	10	

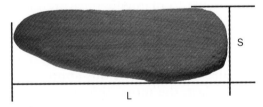

Figure 5.1 Measurement of a clast.

Bulk samples of fines (silts and clays) cannot generally be analyzed this way because the mesh required would be too fragile, and the tiny particles being analyzed have a tendency to stick to everything anyhow. The particle size distribution of fines is most often assessed on the basis of settling velocity, or the time it takes the particles to settle through a column of water at a specific temperature. The most common approach is to use *pipette analysis*, though hydrometer and decantation methods are also popular under some circumstances (see Chapter 7).

Grain size separation by the use of *density gradient columns*, a method that has been commonly used for forensic comparison, may in fact be of only limited utility. A discussion of this method can be found in Murray and Tedrow (1992). Chaperlin and Howarth (1983) established that the method is highly subjective, prone to error, and indicate that is should not be used as a sole measure of comparison, especially for the examination of sand samples. Similarities between density gradient column distributions of sands are generally not because of similarity in source but are instead because the bulk of the sample will be composed of minerals of approximately the same density (Fitzpatrick and Thornton, 1975).

There are also several more sophisticated tools for measuring grain size, especially for fines. *Laser-diffraction size analysis* is based on the idea that it is possible to determine a particle's size by determining the angle at which it deflects a specific wavelength of light. A *photohydrometer* calculates particle size based on changes in a beam of light as it passes through a column of settling suspended sediment. *Image analysis* techniques use a camera to capture digital images of sediment grains, which are then analyzed using computer software.

Regardless of the method used, the result is a summary that lists the proportion of grains in each of the grain size classes (Table 5.2), usually in terms of a percentage. For example, the percentage of the bulk sample that is contained in a particular sieve, or the number of grains on a microscope slide, out of the total number of grains measured, that can be considered very fine sand. This information is best conveyed in either statistical terms or graphical form. When performing statistical analysis, it can be useful to work in terms of the *phi scale* rather than millimeters (or micrometers), which allows you to plot sizes in units of equal value. In this scale, the grain diameter in phi units (ϕ) = $-\log_2$ of grain diameter in millimeters. Thus, a grain with a diameter of 8 mm has a phi value of -3ϕ, and a grain with a diameter of one-eighth of a millimeter (0.125 mm) has a value of 3ϕ.

The most common graphical methods for presenting grain size data are by using a histogram, a frequency curve, or a cumulative curve. A *histogram* is a type of bar diagram with grain size, or size classes, listed along the x-axis by convention, decreasing to the right. Each vertical bar represents a single size class delineated by the grain sizes listed on the left and right of the base of the bar. The height of each bar in the y-direction corresponds to the proportion or percentage of grains in that size class (Figure 5.2a). Though there are more accurate methods of construction, in general a *frequency curve* is a smooth curve that can be generated by connecting the midpoint of each size class bar (Figure 5.2b).

To create a cumulative curve, data are converted into cumulative percentages of grains of that size class or larger (% coarser). Start from the left, with the largest size class, and move to the right adding the percentage (or amount) of all coarser

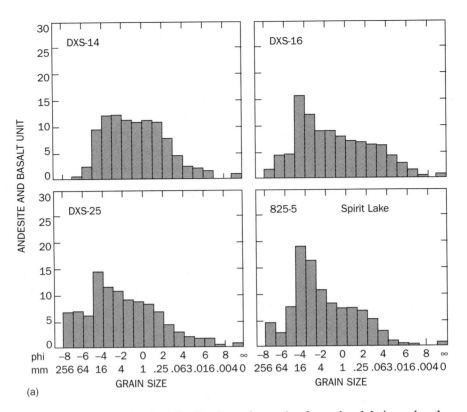

Figure 5.2a Histograms of grain size distribution of samples from the debris avalanche and blast deposits around Mount St. Helens volcano, Washington.
Source: Courtesy of the United States Geological Survey.

Figure 5.2b Histogram of grain size with overlain frequency curve for Spirit Lake.
Source: Courtesy of the United States Geological Survey.

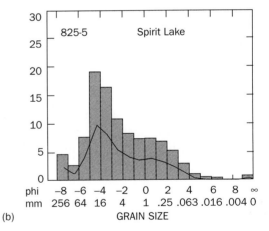

material to each subsequent size class. Grain size is plotted on the x-axis (using a semi log scale if the data are in millimeters) and cumulative weight percentage is plotted on the y-axis (Figure 5.2c). The slope of a cumulative curve reflects the degree of sorting. A very steep slope indicates good sorting, while a very gentle slope indicates poor sorting (Figure 5.3).

There are also several statistical measures of grain size that can be used to characterize a sediment sample. Average grain size, more correctly termed *central tendency*, is reported in a number of ways. The *mean* grain size is technically the

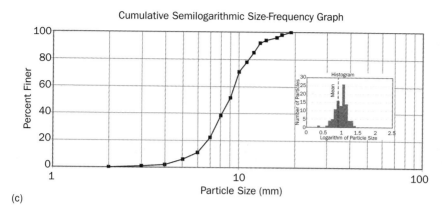

Figure 5.2c Grain size frequency plot.
Source: Courtesy of the United States Geological Survey.

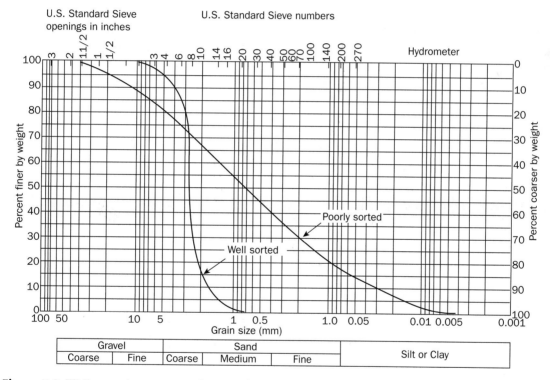

Figure 5.3 Well-sorted versus poorly sorted grain size frequency curves.

arithmetic average of all of the particle sizes in the sample. Since you usually will not actually know the total number of grains in your sample, nor will you actually know the size of every individual particle, mean is approximated by picking selected percentile values from the cumulative curve and averaging those values. Most commonly, sedimentologists take the sum of the particle sizes at the 16, 50, and 84% marks, read off of a cumulative curve, and divide by 3 to get a reportable mean. In phi values mean = $[(\phi_{16} + \phi_{50} + \phi_{84})/3]$.

The *median* size is the midpoint of grain size distribution, which corresponds to the 50th percentile particle size diameter read from a cumulative curve (median = ϕ_{50}). Thus, half of the particles in the sample are smaller than the median size and half of the particles in the sample are larger than the median size. The *mode* is the most frequently occurring particle size in a population of grains. This corresponds to the diameter of the grains at the steepest point (inflection point) on a cumulative curve, or the highest point on a frequency curve. Most sediment samples have a single modal size, but it is possible for samples to be bimodal or even polymodal (have three or more distinct clusters).

The mean, median, and mode sizes for a sample may be the same, but much more commonly they will be different. When the mean, median, and mode are different, grain size frequency curves will show some degree of asymmetry, or *skewness*. Skewness is reported numerically as:

$$S = \frac{\phi_{84} + \phi_{16} - 2\phi_{50}}{2(\phi_{84} - \phi_{16})} + \frac{\phi_{95} + \phi_{5} - 2\phi_{50}}{2(\phi_{95} - \phi_{5})}$$

Table 5.3 lists how skewness is described verbally.

Sorting describes the range of grain sizes within a sample (Figure 5.4). If all the grains in a sample are nearly the same size, the sample is called *well sorted*. If the sample contains gravel, sand, silt, and clay it would be called very *poorly sorted*. Examination of the degree of sorting tells you something about the manner in which the sediments have been transported. Running water is one of the best erosional agents at sorting particles by size (and density; think about panning for gold). Thus, different stretches of a river will be characterized by different sizes of well-sorted particles. Wind is also an excellent sorter, but only for the finer

Table 5.3 Verbal descriptions for skewness.

Skewness	Verbal description of skewness
>+0.30	Strongly fine-skewed (lots of fine particles)
+0.30 to +0.10	Fine-skewed
+0.10 to -0.10	Near-symmetry (unskewed)
-0.10 to -0.30	Coarse-skewed
<-0.30	Strongly coarse-skewed (lots of coarse particles)

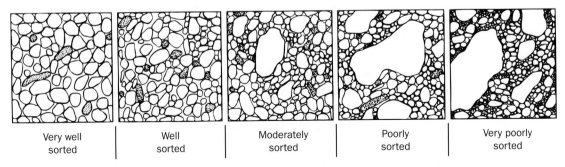

Very well sorted   Well sorted   Moderately sorted   Poorly sorted   Very poorly sorted

Figure 5.4 Visual reference chart for sorting.
Source: After Compton, 1985.

(smaller) grain sizes. Glaciers, on the other hand, are extremely poor at sorting, since the ice basically picks up everything, moves it, and then drops everything when it melts, acting more or less like a bulldozer. Glacial deposits tend to include sediments of all shapes, sizes, and conditions.

On a beach, sediments are sorted by the waves as they crash against the shore. The finer particles get carried inland to the onshore portion of the beach (sometimes into back-beach eolian dunes), while the larger and heavier particles remain closer to the water's edge. In addition, the waves also sort the particles by density. Quartz has a specific gravity (SG) of 2.65 grams per cubic centimeter (g/cm^3) and the SG of feldspars is similar, ranging from 2.5 to 2.6 g/cm^3. Minerals that have a significantly higher SG than average will be found mixed in with much larger clasts. On beaches that have a large fraction of colored *heavy minerals* (minerals with an SG greater than 2.9), there will often be a discernible band (called a *strand-line*) along the top of the swash zone, which is the area at the top of the beach where high tide deposits material (Figure 5.5). This is where the heavier grains will accumulate. In addition, the strong wind that blows in across the water also carries lighter particles inland, to the upper area of the beach or to inland dunes. For this reason, when collecting comparison samples, it is important to be working in the correct beach zone, or stretch of river.

Sorting can be estimated by reference to a visual chart (Figure 5.4), or expressed in terms of a mathematical expression called *standard deviation*. Sedimentologists typically use a formula for determining the inclusive graphic standard deviation that is based on phi values:

$$\sigma_i = \frac{\phi_{84} - \phi_{16}}{4} + \frac{\phi_{95} - \phi_5}{6.6}.$$

In verbal terms, Table 5.4 lists phi standard deviations.

In theory, grain size data can be used to identify the agents of sediment transportation and to determine depositional environments. For example, the mean,

Figure 5.5 Photograph of heavy minerals (dark) in the strand-line of a quartz-sand beach (Chennai, India).
Source: Photograph taken by Mark A. Wilson.

Table 5.4 Phi standard deviations and sorting terms.

Standard Deviation	Sorting
$<0.35\phi$	Very well sorted
0.35–0.50ϕ	Well sorted
0.50–0.71ϕ	Moderately well sorted
0.71–1.00ϕ	Moderately sorted
1.00–2.00ϕ	Poorly sorted
2.00–4.00ϕ	Very poorly sorted
$>4.00\phi$	Extremely poorly sorted

median, and modal sizes of sediments that have been transported by wind will be much finer than those of sediments that have been transported by glaciers. Beach sands generally have lower standard deviations than river sands do, and dune sands are extremely well sorted. However, despite decades of work, and increasingly sophisticated statistical analyses, grain size data alone cannot uniquely link a sediment sample to a specific environment. Such data can, however, be extremely useful in forensic comparisons (see, for example, Pye and Blott, 2004). A more complete discussion on the acquisition and analysis of grain size data can be found in Folk (1974).

Grain shape is actually more important than size for gathering clues to sediment origin. Sediment grains can be described in terms of their roundness, sphericity, and surface texture. *Roundness* is the term used to describe the smoothness or angularity of a particle. The further a grain is transported, the more it is bumped and banged into other particles and into the ground. This causes the edges of the particle to get broken off and smoothed out, resulting in an increasingly smoother shape. So a well-rounded particle has generally traveled a great distance from its point of origin, while an angular grain is more likely closer to its parent material. Each form of transport leaves its signature on the individual grains it moves. Wind action is far more abrasive than transport by water, so it creates grains that are more angular, while water transport will create smoother grains. Glaciers indiscriminately transport everything, so no sorting occurs and there is a jumble of smooth and angular particles.

Roundness (R) can be described subjectively by *roundness class* (modified after Powers, 1953: 118):

- **Well-rounded**: Original faces, edges, and corners have been destroyed by abrasion and whose entire surface consists of broad curves without any flat areas.
- **Rounded**: Round or curving in shape; original edges and corners have been smoothed to rather broad curves and whose original faces are almost completely removed by abrasion. Some flat areas may remain.
- **Sub-rounded**: Partially rounded, showing considerable but not complete abrasion, original form still evident but the edges and corners are rounded to smooth curves. Reduced area of original faces.

- **Sub-angular**: Somewhat angular, free from sharp edges but not smoothly rounded, showing signs of slight abrasion but retaining original form. Faces untouched, while edges and corners are rounded off to some extent.
- **Angular**: Sharp edges and corners. Little or no evidence of abrasion.
- **Very Angular:** Reserve this term for those few particles whose edges and corners look so sharp that they could cut you. A sample that looks freshly broken.

This is a bit easier to understand if you refer to Figure 5.6. It is important to remember that roundness really means smoothness, not how round a particle is, i.e. spherical. Thus, rod-shaped particles that are very smooth are also called well rounded (see Figure 5.6, bottom right). Comparing samples to a sand gage or photographic examples can also be helpful (Figure 5.7).

Another term used to describe a sand grain's shape is *sphericity*, which is a measure of the degree to which the shape of a sedimentary particle approaches that of a sphere. Sphericity can be thought of as the degree of similarity of a particle's length, width, and thickness. In a perfect sphere, the length (L), width (I), and thickness (S) are the same. The more dissimilar the dimensions are, the less spherical the particle. Note that, by this definition, a cube also has a high degree of sphericity; thus, it is still necessary to characterize a particle's roundness.

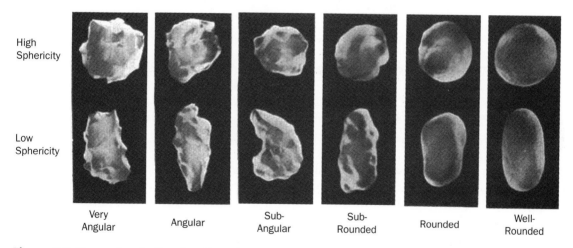

Figure 5.6 Powers' scale for visually estimating roundness.
Source: Powers, 1953; reprinted by permission of the Society for Sedimentary Geology.

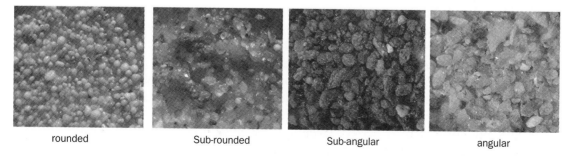

Figure 5.7 Photograph of rounded, sub-rounded, sub-angular, and angular sand grains. Please see Color Plate section.

Table 5.5 Terms and examples of sphericity.

Shape	Diameters	Example
Bladed	L > I > S	Knife blade
Rod-like	L > I = S	Rolling pin
Discoidal	L = I > S	Compact disc or DVD
Spherical	L = I = S	Ball

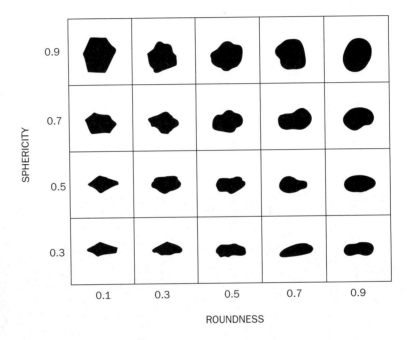

Figure 5.8 Krumbein visual scale for roundness and sphericity.
Source: Krumbein and Sloss, 1963.

Sphericity (S) can be reported as a value between 0.0 (non-spherical) and 1.0 (perfect sphere), though most sediment grains fall in the range from 0.3 to 0.9. Technically, to determine the sphericity of a clast, the length of each of its axes (L, I, S) should be measured with a caliper or some other device, and sphericity = (LIS ÷ L^3). This can be quite difficult to do in practice, especially on small grains, so the following descriptive names are applied based on a visual examination in Table 5.5.

A useful tool for combining these two characteristics, roundness and sphericity, is the *Krumbein visual scale for roundness and sphericity* (Figure 5.8). To use the scale, simply compare the general shape of the particles in your sample to the images in the diagram, and use the numerical scale listed on the sides. Just like sphericity, roundness can also be given a numerical value ranging from 0 (which is not physically possible) to 1 (a perfect circle), and is determined by dividing the radius of the curvature of a circle inscribed around the largest portion of the clast divided by the radius of the curvature for the smallest inscribing circle. In practice, values below 0.1 are not used. For additional examples of roundness, see Figure 5.9. By using the Krumbein visual scale, you can get approximate values for roundness and

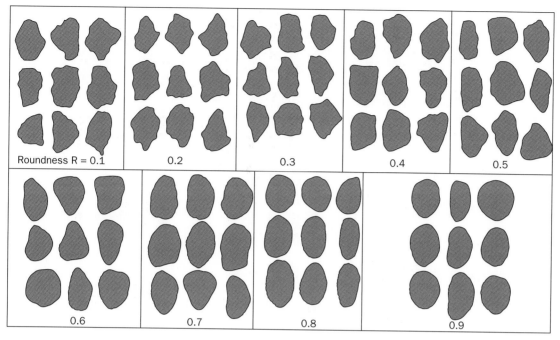

Figure 5.9 Numerical visual scale for estimating roundness.
Source: Modified from Powers, 1958; reprinted by permission of the Geological Society of America.

sphericity without attempting to physically measure all of the individual clasts in your samples. For example, particles shaped like smooth rolling pins would have S = 0.3 and R = 0.9, while nearly perfect spheres would have S = 0.9 and R = 0.9. For additional help with visual determination of sphericity, see Figure 5.10.

If the particles being examined bear a strong resemblance to a common shape, such as a hexagon, a triangle, a cross (X), a tetrahedron, or an octahedron, this should also be reported along with roundness and sphericity. For particles with highly irregular or complex shapes, roundness and sphericity might not be sufficient descriptors. In that case, additional terms can be employed to describe particle morphology, such as *form, elongation,* and *irregularity.* A more detailed discussion can be found in Blott and Pye (2008) and Pye (2007).

Besides the simple techniques described above, there are several more sophisticated ways of measuring and reporting grain shape, including digital imaging and laser particle size analysis, but these methods usually require expensive equipment, significantly more training, and the investment of much more analysis time. There are cases where such an analysis can be highly fruitful, but they are not usually necessary for most comparisons.

When performing an analysis of sedimentary particles, other factors must be kept in mind. First, there is a correspondence between roundness and grain size. Roundness tends to increase with grain size because the larger grains have larger areas of potential contact with other grains. The impacts between larger grains also tend to be more energetic, resulting in greater roundness (Krumbein and Pettijohn, 1938). The smaller the particle, the more likely that cleavage is a controlling factor on particle shape, resulting in more platy particles. In addition, because different minerals have differing levels of susceptibility to weathering processes, the mineralogy of a sample will vary by size fraction. Thus, it is essential to compare the same size fraction of all samples analyzed and to make sure that the same minerals out of each sample are being examined.

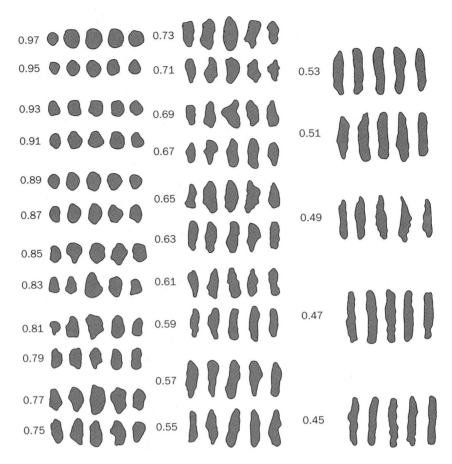

Figure 5.10 Visual images for estimating sphericity.
Source: After Rittenhouse, 1943; reprinted by permission of the Geological Society of America.

The Murder of Italian Prime Minister Aldo Moro (Lombardi, 1999)

In 1978, the body of former Italian prime minister Aldo Moro was found in the trunk of a car parked on a street near the center of Rome. His kidnapping, two months earlier by the *Brigate Rosse* (Red Brigades), a militant terrorist group, initiated a period of serious political turmoil in Italy. At first, the Red Brigades indicated that they wanted to exchange Mr Moro for the safe release of certain prisoners. During his captivity, Mr Moro sent letters to his family, political friends, and other parties urging them to negotiate. Then, on the morning of May 8, authorities received a telephone call informing them that the last message of the Red Brigades was in a car parked midway between the headquarters of the Christian Democratic Party (of which Moro was a member) and the Communist Party.

Mr Moro had been shot only a few hours earlier. His body, dressed in a well-tended blue suit and black leather shoes, rested on a blanket. Mr Moro's trousers had turned cuffs, inside of which a tiny amount of sand was found. More sand was found adhering to his shoes, on the blanket, and inside the car where the body rested. Samples were also collected from the car's fenders and from inside of the tire treads. A forensic geologist was brought in to analyze the material collected and, if possible, indicate the source of the sand.

The samples were examined microscopically, loose and in thin-section, sorted by grain size and analyzed using X-ray diffraction, scanning electron microscope, and microprobe. The sand was very well sorted, with over 92% of the grains falling into the 500–125 µm range, and the individual grains were dominantly rounded to sub-rounded. There was also an absence of fines. The characteristics all pointed to a supratidal marine environment (a zone that extends inland from the high-water line of the mean tides). Compositionally, the sand samples contained a mixture of minerals (quartz, feldspars, amphiboles, muscovite and biotite micas, chlorite, calcite, dolomite, apatite, magnetite, titanite, and garnet) and rock fragments (carbonates, sandstones, chert, quartzite, phyllites, schists, and dominated by a variety of volcanics). The sands also contained a variety of shells from marine organisms, pollen and plant fragments that were also classified.

Taken together, given the mixture of volcanics, carbonates, and metamorphic materials, the evidentiary sand was determined to be beach sand from somewhere along the Tyrrhenian coast near Rome. This corresponded to roughly a 150 km stretch of coastline running from Tarquinia to Terracina. However, there was no published data on sand composition that could allow researchers to narrow the source area down further. Reference samples had to be collected from all of the beaches that were reasonably remote, had road access, and where there was an appropriate source of volcanic material inland. Because the Red Brigades were still very active, Dr Lombardi (the forensic geologist on the case) collected the necessary samples (from 92 sites) by traveling there with his wife, Patrizia, as if on vacation. Patrizia would wander around the beach looking at the plants and gazing romantically at the sea, while her husband discreetly examined the sand. If the sand was a possible match, the couple would sit down together and Dr Lombardi would surreptitiously fill a plastic bag with a reference sample.

The reference samples were initially screened microscopically and those that were clearly different were excluded. Thin sections were prepared of the 70 remaining samples, which were then subjected to a more detailed mineralogical examination. The reference samples that had the most similarity with the evidentiary sand came from an area between Focene Nord and Marina di Palidoro, a stretch of coastline only 11 km long. A few weeks after the killing, police searched the area but found no sign of a terrorist hideout. Years later, in a confession, Mr Moro's captors claimed that they had planted the sand and plant material in order to mislead the police. For a variety of reasons, including the layering of the trace material, the types of pollen found, and the widespread media coverage of the existence of the sand, this seems unlikely, though not inconceivable. Even with the arrests that were eventually made, there are still several unanswered questions about the original kidnapping (during which five bodyguards were killed) and the 55 days that Mr Moro spent in captivity.

Surface Features

The basic morphology of sediment particles is not the only important issue, so is the surface condition of the individual grains, or their *surface texture*, which can reveal important information about source area. In general, the overall morphology (shape, roundness, and sphericity) of a grain is the result of multiple cycles of transport, while the textural features on the surface of the grain are most likely a record of only the most recent transport cycle. Theoretically, one should be able to use surface textures to determine with some precision the environment from which a sample came.

Regrettably, surface texture is less easy to codify than morphology, as there is no well-accepted scheme for its description or classification. For example, the number and type of surface texture features identifiable on quartz-sand grains varies depending on the reference consulted. Krinsley and Doornkamp (1973) and Margolis and Krinsley (1974) recognize 22 distinct surface features, Le Ribault (1977) recognizes 73, Higgs (1979) recognizes 30, Culver *et al.* (1983) recognize 32, Hill and Nadeau (1984) recognize only 6, and Mahaney (2002) recognizes 39. Bull and Morgan (2007) proposed a forensic technique for comparing soil and sediment samples based on quartz-sand grain surfaces that uses 41 identifying characteristics. The key to performing a comparison that involves surface texture characterization is to consistently use a standard operating procedure based on one (or more) standard reference(s).

There are several useful terms that can be applied during a visual or optical microscopic examination. For example, the first surface texture characteristic that should be noted is the overall *luster* of the grain. Most grains will have an average, somewhat dull surface, but some will be either extremely dull or extremely bright. A *frosted* particle is a particle that is basically lusterless and has the appearance of sandblasted glass. This is because the surface of the grain has been broken up by innumerable impacts with other grains while being transported by the wind. The cracking and crazing of the grain surface can render normally transparent grains of quartz translucent. In fact, very mature (i.e. highly weathered) grains can appear nearly opaque. It is important not to confuse *frosting*, which is a surface texture, with true *opacity*. A frosted grain will have numerous small pits and cracks. Frosted sand grains are usually associated with environments such as the desert or near shore sand dunes. Scanning electron imaging of such particles can reveal additional information about the types of impacts suffered.

A dull *matte* surface can also be the result of surface deposition of chemical weathering products or precipitates, like iron oxides or hematite (salt) crusts. This type of surface is theoretically produced by alternating wet and dry cycles, when a thin film of water, perhaps only dew, repeatedly forms and then evaporates, leaving behind tiny crystals. Since frosting is the result of abrasion, it usually produces a rough surface with features greater than $2\,\mu m$ in size, and primarily affects the topographic highs around the grain. A chemically produced matte surface will have a finer texture, with features $2\,\mu m$ or less in size, and be found even in the recesses of the grain.

Grains that appear to be very shiny and strongly reflect light are said to have a *polished* surface. Rather than being dulled by having numerous impacts, or chemical precipitates, the surface is very glossy. Highly polished surfaces are usually the result of abrasion by very fine particles and are often related to water transport of some type, though not always. There are other types of polish, including glacial polish and desert polish. The surface of a gastrolith, or a stomach stone used by some birds and dinosaurs to assist in grinding food, has a very high polish thought to be the result of long exposure to stomach acid.

It is also possible for the true surface luster of a grain to be hidden by a light coating of colloidal (very fine) particles (i.e. it's dirty). Under these circumstances, the actual color and surface texture of a grain might be quite different than it first appears. These grains need to be examined both before and after washing to gather as much information as possible.

Once the overall luster of a grain has been characterized as *average, frosted, matte,* or *polished*, any additional *microrelief features*, or scattered markings on the surface, which are usually irregularly distributed, should be noted. There are several microrelief features that can be seen with the unaided eye or under an optical microscope. For example, *striations* are multiple narrow lines or scratches, generally running parallel or nearly parallel, that are cut deeply into the surface of the grain. Such marks are usually the result of glacial action. *Impression marks,* such as *imbricated grinding features,* are created when a grain has been mechanically ground into adjacent grains.

A *percussion mark* is a divot produced by a violent impact, often associated with a high-velocity flow, like a rock hitting your windshield. There are many different types of percussions marks: *conchoidal fractures* are smooth, curved fracture patterns that commonly occur on quartz, glass, and sometimes feldspars. Abundant conchoidal fracture is considered characteristic only of glacial environments or sediments that are still very close to their source. *Crescentric,* or *arc-shaped, gouges* are deep, crescent-shaped depressions. *Pits* are irregular depressions, while *V-shaped pits* have a distinctive triangular shape.

If additional factors are necessary for comparison, a scanning electron microscope (SEM) would be needed, and one of the references concerning quartz surface textures mentioned above should be employed.

Some Very General Notes About Sedimentary Environments

Beach sands are often characterized by moderately to well-sorted and angular to sub-angular grains that are often highly polished. The grains usually show signs of small V-shaped impacts with other grains or scratches caused by one grain scraping over another (Figure 5.11c, d, e).

Eolian (desert) sands are typically very well-sorted and well- to very well-rounded. Constant high-impact collisions between windblown grains result in a dull to opaque frosted appearance with lots of small surface pits.

Dune sands found inland from beaches will usually display a combination of beach and eolian characteristics. Most of the sand grains will be relatively polished, but a portion will be frosted, though not as highly frosted as eolian sands. Dune sands are also not as rounded as true eolian sands.

Glacial sands are generally very poorly sorted and contain a range of grain shapes, from very angular to rounded. Parallel striations (parallel rows of scratches) are relatively common.

River sands are typically well- to very well-sorted and the grains tend to be well- to very well-rounded. River sands tend not to have the impact marks common to beach or desert sands because the water cushions the grains (Figure 5.11a, e).

Along passive continental margins, such as the eastern coast of North America, extensive chemical weathering produces many quartz-sand beaches. Active continental margins, such as the western coasts of North and South America, have

Figure 5.11 Images of different types of sediments. The black bar in each image represents 1 mm: (a) river sand, (b) inner shelf sand, (c) beach sand, (d) beach gravel, (e) river gravel with cobbles, (f) beach gravel and cobbles. Please see Color Plate section.
Source: Buscombe, Rubin and Warrwick, 2010. Courtesy of the United States Geological Survey.

more diverse characters, combining volcanic, plutonic, and metamorphic source rocks with sedimentary materials, resulting in sands with quartz, feldspars, other silicates, and rock fragments. Volcanic ocean islands, such as Hawaii, are almost entirely composed of basalt, producing black- and green-sand beaches. Island arc beaches also tend to be dominated by dark ferromagnesian minerals.

Beach Sand: Here Today, Gone Tomorrow

This might sound like the plot of a bad television show, but in the past, at least two beaches have been stolen. Really, the entire beach, including all of the sand, was stolen. The first incident occurred in December 2007, when thieves stole the only beach in Hungary. The land-locked country has no natural beaches, so in order to have a beach as near to the real thing as possible several tons of sand were imported to Mindszent in eastern Hungary to form a riverside beach. The beach had been cleaned and covered up for winter when, in December, managers found that all of the playground rides, huts, wooden vendor's stalls, and all 6000 cubic meters of sand were gone.

Unbelievably, the same thing happened again in July 2008, when an estimated 500 truckloads of sand were removed from Coral Beach in Trelawny, Jamaica. The white sand had been imported for a planned resort, and covered 400 m (1300 ft) of strand. Though some of the sand was apparently found on beaches on the northern coast of Jamaica, no arrests were made.

On a much smaller scale, people often steal sand from beaches that have unique characteristics. While this usually just amounts to a zip top bag, occasionally it involves suitcases or even entire truckloads. And, sometimes it even involves shipping containers . . .

Sample Collection

The most common approaches to collecting moderately sized particulate samples, such as sand, from a crime scene include handpicking, brushing, scraping, tape sampling, and vacuum collection. The method used depends on local conditions and the original disposition of the material. Each forensic worker will have his or her own standard operating procedures but a general description of each is as follows.

Handpicking involves literally picking each individual particle up using stainless-steel forceps or a similar device. *Brush sampling* is performed using a brush with synthetic bristles to gently sweep particles into a re-sealable plastic bag. Just prior to sealing the bag, the brush used for collection should also be included. To collect a *scrape sample*, a stiff-edged card made of plastic or a razor blade is used to carefully scrape particles off of a surface and push or scoop them into a plastic or glass container. The scraper is also usually placed into the sample bag.

Tape samples are collected by pressing a short strip of clear adhesive tape against a surface (Figure 5.12). The tape is then adhered to a clean glass slide or clean piece of clear Mylar. The entire glass slide or Mylar sheet is placed into a plastic

Figure 5.12 Example of a tape sample.

sample collection bag. *Vacuum sampling* procedures depend in large part on the type of vacuum sampler used for collection. In general, the surface area is vacuumed until there is no visible dirt in the area or for a fixed duration. At the end of collection, the sampling bag, filter, or entire collection head of the vacuum is placed into a sample bag. For sand samples, it is usually best to use plastic collection bags and bottles, preferably antistatic, rigid plastic containers. Many of the mineral components of sand are hard enough to chip out fragments from glass sampling bottles, complicating the subsequent analysis.

Control, reference, and alibi samples should be collected using stainless-steel tools such as small trowels or pallet knives. A sample measuring 10 cm × 10 cm × 2 cm is usually sufficient, unless the material is very poorly sorted or contains lots of gravel. The more heterogeneous the material, the larger the collected sample must be in order to be representative of the material. For highly heterogeneous materials, 1–2 kg should be collected, if possible. The most important aspect of sample collection is ensuring that the material being collected is appropriate as comparison material. If the questioned sample is from shoe treads, then surface collection is of primary importance, but if the questioned sample is from a shovel blade, then profile material from various depths must be collected. All samples should be delivered without alteration, such as drying or freezing, to a laboratory as soon as possible after collection.

Sample Preparation

Sample preparation procedures can vary greatly, depending on the size of the sample and the types of information needed. Only one sample is examined at a time and the first step is to weigh the sample in its original condition. Commonly with sand samples, one would next examine the whole sample as it was received to determine the types of minerals and other materials present, such as plant fragments, glass, plastic, or fibers. If the sample appears to contain a significant amount of thin plastic, such as tinsel or fragments of plastic wrapping material, these items need to be removed *before* drying, and placed in separate labeled containers. Next, determine the overall color and approximate moisture level of the sample. Then, place the sample into a drying oven at 100 °C (or less, depending on the fragility of the sample) for between 4 and 24 hours, depending on the material and the protocol being observed. After drying, *debride* the sample by removing visible fibers, leaves, roots, and any other plant material and placing it to the side. These items should also be placed separately in appropriately labeled containers.

If the sample is fairly small, a mineral count, analysis of morphology and surface texture, and a basic grain size distribution can be performed without further treatment. For larger or more complex samples, a range of additional treatments may be required. These can include sieving the sample into discrete size fractions, careful washing with water or another solvent, or even disaggregation to break clumps up into smaller pieces, though this last procedure should normally be performed under specific conditions. One classic paper describing soil preparation (which also applies in large part to sand samples) is Graves (1979). One note of caution though: many carbonates and some other sand components will be affected by the indiscriminate use of propellers and ultrasonic cleaners, thus their use is not advocated in cases where surface texture information is of significance, or where fine-grained carbonates make up a significant proportion of the sample. Unlike soil samples, many sand samples are "clean" enough that additional treatments are not required.

The Stereomicroscope

Questioned samples are usually quite small, which means that it is necessary to analyze the samples with some kind of magnification. A *stereomicroscope* is a type of compound microscope that has two eyepieces (oculars) with separate sets of lenses that are aligned through a single primary magnification lens, called an *objective*, to create a three-dimensional view. The total amount of magnification is determined by the power of the eyepieces multiplied by the power of the objective lens, usually between 10× and 40×.

Samples can be examined using *transmitted light*, where the light source is beneath the sample, or with *reflected light*, where the light source is above the sample. Use of transmitted light requires most, if not all, of the sample to be transparent. Some stereomicroscopes are outfitted with polarizing filters, allowing samples to be examined using transmitted polarized light, which can be useful for mineral identification (see Chapter 6). If the components of the sample are predominantly opaque, little data can be obtained using transmitted light. Reflected light is useful for highlighting the surface features of a sample and by changing the angle of the incident light the three-dimensional character of a sample can be revealed. Samples should be viewed against both white and black backgrounds when using reflected light. Transmitted and reflected light provide complimentary data, thus each should be used for sand samples.

Regardless of the entire sample preparation protocol used, at some point in the procedure chances are that at least one, more probably several, glass slides will be prepared for detailed microscopic examination. A brief description of how to make a *smear slide* using Cargille Meltmount is available on the companion website.

Forensic Examination of Sand

To perform a simplified forensic analysis, you will want to examine each sample (1) with the unaided eye, (2) as a smear slide using a stereomicroscope with transmitted light, (3) as a smear slide using a stereomicroscopic under reflected light on a white background, and (4) reflected light on a black background. For each sample, note the following:

- **composition** of the grains in the sample and the relative amounts of each of the different components of the sample (usually reported as a percentage). A rapid counting procedure (Graves, 1979) is often the most expedient approach, but on top of the normal rotating stage requires a mechanical stage that can be used to position the glass slide north–south and east–west. Basically, the viewer methodically tracks over the entire surface of the slide, looking for a specific list of minerals or particles, keeping track of them on a counting sheet.
- **grain size** and **grain size distribution**
- the **surface texture** (polished, frosted, etc.) and any **surface coatings**
- **relief** (the degree to which mineral grains stand out from the mounting medium) (Figure 5.13)
- the **roundness** and **sphericity** of the grains (use the Powers' scale for visual roundness and the Krumbein visual scale for roundness and sphericity)

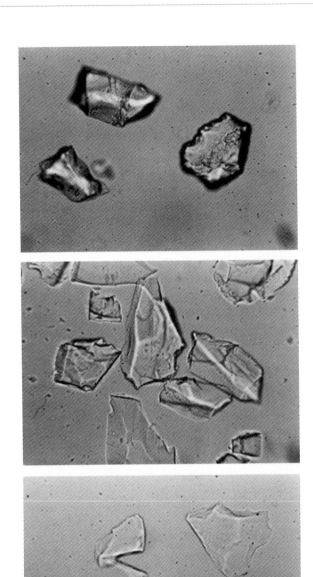

Figure 5.13 Relief: (a) high relief, (b) moderate relief, (c) low relief.
Source: Nesse, 1991.

- the **magnetic properties** of the grains
- the presence of **organic materials** such as plant parts, insect parts, twigs, etc.
- the presence of any **marine carbonates**, such as shells or coral
- the presence of any **synthetic materials**, such as glass, aluminum foil, fibers, etc.

It cannot be overemphasized that the goal of a forensic analysis is not to achieve a "match" between two samples but is instead to *exclude* samples that are clearly different.

Geologists are most accustomed to studying thin-sections of rocks, or slices of rock that are uniform in thickness (around 30 µm usually), to make a precise comparison of the optical properties of minerals. The study of sand grains in a smear slide is a bit different, since you are looking at particles that vary in thickness. The listing below will help with the identification of common minerals, though in some cases polarized light is needed to make a more definite identification. Chapter 9 discusses the microfossil component of sand. If this style of analysis provides insufficient information, a traditional thin-section should be created, and an examination performed using a petrographic microscope (see Chapter 6).

Common Minerals

Quartz is extremely common. Sand-sized fragments are generally colorless in transmitted light, regardless of the color they might have been in the hand sample. Under reflected light, quartz is slightly lighter than the background. Polished to highly frosted and angular to rounded, though most commonly angular to sub-angular. Grains have conchoidal fracture and lack cleavage. The lack of cleavage is a key characteristic that differentiates quartz from feldspar. Under plane polarized light, quartz has a low relief (0.004 to 0.013), which means that it kind of blends in with the Meltmount.

Potassium and alkali feldspars are colorless in transmitted light, though weathered areas can be light yellow to light brown. Grains show a range of morphologies, though most show good cleavage, distinguishing them from quartz. Feldspar grains look more like a step mesa, while quartz looks more like a glacial mountaintop or hill. Under plane polarized light, these feldspars have very low relief and might be hard to distinguish from Meltmount. *Twinning*, repetitions of crystal shapes, is also a distinctive feature, as quartz does not twin.

Plagioclase feldspar is also colorless in transmitted light, with weathered portions that might appear light yellow to light brown. Grains are typically sub-angular and show good cleavage in one or two directions. *Polysynthetic twinning*, i.e. lots and lots of twins, is common.

Biotite, phlogopite, and *muscovite mica* appear as plates, flakes, or long strands, basically the same way they appear in hand samples, only smaller. In general, under plane-polarized light, biotite appears brownish to brownish-green, phlogopite is orangey-brown to brownish, and muscovite is colorless. Relief is low to moderate (+0.04 to +0.11). Since mica has only one cleavage plane, which is responsible for creating all of the flakes, no cleavage traces will be apparent.

Calcite and dolomite are colorless in transmitted light, though if the crystals are large enough fine bands of iridescent colors are sometimes visible. Grains range from rounded to angular. Relief is variable (−0.05 to +0.14) and will change as the sample is rotated.

Rock fragments can also often be identified in sand samples. If you see small fragments of material that have the same appearance as rock hand samples, chances are that you have found a small fragment of a rock. Schist, shale, and limestone are commonly found in sand.

Volcanic glass shards are common in volcanically active areas. They can be angular fragments (if fresh) to well-rounded. Mafic *obsidian* shards tend to be brown, while felsic shards are colorless. *Pumice* and *tuff* fragments can have a frothy appearance with many vesicles and visible inclusions.

Less Common Minerals

Garnet varies in color from light red to purple, pale violet, orange or even green (though this is very rare). Really fine particles might be clear. Garnet commonly gives sand a pink tint. Grains are usually sub-rounded and show conchoidal fracture. Relief is very high (+0.26 to +0.46), making the grains stand out against the background, usually with a pronounced black band around the mineral.

Clinopyroxene grains are colorless to bright or pale green in transmitted light. Stubby sub-angular fragments showing cleavage in two directions at nearly 90° are common. Relief is moderate to high (+0.06 to +0.20).

Orthopyroxene varies in color from pale pink, pale red and yellow to colorless. Sometimes the grains will appear to change hue or color as the sample is rotated (a property called *pleochroism*). Stubby sub-angular fragments showing cleavage in two directions at nearly 90° are common. Relief is high (+0.12 to +0.19).

Amphiboles range in color from green and bluish-green to olive and brown. Colors are deeper and more intense than the pyroxenes. Sometimes the grains will appear to change hue or color as the sample is rotated (pleochroism). Large grains might appear nearly opaque except at their thinnest points. Grains are moderately elongated, with good cleavage at roughly 60° and 120°. Cleavage traces are common. Relief is moderate to high (+0.07 to +0.22).

Olivine is pale yellow to light green under transmitted light. Very small grains might appear colorless. Grains are usually at least somewhat rounded, and conchoidal fracture is common. No cleavage occurs. High relief.

Apatite is colorless under transmitted light. Most grains are *equant* (equal dimensions in all directions) and rounded. The surface is commonly pitted. Grains lack cleavage. Relief is moderate (+0.003 to +0.005), so it appears to stand out more than quartz.

Gypsum is colorless under transmitted light. Grains are typically rhombic and show cleavage. Irregular shapes with ragged-looking edges are common as well. The interior can appear fibrous. Grains are very soft. Relief is very low (−0.02 to −0.01).

Zircon is transparent to faint yellow. Grains are commonly stubby rounded prisms or irregular in shape. Inclusions are very common. Zircons have very high relief (+0.38 to +0.45).

Opaque Minerals

Pyrite grains appear to be a dull metallic brass color under reflected light. Many shapes are common, and pyrite commonly appears as infill in some fossils and tests.

Magnetite is grayish-, brownish-, or bluish-black under reflected light with a metallic luster. Irregular shapes and ragged edges are common. Magnetite's reaction to a magnet is its distinctive property, which must be tested on loose particles, not on a completed slide.

Hematite is usually dark reddish-brown (rust-colored) under reflected light. Irregular shapes and ragged edges are common. Nonmagnetic.

Limonite is usually earthy to metallic, yellow, brown, or orange. Nonmagnetic.

Anthropogenic Materials

Glass is possibly the most commonly found synthetic material in samples. Angular shards of glass that show conchoidal fracture are either freshly broken or were buried very quickly after being broken. Through time, glass shards will behave somewhat like quartz, getting more rounded and possibly even becoming frosted. Thus, glass fragments found in beach sand are usually translucent, sub-angular, and possibly frosted, but will still have remnants of the parallel planer curved surfaces of the bottle from which they came. Decorative glass is often brightly colored and may contain swirls of pigment or fragments of glitter. Unless the shard is obvious (i.e. clearly an identifiable fragment of a bottle, headlight, etc.), polarized light might be necessary to distinguish glass from quartz. Glass is also commonly found in the form of small, often hollow, beads that are the byproduct of smelting or other combustion processes. Waste glass from combustion processes, and glass melted by fire, is often black and might have bands of frothy bubbles.

Plastic pieces are usually recognizable by their low density, high shine, and unnatural color, though small fragments of high-density plastics, especially those with mineral fillers, can be harder to distinguish.

Concrete is a construction material composed of gravel, sand, and/or other aggregate material held together by cement. There are many different types of concrete, but it is usually a fairly uniform gray or off-white color. Pigments can be added to change the color, as can other components that change the basic properties, such as plasticizers. Flat surfaces and decorative textures are common.

Asphalt (blacktop) is a mixture of aggregate (gravel, sand, and/or other solids) and a dark bituminous substance that is usually obtained from petroleum processing. Fragments of asphalt will have a distinctive black color and tend to be somewhat malleable, flattening rather than breaking.

Bricks are made of sun-dried or fired clay. They are usually red, brown, yellow, or tan and may contain sand or other fragments within them. Decorative features on the surface are common. Brick is usually dull in appearance and more uniform in appearance than naturally occurring materials.

Tile is a generic term for any thin, ceramic building material, including glazed floor tiles, brick-like roofing tiles and vitreous plumbing tiles, which are shiny on the surface and usually curved. These items are usually uniform in thickness and color, and may show rough, decorative surface textures and colorful glazing.

Slag is a byproduct formed in the course of steel production and from smelting (heating to a very high temperature) of various ores to purify metal. It is a silicate that will contain a variety of different metal oxides, depending on the source ore and production processes involved. Slag always contains some amount of sulfur. Often similar in composition and appearance to natural igneous rock, the outer surface of slag tends to be ropy and irregular. If the slag is relatively fresh, the surface is usually shiny or glassy and may be iridescent. The inner surface is usually dull and contains numerous voids. Slag is commonly used around railroad tracks and in asphalt and concrete. Slag that is a byproduct of steelmaking is largely carbonate that has absorbed a significant amount of phosphate, which makes it useful as a soil amendment.

Clinker is a byproduct of burning coal, similar to slag. Clinker has lots of bubbles or vesicles and will often contain inclusions of brick fragments, pebbles, and other debris. Colored tarnish or films on surfaces are common. Naturally forming clinker can be found where burning underground coal seams melt adjacent rocks.

Metal fragments can usually be distinguished by their metallic luster and reflectivity. Shapes are often irregular, and most of the time metal pieces are soft and malleable.

Another Geologic Substitution (Murray, 2004)

For example, consider a $2 million shipment of exotic perfume from Paraguay that arrived in Miami, Florida as several tons of sand. An investigator, forensic geologist Fred Nagle, was brought in to establish liability, as different insurers covered different portions of the shipment's journey. Nagle found that the shipping containers were filled with a mixture of burlap bags with green and yellow threads sewn into their sides and modern, white, synthetic bags. A student of Nagle's identified the burlap bags as Brazilian coffee bags and pointed out that the white bags were also used for coffee. Each bag was filled with mature, fine-grained, brownish sand. Based on the features observed, the sand was identified as a river deposited granitic sand rich in rutilated quartz (i.e. quartz grains that contain fine needles of the mineral rutile).

Nagle began comparing the evidentiary sand with samples in the University of Miami's collection, and with some other collections, but found nothing similar. With no other available references for comparison, Nagle decided to go to Brazil and check each of the places where the freighter stopped, to look for the sand source. The first place he visited was a small port south of Rio, where the freighter had apparently made an unexplained, unscheduled three-day stop. The port was unguarded and the shipping containers would have been readily accessible. However, the sand was not right. With the most obvious anomaly out of the running, Nagle traveled for two weeks, still without locating similar sand.

Then Nagle flew to Iguaçu Falls, on the border between Brazil and Paraguay, to see the falls and visit the new hydroelectric dam under construction there. While he was there, he asked where they got the sand for the dam. Most of it had been produced by grinding up local basalt, but some of the sand was dredged from the Paraná River. Nagle tested the river sand and bingo: it had the same characteristics as the evidentiary sand. In addition, as he crossed the top of the dam into Paraguay, Nagle noticed a mix of coffee bags filled with sand being used to block access to the construction site. Laboratory examination of the sand from the river and the bags demonstrated that they were almost identical to the sand from the shipping containers. In this case, unusual minerals helped make it possible to make the identification with an uncommonly high level of certainty. Not only did the sand have the rutilated quartz, it also contained fragments of blue turquoise, which was commonly found filling the cavities of the basalts in that area.

With this information, Nagle was able to determine that the substitution took place at the very start of the journey, which explained why the container seals had looked undisturbed and why the weight of the containers was correct. They had been filled with sand all along. Oh, and it turns out that the unexplained stop south of Rio was because the captain stopped there to visit a girlfriend.

Sand from the Bottom of a Shoe (Petraco, Kubic, and Petraco, 2008)

In early spring one year, the body of a young adult female was seen floating down the East River near midtown Manhattan, New York City. The woman was later identified as a missing corrections officer and autopsy revealed that she had been shot with a handgun. Investigators learned that she had been in the process of obtaining a divorce from her husband, who became the prime suspect.

A search warrant was issued for the husband's residence, where officers discovered a water-stained right shoe in the bedroom closet. On the underside of the shoe, there was a small quantity of sand adhering to the inside of the heel area. Forensic investigators examined the shoe under a stereomicroscope and were able to recover 10 mg of sand. This was not a large enough sample to perform a sieve analysis, but color was determined (5Y 7/2). The sand was also analyzed using X-ray diffraction (XRD) and found to consist predominantly of quartz (88%), with smaller amounts of hornblende, garnet, hematite, magnetite, limonite, tourmaline, and shell fragments. Both the water-stained area of the shoe and the sand sample tested positive for sodium and chloride ions, consistent with exposure to salt water.

Based on the information gathered, investigators hypothesized that the man had taken his wife to a local beach where he pulled her into the water and shot her with her own revolver. Sand samples were collected from that local beach, as well as from various other locations along the East River. The beach sand had similar coloration, morphology, and mineralogy to the sand recovered from the suspect's shoe, thus the samples could have shared a common origin. The rest of the samples collected were dissimilar to the questioned sand and therefore excluded from sharing a common origin. This information was later used at the trial, where the husband was found guilty of second-degree murder.

Summary

This chapter introduced you to some of the complexities of characterizing sand, a common form of geologic trace evidence. Given its prevalence in a variety of types of crime, it could be worth knowing something about some of the more interesting sands found around the globe (Figure 5.14). A very short list of some distinctive sands can be found on the companion website. Sand particles are small enough that they require microscopy for identification but retain many of the recognizable characteristics of hand samples. This is not generally true of soil, which we will investigate in Chapter 7. Learning to work with and characterize sand-sized material is a good intermediary step toward learning to work with a wide range of soils and other forms of geologic evidence.

(a)   (b)   (c)

(d)   (e)   (f)

Figure 5.14 Examples of some of the wide variety of distinctive sand types from around the world: (a) White Sands, New Mexico desert sand composed of small soft grains of gypsum; (b) Texas City, Texas beach sand composed of coarse rock fragments and shell pieces with almost no fines; (c) Smith Island, Baja California, Mexico, beach sand composed of shiny metamorphic rock fragments with some organic material; (d) Oak Creek, Nevada desert sand composed of sub-angular grains of quartz with some feldspar; (e) Indiana Dunes, Indiana, dune sand composed of pitted, frosted, rounded quartz grains that will make a squeaking sound when shaken; (f) Half Moon Cay, Bahamas, white oolitic beach sand with small, rounded grains of pink coral; (g) Big Island, Hawaii, green beach sand formed predominantly of olivine with some obsidian and shell fragments; (h) Oahu, Hawaii, black basalt beach sand; (i) Fort Pierce, Florida, mature beach sand composed almost solely of clear, well-polished quartz grains; (j) Antelope Island, Great Salt Lake, Utah, oolitic lake sand; (k) Lake Bratan, Bali, Indonesia, lake sand from a lake that fills a volcanic crater; (l) Vanua Levu, Fiji, carbonate beach sand composed almost entirely of shells, sea urchin spines, and other organics; (m) Coral Pink Sand Dunes State Park, Utah, frosted dune sand; (n) Pismo Beach, California, beach sand; (o) Third Beach, Vancouver, Canada, beach sand; (p) Hoshizuna-no-hama, Iriomote, Okinawa, Japan, star sand; (q) Kalalau, Kauai Island, Hawaii, beach sand; (r) Perissa, Santorini, Greece, black volcanic sand. Please see Color Plate section.
Source: Photograph m. and n. by Mark A. Wilson; o. by Bobanny; p. by Geomr; q. by Psammophile; r. by Stan Zurek.

Figure 5.14 *continued*

Further Reading

Greenberg, G. (2008) *A Grain of Sand: Nature's secret wonder*. Voyageur Press, Minneapolis, MN. [Excellent microphotographs of sands from around the world.]

McPhee, J. (1998) *Irons in the Fire*. Noonday Press, Farrar, Straus and Giroux, New York. [This book includes The Gravel Page article mentioned.]

Prothero, D. R. and Schwab, F. (2004) *Sedimentary Geology: An introduction to sedimentary rocks and stratigraphy*, 2nd edn. W. H. Freeman, New York.

Stow, D. A. V. (2005) *Sedimentary Rocks in the Field: A color guide*. Elsevier Academic Press, Burlington, MA. [Excellent photographs.]

Webber, B. (1992) *Silent Siege III: Japanese attacks on North America in World War II: Ships sunk, air raids, bombs dropped, civilians killed*. Webb Research Group Publishers: Central Point, OR. [Book that has the most detailed information.]

The American Public Broadcasting System (www.shoppbs.org) has released a DVD entitled *On a Wind and a Prayer*, a 2005 documentary by Michael White Films about the Balloon Bombs (http://www.onawindandaprayer.com/).

References

Basu, A. (1976) Petrology of Holocene fluvial sand derived from plutonic source rocks: Implications for paleoclimatic interpretation. *Journal of Sedimentary Petrology* **46**(3): 694–709.

Basu, A. and Molinaroli, E. (1989) Provenance characteristics of detrital opaque Fe-Ti oxide minerals. *Journal of Sedimentary Petrology* **59**(6): 922–34.

Blott, S. J. and Pye, K. (2008) Particle shape: A review and new methods of characterization and classification. *Sedimentology* **55**: 31–63.

Bull, P. A. and Morgan, R. M. (2007) Sediment Fingerprints: A forensic technique using quartz sand grains. A response. *Science and Justice* **47**(3): 141–4

Buscombe, D., Rubin, D. M., and J. A. Warrick. (2010) A universal approximation of grain size from images of noncohesive sediment. *Journal of Geophysical Research* **115**, F02015, doi:10.1029/2009JF001477.

Chaperlin, K. and Howarth, P. S. (1983) Soil comparison by the density gradient method: A review and evaluation. *Forensic Science International* **23**: 161–77.

Chicago Daily Tribune (1945) Disclose How Japs' Paper Balloons Drop Bombs In U.S. Chicago Daily Tribune, May 23: 1.

Christian Science Monitor (1944) Bomb Balloon in West Bore Nippon Marks. Christian Science Monitor, December 19: 8.

Compton, R. R. (1985) *Geology in the Field*. John Wiley & Sons Ltd, New York.

Culver, S. J., Bull, P. A., Shakesy, R. A., and Whalley, W. B. (1983) Environmental discrimination based on quartz grain surface textures: A statistical investigation. *Sedimentology* **30**: 129–36.

Fitzpatrick, F. and Thornton, J. I. (1975) Forensic science characterization of sand. *Journal of Forensic Science* **20**(3): 460–475.

Folk, R. L. (1974) *Petrology of Sedimentary Rocks*: Hemphill, Austin, TX.

Graves, W. J. (1979) A mineralogical soil classification technique for the forensic scientist. *Journal of Forensic Science* **24**: 323–38.

Higgs, R. (1979) Quartz grain surface features of Mesozoic-Cenozoic sands from the Labrador and Western Greenland continental margins. *Journal of Sedimentary Petrology* **49**: 599–610.

Hill, P. R. and Nadeau, O. C. (1984) Grain surface textures of the late Wisconsians sands from the Canadian Beaufort shelf. *Journal of Sedimentary Petrology* **54**(4): 1349–57.

Krinsley, D. H. and Doornkamp, J. C. (1973) *Atlas of Quartz Sand Surface Textures*. Cambridge University Press, Cambridge.

Krumbein, W. C. and Pettijohn, F. J. (1938) *Manual of Sedimentary Petrography*. Appleton Century Crofts Inc., New York.

Krumbein, W. C. and Sloss, L. L. (1963) *Stratigraphy and Sedimentation*, 2nd edn. W. H. Freeman and Company, San Francisco.

Le Ribault, L. (1977) *L'exoscopie des quartz*. Editions Masson, Paris.

Lombardi, G. (1999) The contribution of forensic geology and other trace evidence analysis to the investigation of the killing of Italian Prime Minister Aldo Moro. *Journal of Forensic Sciences* **44**(3): 634–42.

Los Angeles Times (1944) Jap Balloon Found in Montana Raises Sabotage Fears. Los Angeles Times, December 19: 1.

Mahaney, W. C. (2002) *Atlas of Sand Grain Surface Textures and Applications*. Oxford University Press, Oxford.

Margolis, S. V. and Krinsley, D. H. (1974) Processes of formation and environmental occurrence of microfeatures on detrital quartz grains. *American Journal of Science* **274**: 449–64.

McCrone, W. C. (1982) Soil comparison and identification of constituents. *The Microscope* **40**: 109–21.

Murray, R. C. (2004) *Evidence from the Earth*. Mountain Press Publishing Co., Missoula, MT.

Murray, R. C. and Tedrow, J. C. F. (1992) *Forensic Geology*, 2nd edn. Prentice Hall Inc., Englewood Cliffs, NJ.

Nesse, W. D. (1991) *Introduction to Optical Mineralogy*, 2nd edn. Oxford University Press, New York.

Newsweek (1945) Trial Balloons? Newsweek, January 15: 40–41.

Petraco, N., Kubic, T. A., and Petraco, N. D. K. (2008) Case studies in forensic soil examinations. *Forensic Science International* **178**: 23–7.

Powers, M. C. (1953) A new roundness scale for sedimentary particles. *Journal of Sedimentary Petrology* **23**: 117–119.

Powers, M. C. (1958) Roundness of sedimentary particles: Comparison chart for visual estimation of roundness: AGI data sheet. *Geotimes* **3**(1): 15–16.

Pye, K. (2007) *Geological and Soil Evidence: Forensic applications*. CRC Press, Boca Raton, FL.

Pye, K. and Blott, S. J. (2004) Particle size analysis of sediments, soils and related particulate materials for forensic purposes using laser granulometry. *Forensic Science International* **144**: 19–27.

Rittenhouse, G. (1943) Transportation and deposition of heavy minerals. *Geological Society of America Bulletin* **54**: 1725–80.

Seattle Times (1945) Jap Balloons Boomeranged. Seattle Times, October 3, http://www.stelzriede.com/ms/html/mshwfug4.htm.

Washington Post (1945) Jap Balloons: "Balls of Fire" Stump Experts. The Washington Post, January 3: 5.

Chapter 6
Gems and Gemstones: Those Most Precious of all Minerals

Sometimes crimes are actually committed over the possession of small amounts of minerals, though the people committing these crimes certainly do not think of it this way. Gems and gemstones are just particular minerals, or rocks, that people have taken a special fancy too. There is nothing intrinsically different about them. They have the same properties as garden-variety minerals and can be identified using the same procedures. The big difference is that because people find gems and gemstones especially attractive there is a large monetary value associated with them . . . and where money goes, crime follows.

The Federal Bureau of Investigation (FBI) has also become very interested in jewelry, gem, and gemstone theft because, increasingly, these crimes are being committed by organized criminal enterprises, or "theft groups," that are also involved in activities such as terrorism and drug smuggling. The high value, small physical size, and relative untraceability of gems and gemstones make them very appealing to criminals. Because of national and international governmental currency tracking, criminals have moved to alternative monetary systems. Gems, gemstones, jewelry, and artwork are commonly used for money laundering, international transactions, and the storage of crime proceeds. From our perspective, gems and gemstones are also good tools for learning about the optical properties of minerals and some of the more rarely used physical properties.

An Introduction to Forensic Geoscience, First Edition. Elisa Bergslien.
© 2012 Elisa Bergslien. Published 2012 by Blackwell Publishing Ltd.

The Thailand Gemstone/Jewelry Scam

An example of large-scale fraud is the Thailand gemstone/jewelry scam that has gotten quite a bit of press in the United Kingdom, though much less so in the United States. This has gone on for over 20 years and the basic scam goes something like this: tourists to Bangkok will "accidentally" meet a respectable-looking man who chats with them, points out some local temples, or discusses other cultural activities of interest to a tourist. During the conversation, the man casually mentions an annual wholesale gem or jewelry sale going on that week. He explains that the sale is to allow Thai students to buy jewelry or gems cheaply so that they can sell them abroad to help pay for their education, or some other similar story. Tourists are also allowed to take advantage of this sale because the government wants to encourage tourism. If the tourists show no interest at this point, they are usually hit up again during another "chance" encounter with another respectable-looking person who also mentions this wonderful sale.

The tourists' friendly *tuk tuk* (motorbike taxi) driver, who has helpfully escorted them around the city, will take interested parties to a "Thai Export Office" (the actual name of the store varies), where they meet with well-dressed salesmen and are often given VIP treatment. The tourists are shown a variety of items, but the sale is always ending that day, so they must decide to make a purchase immediately if they want to take advantage of the sale. The other catch is that the jeweler does not accept credit cards and must be paid in gold or cash only. The jeweler will even assist the tourist in finding a gold merchant. After the sale, the newly purchased merchandise is often carefully packed in an international courier bag for shipment back home, so that the tourist does not have to worry about it during the remainder of their holiday.

At this point, the tourists are usually treated to a sightseeing tour, often some hours away from Bangkok. It is only on the next day, or later, that the tourists realize they have been scammed. Their *tuk tuk* driver actually arranged all of the "chance" meetings and has carefully stage-managed events. By the next day, the name of the jewelry store where the purchases were made has been changed, or the store has closed down, moving to an entirely new location. If the tourist manages to take the merchandise with them, which is apparently rare, it is often stolen from their hotel room. An estimated 15,000 people are affected each year.

Apparently, if you manage to keep hold of the jewelry or stones you were sold, which are worth only a small fraction of what you paid for them, and catch the store with the name it had when you made the purchase, it is occasionally possible to get some percentage of your money back through the Thai government's Department of International Trade. If the merchandise was couriered out of the country, there is generally no recourse. Tourists who attempt to pursue their cause through the tourist police, or local Thai government, often become the victims of assaults and/or robberies. There are also several stories of violence against tourists who return to the shops to attempt to get their money back.

These stories have the same general themes of the Ralston Fraud and the Bre-X affair discussed previously: the chance of making a lot of money for a small upfront investment and decisions that are made with limited information and a lack of time for adequate deliberation or research. As is often said, if a deal seems too good to be true, it usually is.

An Introduction to Gemstones

A precious stone, or gemstone, is a natural object (mineral, rock, or organic material) that is rare, hard, chemically resistant, and beautiful. *Gemstone* is the term used for the raw material. A *gem* is a gemstone that has been cut and polished or otherwise treated to enhance its beauty. *Semiprecious* stones are usually either more common or less durable than true gemstones, though this distinction is purely subjective. Diamond, corundum (ruby and sapphire), beryl (emerald and aquamarine), topaz, and opal are generally classed as *precious stones*. All other gemstones are usually classed as semiprecious.

A *mineralogist* is a person who studies the formation, occurrence, properties, composition, and classification of minerals. A *gemologist* is a person who has successfully completed recognized courses in gemology (the science and study of gemstones) and has proven skills in identifying and evaluating gem materials. A *lapidary* is a cutter, polisher, or engraver of precious stones.

When discussing gems, or cut stones, certain particular descriptive terms are used. The weight standard for gems and pearls is the *carat*, which is now defined as equaling exactly 200 mg. The term originates from a Greek word for carob seeds, which were historically used as weights on precision scales. *Cutting* refers to the way a stone is shaped. There are basically two major categories of cuts: *non-faceted* (or plain) *cut* and *faceted cut*. Historically, most gems were simply rounded or smoothed in what is called a *cabochon*. Such stones can be completely rounded, or can have a flat base. This style of non-faceted cut is still popular with opaque stones, like opal, and is used to create effects such as cat's eyes. Nowadays, most transparent stones are faceted. A *facet* is simply an artificially created, polished, planer surface.

There are a variety of different faceting styles, but the basic terminology used is the same for all of them (Figure 6.1). The midline of a faceted gem is called the *girdle* and may or may not be faceted. The area above the girdle is called the *crown*. The top facet is called the *table*. Surrounding the table are star facets, bezel (or kite) facets and a series of upper girdle facets. The area below the girdle is called

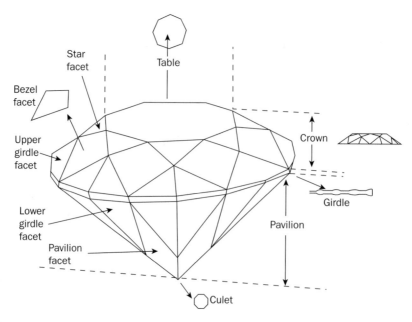

Figure 6.1 Terminology used to describe gems.

the *pavilion*. The bottom of the stone is called the *point*, or if it is cut flat it is called the *culet*. The culet, or point, is surrounded by a series of *pavilion facets* and lower girdle facets. Faceting results in the loss of up to 60% or more of the original stone, depending on the shape and number of imperfections. The Cullinan, the largest diamond gemstone ever found, started out at 3106 carats. It was cut into nine major gems, ranging in size from the Cullinan I (or Star of Africa) at 530.2 carats to the Cullinan IX, a 4.39-carat pear-shaped diamond, plus 96 smaller gems, with a loss of 65% of the original material.

In many ways, rarity is a gemstone's key characteristic. Gemstones do not form large-scale ore-type deposits. Instead, they tend to be scattered sparsely through a large body of host rock, or isolated in small cavities, vugs, or veins. Diamondex Resources Ltd has on its website a very nice animation of a kimberlite pipe forming, which helps to explain how diamonds can be distributed (http:// www.bcminerals.ca/i/video/kimberlite-anim.swf). In the richest diamond kimberlite pipes in Africa, there is only about 1 part diamond per 40 million parts rock. Much of that diamond is not of gem quality, so the average stone in an engagement ring is actually the product of the removal and processing of 200 to 400 million times its volume of rock. Most gemstones are found in igneous rocks and alluvial gravels, but sedimentary and metamorphic rocks may also contain gemstones. Each environment tends to have a characteristic suite of gemstones, but many kinds of gemstones occur in more than one environment.

Beauty, one of the key characteristics of a gemstone, is very subjective in nature, and difficult to impossible to quantify in any meaningful way. Therefore, when examining gemstones as a scientist, one must concentrate on the quantifiable characteristics, such as hardness and chemical resistance, both of which are important with respect to the durability of a gem during the course of everyday use, and on other properties used in mineralogy, like crystal form, cleavage, and luster.

The vast majority of precious stones are *crystalline*, which means that they possess a regular internal arrangement of atoms that is repeated over and over. This internal arrangement determines the external shape, or crystal form, of a gemstone and its cleavage, or zones of breakage, which are key properties that can be used for identification. However, not all gemstones technically qualify as minerals. Some gemstones, such as amber and opal, are amorphous (glass) instead of crystalline, which means that they lack this internal order, and so other properties must be used for their identification.

Because one of the key requirements for the identification of a gemstone is that the examination be nondestructive, some of the commonly used tools of a mineralogist, like the streak test and acid, are generally forbidden. As already mentioned, crystal shape and cleavage prove helpful with uncut stones. Hardness testing is also sometimes necessary, though almost never applied with worked stones, as scratching up valuable gems is frowned upon. When examining cut gems and small gemstones, refractive index (RI) and specific gravity (SG) are the major characteristics most commonly assessed.

Crystal Forms

Many of the important properties used to identify gems and gemstones, such as SG, hardness, luster cleavage, and crystal habit, have already been described. Remember that it is possible to identify cleavage, and fracture, without damaging a

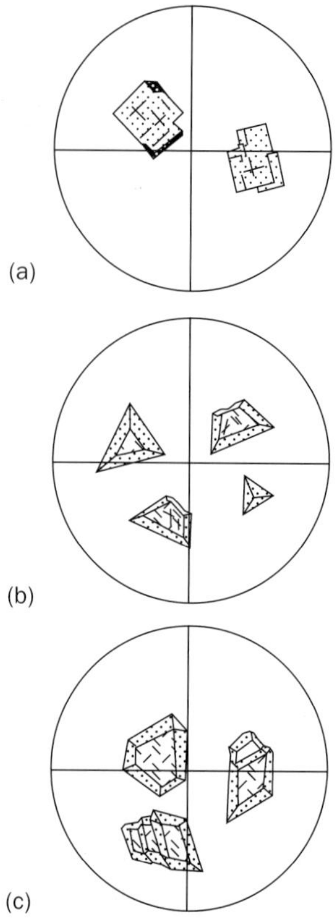

Figure 6.2 Cleavages in isometric minerals as seen in grain mount: (a) cubic cleavage, (b) octahedral cleavage, (c) dodecahedral cleavage. *Source:* Nesse, 1991.

gemstone or gem, by looking carefully inside of it under a microscope (Figure 6.2). Narrow cracks, called *cleavage traces*, that run parallel to planes of symmetry and correspond to incipient cleavage zones, can often be seen. Smoothly curved surfaces, with or without conchoidal rings, as well as irregular, jagged or splintery surfaces are all usually signs of fracture. Cleavage plays a very important role in fashioning gems, often determining the placement of facets.

Remember that *crystal habit* is the general term for the external shape of a mineral. This is determined from uncut gemstones, not from faceted crystals. *Crystal form*, on the other hand, has a more restrictive meaning. Technically speaking, a *crystal* is a solid that possesses a repetitive internal structure and is bounded by flat planes called *crystal faces*. These crystal faces, and the angles between them, are constant for a given mineral type. So the term crystal form refers to the manner by which the crystal faces are related to each other by symmetry. In this context, *symmetry* means the periodic repetition of structural components. Crystal faces have a symmetrical arrangement that is the same for all crystals of the same mineral type.

Externally there are three types of symmetry (*symmetry elements*): symmetry across a plane (think of a mirror, that is one type of plane symmetry), symmetry around a line (axial symmetry), and symmetry about a point (Figure 6.3). There are also only a few methods, called *symmetry operations*, by which repeating patterns can be generated. You can think of a symmetry operation as a way of moving an object such that it appears the same before and after the motion. An easy one to

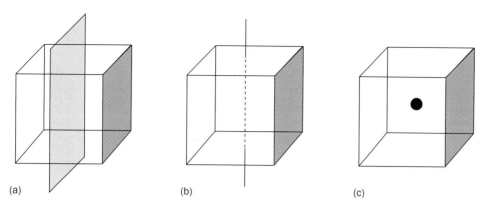

Figure 6.3 Elements of symmetry for a cube: (a) plane symmetry, (b) axial symmetry, (c) point symmetry.

understand is *rotation*, or the repetition of an element around an imaginary circle at a fixed angle. For example, if you make a mark at 90° intervals around a fixed point, you end up with four marks and it takes four rotational motions to return to the original position. This is called *four-fold rotation*. A six-sided die has four-fold rotational symmetry as long as you ignore the pips. If you place a die on a table and look at it top-down, and turn the die 90°, it looks the same. If you rotate the die four times, it will return to its original orientation.

Other symmetry operations include reflection, inversion, and translation. When a form is replicated across a plane, like what happens with a mirror, it is called *reflection*. Slightly more complex is *inversion*, or the projection of all points along your imaginary circle through a single point while turning them inside out. *Translation* is the periodic repetition of an element at a fixed linear interval, kind of like the dashed lines in the center of a road, or the line of footprints you would leave if you hop on one foot. Crystals are defined based on the type and number of symmetry operations necessary to create (or replicate) their three-dimensional structure.

Crystallographers group mineral crystals into 32 classes based upon the types of symmetry they possess. The full descriptive system used in mineralogy is fairly complex but, fortunately for gemology, classification by *crystal system*, rather than class, is usually sufficient. While looking at a crystal, you need to image three axes along the exterior. Starting at the rear, lower left corner of the crystal, called the *origin*, the x-axis will point forward along the left-hand edge of the crystal, the y-axis will point to the right along the back edge of the crystal, and the z-axis will point up along the back edge of the crystal. You can determine which crystal class a sample belongs to by looking at the length of the crystal's sides (a, b, c) and the angles (α, β, γ) at which those sides intersect (Figure 6.4).

There are six (or seven) crystallographic systems (depending on the reference you consult). Going from most to least symmetrical, they are (Figure 6.5):

- **Isometric** (or cubic) – three crystal axes are all at right angles to each other ($\alpha = \beta = \gamma = 90°$) and the sides are all of equal length (a = b = c). This system displays the highest degree of symmetry with three four-fold axes of rotation, four three-fold axes of rotation, six two-fold axes of rotation, and a center (play with a die to see if you can find them all). Common examples include the cube, octahedron, and dodecahedron.

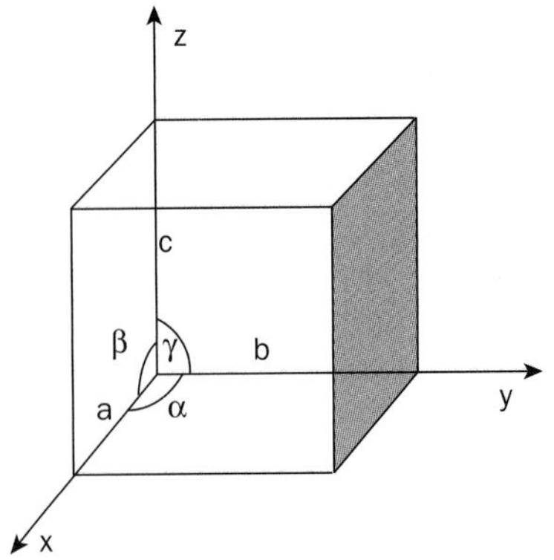

Figure 6.4 Establishment of a coordinate system.

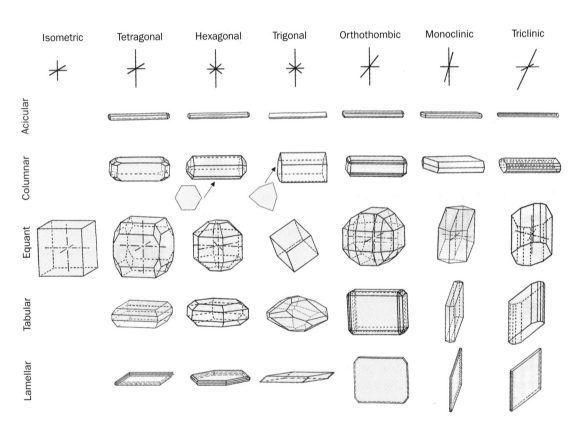

Figure 6.5 Crystallographic systems.
Source: Modified from Winchell, 1949.

- **Tetragonal** – three crystal axes intersect at right angles to each other
  ($\alpha = \beta = \gamma = 90°$), and the horizontal axes are equal in length, but the vertical axis
  is longer or shorter than the other two ($a = b \neq c$). Common examples include
  four-sided prisms and pyramids, trapezohedrons, and eight-sided pyramids.
- **Hexagonal** – is a little trickier. You need to imagine the origin at the center of
  the bottom of the crystal with the z-axis shooting up through the middle of the
  crystal, and three horizontal axes, each going out through every other point on
  the base of the crystal. Thus, you have three segments that are equal in length
  and lie in the horizontal plane at 120° to each other, while the fourth vertical axis

(perpendicular to the other three) is either longer or shorter than the other three ($a_1 = a_2 = a_3 \neq c$) and ($\alpha = \beta = 90°$, $\gamma = 120°$). Hexagonal prisms and pyramids are examples.

- **Trigonal** – also uses a system of four axes, with three axes of equal length in the same plane ($a_1 = a_2 = a_3 \neq c$) intersecting at 120° while the fourth axis lies perpendicular to the other three and is either longer or shorter than the other three ($\alpha = \beta = 90°$, $\gamma = 120°$). The trigonal system is often included in the hexagonal system as a division, called the *rhombohedral division* with the other hexagonal classes grouped into the *hexagonal division*. The major distinction to keep in mind is that hexagonal crystals have a six-sided cross-section, while trigonal crystals have a three-sided cross-section. Three-sided pyramids and rhombohedra are examples.
- **Orthorhombic** – three crystal axes of unequal length ($a \neq b \neq c$) all at right angles to each other ($\alpha = \beta = \gamma = 90°$). Shapes include basal pinacoids and rhombic prisms.
- **Monoclinic** – three axes of unequal length ($a \neq b \neq c$), two of which are inclined to each other at an oblique angle and the third is perpendicular to the plane of the other two (the vertical z-axis is at 90° to the left–right y-axis, which is at 90° to the front–back x-axis, which is inclined to the vertical z-axis) ($\alpha \neq \beta = 90° \neq \gamma$).
- **Triclinic** – all three axes are of unequal length ($a \neq b \neq c$) and at oblique angles to each other ($\alpha \neq \beta \neq \gamma$).

The Petrographic Microscope

The remaining properties we will use to characterize gemstones and gems are all optical properties, which means that we are going to need to add another tool to our forensic toolkit. A *petrographic microscope* is a particular type of polarized light microscope, similar to a stereomicroscope, but it is capable of higher magnification and has additional tools necessary for optical mineralogy. A variety of different models are available, but all of them will have the same basic components. Starting at the top of the microscope is the *eyepiece* (or ocular), the lens you look through at the top of the microscope. The eyepieces magnify the microscope image, usually between 5 and 12×, and focus it for the human eye. The most common magnification is 10×. Eyepieces will usually have cross-hairs that should be oriented N–S and E–W, and can be brought into focus by repositioning the eyepiece in the microscope tube or by rotating the top of the eyepiece.

Some eyepieces contain a scale instead of cross-hairs. To use an ocular scale, you need to calibrate it using a stage micrometer (a glass slide that is etched with distances in micrometers) to determine the actual distance represented by the markings in the ocular. Most eyepieces are designed for use without eyeglasses. If the user wears eyeglasses, it is usually best to remove them before using the microscope, unless the wearer has significant astigmatism. It is possible to focus images without wearing your glasses, and removing them keeps you from scratching your lenses. There are high eye-point eyepieces with rubber cups specially designed for use with eyeglasses, but they are not commonly found in most laboratories.

Beneath the eyepiece is a *Bertrand lens*, which is mounted on a pivot so that the lens can be swung into or out of the light path in the microscope tube. It is used to create interference figures. If your microscope lacks a Bertrand lens, you can achieve a similar effect by removing the ocular and looking down the microscope tube.

Next is the *analyzer* (or upper Nicol), the first of two polarizing filters. The analyzer can be inserted or removed from the field of view usually by the use of a lever or a slider. If the analyzer is in the light path, you are viewing your samples with cross-polarized light. If the analyzer is out of the light path, you are viewing your samples in plane-polarized light, which is in effect like normal light.

There is also usually an *accessory slot* in the microscope tube, oriented at a 45° angle to the analyzer. The slot should be covered if no plates are present in the slot, to present dust from entering the microscope tube. The most commonly used accessory plates are the quarter-wave plate (or mica plate), the full wave plate (also called a first-order red plate or gypsum plate) and the quartz wedge.

At the base of the microscope tube is a rotating turret that holds three or four objectives. An *objective* is a unit that contains a set of lenses for magnifying images. Objectives are usually inscribed with several important pieces of information. The most prominent marking is the magnification power (usually 4×, 10×, 40× on student geology microscopes). The term refers to the ratio of the image size to the object size, for example 10× means that the image is ten times larger than the actual object being viewed. Many objectives on newer microscopes are color-coded, a red ring for 4×, yellow for 10×, green for 20×, blue for 40× and white for 100×. The total magnification of a microscope is equal to the magnification of the objective multiplied by the power of the objective used. For example, using a 10× objective with a 6× eyepiece gives a total magnification of 60×.

Objectives are also labeled with their *numerical aperture* (NA), a parameter that ranges from 0.1 to 1.0 (in dry air) and describes the ability of the lens to resolve details. For example, if you are looking at an image with a series of closely spaced dots, the higher the NA, the better you can differentiate the individual dots. This information becomes important if you use oil-immersion lenses.

The objectives on a turret are set up so that little re-focusing is needed when you switch from one to another. You should always start by surveying a sample at lower power and locating areas of interest before switching to higher magnification. The higher the magnification employed, the narrower the field of view and the dimmer the image. You should always rotate the objectives by using the ring on the turret and never by grabbing onto the objectives. Achromatic objectives are the least expensive type and most commonly found on student microscopes. Lenses with higher levels of correction are fluorites (semi-apochromats) and apochromats.

Directly beneath the objective is the *stage*, a rotating plate upon which you place your sample. Stages can be rotated 360° and are calibrated with a scale along the outside perimeter. Most stages will have two metal spring clamps meant to hold down glass slides and a screw that is used to fix the stage in place. You should check that your stage is aligned properly by placing a slide on it, viewing it through the eyepiece, and centering a particular feature under the cross-hairs. Rotate the stage. The feature should stay in position under the cross-hairs. If it swings out of view, the stage is out of alignment and needs to be repositioned.

Under the stage is a knob for rotating a condensing lens into the light path. A *condensing lens* is used to convert the incident parallel light rays into a convergent cone of light (i.e. the light is angled in so that the rays will meet at a point). This is used with the Bertrand lens to create interference figures.

Somewhere along the light source will be a little slide or lever that works the *diaphragm*, an iris that restricts the amount of light reaching the stage. In most cases, you want the iris to be fully open, unless you are creating interference figures or trying to determine RI.

The *polarizer* (or lower Nicol) is usually fixed in position beneath the diaphragm. It converts the light from the light source into plane-polarized light and should be positioned 90° to the analyzer. You do not normally have to worry about the polarizer, but in some inexpensive microscopes it is possible to accidentally re-position it. To check that the analyzer and polarizer are oriented appropriately, turn on the light source, leave the stage empty, and flip the analyzer into position. The view through the eyepiece should be totally black. If it is not, something is out of alignment.

Beneath the sub-stage assembly, the *light source* is located in or on the base of the microscope. It is normally a low-powered bulb that may be covered with a blue filter. More sophisticated microscopes have a range of different possible illumination sources for specialized techniques.

Most microscopes will have two *focusing knobs* located on the side of the body of the microscope: a *coarse focus* and a *fine focus*. They work by raising and lowering the stage assembly, changing the distance between the base of the objective and the sample on the stage.

To focus a microscope, start by looking at it from the outside, with your eye level to the stage. Move the focusing knob so that the distance between the objective and your sample decreases until there is just a thin sliver of light between the sample and the objective. Never let the objective actually touch the sample. Next, look through the eyepiece and turn the focusing knob so that the stage moves down, away from the objective, until the sample is clearly in focus. This procedure will help ensure that you never slam the objective into the sample, possibly damaging or destroying both. This is especially important when you are working with high-power objectives, like the 40×. There is usually very little space between the lens and the sample, making it incredibly easy to accidentally destroy (crunch!) your sample.

Light and the Optical Properties of Minerals

Visible light, energy that travels in waves at wavelengths visible to our eyes, is actually just a type of electromagnetic radiation. When a ray of light encounters an *interface* between two different media, such as air and a gemstone, part of that ray is *reflected* (sent back into the first medium, or the air in this example), part is *refracted* (enters the second medium, the gemstone), and part is *absorbed* (Figure 6.6). Opaque materials will either absorb most of the incident (incoming) light or reflect most of it, so that no light passes through. The color that an object appears to be is determined by the wavelengths of light that are reflected from its surface to your eye. Thus, black objects appear black because they absorb all of the incident white light, while red objects, like an apple, reflect red light to your eyes and absorb all of the other wavelengths. Reflection is symmetrical; the incoming rays of light are reflected from the surface at the same angle at which they arrived, or in other words the angle of incidence equals the angle of reflection.

In transparent (clear) materials, the majority of the light is refracted, with little reflection or absorption. The light passes through, allowing you to see objects on the other side. Refraction is not symmetrical. When incident light rays pass into a new medium, the change in density causes the light rays to slow down (or speed up), which also causes them to change direction (or "bend"). Most people are familiar with this phenomenon as it applies to water; when you immerse a straw into a drink at an angle and look at it through the side or top of the glass, the straw appears to be bent or even broken (Figure 6.7). The difference in the speed and

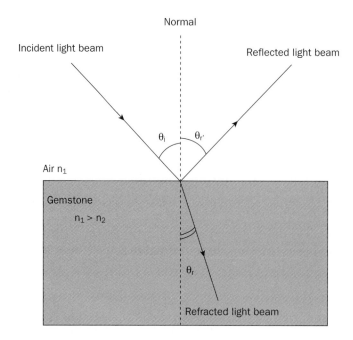

Figure 6.6 Interaction of a light ray at a change in medium.

Figure 6.7 Picture of a straw in a beaker of water.

orientation of the refracted light ray is a function of the refractive indices of each of the mediums involved.

When light enters a denser material, it slows down and the light ray bends inward toward *normal* (think of this as a line perpendicular to the boundary between two substances) (Figure 6.6). The denser the medium, the more light bends. Because this property is related to the internal arrangement of the atoms, it is different for different minerals. Light travels 1.54 times faster through air than it does through quartz and 2.41 times faster through air than through diamond. This means that light is refracted (bent) to a greater degree in diamond than in quartz.

The *refractive index* (RI) is the ratio of the velocity of light in a vacuum (in practice, air) and in another medium. For example, the speed of light in a vacuum is ~300,000 km/sec (~186,000 miles/sec), while the speed of light in a diamond is 124,120 km/sec (77,125 miles/sec). A diamond's RI is calculated as follows:

$$RI_{diamond} = \frac{V_{air}}{V_{diamond}} \frac{300\,000\,km/sec}{124\,120\,km/sec} = 2.415$$

This is also measurable in terms of the ratio of the sines of the angle of incidence and the angle of refraction at the interface between the two media (Figure 6.8). This is called *Snell's law*:

$$RI_{mineral} = \frac{\sin\theta_{incident\ on\ mineral}}{\sin\theta_{refracted\ in\ mineral}}$$

(treating the RI of air as unity). RI is always measured relative to the *normal* of the surface of the gem (i.e. perpendicular to the interface between the two media). Gemologists usually measure RI using a *refractometer*, a small instrument that allows the user to read values directly from a scale. In practice, these instruments only work for RIs between around 1.40–1.81.

Becke Line Method

Another method of determining the approximate RI of a gemstone is to use immersion liquids of known RI. For precision work, sets of calibrated mixed

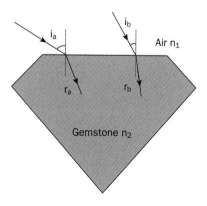

Figure 6.8 An example of Snell's law. Light rays incident to the surface of a diamond at different angles are also refracted at different angles: incident $i_a \approx 52°$, incident ray $i_b \approx 30°$, refracted ray $r_a \approx 19°$, and refracted ray $r_b \approx 12°$. While the angles are different, the ratio of sin i:sin r is constant at 2.415.

immersion liquids of known RI created by Cargille Laboratories can be purchased from a variety of resellers. To determine RI you need a petrographic microscope (or compound light microscope) adjusted for *Köhler illumination*. Do this by closing the *diaphragm* (located in the base of the microscope next to the illumination source) down until the edges are clearly visible in the field of view. Use the *sub-stage condenser* (the assembly mounted just below the stage) control knob (which moves the condenser assembly up and down independently of the stage) to bring the edges of the diaphragm into crisp focus. Ensure that the closed diaphragm is centered in the field of view by adjusting the centering screws of the sub-stage condenser. Now open the *diaphragm* just enough so that the edges are just out of the field of view.

The image contrast of samples can now be adjusted using the *condenser aperture diaphragm* (this diaphragm is located in the condenser assembly and is usually controlled by a lever, a swing arm or a collar on the condenser housing), so that the light from the condenser fills the back focal plane of the objective. Remove one eyepiece and look down the tube of the microscope. Watch the image of the aperture diaphragm as it is opened and closed. If the microscope is equipped with a diffuser, an evenly light circle will be visible; otherwise, you will see the image of the lamp filament. You want to set the aperture diaphragm so that contrast is maximized. This is usually such that somewhere between 60 and 90% of the size of the entire light disc is visible in the tube; in practice, you will have to play around with this adjustment until you are satisfied with the contrast. Also, if the level of magnification is altered (i.e. you change objectives) the optical components must be realigned to maintain Köhler illumination, so you must start over from the beginning.

Once Köhler illumination has been established, use only *neutral density filters* placed in the light path to alter the intensity of the illumination. As a general rule, you should not significantly alter illumination by changing the supply of power to the light source. Use the manufacturer's specifications to set the light source at its optical voltage (usually 6–10 volts) and leave it at that setting. If you are using a polarizing microscope, make sure that the analyzer is out and that you are using plane-polarized light.

Place the stone into a glass dish filled with a liquid of known RI (Table 6.1) and place the dish onto the microscope stage. Close the diaphragm of the light down to approximately the size of the stone, and focus to a point just above the stone. When there is a difference between the RI of the sample and the medium in which it is immersed, a bright halo, called a *Becke line*, will appear around the sample; the brighter the line, the greater the difference in RI. If you look carefully, there are actually two Becke lines: the bright line that is usually easily visible and a companion dark line created because the incident light was redirected. Now, raise the plane of focus by slowly increasing the distance between the stage and the objective. As you change focus, the bright Becke line will move toward the medium with the higher RI.

Now, focus just above your gemstone or gem sample. If the edges and facets of the stone appear black, and the stone has a white halo around it, the immersion liquid has a higher index of refraction than the gem. The larger the bright halo around the stone appears, the larger the difference between the refractive indices. As you focus carefully down into the stone, the facet edges will start to brighten.

If the stone appears black with white facets, there is no white halo around the stone when you are focused just above the stone, and the facets start to darken

Table 6.1 Immersion liquids of known refractive index (20–22 °C).

RI	Liquid
1.0003	Air
1.33	Methyl alcohol
1.333	Water
1.36	Acetone
1.36	Ethyl alcohol
1.38	Hexane
1.38	Isopropyl Alcohol
1.40	Silicon Oil
1.41	Decane
1.43	Ethylene glycol
1.43	Xylene
1.44	Chloroform
1.45	Kerosene
1.45	Coconut Oil
1.46–1.47	Peppermint Oil
1.47	Corn Oil
1.47	Glycerin
1.47	Olive Oil
1.48	Cod-liver Oil
1.48	Castor Oil
1.49	Linseed Oil
1.5	Benzene
1.51	Ethyl iodide
1.53–1.54	Clove Oil
1.57–1.60	Cinnamon Oil
1.6	Bromoform

when you focus down into the stone, then the gem has a higher RI than the immersion liquid.

If the stone seems to disappear into the liquid, then the indices of refraction of the gemstone and the immersion liquid are nearly the same. Note that unless you have oil immersion objectives on your microscope, you need to take care *not* to submerge the end of the objective into the immersion liquid! This same basic technique is used to determine the RI of any mineral, and the RI of glass fragments as well, though in the latter case much higher precision is necessary.

Once the relative indices of refraction for the liquid (with a known RI) and the gemstone (unknown) are understood, the sample is removed from the immersion liquid, cleaned, and then placed into a new immersion liquid in a clean glass dish. Based on the results of the previous test, a new immersion liquid, with an RI closer to that of the sample, should be selected. The idea is to find a *match point*, or the point at which the gems or gemstone will almost disappear in the immersion liquid and no Becke line appears. At this point, the RI of the immersion fluid and the sample fragment are the same, at that temperature and wavelength. Ideally, you want to find the match point, but more often two liquids will bracket the match point.

If you do not have access to immersion liquids, you can use a less precise method of estimating RI using a microscope, called the Duc de Chaulnes' Method. For this, you need to secure the stone on a glass slide with the culet (bottom facet) of the stone touching the glass slide and the table (top facet) parallel to the slide. Now, focus the microscope on the glass slide and record the reading on the fine-adjustment focusing knob. Move the slide so that the table of the stone is in the field of view and adjust the fine focus until the table is sharply in focus. Then record the reading. The difference between these two readings is the total thickness (T) of the stone. Now, without moving the slide, focus the microscope downward through the table of the stone until the cutlet (or the point at the bottom of the stone) is in focus and record the reading. The difference in the reading on the focusing drum between this measurement and the second one is the apparent thickness (t) of the stone. The RI of the stone is approximately equal to the actual thickness divided by the apparent thickness (RI = T/t) (Figure 6.9).

The Forensic Identification of Glass

Glass is one of the most common types of trace evidence analyzed. The name *glass* is a generic term for the brittle, transparent, or translucent solids that are employed for a wide variety of applications, including as windows, bottles, and light bulbs. Various forms of glass are also commonly used as substitutes for gems. Glass is technically defined as the "inorganic product of fusion which has cooled to a rigid condition without crystallizing" (ASTM C162-05, 2010). This means that, unlike

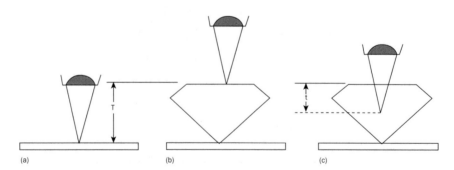

Figure 6.9 Using the Duc de Chaulnes' Method, three measurements are taken for an approximate measurement of refractive index: (a) focus on the top of a glass slide and note the reading on the fine-focus knob; (b) affix the gem to the glass slide resting on the culet and focus on the table of the gem. Note the reading on the fine focus knob. The difference between the first reading and the second is T; (c) focus through the gemstone on the culet and note the reading on the fine-focus knob. The difference between the second and third reading is t. *Source:* After Hurlbut and Switzer, 1979.

minerals, glass is *amorphous*, i.e. lacking in internal structure, which gives glass some distinctive optical properties.

There are several different materials that are commonly referred to as glass. These include naturally occurring materials such as obsidian or volcanic glass and, more commonly in forensic work, manufactured materials such as *aluminum oxynitride optical ceramic* (ALON or AION, used for the windows of armored vehicles); *borosilicate glass* (with > 5% boric oxide, commonly used for laboratory glassware); *alumino-silicate glass* (used to make glass fibers and stovetop cookware); *lead-alkali-silicate glass* (with a high RI; it is used for costume jewelry and chandeliers); and *soda-lime silicate glass* (used for windows and containers).

The manufacture of the most common form of glass (soda-lime silicate glass) begins with large quantities of silica sand (silicon oxide), which is sieved, cleaned, and crushed. Then other geologically derived materials are added such as lime (CaO), which is derived from limestone ($CaCO_3$) or dolostone [$MgCa(CO_3)_2$] and improves durability. Color is controlled by the addition of elements such as manganese for purple/violet, copper to create blue or green, selenium for pink, gold to create red, and chromium for green. Iron may be added to create a yellowish-brown (Fe^{2+}) or bluish-green (Fe^{3+}) color, though it is also an impurity that is usually found in most glass. Metal oxides are added to improve durability (lead), resistance to chemical corrosion (alumina), and increase heat resistance (boron). *Cullet*, or recycled broken glass, and sometimes soda ash (thus the term soda-lime) or potash, are added to act as *flux* (lowers the melting point and promotes glass formation). Everything is ground into powder together to form a *batch* and then melted in a furnace to produce liquid glass.

For the large-scale production of flat glass, raw materials are constantly being fed into a huge furnace to produce molten glass that is continuously poured onto a bed of molten tin. Liquid tin is used because it is denser than the molten glass, does not bond with it, and gives the glass a very smooth finish. The resultant glass product, with nearly perfectly parallel surfaces and uniform thickness, is called *float glass*. After the glass is formed, it must be cooled in a controlled fashion, called *annealing*, to relieve internal stresses and prevent cracks from developing at lower temperatures.

Float glass is commonly used for windows and flat panel displays. Glass containers, including bulbs and headlamps, on the other hand, are usually created from smaller batches using either a ribbon and blow method or a drop and press method. To create light bulbs, molten glass pours out of a furnace onto rollers and is directed into a ribbon machine onto a series of plates. There, air nozzles blow the glass through holes in the plates and into molds that shape the glass. Bottles and jars are usually made using an automated process during which precise lengths of molten glass (called *gobs*) are sheared off as they flow out of a furnace. The gobs then drop into molds and are pressed (or blown) into shape. Optical glass, used for creating items such as lenses for digital cameras, telescopes, and medical equipment, is created using a complex precision-molding process. And this is just a sample of the variety of different ways that modern glass products are made. Plus, hand-blown glass products are still manufactured as well.

Secondary processes are also employed to achieve certain characteristics. *Tempered glass*, which is more resistant to breaking than annealed glass, is created either through a thermal process, where the glass is heated and then rapidly cooled, or through a chemical process, where the glass is plunged into a bath of molten potassium nitrate. This causes large potassium ions to exchange for some of the sodium ions in the glass, changing the stress state on the exterior layer of the glass panel. Tempered glass does not break into sharp shards. Instead, when compromised, the whole panel will shatter at once into small square-shaped pebbles (dicing), which is why it is often called *safety glass*. Tempered glass is used for the side and rear windows of automobiles, as well as for computer displays, skylights, shower/tub enclosures, refrigerator shelves, oven doors, storm windows, and sliding glass doors. Building codes in many areas also often require the use of tempered glass in certain places in private homes (such as glass doors, interior partitions, and windows that are less than 46 cm (18 in) above the floor) and on many public structures.

Laminated glass is created by sealing one or more thin layers of flexible clear plastic, usually polyvinyl butyral (PVB), between two or more pieces of heat-treated glass. When laminated glass breaks, the plastic film holds the glass pieces in place, usually creating a characteristic spider's web crack pattern. Bullet-resistant glass is often constructed by sandwiching together multiple layers of laminated glass. In the United States, the front windscreen on all automobiles must be laminated glass. It is also used for large skylights, exterior storefronts, and where hurricane-resistant construction is required.

Modern glass manufacture results in a very uniform product; however, the use of natural materials, combined with the differing compositions of batches and differing means of production, means that the elemental composition, and in consequence the physical properties, of glass can vary considerably from manufacturer to manufacturer and from batch to batch. These small but measurable differences in physical, chemical, and optical properties allow forensic workers to differentiate samples and to determine whether a fragment could have come from a particular source.

When glass shatters, fragments can travel a remarkably great distance in all directions. In experimental studies, fragments have been found up to four meters away from a breaking object and recovered from the clothing of people standing up to one meter away (Locke and Unikowski, 1991). Fragments are frequently found in shoe or tire treads, on clothing, and on tools, even in hair. The most common forms of glass confronted in forensic work are typically window glass and automobile glass (windshield, side window, and headlamp). Only by the physical matching of two or more glass fragments can a forensic worker conclusively demonstrate the *association* of samples to the exclusion of all other possible sources. However, since most glass fragments received for forensic examination are from transfer and smaller than 0.5 mm (usually around 0.2 mm), physical matching is usually not possible (Zadora, 2009).

The first step in any forensic examination of glass is to check that the sample really is glass. Plastic can usually be distinguished from glass by carefully pressing the sample with a needle. If the sample flexes, deforms, dents, or easily scratches, it is plastic. This includes products like *acrylic glass*, which is not technically glass at all. A common example is poly(methyl methacrylate) (PMMA), which is sold under a wide range of trade names, such as Plexiglas, Lucite, and Polycast. PMMA is a transparent thermoplastic that is commonly used as a glass substitute. With practice, it can easily be distinguished by its low SG (around 1.2), flexibility, low RI (approximately 1.33), and almost total lack of scratch resistance.

Glass fragments will also often show conchoidal fracture. Glass is technically considered isotropic, so when observed under cross-polarized light, it should stay extinct through a 360° rotation. However, glass that has been heat-treated or subjected to thermal or mechanical stress can show interference colors. In practice, some samples of glass stay an even shade of gray throughout rotation, while scratches, pits, and abrasion features can cause strange effects that can confuse the novice microscopist.

The overall condition of each sample should be observed. Freshly broken glass will have sharp edges. Pits, scratches, and other abrasion features should also be noted. If the glass samples are large enough, color can be useful for comparison. The samples must be placed on a white background and illuminated using natural light. Sample size affects apparent color, so side-by-side comparisons should only be made on fragments that are the same size. With small particles, however, color density is too low to be useful and it is not possible to reliably use this characteristic.

Many types of glass will fluoresce when exposed to *ultraviolet* (UV) light. Float glass exposed to short-wave UV light (~254 nm) will fluoresce on the side that was in contact with molten tin (usually a yellowish color). Glass that has certain additives or impurities, such as uranium, will fluoresce throughout its body. Surface coatings can also cause glass to fluoresce in different colors. Fluorescent comparisons should be performed side by side, on particles of similar size and condition.

Initial comparisons can also be made on the basis of physical features. Curvature, thickness, surface coatings, such as thin films and mirrored backing, and manufacturing features, like etching, texture and frosting, can all be used to exclude fragments from a given source. The caveat is, of course, that the samples under examination must be large enough to be fully representative of the feature being investigated. For example, in order to determine thickness, both original parallel surfaces must be present. Thickness can be a helpful feature to establish because this property is tightly controlled by manufacturers (often within fractions of a millimeter). Plus, thickness can be related to product type (see ASTM C 1036-11, 2011). A caliper or micrometer capable of +/- 0.02 mm precision or better must be used for forensic comparison. When the thickness of a piece of flat glass is measurably different from the range expressed in a set, it can be excluded as having come from the same source.

Most commonly, the next step to take for examination of small glass samples is to determine RI. First, the some particles of the glass sample are isolated and cleaned in an ultrasonic bath. A solvent may also be used to remove any surface coatings. Next, the glass is dried thoroughly, placed on a microscope slide, and secured using a few drops of an appropriate calibrated immersion liquid (Table 6.2). The glass is then carefully crushed using a metal spatula to produce multiple fragments approximately 150 μm in diameter with fresh, sharp edges. The fragments are mounted in more of the same immersion liquid and topped with a cover slip. The cover slip must be as close to parallel as possible. If it sits at an angle, re-orient or remove some of the glass fragments.

Adjust your microscope for Köhler illumination as described previously and make sure to use plane-polarized light. Now, place a slide with a glass sample under the microscope and focus on the edge of a representative particle. You might want to adjust the aperture diaphragm a bit to improve the contrast. Due to the difference between the RI of the glass sample and the medium in which it is immersed, a Becke line will appear around the glass. As before, you want to raise the plane of focus by slowly increasing the distance between the stage and the objective. As you change focus, the bright Becke line will move toward the medium

Table 6.2 Refractive indices of some commercially available testing liquids.

Liquid	RI at 25 °C	Use for
Dow Corning 710 Silicone Oil	1.533	Soda-lime silicate glass
Dow Corning 550 Silicon Oil	1.4935	Borosilicate glass
Dow Corning F/6/7024	–	High RI glass (lead glass)
Cargille (www.cargille.com)	1.3–2.3	Sets of certified refractive index liquids (some of these liquids will break down under high heat)

Table 6.3 Average refractive indices of some common types of glass.

Type of Glass	RI
Headlamps	1.47–1.49
Television glass	1.49–1.51
Window glass	1.51–1.52
Bottle glass	1.51–1.52
Flat glass	1.51–1.53
Vehicle float glass	1.5
Fused silica quartz	1.459 (avg.)
Borosilicate	1.479 (avg.)
Soda-lime silicate glass	1.512 (avg.)
Alumina-silicate glass	1.530 (avg.)
High lead glass	1.693 (avg.)

with the higher RI. Once the relative indices of refraction for the liquid (with a known RI) and the glass (unknown) are understood, the glass fragments are removed from the immersion liquid, cleaned, and then placed into a new immersion liquid on a clean slide. Based on the results of the previous test, a new immersion liquid, with an RI closer to that of the glass, is selected. The idea, as before, is to find the match point, or the point at which the glass fragment will disappear in the immersion liquid and no Becke line appears. However, there is an additional difficultly when working with glass.

The indices of refraction for natural materials have a broad range, so it is usually possible to narrow down RI sufficiently for mineral identification using the liquids listed in Table 6.1. For manufactured materials like glass, however, RI is narrowly controlled and there might be only very small differences between similar types of glass (Table 6.3). This is especially true for float glass, where the variability in RI across a flat panel of float glass was found to be approximately +/− 0.00004 for annealed glass and +/− 0.0016 for tempered glass (Underhill, 1980). Significantly, the float surface of glass may be 0.0013 to 0.0047 higher than the bulk glass, while the non-float surface may be 0.0011 to 0.0039 lower (Underhill, 1980). The Becke Line Method of determining RI is only accurate to +/− 0.001 under the best of conditions, so in general it is only useful as a screening method for glass fragment analysis.

Most laboratories use more complex methods to establish RI, such as the Emmons Double Variation Method (Association of Analytical Chemists Method, 1990), or manual or automated temperature variation, all of which require some specialized equipment. Each of these methods relies on the fact that the RI of a material is temperature- and wavelength-dependent. Thus, one way to improve the accuracy of RI measurements is to determine values either a range of temperatures at a fixed wavelength or for a range of wavelengths at a fixed temperature. Because of thermal lag, it is usually easier to do the latter.

Many forensic laboratories utilize the Glass Refractive Index Measurement system (GRIM 3). The GRIM 3 uses an oil-immersion/temperature variation method and operates through a standard laboratory microscope. By varying the temperature to alter the RI of a calibrated oil, the RI of an immersed sample can be determined at the point of null refraction, when the refractive indices of the sample and oil match. Samples as small as 50 μm can be analyzed, and the RI of a sample can be determined to five decimal places.

If you are interested in more information on forensic glass examination, see Bottrell (2009) and Koons *et al.* (2002).

More Optical Properties

A property related to RI is *relief*, defined as the contrast between the mineral and its surroundings (such as an immersion oil or binder for a glass mount). Grains with low relief are barely visible (i.e. close to the RI of the immersion medium), while those with high relief stand out clearly (i.e. have a very different RI). Some minerals, like calcite, can even show variable relief with state rotation.

RI also determines the play of light in a gem or gemstone. This involves the *critical angle*, or the minimum angle at which *total internal reflection*, where all of the light from the incident ray travels through the medium to a second interface and then is reflected inside of the second medium, is achieved (Figure 6.10). Light rays

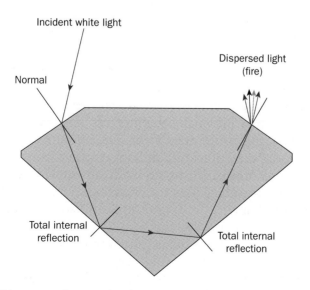

Figure 6.10 Light striking a medium with a lower index of refraction will be totally internally reflected if it intersects the interface at or above the critical angle.

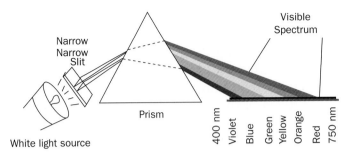

Figure 6.11 Dispersion of white light into its component wavelength by a prism.

striking the interior surface of a facet at an angle less than the critical angle are refracted through the surface interface and exit the stone. Light rays striking the interior surface of a facet at greater than the critical angle will be reflected back into the gemstone, making it shine brightly. The RI of a gemstone is used to determine the placement and angle of facets in order to control the path of light in a gemstone. Gems appear at their most brilliant when light is reflected back out through the top facets and not lost through the lower portion of the gem (the pavilion).

Depending on their indices of refraction, some stones will also cause the dispersion of light. *Dispersion* is the separation of white light into its component wavelengths, as by a prism (Figure 6.11). Visible light is commonly divided into seven colors: red, orange, yellow, green, blue, indigo, and violet. Violet light has the shortest wavelength (380–450 nm) and red light has the longest wavelength (620–750 nm). All light travels at the same speed in a vacuum, but within a material light will slow down to a varying degree depending on its wavelength. The shorter the wavelength, the more the light will be slowed down or refracted, thus blue light is bent more than green light, which is bent more than red light.

If dispersion in a mineral is low, then white light can travel through the mineral nearly unaffected and emerge again as white light. But if dispersion is high, the white light will increasingly be separated into its component wavelengths, or colors, through increasing refraction. This is what causes the flashes of color, called *fire*, in cut gemstones. Diamond has a high degree of dispersion, which is why a well-cut diamond will create dozens of flashes of rainbow light. This property is often used to recognize diamond stimulants such as strontium titanate (ST), cubic zirconia (CZ), and yttrium aluminum garnet (YAG), which all have high dispersions, but not as high as a real diamond. Dispersion is usually expressed numerically as the difference between the red and violet refractive indices.

Depending on the manner in which they interact with light, crystals can be divided into one of two categories: *isotropic* or *anisotropic*. Isotropic substances have the same properties in all directions. When light passes through an isotropic mineral, it slows down as previously described. Light moves through isotropic crystals at the same velocity, regardless of the orientation of the mineral, thus they can be characterized by a single RI (*n*). Examples of isotropic minerals are opal and all minerals in the *isometric* crystal system.

Minerals in all of the rest of the classes (all classes except the isometric) are anisotropic substances, with properties that vary by direction. This means that the velocity at which light travels through the crystal varies depending on the crystal's orientation (*crystallographic direction*). Such substances have a range of values for RI. With a few specific exceptions, all light entering an anisotropic crystal is broken

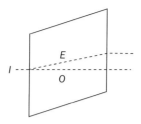

Figure 6.12 The light that enters an anisotropic material (I) is split into two rays with different velocities (O and E). As the resultant rays travel along different paths, this phenomenon is called *double refraction*.

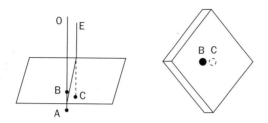

Figure 6.13 Ordinary (O) and extraordinary (E) rays as they move through a crystal of calcite, resulting in double refraction. A is a dot on a page, B is the fixed image produced by the O-ray, and C is the image produced by the E-ray, which revolves around B when the calcite crystal is rotated.
Source: Modified from Winchell, 1949.

into two rays (Figure 6.12). Thus, in general, for a random orientation, an anisotropic crystal will have two indices of refraction, one associated with each light wave. This is known as *double refraction*.

All anisotropic minerals exhibit the phenomenon of double refraction, but it is only when a crystal's *birefringence*, the difference between the highest and lowest index of refraction in a mineral, is very high that it is apparent to the unaided eye. This phenomenon is best observed in nice clear crystals of calcite, but can also be seen in zircon, sphene, tourmaline, and peridot, though you might need a magnifying glass. When you place a large, clear piece of calcite that has been cleaved into a parallelogram (called Iceland spar or optical calcite) onto some text, and look through it, there appears to be a double image.

The light reflected from the text is split by the calcite into two wavefronts that are traveling at two different velocities. Remember that when light changes velocity, it also bends or changes orientation (refraction), thus some of the light is still moving along the original path while the rest of the light bends because it is traveling at a slightly different velocity. If you rotate the calcite crystal, one image will stay fixed while the other image will appear to float around. The light that travels straight through the crystal, creating the stationary image, is termed the *ordinary ray* (O or ω), while the refracted light, which creates the floating image, is termed the *extraordinary ray* (E or ε) (Figure 6.13).

So far, the properties described can all be seen with normal incident light, but to really understand double refraction it is necessary to use polarized light. Ordinarily, light waves vibrate in all directions perpendicular to their direction of propagation (i.e. the wave crests and troughs are at 90° to the direction the light ray is moving). This means that incident light rays come at an object with their "wave peaks" oriented in all directions. If light is constrained to vibrate in a single plane, i.e. all of the waves are oriented parallel to each other, it is called *plane-polarized light* (Figure 6.14). When light passes through a calcite crystal, not only is it split into two rays that travel different paths but also the rays are polarized so that they vibrate in mutually perpendicular planes (Figure 6.15). Use of plane-polarized light allows for the characterization of many crystallographic properties and is an essential tool in mineralogy and gemology.

The best way to determine the crystallographic properties of an unknown mineral is by using *cross-polarized light*. To do this, gemologists use an instrument called a *polariscope*, while mineralogists use a petrographic microscope. In both cases, the instrument contains two polarizing plates, called *polars*, one above the specimen and one below. The plates are oriented such that the lower one (usually called the *polarizer*) is oriented to transmit light vibrating E–W (side to side) and the upper polar (usually called the *analyzer*) is oriented to transmit light vibrating N–S (front to back). The reverse orientation is often found on older microscopes. In either case, the polars are *crossed*, or at 90° to each other, effectively blocking the transmission of light (Figure 6.16).

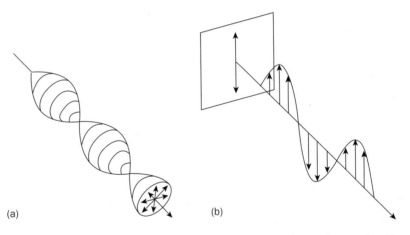

(a)   (b)

Figure 6.14 (a) Unpolarized light vibrates in all directions at right angles to the direction of propagation, (b) plane polarized light means that the light vibrates in a single plane. *Source:* Modified from Nesse, 1991.

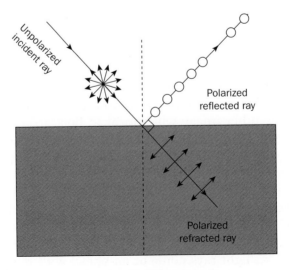

Figure 6.15 Light split into two rays that travel different paths.

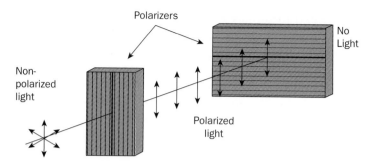

Figure 6.16 Cross-polarized light and extinction.

Isotropic versus Anisotropic Minerals

When viewing an isotropic crystal through crossed polars, it appears dark at all orientations, because the light that travels through isotropic materials continues to vibrate in the same plane it entered. Thus, light that travels through a polarizer and enters a crystal stays polarized with the same orientation as it travels through the crystal and then exits to be blocked by the analyzer. This is called *extinction* (Figure 6.16). One can quickly tell if a crystal is isotropic by rotating it on a microscope stage between crossed polars. At all orientations light is blocked so the sample will remain dark through an entire 360° rotation of the stage. This allows gemologists to quickly distinguish many synthetic glass materials from all gems except those in the isometric system. Glass is technically amorphous (i.e. lacking a repeating atomic structure), so the light traveling through is not split into two beams, thus glass tends to look black (or dark gray) through a 360° rotation, similar to an isotropic crystal.

Anisotropic crystals will have either one or two orientations through which there is no double refraction. Such orientations are called *optic axes*. Crystals with one optic axis are called *uniaxial*, and crystals with two optic axes are called *biaxial*. If such a crystal is mounted on the microscope stage such that the optic axis is exactly perpendicular to the stage, the crystal behaves just like an isotropic mineral. Light polarized in the E–W direction as it enters the crystal remains oriented in the E–W direction as it leaves the crystal, so through crossed polars (with the analyzer pushed in) the crystal goes extinct (light is blocked) at all orientations, through a complete 360° rotation of the microscope stage, just as occurs with isotropic crystals.

However, at all other orientations, when viewed through crossed polars, anisotropic minerals will only go extinct every 90° as you rotate the microscope stage. Most of the time, the light you see as the crystal is rotated from one extinction point to the next is going to be a combination of light from the ordinary and the extraordinary rays. But remember that these two rays are polarized perpendicular to each other. As you rotate the stage, every 90° the light passing through an anisotropic crystal will be either all from the ordinary ray or all from the extraordinary ray, and vibrating in a plane parallel to the direction of the polarizer. Thus, when the light reaches the analyzer it will be completely blocked.

If you start with a crystal at extinction, where all of the light from the extraordinary ray is being blocked, and then start to rotate the crystal, initially most of the light passing through the analyzer will be from the ordinary ray, and only a little will be from the extraordinary ray. As the crystal is rotated from the extinction position, progressively more light passes through the analyzer, reaching a maximum brightness, which occurs at 45° from extinction for most anisotropic minerals, though many monoclinic and all triclinic minerals have their maximum brightness at a point inclined from 45°. At maximum brightness, the amount of light from each ray will be equal. As you rotate further, there will be more and more light from the extraordinary ray, while there will be less and less from the ordinary ray. At 90° from the previous extinction, all of the light from the ordinary ray will be blocked.

As light passes through the crystal, the two rays also move at different velocities, therefore when they emerge they will be out of phase (i.e. the wave peaks and troughs will not light up with each other), causing interference. This interference produces an effect by which certain wavelengths of light are eliminated, and others are strengthened, resulting in the appearance of a particular color. The colors produced are called *interference colors* and depend on the orientation, thickness, and birefringence of the crystal examined. With a continuous rotation of the stage, there is a continuous change in the apparent color of the crystal.

For minerals with low birefringence, the interference color will be white or gray, so as you rotate the microscope stage the mineral goes from white (or gray) to black every 90°. Quartz, for example, usually shows up as first-order white or gray. For minerals with a slightly greater birefringence, yellow, orange, or red interference colors will appear (still called *first-order colors*). For minerals with high birefringence, more colors will be produced, shifting through violet, indigo, blue, and green (called *second-order colors*). The pattern of colors repeats every 550 nm (yellow, orange, red, violet, indigo, blue, and green), becoming more and more washed-out-looking. Interference colors are shown in a diagram called the Michel-Lévy Color Chart (copies of which are available from most major microscope manufacturers), and can be used as a distinguishing characteristic for some minerals. When describing interference colors, both the color and order (first, second, third, etc.) must be stated.

Remember that the birefringence of isotropic crystals is always zero. The birefringence of anisotropic minerals in a grain mount can be estimated as long as you can estimate the thickness of the sample and the interference color can be recognized. The easiest way, for samples less than or equal to 0.05 mm in thickness, is to use the Michel-Lévy Color Chart. Find a grain of the mineral under examination that displays the highest-order interference color under cross-polarized light. On the Michel-Lévy Color Chart find the approximate thickness of your sample on the y-axis on the graph and follow the horizontal line to the right until you reach a vertical line corresponding to the correct interference color. At the point where these two lines intersect, find the nearest diagonal line and follow it to the right. Where the diagonal line exits the graph, there will be a numerical value corresponding to birefringence. For example, if you are looking at a sample under cross-polarized light that appears to be an intense violet/indigo color, corresponding to a wavelength of 600 nm on the x-axis of the chart (which lies in the second-order region), and are looking at a standard mineral thin-section with a thickness of 0.03 mm (thickness is along the y-axis), you would follow the diagonal intersecting that x, y point to the edge of the chart, which gives you a birefringence of 0.020.

For samples that are thicker than 0.05 mm, the relationship $\Delta = d(n_s - n_f)$ can be used, where Δ is retardation (which can be translated as the wavelength of the interference color), d is the thickness of the sample and $(n_s - n_f)$ is birefringence.

Again, you look for a grain (or for the orientation) with the highest-order interference color. Look at the color chart and determine the approximate wavelength of the interference color. Wavelength is in nanometers, so divide it by 1000 twice to convert to mm. Now divide your converted wavelength by the approximate thickness of the sample (also in mm) to obtain the birefringence. So if you are looking at a rich third-order green color that corresponds to a wavelength of approximately 1250 nm, convert that to 0.00125 mm and then divide by sample thickness. Assuming that our sample is 0.05 mm thick, this corresponds to a birefringence of 0.025. For a more detailed discussion of the Michel-Lévy Color Chart see Delly (2003).

Because these methods are subjective and you will not necessarily be looking at a grain oriented to display maximum birefringence, it is often better to use terms such as "low," "moderate," "high," and "extreme" rather than a number when describing birefringence. Both of these procedures can be used to obtain rough approximations of a sample's birefringence but, in practice, for samples more than 0.1 mm thick it can be very difficult to interpret interference color. If detailed information on birefringence is required, axially oriented samples and immersion oils must be used to determine it directly.

Anisotropic Crystals

In uniaxial crystals, crystals from the hexagonal, trigonal, and tetragonal systems, the only direction that light can travel in all directions at equal velocity is along the c-axis. So if you are using cross-polarized light to look at a uniaxial mineral oriented such that you are looking along the c-axis, it will be extinct at all angles as you spin the microscope stage. At all other crystal orientations, incident light is split into two rays: an ordinary ray (symbolized ω or O) and an extraordinary ray (symbolized ε or E) (Figure 6.17). The ordinary ray continues to travel as predicted

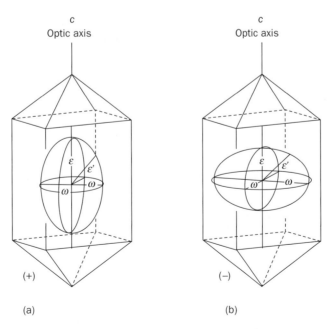

Figure 6.17 The uniaxial indicatrix and orientation in a tetragonal crystal: (a) positive indicatrix, (b) negative indicatrix.
Source: Hurlbut and Switzer, 1979.

by Snell's law, while the extraordinary ray does not. This means that the refractive indices of uniaxial minerals varies between two extreme values, ω and ε. Intermediate values between ω and ε are designated ε'. This can be visualized by looking at a uniaxial indicatrix (Figure 6.17). An *indicatrix* is simply a three-dimensional shape with its axial lengths drawn in proportion to a crystal's RI for light vibrating in each vector direction. An isotropic indicatrix, for example, would simply be a sphere.

If you look through your microscope at a uniaxial mineral that is randomly oriented (i.e. not looking along the c-axis), as you rotate the stage extinction only occurs in a few places, and the rest of the time you will see light from the extraordinary ray shining through the mineral. Just how extinction occurs can also be a useful property for identifying minerals. If you place a uniaxial crystal on your microscope stage so that it is oriented with the c-axis parallel to the microscope stage, you will find that the crystal will go extinct twice, at positions that are 180° apart.

Uniaxial minerals can be further divided into two classes. If the velocity of the ordinary ray is greater than the velocity of the extraordinary ray ($\omega > \varepsilon$) the mineral is said have a *negative optic sign* or is *uniaxial negative*. If the velocity of the extraordinary ray is greater than the velocity of the ordinary ray ($\varepsilon > \omega$) the mineral is said to have a *positive optic sign* or is *uniaxial positive* (Figure 6.17). The absolute birefringence of a uniaxial mineral is defined as $|\omega - \varepsilon|$ or the absolute value of the difference between the extreme refractive indices. The optic sign of a mineral is most easily determined by using creating *interference figures* on a polarized light microscope.

Uniaxial Interference Figures

An idealized summary of this procedure is as follows:

1. Set up your crystal so that you are looking at it along the c-axis through the microscope (i.e. the c-axis is perpendicular to the microscope stage and parallel to the microscope tube) and focus on it using a medium-powered objective. (You can obtain interference figures from crystals that are not oriented along the c-axis, but they are more complicated to interpret.)
2. Flip the condenser in to create *convergent light* (a cone of light that radiates out in all directions).
3. Now choose a high-powered objective lens and re-focus on the crystal.
4. Using cross-polarized light, check that you are oriented properly by rotating the microscope state and verifying complete extinction of the mineral.
5. With the analyzer still in, flip in the Bertram lens on your microscope (or remove the ocular and look down the microscope tube).

You should now be looking at an image of concentric circles of color (interference colors) with a black cross through the middle (Figure 6.18). This is a uniaxial interference figure. The black bars forming the cross are called *isogyres* and the point in the center, where the isogyres cross, is called the *melatope*. A melatope marks the location of an optic axis. Each ring of color is called an *isochrome*. The lowest-order (i.e. first- or second-order) *isochrome* will be nearest the melatope and

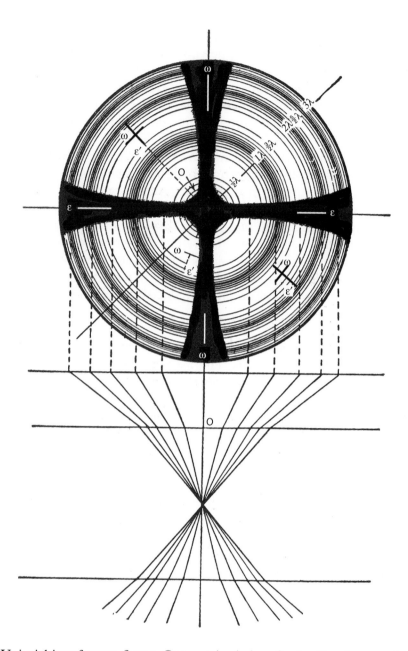

Figure 6.18 Uniaxial interference figure. Concentric circles of color (*interference colors*) with a black cross (the black lines are called *isogyres*) through the middle.
Source: Modified from Winchell, 1949.

the interference colors will increase in order outward, as shown on a Michel-Lévy Color Chart.

If the mineral has a low birefringence (like quartz), there will be few or no concentric colors, while minerals with a high birefringence (like calcite) will have many rings of colors. To determine the crystal's optic sign you need to insert a quartz-wedge accessory plate into the slot above the objective and observe the interference colors. If the colors in the bottom-left (SW) and top-right (NE) move inward, while the colors in the other two quadrants move outward, the crystal is optically positive (+). If the opposite happens, then the crystal is optically negative (–). For an explanation of why this happens, consult an optical mineralogy book. You can also use other types of accessory plates to determine optic sign, but interpreting the results is a little more complex. Basically, when you insert a half-wave or full-wave plate, the mineral is optically positive if the colors in the NE and SW quadrants shift to higher-order interference colors, while the colors in the NW and SE quadrants shift to lower-order interference colors.

All crystals in the orthorhombic, monoclinic, and triclinic systems are *biaxial*. Biaxial crystals have two optic axes along which light behaves as it would in an isotropic medium. Light moving through the crystal in any other direction is broken up into two rays with mutually perpendicular directions of vibration, both of which are usually extraordinary rays. Biaxial crystals have refractive indices that vary between two extremes, like uniaxial crystals, but unlike uniaxial crystals the RI varies in all three axial directions. For biaxial crystals, the smallest RI is given the symbol α, the intermediate RI is given the symbol β, and the largest RI is given the symbol γ (Figure 6.19). The absolute birefringence of a biaxial mineral is defined as $(\gamma - \alpha)$.

The orientation with the lowest RI (α), along which light moves at the highest velocity, is designated the X-axis. The direction with the highest RI (γ) is located at 90° to X, and is designated the Z-axis. The axis perpendicular to the plane defined by XZ is designated the Y-axis, and has an intermediate RI (β). Light that enters a biaxial crystal from a random orientation is split into two rays, neither of which is oriented parallel to one of the principal axes, so their refractive indices will have values between α and γ. At two of these random orientations within the crystal, the intermediate value of the RI will be equal in all directions and no double refraction will occur. These are the optic axes of the mineral, with refractive indices equal to β, and the only time that a ray will be "ordinary."

In orthorhombic crystals the optical directions correspond to the crystallographic axes, i.e. the X direction and its corresponding RI, α can be aligned with the a, b, or c crystallographic direction, bringing the Y (β) and Z (γ) directions parallel to the other two crystallographic axes. In monoclinic crystals, only one of the X (α), Y (β), or Z (γ) directions is parallel to the b crystallographic axis (y-axis), and the

Figure 6.19 Biaxial indicatrix. Principal optical directions OX, OY, and OZ are at right angles and proportional respectively to the refractive indices α, β, γ. *Source:* Hurlbut and Switzer, 1979.

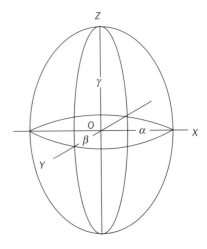

Table 6.4 Indices of refraction and axes.

	Principal indices of refraction	Index of refraction for light traveling parallel to an optic axis	Indices of refraction in a random direction	Birefringence in a random direction	Maximum possible birefringence
Isotropic crystals	n	n	n	0	0
Uniaxial crystals	ε, ω	ω	ω, ε'	$\delta' = \lvert \omega - \varepsilon' \rvert$	$\delta = \lvert \omega - \varepsilon \rvert$
Biaxial crystals	α, β, γ	β	α', γ	$\delta' = \gamma' - \alpha'$	$\delta = \gamma - \alpha$

other two do not coincide with crystallographic directions. In triclinic crystals, none of the optical directions coincides with crystallographic directions. The relationship between RI and axes for all crystal types is shown in Table 6.4.

Biaxial Interference Figures

To determine the optical sign of a biaxial mineral, obtain the interference figure as follows:

1. Focus on a grain using a medium-powered objective. You want to select a grain with low birefringence (i.e. shows the lowest-order interference colors or goes completely extinct under cross-polarized light).
2. Flip the condenser in to create convergent light.
3. Now choose a high-powered objective lens and re-focus on the crystal (always remember to be very careful when focusing at high power!).
4. Push the analyzer in, rotate the crystal so that it is extinct or as dark as possible, and flip in the Bertram lens on your microscope (or remove the ocular and look down the microscope tube).
5. For biaxial minerals, four types of interference figures are possible (Figure 6.20):
   a. an optic axis figure (OA) with only one isogyre (curved black line) in the field of view (or the other isogyre is at the edge of the field of view)
   b. the acute bisectrix (Bxa), which looks like a black cross (uniaxial interference figure) if the grain is extinct under cross-polarized light, but if you rotate the stage, the isogyres (black curved lines) move away from each other
   c. an optic normal figure (ON), which looks like the Bxa but breaks apart or dissolves with just a slight rotation of the stage
   d. the obtuse bisectrix figure (Bxo), which also starts out looking like a Bxa, but the isogyres will always leave the field of view when you rotate the stage.
6. Ideally, you want to use a Bxa figure to determine optic sign. With the Bxa centered (or as nearly centered as possible), rotate the stage so that the isogyres are located in the southwest and northeast quadrants (for some minerals the isogyres will move completely out of the field of view, in which case it is better to use the OA method of determining optic sign). Insert a quartz wedge and watch as the color rings move. If the color rings move outward from the center

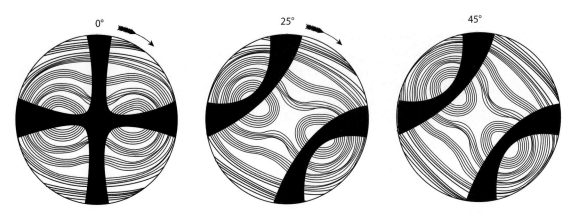

Figure 6.20 Centered acute biaxial interference figures.

of the field of view away from the isogyres, but inward toward the concave sides of the isogyres from the sides, the crystal is optically positive. If the color rings move from the center outward toward the isogyres, and from the concave sides of the isogyres outward, the crystal is optically negative.

7. If a good Bxa figure cannot be obtained, use an OA figure instead. Look for a grain showing very low or zero retardation (shows the least variation or goes completely extinct when rotated under cross-polarized light). You want to obtain a centered or nearly centered OA figure (one isogyre in the center of the field of view). Rotate the stage so that the isogyre is concave to the northeast. Insert a quartz wedge and watch the colors move. If the colors on the convex side of the isogyre (southwest) flow outward while the colors on the concave side of the isogyre (northeast) flow inward, the crystal is optically positive. If the reverse happens, i.e. the color on the convex side of the isogyre flow inward while the colors on the concave side flow outward, the grain is optically negative.

Another useful property of biaxial minerals is 2V°, the acute angle between the optic axes. This property can be estimated in a couple of ways. First, you can estimate 2V° from the Bxa figure by looking at the amount of separation between the isogyres. Start with the stage at the point where the isogyres appear as a cross (like a uniaxial interference figure). Now, rotate the stage so that the isogyres move away from each other. After a rotation of 45°, the isogyres are at maximum separation. The 2V° can then be estimated using Figure 6.21. For standard lenses, if the isogyres just leave the field of view at their point of maximum separation, then 2V° is around 60 to 65°. If the isogyres barely more apart from each other, then 2V° is around 10°.

The 2V° can also be estimated simply from the curvature of the isogyre in an OA figure. If 2V° is less than 10–15°, the isogyre will form a 90° bend, while if the isogyre is almost straight, the 2V° will be around 90°. Figure 6.22 can be used to estimate 2V° values between these endpoints.

This probably seems overwhelming in text form, but when you get some practice using a microscope it starts to make more sense. The best way to get practice looking at biaxial interference figures is to put a thin sheet of muscovite mica on a glass slide without a cover slip. By virtue of its cleavage, mica will automatically be

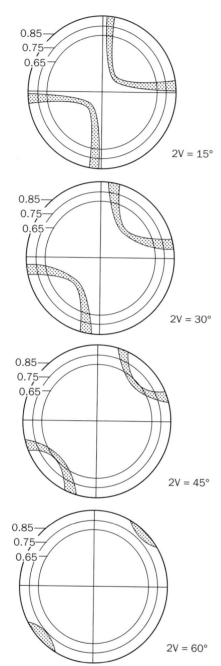

Figure 6.21 Visual estimation of 2V° from the Bxa figure by looking at the amount of separation between the isogyres. The positions of the isogyres are constructed for a mineral with a β (intermediate RI) equal to 1.60 and an objective lens with the numerical aperture of 0.85. The two inner circles in each diagram show the field of view for lenses with numerical apertures of 0.75 and 0.65.
Source: Nesse, 1991.

aligned to produce the acute bisectrix (Bxa) so that you can become accustomed to determining optic sign and 2V°. If you are working with an unknown grain and have an uncentered interference figure, rotate the stage. If you see movement parallel to the isogyre, the mineral is uniaxial, while if you see a change in the curvature of the isogyre, the mineral is biaxial.

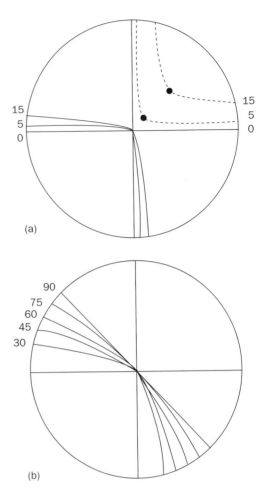

Figure 6.22 Visual estimation of 2V° from the curvature of the isogyre in an OA figure: (a) both melatopes in the field of view. The position of the second (dashed) isogyre depends on the value of β and the NA of the objective lens, (b) one melatope in the field of view.
Source: Nesse, 1991.

Conflict Diamonds

Technically speaking, *conflict diamonds*, or *blood diamonds*, are diamonds illegally traded by forces that oppose legitimate and internationally recognized governments in order to fund the procurement of arms and support armed conflict. Some of Africa's most war-torn countries – Sierra Leone, Angola, Liberia, the Ivory Coast, and the Democratic Republic of Congo – are the most diamond-rich on the continent. They are also among the poorest and most under-developed countries in the world. Warlords often used children as slave labor to work in diamond mines, where they shoveled the stones by hand from alluvial deposits. The raw diamonds were then traded for weapons or cash in order to support war. Meanwhile, villages were burned, crops destroyed, and the local population displaced or murdered. In Sierra Leone's long civil war, explored in the 2006 movie *Blood Diamond*, the fiercest fighting was over possession of the diamond fields. Atrocities such as amputation, conscription of child soldiers, and mass murder were common.

Conflict diamonds have also been linked to al-Qaeda and Hezbollah. At the time of the 1998 bombings of United States embassies in Kenya and Tanzania, al-Qaeda allegedly transferred vast sums of cash into diamonds and other gemstones as their assets were being frozen. Many of the diamonds were reportedly mined by the Revolutionary United Front (RUF) in Sierra Leone, a group whose trademark was to hack off the hands of their victims. And there are indications that smuggled gemstones are still being exchanged for weapons by a variety of terrorist groups.

To prevent the trade of conflict diamonds, the United Nations-backed Kimberley Process (KP) was launched in 2000. It is a voluntary international certification process for rough diamonds meant to ensure that diamonds available for purchase on the open market are "conflict-free." The uncut gems have to be transported to market in tamper-proof containers and diamonds from rebel-held areas are not accepted for trade. As of November 2008, the KP 75 had 49 members representing 75 countries. However, because the process is voluntary there are still some unresolved issues. Plus, terrorists and militant groups are not particularly concerned about the difficulties of the legitimate diamond trade, so significant smuggling still occurs.

Other Important Properties of Gems and Gemstones

A mineral whose color changes with its orientation is said to be *pleochroic* ("many colors") or to exhibit *pleochroism*. For example, when viewing a crystal from one direction it appears to be yellow, but when rotated it appears green. More commonly, the mineral will change from a strong shade to a light shade of the same color. Pleochroism is caused by the differing absorptions of the wavelengths of light that are traveling different directions in a crystal. Sometimes this effect can be quite stunning, with distinct color variation, but more often the effect is subtle and might be impossible to detect with the naked eye. To determine pleochroism, do not use cross-polarized light, just view the crystal at various angles under normal room light, or rotate in plane-polarized light on a microscope.

Isometric minerals cannot be pleochroic, because the light traveling through such crystals is the same in all directions. Tetragonal, trigonal, and hexagonal minerals can be *dichroic* or "two-color," since they have one set of optical properties along the major symmetry axis (c-axis), while the other two/three axes have the same properties. Orthorhombic, monoclinic, and triclinic minerals can have the largest color-change effect, which is *trichroic*, or "three-color." Each of these kinds of crystals has different properties along all three of its axes; therefore, the light behaves differently with different crystal orientation. Strong pleochroism is a rare phenomenon, and can be a diagnostic property.

Luminescence is a general term that refers to the emission of visible light under a given set of circumstances, as a reaction to a particular wavelength of incoming light, as a result of a chemical reaction, or as a response to heat. *Fluorescence* is the emission of visible light under exposure to UV light. Some gems/gemstones react to any UV wavelength, some react only to short-wavelength UV light (254 nm), while others only react to long-wavelength UV light (366 nm). This property can be helpful for identification, but it is not diagnostic, since some specimens of the same gemstone can produce different colors, or may not fluoresce at all. It does help differentiate high-quality stones, which generally will fluoresce, from cheaper stones, which will not. UV light is also commonly used to help detect synthetic stones and alterations to natural stones. If a substance continues to emit light after illumination has ceased, the effect is called *phosphorescence*.

Adularescence is a bluish-white sheen that plays over the surface of cabochon cut stones. *Asterism* is the formation of distinct stars of light with four, six, or, rarely, twelve rays, caused by the reflection of light off of thin, needle-like inclusions. *Chatoyancy*, or iridescent luminosity, is the production of a thin bright silver line similar to that of a cat's pupil, caused by the presence of threadlike parallel inclusions. *Iridescence* is the term for the rainbow-like sheen of colors seen on the surface or inside of some gems. *Labradorescence* is metallic iridescence, commonly a bluish-green, but whole spectrums can sometimes be observed.

Play-of-light is a term of art that refers to the movement of light and rainbows of color that occur in a stone as it is rotated in the light. Some secondary classification terms based on *play-of-color* are: *pinfire* or *pinpoint*, very small, closely spaced, color flashes; *mosaic* or *harlequin*, large (approx. 2 mm or greater), regular, angular, patches of color; *flame*, sweeping reddish bands, like a wind-blown flame; and *flash*, sudden play-of-color as stone is moved.

Scams, Cons, and Thievery

Scams involving gemstone, gems, and jewelry come in many shapes and forms. The most garden-variety frauds involve buying them sight unseen over the Internet from places like eBay. At least one person, a man from Springfield, Missouri, was arrested for this kind of activity. The basic ploy works something like this: stones are offered for sale online, accompanied by pictures of real gemstones. Once purchased, the seller ships fake gems, low-grade gems, or occasionally nothing at all. Often these stones arrive sealed in plastic with fake grading reports from "Gem Information Laboratory, Inc." or "Gemstone Identification Laboratory, Inc." neither of which is a legitimate agency. This should set off alarm-bells for anyone conversant in the gem trade since stones from legitimate dealers would never be sealed in plastic. There are at least 83 victims in various states and countries who attempted to buy the stones from the Springfield, Missouri suspect. No doubt this type of fraud still flourishes.

Identifying Gems and Gemstones

Sixteen minerals and mineral groups are listed in alphabetical order in Tables 6.5 and 6.6, with additional notes here and on the companion website (www.wiley.com/go/bergslien/forensicgeoscience). They are highly prized for their beauty, durability, and rarity. Following that are some common organic gemstone groups that are highly prized.

Beryl ($Be_3Al_2Si_6O_{18}$), beryllium aluminum silicate, is found as an accessory mineral in granite pegmatite and in metamorphic rocks. Nearly all stones are *included*, i.e. they have solid, liquid, gaseous, or multiphase material that was incorporated into them while they were forming. Some highly prized stones have a

Mineral	Gem Variety	Crystallography	Cleavage	Fracture	Luster	Mohs	SG
Beryl ($Be_3Al_2Si_6O_{18}$) Beryllium aluminum silicate	*Emerald; Aquamarine; Morganite (Pink Beryl); Heliodor (Golden Beryl); Yellow beryl; Red beryl (Bixbite); Goshenite (Colorless)*	Hexagonal	Lacks good cleavage, imperfect parallel to basal plane	Conchoidal	Vitreous	7.5–8; fairly brittle	2.67–2.91; Synthetics 2.65
Chrysoberyl ($BeAl_2O_4$) Beryllium aluminum oxide	*Alexandrite; Cat's eye (Cymophane); Chrysoberyl; Golden Chrysoberyl*	Orthorhombic; uncut crystals are commonly found in the shape of cyclic twins	Weak prismatic	Conchoidal	Vitreous	8.5 (third-hardest gemstone)	3.68–3.78
Corundum (Al_2O_3) Aluminum oxide	*Ruby; Sapphire; Padparadscha*	Trigonal (Hexagonal-rhombohedral)	None; basal and rhombohedral parting	Conchoidal to uneven	Adamantine to vitreous	9 (second-hardest gemstone)	3.96–4.05 (most often ~4)
Diamond (C) Carbon	Diamond	Isometric	Perfect octahedral	Conchoidal to splintery	Adamantine, uncut greasy	Tenth-hardest gemstone, much harder than 9	3.50–3.53
Alkali feldspars ($KAlSi_3O_8$)	*Microcline:* ★*Amazonite or Amazonstone*	Triclinic	Perfect, 2 directions at nearly 90°	Irregular to Conchoidal	Vitreous	6–6.5	2.54–2.58
	Orthoclase: ★*Adularia* ★*Moonstone* ★*Noble orthoclase*	Monoclinic	Perfect, two directions, distinct 90°	Irregular to Conchoidal	Vitreous	6–6.5	2.56–2.58
	Sanidine						2.56–2.62
Plagioclase Feldspar Series solid solution series from Albite ($NaAlSi_3O_8$) to Anorthite ($CaAl_2Si_3O_8$)	*Albite: Moonstone;* ★*Peristerite*	Triclinic	Perfect, 2 directions, distinct 90°	Irregular to splintery	Dull to vitreous	6–6.5	2.62–2.76
	Oligoclase: ★*Aventurine, also called* ★*Sunstone*						
	Andesine: ★*Labradorite*						
	Bytownite and Anorthite						

continued

Table 6.5 *continued*

Mineral	Gem Variety	Crystallography	Cleavage	Fracture	Luster	Mohs	SG
Garnet [isolated tetrahedra, general formula $A_3B_2(SiO_4)_3$, where A = Ca, Mg, Fe^{2+}, or Mn and B = Al, Fe^{3+}, or Cr, with lots of solid solution between members]; neosilicate mineral group	*See list of varieties in notes section*	Isometric (commonly in a distinctive well developed dodecahedra)	None	Conchoidal; splintery, brittle	Vitreous to resinous; transparent to opaque	6.5–7.5 (reds 7–7.5; others softer)	3.62–4.32; see note for details
Jadeite ($NaAlSi_2O_6$) Sodium aluminum silicate (pyroxene group)	*Jade*	Monoclinic	prismatic at 87° and 93° (very difficult to break)	Splintery	Vitreous. Pearly on cleavage surfaces	6.5–7	3.3–3.5
Nephrite [$Ca_2(Mg, Fe)_5Si_8O_{22}(OH)_2$] Calcium magnesium silicate (amphibole group)		Monoclinic	prismatic at 56° and 124° (very difficult to break)	Splintery	Vitreous	6–6.5	2.90–3.02
Lapis lazuli (Rock composed mainly of the mineral lazurite with variable amounts of pyrite and white calcite)	**Lapis lazuli** (see pigments Chapter 9 for more about lapis)	Isometric (blue mineral); dense aggregate	Indistinct	Conchoidal	Vitreous	5–5.5	2.4–2.45
Opal ($SiO_2 \cdot n\, H_2O$) Hydrated silica	*Black Opal; Crystal Black Opal; Crystal Opal; White Opal; Boulder Opal; Transparent or Semi-transparent Opal; Common Opal*	None; amorphous	None	Conchoidal; splintery, brittle	Vitreous to waxy	Varies on water content; typ. 5.5–6.5	Depends on water content; typ. ranges from 1.98–2.20
Olivine [(Mg, Fe)$_2$SiO$_4$] Magnesium iron silicate	*Peridot (Chrysolite)*	Orthorhombic	Indistinct, parallel to c-axis	Conchoidal; brittle	Vitreous to greasy	6.5–7	3.28–3.48

Mineral / Composition	Varieties	Crystal system	Cleavage	Fracture	Luster	Hardness	Specific gravity
Quartz (SiO_2) Silicon dioxide or silica; coarsely crystalline varieties	Rock crystal; Amethyst; Citrine; Morion; Smoky quartz or cairngorm; Rose quartz; Green quartz or prasiolite; Chatoyant Quartz (Tiger's-Eye and others)	Hexagonal (Rhombohedral)	Extremely poor	Conchoidal	Glassy to vitreous	7	Typ. 2.6–2.7; can be 2.46–2.71
Quartz (SiO_2) Silicon dioxide or silica; cryptocrystalline varieties of silica	Chalcedony; Bloodstone or heliotrope; Carnelian; Chrysoprase; Sard; Jasper; Agate; Onyx; Sardonyx; Plasma	Microcrystalline hexagonal (Rhombohedral)	None	Conchoidal	Vitreous to waxy	6.5–7	2.59–2.61; agate 2.55–2.64
Spinel ($MgAl_2O_4$) Magnesium aluminum oxide	Balas spinel (Balas ruby); Almandine spinel; Rubicelle or Flame spinel; Sapphire spinel and gahnospinel; Chlorospinel; Ceylonite (Pleonaste)	Isometric (Cubic)	None	Conchoidal	Vitreous	8	3.58–3.61; rare up to 4.40
Topaz [$Al_2(F,OH)_2SiO_4$] Aluminum silicate fluoride hydroxide	Topaz	Orthorhombic	Perfect parallel to c-axis	Conchoidal to uneven	Vitreous	8	Varies with color: 3.49–3.57
Tourmaline [Extremely complex group of hydrous minerals containing Li, Al, B, and Si, plus varying quantities of alkalis (K, Na) and metals (Fe, Mg, Mn)] Complex aluminum borosilicate	Achorite; Brazilian emerald; Dravite; Elbaite; Indicolite; Rubellite; Siberite; Schorl; Uvite; Verdelite; Watermelon	Hexagonal (Trigonal)	None	Conchoidal; sometimes forming globular or spherical lumps free of inclusions	Vitreous	7–7.5	2.9–3.15

continued

Table 6.5 *continued*

Mineral	Gem Variety	Crystallography	Cleavage	Fracture	Luster	Mohs	SG
Turquoise (Hydrous copper aluminum phosphate)	*Turquoise*	Aggregate of crystals; Triclinic (rare)	None	Uneven, conchoidal	Waxy, vitreous if crystalline	5–6	2.6–2.8
Zircon ($ZrSiO_4$) Zirconium silicate	*Jargon; Matura diamond; Starlight*	Tetragonal	1, indistinct	Conchoidal	Vitreous to adamantine	6.5–7.5; gem grade 7–7.5	3.93–4.73; gem grade 4.6–4.73
Organic Gemstones							
Amber (organic)	*Amber*	Amorphous	None	Conchoidal; brittle	Resinous or greasy; polishes to vitreous	2–3	1.05–1.10
Coral (mainly calcite, $CaCo_3$)	*Coral*	Cryptocrystalline Trigonal	None	Irregular	Varies	3–4	2.60 –2.70
Ivory [mainly Ca_5 $(F,OH,Cl)(PO_4)_3$] Organic calcium phosphate	*Ivory*	Cryptocrystalline	None	Fibrous or splintery	Greasy	2–3	1.7–2.0
Jet (organic carbon)	*Jet*	Amorphous	None	Conchoidal	Velvet	2.5–4	1.19–1.35
Pearl (organic)	*Pearl*	Amorphous	None	Uneven	Pearly to dull	2.5–4.5	2.60–2.85

Table 6.6 Optical properties of gems and gemstones.

Mineral	Gem	Optical System	RI	Birefringence	Dispersion	Pleochroism	UV Fluorescence
Beryl	*Emerald*	Anisotropic: Uniaxial (–)	1.57–1.58	Low (0.005–0.009)	Low (0.014)	Weak; green to bluish-green	Emerald – none to weak orangey-red or green; others very weak to none; synthetic emerald may fluoresce a weak dull red and appear opaque under long wavelength UV light
	Aquamarine		1.56–1.59			Weak; blue to darker blue	
	Morganite (Pink beryl)		1.57–1.6			Weak; light red to light violet	
	Heliodor, Golden beryl or Yellow beryl		1.56–1.59			Weak; greenish-yellow to yellow	
	Red beryl (Bixbite)		1.57–1.6			Distinct purplish-red/orangey-red to red	
	Goshenite		1.56–1.59			Transparent so not discernible	
Chrysoberyl	*All varieties (see full list in notes)*	Anisotropic: Biaxial (+)	$\alpha = 1.745$ $\beta = 1.746$ $\gamma = 1.755$	0.010	0.015	Weak in poorly colored stones; strong in deeply colored stones; very strong in alexandrite	Alexandrite: weak red under both short and long wavelength UV; none for the others
Corundum	*All varieties (see full list in notes)*	Anisotropic: Uniaxial (–)	$\omega = 1.769$ $\epsilon = 1.760$ (reported range 1.757–1.778)	Low (0.008–0.009)	Low (0.018)	Strong, with deepest color perpendicular to c-axis (see notes)	Varies with place of origin (Fe content) (see notes)
Diamond (C)	Diamond	Isotropic with *anomalous double refraction*	2.417–one of the highest of any gemstone	None, isotropic	Very high (0.044)	None	Many, but not all, fluoresce

continued

Table 6.6 *continued*

Mineral	Gem	Optical System	RI	Birefringence	Dispersion	Pleochroism	UV Fluorescence
Alkali feldspars	*Microcline:* ★*Amazonite or Amazonstone*	Anisotropic: Biaxial (–)	$\alpha = 1.519$ $\beta = 1.523$ $\gamma = 1.525$	Low (0.008)	None to low (0.012)	None	Weak; olive green or red
							Weak red to moderate blue
	Orthoclase: ★*Adularia* ★*Moonstone* ★*Noble orthoclase* Sanidine	Anisotropic: Biaxial (–)	$\alpha = 1.520$ $\beta = 1.525$ $\gamma = 1.527$	Weak to medium (0.005–0.015)	None to low (0.012)	None	Weak; bluish or reddish, orange
Plagioclase Feldspar Series	*Albite:* Moonstone ★*Peristerite*	Anisotropic: Biaxial (+) or (–)	$\alpha = 1.527–1.577$ $\beta = 1.5231–1.585$ $\gamma = 1.534–1.590$	Low to medium (0.007–0.013)	Very low (0.012)	None	White, green
							Very weak brown
	Oligoclase: – ★*Aventurine, also called* ★*Sunstone*						Pale white, pale brown; dark brownish-red
	Andesine: ★*Labradorite*						Blue or pinkish-orange, yellow striations
	Bytownite						Cream, pale brown
	Anorthite						Cream, pale brown
Garnet Group	*See full list of varieties in notes*	Isotropic, often with anomalous double-refraction	Almandine 1.76–1.82 Andradite 1.88–1.89 Demantoid 1.85–1.89 Grossular 1.73–1.76 Tsavorite 1.74 Hessonite 1.74–1.75 Pyrope 1.71–1.77 Malaia 1.742–1.78 Rhodolite 1.75–1.77 Spessartine 1.79–1.814	None, isotropic	0.024–0.057; varies with composition	None	None, except in some green garnets (grossular and tsavorite), which may show a weak orange in long UV and weak yellow in short UV

Jadeite	*Jade*	Anisotropic Biaxial (+)	$\alpha = 1.65–1.67$ $\beta = 1.66–1.68$ $\gamma = 1.67–1.69$	0.013–0.020	None	None	Weak, greenish
Nephrite		Anisotropic Biaxial (−)	1.60–1.63 average	0.022–0.027	None	None to strong(?)	None
Lapis lazuli	*Lapis lazuli*	Isotropic; sparkles of calcite and pyrite	Near 1.50	None	None	None	Strong; white, orange, copper
Opal	*All varieties (see full list in notes)*	Amorphous	Depends on water content 1.42–1.47	None	None	None	White, bluish, brownish, greenish; can be strong
Olivine	*Peridot (Chrysolite)*	Anisotropic: Biaxial (+) or (−)	1.650–1.703	Moderate 0.0036	Medium 0.02	Distinct; variety of greens	None
Quartz Coarsely crystalline varieties	*All varieties (see full list in notes)*	Anisotropic: Uniaxial (+) *Wavy extinction is common in some crystals*	1.54–1.55; fine-grained varieties as low as 1.53	Very low	Very weak	Usually very weak; Amethyst – blue to violet; Smokey quartz – pale to dark brown/ brown to black	Usually none
Quartz microcrystalline varieties of silica	*All varieties (see full list in notes)*	Amorphous	1.54–1.55; fine-grained varieties as low as 1.53	None	None	None	None to blue white
Spinel	*All varieties (see full list in notes)*	Isotropic	About 1.72 for most natural gem material; red spinel may be as high as 1.74; Zn-rich as high as 1.805.	None, isotropic	Medium 0.02	None	Strongest under long wavelength UV light; weak to absent in short wavelength UV; color varies with body color

continued

Table 6.6 *continued*

Mineral	Gem	Optical System	RI	Birefringence	Dispersion	Pleochroism	UV Fluorescence
Topaz	*Topaz*	Biaxial (+)	Usually 1.61–1.64; reported range is 1.607–1.64; pink, yellow, red, 1.62–1.63; colorless and blue, 1.61–1.62	Low (0.005–0.009)	Low (about 0.014)	Weak to none, except in pink stones; some yellow varieties may show a weak yellow to pink	Weak
Tourmaline	*All varieties (see full list in notes)*	Uniaxial (–)	1.62–1.64; large reported range 1.610–1.675	High; average about 0.018	Medium 0.017	Strong, diagnostic, visible to the naked eye	Weak or none
Turquoise	*Turquoise*	Biaxial (+)	1.61–1.65	0.04	None	Weak to absent	Weak to absent
Zircon	*All varieties (see full list in notes)*	Uniaxial (+)	1.81–2.02	High; 0.04–0.062	Strong; 0.022–0.039	Depends on color, see text	Blue: very weak orange; red/brown: weak yellow
Organic Gemstones							
Amber	*Amber*	Amorphous	1.53–1.55	None	None	None	Bluish-white to yellowish-green
Coral	*Coral*	Microcrystalline	1.48–1.66		None	None	Some weak; violet
Ivory	*Ivory*	Microcrystalline	1.53–1.57		None	None	Various blues
Jet	*Jet*	Amorphous	1.64–1.68	None	None	None	None
Pearl	*Pearl*	Amorphous	1.52–1.69	None	None	None	Usually weak

large number of visually interesting inclusions. Synthetics, however, will not show inclusion. Varieties include *emerald* (intense green or bluish-green from chromium impurity); *aquamarine* (light blue or light greenish-blue from trace-iron impurity); *morganite* (pink, purplish-pink, peach or salmon from manganese impurity); *heliodor* or *golden beryl* (golden yellow to golden green from trace-iron impurity); *yellow beryl* (yellow to yellowish-green from trace-iron impurity); *red beryl* or *bixbite* (raspberry-red from manganese impurity); and *goshenite* (colorless, very clear, sometimes used as a diamond substitute).

Chrysoberyl ($BeAl_2O_4$), beryllium aluminum oxide, is found in granite pegmatite and mica schists. Chrysoberyl is *not* the same as beryl. It is among the world's rarest gems. Varieties include *alexandrite*, which is green, rarely almost emerald colored, more often yellowish or brownish green. The color changes to red when exposed to candlelight or incandescent light and to green in daylight. Extremely rare, it is one of the most expensive gemstones; *cat's eye* or *cymophane* is usually yellowish or greenish chrysoberyl that displays phenomena of chatoyancy or iridescent luminosity, which produces a thin bright silver line similar to that of a cat's pupil, caused by the presence of threadlike parallel inclusions; *chrysoberyl*, transparent yellowish-green to greenish-yellow and pale brown; and *golden chrysoberyl*, yellow to greenish-yellow.

Corundum (Al_2O_3), aluminum oxide, is found in nepheline syenite, nepheline syenite pegmatite, and in metamorphic rocks. Varieties include *ruby*, which is intense red; *sapphire*, a term which technically applies to any non-red corundum, most commonly blue but also pink, yellow, green, purple, orange, or clear; and *padparadscha* ("lotus blossom color"), a very rare pinkish-orange sapphire. Ruby is distinguished from red spinel, garnet, or glass (all isotropic) with cross-polarized light and RI is notably different from other red stones. Sometimes rubies show the property of *asterism*, luminosity in the form of a six-pointed (more rarely twelve-pointed) star due to oriented included rutile needles. Synthetic ruby lacks the natural mineral or fluid inclusions that are distinctive for all sources of natural rubies. See companion website for additional notes on pleochroism and UV fluorescence characteristics of the different varies.

The Millennium Sapphire

The largest carved sapphire in the world, the Millennium Sapphire, weighs 61,500 carats, or 12.3 kg (about 27 lb). It was discovered in central Madagascar in 1995, weighing 89,850 carats (almost 18 kg or 40 lb) in its rough form. A French mining group acquired the stone from its discoverer and shipped it to Thailand for an initial laboratory certification. With the appropriate certifications of authenticity in hand, a publicity campaign was launched and the stone was put up for sale. The sapphire's sale and exportation caused significant political controversy inside Madagascar, eventually leading to the impeachment of the then-president.

In 1998, the raw sapphire was purchased by an Asian consortium that decided against cutting the stone into smaller pieces, choosing instead to have it carved. They commissioned Italian artist Alessio Boschi to oversee the design. Boschi selected 134 representations of historically important individuals and images including Mahatma Gandhi, the Great Wall of China, Gutenberg's Printing Press, Albert Einstein, Martin Luther King, and Alexander the Great. The project took two years.

Diamond (C), carbon, is composed of almost pure carbon that has been compressed under extremely high pressures and at high temperatures in the Earth's upper mantle. Diamonds can be perfectly colorless (rare) to yellow-tinged; rarely shades of red, orange, deep yellow, pink, violet, green, blue, brown, and black. Stones with intense color, called *fancy diamonds*, are very rare. Most have at least a tint of yellow. Habit can range from octahedra and cubes, to elongated or flattened crystals, or flat twinned crystals. High luster, extreme hardness, single refraction, and SG of 3.50–3.53 generally distinguish diamond from all natural and synthetic substitutes. Sharp facet junctions and the absence of scratches can be diagnostic. For additional information, see the companion website.

The Importance of Knowing Your Gems

In April 1999, the FBI arrested Florida jeweler Jack Hasson, who was accused of bilking more than $83 million from a variety of people including Aben Johnson, retired founder of a Detroit TV station, and golfers Jack Nicklaus and Greg Norman. In May 1997, Johnson paid $17 million for a collection of red, blue, green, and yellow diamonds that supposedly belonged to "Wal-Mart Heiress, Sylvia Walton." Part of the collection was the so-called Streeter Diamond, which had allegedly been won in a poker game by Wal-Mart founder Sam Walton. The stone was sold as a 418-carat natural blue diamond. Briefly consulting the Internet, or any book on famous gems, reveals that the largest deep-blue diamond in the world is the 45.52-carat Hope Diamond, currently housed at the Smithsonian Natural History Museum in Washington, DC.

It is extraordinarily unlikely that a blue diamond almost ten times the size of the Hope could be privately sold, with no fanfare, by an unknown jeweler in Florida. The Streeter was actually a 300-carat blue topaz. Johnson paid about $3 million for the topaz, or $10,000 per carat for a stone worth about $10 per carat. The collection that Johnson purchased was formed of low-grade sapphires or blue topazes rather than blue diamonds, and synthetic rubies rather than red diamonds. The green diamonds were tourmalines and the yellow diamonds were citrines (a type of quartz) or colored cubic zirconium. To add insult to injury, Wal-Mart founder Sam Walton does not even have a daughter named Sylvia. Hasson is currently serving a 40-year sentence following his conviction in 2000 on fraud, money-laundering and obstruction of justice charges.

Feldspar (framework silicates, chemistry varies) is the most abundant mineral in the Earth's crust; however, a few forms of feldspar have found favor for use in jewelry and artwork. This list just includes the most common gemstone varieties; see the companion website for more information.

Alkali feldspar varieties include *amazonite* or *amazonstone* ($KAlSi_3O_8$) a bluish-green, or yellowish- green to greenish-blue, form of microcline; *adularia* is clear, colorless orthoclase ($KAlSi_3O_8$) with a bluish-white sheen; *moonstone* is translucent or nearly colorless orthoclase with adularescence, i.e. slightly cloudy with a mobile silver reflection like a small hazy moon. It can also be white to yellowish- or reddish- to bluish- gray; *noble orthoclase* is transparent yellow to golden yellow.

Plagioclase Feldspar Series a solid solution series from albite (100–90% $NaAlSi_3O_8$) to anorthite (100–90% $CaAl_2Si_2O_8$). Gemstone varieties include *albite moonstone*, a colorless, white to yellowish-, or reddish- to bluish-gray albite with adularescence; *peristerite*, an albite with blue white iridescence; *sunstone*, a golden or reddish oligoclase with "spangles" from inclusions of hematite; and *labradorite* (50–30% albite, 50–70% anorthite), a stone with colorful rainbow effects, or labradorescence.

Garnet [isolated tetrahedra, general formula $A_3B_2(SiO_4)_3$, where A = Ca, Mg, Fe^{2+}, or Mn and B = Al, Fe^{3+}, or Cr, with lots of solid solution between members] is actually a neosilicate mineral family that contains several different gemstones with similar crystalline structures and chemical compositions. Garnets are relatively common in metamorphic rock. One of the most distinctive properties of garnets is that they are isotropic, i.e. most will remain black in all orientations under a cross-polarized light, though there are some varieties that will not. Garnets that do not go black under cross-polarized light possess an *anomalous double refraction* (ADR) due to straining of the crystal lattice. There are many varieties, the most important of which are: *almandine* [$Fe_3Al_2(SiO_4)_3$], orangey-red to purplish-red; *andradite* [$Ca_3Fe_2(SiO_4)_3$,], yellowish-green to orangey-yellow to black; *grossular* [$Ca_3Al_2(SiO_4)_3$], brown, green, colorless orange, pink, yellow; *tsavorite* [$Ca_3Al_2(SiO_4)_3$], green to yellowish-green; *pyrope* [$Mg_3Al_2(SiO_4)_3$], colorless if pure (rare), red, reddish-purple, reddish-brown; *melanite* (titanian andradite) [$Ca_3Fe_2(SiO_4)_3$], opaque black; *uvarovite* [$Ca_3Cr_2(SiO_4)_3$], emerald-green; and *color-change garnet*, which goes from bluish-green in daylight to purplish-red in incandescent light; SG 3.62–4.30. Additional varieties are listed on the companion website.

Jade is the gem name of two completely different mineral species: *jadeite* and *nephrite*. Nephrite [$Ca_2(Mg,Fe)_5Si_8O_{22}(OH)_2$], calcium magnesium silicate, is a white, deep-green, or creamy-brown aggregate of fibrous amphibole and is formed in metamorphosed iron and magnesium-rich rocks. It is pretty common and often found in association with serpentine.

Jadeite ($NaAlSi_2O_6$), sodium aluminum silicate, on the other hand, is a member of the pyroxene group that is formed only under special high-temperature and low-pressure conditions. Color ranges from white, leafy-and bluish-green, and emerald green, to lavender, dark bluish-green, and greenish black. Some jadeite displays chatoyancy, yielding fine cat's eyes in rare pieces (Korea, Alaska, and Russia).

It is difficult to impossible to distinguish nephrite jade from jadeite jade by visual inspection. Jadeite jade is quite rare and in its emerald-green, translucent form is referred to as *Imperial jade*, or gem jade. SG determination is the most reliable of the simple ID methods for distinguishing the two. A great variety of materials are offered as imitations, including talc (soapstone), serpentine, amazonite, dyed chalcedony, and others. All can have colors remarkably similar to jade, but physical properties, particularly SG and hardness, can usually be used to distinguish true jade from imitations.

Jaded

The modern jade market has been in a state of unrest for the past couple of decades. Prices for fine jade carvings soared in the late 1990s and early 2000s with some pieces selling for hundreds of thousands of dollars. As a result, the market was almost flooded with doctored stones and modern fakes. Quality ancient jade is very rare, especially since China has long forbidden the export of antiques dating to 1795 or earlier, and they recently pushed the cutoff date up to 1911. Plus, as of January 14, 2009, it is illegal to import Chinese artifacts created before AD 907 into the United States. Jade artifacts are highly coveted and the dearth of legitimate material has created a thriving market in fakes.

Poor-quality jade is often dyed or treated to temporarily enhance its color and therefore its value. Disreputable sellers will also pass off other green minerals such as serpentine or green garnet as jade. Much of the material for sale online is of poor quality, usually not jade and can sometimes even be plastic. Fortunately, most of this material is easily distinguished from real jade. First, jade has a high SG, greater than that of quartz, glass, or soapstone, and should feel quite heavy for its size. Jade is also durable, with a Mohs' hardness of 6–7, so theoretically you should be able to scrape a stainless steel knife blade across the surface of a jade carving without being able to scratch it. Few dealers would be likely to let you perform such a test, however. Real jade will not have bubbles and should feel very smooth and cool to the touch.

A more tricky issue to deal with is that the reproduction of original pieces has become big business in China and their manufacture has become very sophisticated. Obvious pitfalls like drill marks are now avoided with the use of computer-guided-laser carving machines and highly accurate reproductions can be created relatively quickly. Because these items are modern, there are no prohibitions on their export or import. However, unscrupulous dealers can market the items as if they were ancient. Because these items really are made of jade, even experts are starting to have difficulty telling modern reproductions from true Ming dynasty carvings.

Another event has added to the tumult in the market. Jadeite, the rarer mineral form of jade, is only found in a dozen places on Earth, and Myanmar (formerly Burma) had long been the dominant commercial producer of jadeite. This rarity helped keep the production of modern reproductions partly in check. However, in 1998, when Hurricane Mitch hit Guatemala, it caused landslides that uncovered huge new deposits of jadeite. The ancient Olmec people were known to have used jadeite for creating carvings and blades, but when the Spanish wiped out the indigenous cultures of Central America, knowledge of its source was lost. Geologists and archeologists believe that source has now been revealed in a deposit that is described as being roughly the size of Rhode Island, or 2500 square kilometers. The bulk chemistry of the jadeite from Guatemala is the same as the jadeite from Burma, providing a huge potential resource for new carvings. In future, it appears that ever-more sophisticated techniques, such as trace-element analysis, may be needed to tell true jade artifacts from modern reproductions.

Lapis lazuli is not a single mineral but a rock composed mainly of the blue mineral lazurite with variable amounts of pyrite and white calcite. It is deep blue, azure blue, or greenish-blue with flecks of white and gold.

Opal ($SiO_2 \cdot nH_2O$), hydrated silica, has the same chemical composition as quartz but usually has between 4 and 9% (rarely as high as 20%) water incorporated into the structure. Nothing else shows the play-of-color of precious opal. There are a number of varieties, the most important of which are: *black opal* (Australian opal or coal black), which has a black, dark-green, dark-brown or other dark body color, and vivid play-of-color. With flashes and speckles that appear against a black background, it is the most valuable type of opal; *white opal* (Hungarian opal) is translucent to opaque, porcelain-like white material with play-of-color that resembles flashes or speckles; *transparent* or *semi-transparent opal* is transparent opal with slight to no play-of-color and a body color that is yellow, orange, brown, red, or colorless. Colorless varieties are called *water opal, jelly opal, hyalite, contra-luz,* and *hydrophane*. Stones with yellow, orange, and red body color are referred to sometimes as *fire opal* or *cherry opal*; and *crystal opal* is transparent to translucent, having no body color but strong play-of-color.

Olivine [$(Mg, Fe)_2SiO_4$], magnesium iron silicate, has one important gemstone variety called *peridot*, which is yellowish-green to greenish-yellow.

Quartz (SiO_2), silicon dioxide or silica, includes several varieties that see common use as gemstones or gemstone substitutes. Coarsely crystalline varieties of silica include *rock crystal* (colorless), *amethyst* (purple), *citrine* (yellow to amber), *morion* (black), *smoky quartz* or *cairngorm* (smoky-gray to brown), *rose quartz* (translucent pink), *green quartz* or prasiolite (green), and *chatoyant quartz* (tiger's-eye and others), which is fibrous, opaque quartz in which numerous inclusions create a series of cat's eye effects, usually in brown and gold tones, though other colors exist as well.

Common cryptocrystalline varieties of silica include *chalcedony* (generic name by gemologists for cryptocrystalline quartz), *bloodstone* or *heliotrope* (opaque green stone with red spots of color), *carnelian* (translucent yellow, orange, or flesh-colored), *chrysoprase* (green, apple-green, yellow-green), *sard* (reddish-brown to brown), *jasper* (opaque, color varies; all varieties have spots, streaks, or bands of color), *agate* (wavy concentric bands of color), and *onyx* (black and white form of agate, often applied only to black section of stone).

Spinel ($MgAl_2O_4$), magnesium aluminum oxide, is a mineral group with only a few members that are of gemstone quality. The magnesium in the structure can be replaced by iron, zinc, and manganese, while the aluminum can be replaced by iron or chromium, so a range of colors is possible. Spinel is usually formed by regional or contact metamorphism. Some gemstone varieties include *almandine spinel* (purplish-red), *balas spinel* or *balas ruby* (a pale red spinel), *ceylonite* or *pleonaste* (dark green to black, opaque), *chlorospinel* (green), *gahnite* or *zinc spinel* (blue, violet or dark green to blackish); *rubicelle* or *flame spinel* (orange to orangey-red) and *sapphire spinel* or *gahnospinel* (blue to dark blue or green). See the companion website for more information.

Topaz [$Al_2(F,OH)_2SiO_4$], aluminum silicate fluoride hydroxide, ranges in color from wine yellow, to pale blue, green, violet, or red. SG varies with color.

Tourmaline is an extremely complex group of hydrous aluminum borosilicate minerals containing Li, Al, B, and Si, plus varying quantities of alkalis (K, Na) and metals (Fe, Mg, Mn). The tourmaline group probably displays a greater range of colors than any other gemstone group and occurs as well-formed, elongate, trigonal prisms, with smaller, second-order prism faces on the corners. The prism faces are often striated parallel to direction of elongation (c-axis). The rounded triangular cross-sectional shape of tourmaline crystals is diagnostic; no other gem mineral has such a shape. Tourmaline commonly occurs in granite pegmatite or granite that have been infiltrated by boron-rich fluids, and in metamorphic rocks such as schist and marble. A few tourmaline varieties include *achorite* (colorless, very rare), *Brazilian emerald* (green), *dravite* (yellowish-brown to dark brown), *indicolite* (dark blue), *rubellite* (pink to red), *siberite* (lilac to violet-blue), *schorl* (black); *verdelite* (green) and *watermelon* (pink center with outer rim of green: opposite is called *reverse watermelon*). Pleochroism is strong, diagnostic, and visible to the naked eye. See the companion website for more information.

Turquoise (hydrous copper aluminum phosphate) is a nearly opaque sky-blue, bluish-green, greenish-gray, or robin's-egg blue secondary mineral that usually contains some brown, dark-gray, or black veins of other minerals and is formed via chemical weathering by percolation of water through aluminum-rich igneous or sedimentary rock in the presence of copper. Most often, it is found in association with copper deposits in acid or semi-arid environments.

Zircon ($ZrSiO_4$), zirconium silicate, is one of the most widely distributed accessory minerals in felsic igneous rocks. It is also commonly found in metamorphic and sedimentary rocks. Gemstone varieties include *hyacinth* or *jacinth* (reddish-brown, yellow, orange, brown), *jargon* (old term for straw-yellow to almost colorless or gray), *Ceylon* or *Matura diamond* (colorless) and *starlight* (blue variety, usually created by heat-treating other zircons). Other colors are also common. Zircon has marked birefringence (doubling of pavilion facets), and a very high RI (off the scale of a refractometer). See the companion website for more information.

Organic Gemstones

Amber, a mixture of hydrocarbons, is the hardened, fossilized resin of trees that lived millions of years ago. It is amorphous, i.e. lacks crystalline structure, and its chemical composition varies. Amber has a low SG (1.05–1.10) and it will float in a concentrated solution of saltwater. It is also fairly soft (around 2.5), can be polished to a bright vitreous luster, and is used mainly in making beads or other ornaments. It varies from transparent to semi-transparent and generally from light yellow (honey-colored) to dark brown, but can be orange, red, whitish, greenish-brown, blue, or violet. Plus, amber can be dyed in any color.

Jurassic Park Problems

The popularity of the *Jurassic Park* book and series of movies caused an enormous upwelling of interest in amber. So much so that the quantity of fake amber coming onto the market increased significantly, especially material containing insects and other animals. While some of the fakes are quite sophisticated, much of the fake material is actually created out of acrylic or polyester resin. Often, if a piece has a perfectly centered animal, neatly arranged, that is an indication that it is a fake. Remember, amber is fossilized tree resin and the animals trapped in it would usually have been struggling and unhappy in the extreme. The resulting fossils are often contorted and may even show signs of decomposition. Inclusion of modern animals, like honeybees, is a dead giveaway that the piece is a fake.

Some simple field tests for separating the real from the fake include rubbing the sample with acetone (nail polish remover). Real amber should not react, but copal (tree sap that has not yet fossilized) and some modern synthetic fakes will become sticky. Under UV light, true amber should fluoresce pale blue, yellow, green, or orange. If you rub a sample vigorously with soft velvet or a silk cloth, real amber may emit a pine or turpentine scent and will become heavily charged with static electricity, easily able to pick up small pieces of paper. Under the same conditions, copal can actually begin to soften and get sticky. Another simple test is to mix 23 g of table salt with 200 ml of lukewarm water to create a saltwater solution with an SG of around 1.115 g/l. Amber should float, while copal and most synthetics will sink. These tests can help, but they are not definitive. There are also more sophisticated ways of creating fake amber, and differentiating them may require use of a petrographic microscope to measure RI and other optical properties.

Coral ($CaCO_3$) is formed mainly of calcite (with minor amounts of *conchiolin*, a complex protein used as a binder) secreted by coral polyps (tiny marine animals) that extract calcium carbonate from the sea and exude it to build their homes. The coral used as gem material is usually from reef-building species (like *Corallium rubrum*). Such gem coral ranges from semi-translucent to opaque and occurs in a range of hues from white, pink, orange, and red to blue and violet. Also in use as gem materials, black coral and golden corals are predominantly formed of conchiolin, not calcium carbonate. Unworked coral is dull, but it can be polished to a fine vitreous luster (this can occur naturally too, as by the action of waves on a beach). The finest coral is used to make figurines, cameos, carvings, and beads.

Ivory [predominately $Ca_5(F,OH,Cl)(PO_4)_3$], organic calcium phosphate, is mainly obtained from the tusks of elephants, but the term can also refer to the teeth, horns, or tusks of hippopotamus, walrus, narwhal, and wild boar, and even fossilized mammoth. Ivory is usually a creamy white or off-white color (i.e. ivory), though it goes yellowish-brown with age. Ivory is very fine-grained and fairly soft (2–3), so it is often used for fine carvings. The wholesale slaughter of elephants has led to strict laws banning the majority of the ivory trade.

Jet is basically a form of bituminous coal, thus is made predominantly of carbon. It is black ("jet black") in color and can be polished to a fine waxy, velvety shine. In the Middle Ages, polished plates of jet were actually used as mirrors. Jet is often cut into beads, bracelets, and a wide range of decorative and useful objects. It has been used traditionally to make rosaries and mourning jewelry for decades becoming especially popular when Queen Victoria wore jet after the death of Prince Albert in 1861.

Pearl (organic $CaCO_3$) is a form of aragonite. Pearls are typically formed by saltwater mollusks (genus *Pinctada*, called pearl oysters, though they are not true oysters), by some freshwater clams and mussels (*Unio*), and more rarely by other mollusks. Natural pearls form when an irritant, such as a food particle, gets trapped inside the mollusk. The mollusk responds by secreting layers of *nacre* (which is basically aragonite, a mineral form of calcium carbonate with a different crystalline structure than calcite), and conchiolin. The composite material has a high luster and gives pearls their iridescent sheen. Cultured pearls are created by inserting a small bead, usually made of mother-of-pearl, into the mantel of a pearl-producing mollusk. The mollusks are then kept underwater for two or three years and then harvested. Natural pearls come in a wide variety of shapes: round, pear, drop, egg, bean, and others. They also come in various colors, including white, cream, light rose, rose, black, gray, bronze, blue, dark blue, blue green, red, purple, yellow, and violet. Pearls are rather soft (2.5–4.5) but due to their compact, layered structure they tend to be fairly resilient.

Summary

This chapter introduced you to many important optical properties used for the characterization of minerals and to the use of the petrographic microscope, one of the most important tools in mineralogy. While these concepts were introduced in the context of gem and gemstone identification, it is essential to realize that they apply to all minerals. Learning how to interpret optical information on large samples like gemstones makes it much easier to use these same techniques on smaller mineral samples, such as sands and soils.

Further Reading

Cipriani, C. and Borelli, A. (1986) *Simon and Schuster's Guide to Gems and Precious Stones*. Simon and Schuster, New York.

Grimaldi, D. A. (1996) *Amber: Window to the past*. Harry N. Abrams, Inc., New York.

Nesse, W. D. (2003) *Introduction to Optical Mineralogy*, 3rd edn. Oxford University Press, New York.

Oldershaw, C. (2003) *Firefly Guide to Gems*. Firefly Books Ltd, Buffalo, NY.

Perkins, D. and Henke, K. R. (2004) *Minerals in Thin Section*, 2nd edn. Pearson-Prentice Hall, Upper Saddle River, NJ.

Schumann, W. (2006) *Gemstones of the World*, 3rd edn. Sterling Publishing, New York.

References

Association of Official Analytical Chemists (AOAC) (1990) Method 973.65, characterization and matching of glass fragments: Dispersion microscopy (double variation method). In: *Official Methods of Analysis of the Association of Official Analytical Chemists*: Vol. 1, 15th edn. AOAC, Gaithersburg, MD.

ASTM Standard C162-05 (2010) Standard Terminology of Glass and Glass Products, ASTM International, West Conshohocken, PA, DOI: 10.1520/C0162-05R10, http://www.astm.org/Standards/C162.htm.

ASTM Standard C1036-11 (2011) Standard Specification for Flat Glass, ASTM International, West Conshohocken, PA, DOI: 10.1520/C1036-11, http://www.astm.org/Standards/C1036.htm.

Bottrell, M. C. (2009) Forensic glass comparison: Background information used in data interpretation. *Forensic Science Communications* **11**(2), available online at http://www.fbi.gov/hq/lab/fsc.

Delly, J. G. (2003) The Michel-Lévy interference color chart: Microscopy's magical color key. Modern Microscopy, http://www.modernmicroscopy.com/main.asp?article=15, accessed 25.11.11.

Hurlbut, C. S. and Switzer, G. S. (1979) *Gemology*. John Wiley & Sons Ltd, Chichester.

Koons R. D., Buscaglia, J., Bottrell, M., and Miller E. T. (2002) Forensic glass comparisons. In: R. Saferstein (ed.), *Forensic Science Handbook: Volume 1*, 2nd edn. Pearson Education, Upper Saddle River, NJ, pp. 162–213.

Locke, J. and Unikowski, J. (1991) Breaking of flat glass: Part 1: Size and distribution of particles from plain glass windows. *Forensic Science International* **51**: 251–62.

Nesse, W. D. (1991) *Introduction to Optical Mineralogy*, 2nd edn; Figure 4.3. Oxford University Press, New York.

Underhill, M. (1980) Multiple RI in float glass. *Journal of the Forensic Science Society* **20**: 169–76.

Winchell, A. N. (1949) *Elements of Optical Mineralogy. Part I: Principles and methods*. John Wiley & Sons Ltd, Chichester.

Zadora, G. (2009) Evaluation of evidence value of glass fragments by likelihood ratio and Bayesian Network approaches. *Analytica Chimica Acta* **642**: 279–90.

Chapter 7
Soil: Getting the Dirt on Crime

Soil is almost certainly the most common kind of geologic trace evidence. Soil is ubiquitous and we are in almost constant contact with it during all of our daily activities. Everywhere we go, small amounts of dirt are transferred to our shoes, our clothing, even into our lungs. By examining the traces of soil found on pants, in tire treads, or stuck to a shovel, it is possible to establish relationships between suspects, victims, and crime scenes.

Soil is also the most complex material discussed thus far. While the primary constituent is geologic (mineral and rock fragments), soil also contains organic debris, organisms, plant matter, and anthropogenic materials. In addition, the components of soil are usually very, very small. The approaches discussed thus far are no longer sufficient for good characterization and some new tools must be brought to bear.

A Dirty Deed

Doug Kyle, a Texas law enforcement officer, was on vacation hunting in the Uncompahgre Mountains of western Colorado. He was surprised when a couple set up camp just a few yards away. After all, there are millions of acres of mountain range, so it is considered pretty rude to camp next to other hunters. On October 15, 1995, Doug heard several gunshots and about 30 minutes later saw Janice Dodson walk back to her campsite. Shortly afterward, Janice talked to Doug. She said that her husband, John Bruce Dodson, had not returned on time and that she was going to go look for him. During the conversation, Janice denied hearing shots and Doug noticed that she had changed her clothing.

A few minutes later, Doug heard Janice calling for help. He found Janice standing along a fence line in a grassy field, next to the fallen

An Introduction to Forensic Geoscience, First Edition. Elisa Bergslien.
© 2012 Elisa Bergslien. Published 2012 by Blackwell Publishing Ltd.

body of her husband. Janice was hysterical, waving John's orange hunting vest in the air. Doug determined that John was dead and called 911 for help. It turned out that Janice and John were newlyweds, married three months to the day, and it looked like John was the unfortunate casualty of a hunting season accident.

So things would have remained, except the autopsy revealed John had not been hit by a single stray bullet. He had been shot three times, turning what appeared to be an accident into murder. Investigators analyzed the bullet wounds to determine the paths that the bullets took. Based on their reconstruction, it appeared that John had been walking along the fence line when the first shot ripped through his orange vest, grazing his skin. John then seems to have taken his vest off, waving it over his head and shouting to let any nearby hunters know that he was not a deer. The second shot hit John in the chest, exiting under his arm. John fell to the ground, and a final shot, which struck a fencepost and broke, hit John in the back, severing his spine.

Since the third shot left a hole in the fencepost, investigators used a string running from the location of the body toward the hole to determine the trajectory of the bullets and find the spot where they had been fired, which was located approximately 55 meters away. There they found a .308-caliber shell case and a single shoe print in the mud. Neither Janice nor John owned a .308 and there were no obvious reasons for the murder, so the police started looking for suspects. In the course of the investigation, the police found that at the time of the shooting J. C. Lee, Janice's ex-husband, was camping about 1.3 km away, straight down the fence line.

Even more interestingly, Mr Lee reported that someone had stolen a .308 rifle and a box of .308 cartridges from his tent the day before the murder. Mr Lee came in for quite of bit of police scrutiny, but it turned out that he had an alibi. He had been out hunting with his boss, well away from his camp, at the time of the shooting. Despite intensive searching, investigators were not able to find the rifle, and winter soon closed quickly in, making the crime scene inaccessible.

Investigators Bill Booth, Dave Martinez, and Wayne Bryant returned to the scene during the summers of 1996, 1997, and 1998 to continue their investigation. They were especially interested in locating the missing rifle. During one visit, investigators ran a line of fluorescent string from the suspected location of the shooter, through the approximate area where the first shot penetrated John Bruce Dodson's hunting vest, and out into the surrounding terrain. Based on that geometry, they marked off a grid and performed a search using a metal detector. Under a thick pile of brush, the investigators located a .308 caliber bullet characteristic of the rifle that Janice's ex-husband reported stolen and which corresponded to the casing found earlier.

Despite the probable use of his rifle, J. C. Lee seemed to be in the clear. However, in the time since John's murder, Janice's behavior started to attract attention. In the three months following their marriage, Janice took out three insurance policies in which she had been named the primary beneficiary, and made sure that wills had been written and funerals plans laid out. She also ensured that all of John's property was put into both their names and apparently persuaded him to put money into her ex-husband's bank account, purportedly to benefit her children. Due to this foresight, Janice was in line to receive more than $450,000 in death benefits and assets.

Shortly after John Bruce's death, Janice closed out his bank accounts, cashed out his IRAs, and sold his home, his car, and even his horse, Glory. She also started pestering the life insurance companies to pay out. Further, investigators found out that Janice had visited the Uncompahgre Mountains, without her husband, a few weeks before the shooting. But, despite this rather damning list of activities, there was no physical evidence linking Janice to the crime.

Then, after three years of work, during another search of a pond located near J. C. Lee's former campsite, Al Bieber of NecroSearch International, a non-profit organization that specializes in locating clandestine gravesites (www.necrosearch.org), commented that the mud in and around the cattle pond was bentonite. *Bentonite* is a generic term used to describe several types of clay, formed by the weathering of volcanic ash, that consist dominantly of smectite, a clay mineral that swells in the presence of water. Bentonite has some interesting properties that make it very useful in a variety of industrial and manufacturing processes, and large quantities are not native to the Uncompahgre Mountains. The bentonite near Lee's camp had been intentionally brought in by a rancher to stop water from leaking out of the bottom of the artificial watering pond. With a little research, investigators found out that this was probably the only place in the area with bentonite.

On the morning of the shooting, Doug had noticed that Janice had changed her clothing. In the Dodsons' tent, police had found Janice's hunting coveralls, which were covered in mud from the knees down, and taken them as evidence. At the time, Janice told investigators that she had stepped into a mud bog along the fence near her camp. Investigators Booth and Martinez now realized that the dried mud on Janice's coveralls might turn out to be very important. They collected samples from the mud bog near the Dodsons' former campsite, from around a pond near the camp, and from the pond near Lee's former campsite.

The samples, along with the mud recovered from Janice's coveralls, were sent to the Colorado Bureau of Investigation in Denver, Colorado. There, forensic scientist Jacqueline Battles examined the evidence. She concluded that the mud on Janice's clothing was consistent with the mud from the pond near her ex-husband's campsite and was not consistent with the soils of the mud bog or the pond near the Dodsons' camp. Investigators also confirmed that there were no other bentonite-lined ponds in the region, and no natural bentonite deposits in the area. They believed that the only reasonable explanation for the bentonite on Janice's clothing was that she had walked to her ex-husband's campsite, through the bentonite-augmented mud, to get the .308 rifle. In addition, the shoe print discovered earlier at the shooter's suspected location was consistent with Janice's boots.

Three years after the shooting, Booth and Martinez went to Nacogdoches, Texas and served an arrest warrant on Janice. She was extradited to Colorado and the case went to trial in Mesa County in 2000 (*People v. Hall*, 98CR1274). There, she was convicted of first-degree murder and sentenced to life without parole. The Colorado Court of Appeals affirmed the conviction in 2002 (Court of Appeals No. 00CA0845). To this day, the murder weapon has not been found and Janice Hall (formerly Dodson) still insists that she is innocent. A more thorough discussion of this case can be found in the book *Dead Center: The Shocking True Story of a Murder on Snipe Mountain* by Frank J. Daniels (2006), the district attorney on the original case.

Introduction to Soils

While there are a range of technical definitions for it, most simply put *soil* is the unconsolidated material on the Earth's surface that is capable of supporting vegetation. Unconsolidated material that does not support vegetation is simply referred to as *sediment*. These definitions work reasonably well under most circumstances, unless engineering or architectural reports are being used. For the purposes of forensic work, the distinction between soil and sediment is relatively unimportant and the same general procedures would be followed for both. Perhaps unexpectedly, artificial mixtures like potting soil, compost, and even kitty litter are considered soil for forensic purposes.

Most of the Earth's land surface is covered by *regolith*, layers of rock and mineral fragments produced by physical and chemical weathering. Layers of regolith accumulate through time and are altered through chemical and biological activity. Given enough time and the right conditions, the layers become soil. A true soil is a complex combination of regolith, organic material, water, air, and organisms. The relative proportions of each of these components and the chemical composition of the soil determine its ability to support different types of vegetation and give it specific engineering properties.

In general, soil is composed roughly of 50% solid material (rock and mineral fragments, and organic material) and 50% pore space, with air and water filling the voids. The void space in a soil is vital. Without the free movement of air and water, a soil would be incapable of supporting life. Given the importance of the biological aspects, soil can also be considered an ecosystem. Fertile soil contains a wide range of organisms, such as algae, bacteria (nitrogen fixing and others), fungi (decomposers), worms and nematodes (soil processors), plus a wide variety of bugs, slugs, and other microscopic organisms. These complex ecosystems are fairly poorly understood and thus far have not been artificially duplicated, which clearly differentiates natural soils from the mixtures used in greenhouses and well-cultivated gardens.

Natural soils show a huge spatial variation. The composition and texture of a soil is dependent on the regolith parent material, which is usually the local bedrock but could also be glacially deposited or other non-native material, the local climate, and the regional topography. The interplay of these factors through time plus the effects of human activities mean that even over small distances there can be large variations in soil properties. High-quality, fertile soil accumulates at a rate of approximately 1 mm per year under very good conditions, but can take thousands of years to develop under poor conditions. Young or poor-quality soils will be little more than accumulations of sediment with a few organic nutrients. Mature fertile soils will contain a number of characteristic layers, called *soil horizons*, that have distinctive properties (Figure 7.1).

Soil Horizons

Starting at the surface, the *O-horizon* consists largely of organic material, such as plant litter, fecal matter, and other organic debris, with little to no mineral component. The upper portion is mostly un-decomposed plant litter, while the lower portion consists dominantly of *humus*, which is partly to fully decomposed organic material. Because of the high concentration of organics, this horizon is typically very dark to nearly black in color and usually has a slightly slimy or greasy feel. Living organisms and significant amount of biological activity are concentrated

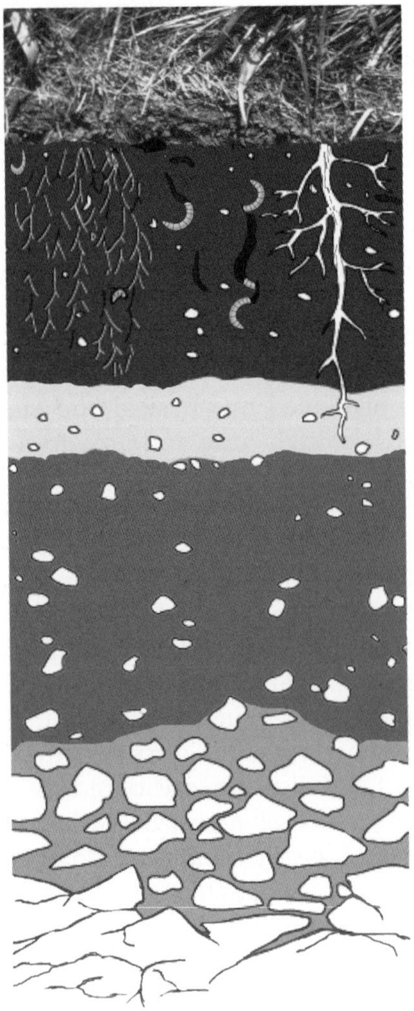

Figure 7.1 Idealized soil profile showing each of the horizons.

in this layer. To qualify as an O-horizon, the layer should contain at least 20% organic matter excluding roots and contain less than 50% regolith. Because O-horizons form from the accumulation of organic debris, they tend to be thickest in cool, dry climates, while in areas with low production they may be missing entirely.

Next is the *A-horizon*, a dark layer where organic matter from the O-horizon is intermixed with >50% mineral grains. This horizon still has high biological activity and significant amounts of humus, ranging up to 30% in highly fertile soils. The O- and A-horizons together are considered the *topsoil*. In cultivated soils, the O-horizon is usually plowed into the A-horizon, which is then designated *Ap* (the lower-case p stands for "plowed").

Beneath the topsoil comes the *E-horizon*, a light-colored mineral layer that contains very little to no organic material. Not all soils have this layer and it is often left out of simplified soil profile models. As water percolates through the E-horizon, it removes soluble minerals and nutrients via *leaching*, and carries away finer silt and clay particles, a process called *eluviation* (thus the E designation). The

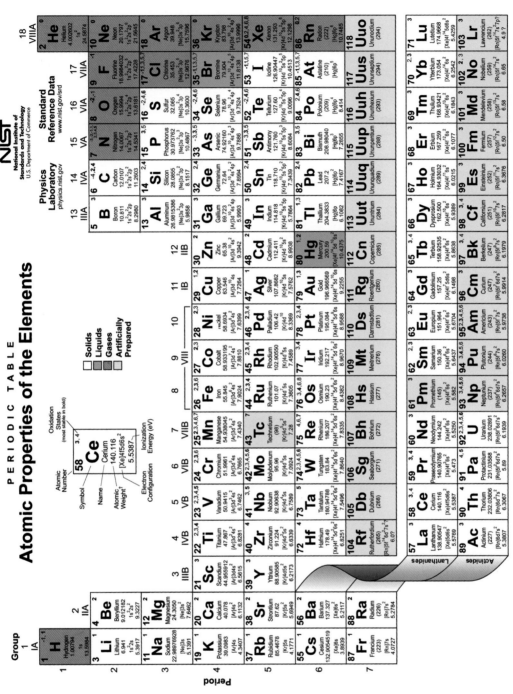

Plate 2.3 Periodic table of the elements.

Source: Modified from Dragoset *et al.*, (2010), Periodic Table: Atomic Properties of the Elements. National Institute of Standards and Technology.

Plate 2.21 Examples of mineral luster, from left to right: metallic luster (two pieces of galena and one cube of pyrite), resinous luster (middle rear copal, middle front orpiment), and vitreous (two pieces of quartz).

Plate 2.22 Examples of the range of possible colors of the mineral fluorite.

(a)   (b)   (c)   (d)   (e)   (f)

(g)   (h)   (i)   (j)

Plate 2.23 Some common crystal habits: (a) acicular, (b) bladed, (c) cubic, (d) dodecahedral, (e) octahedral, (f) tabular, (g) dendritic, (h) fibrous, (i) globular, (j) rosette.

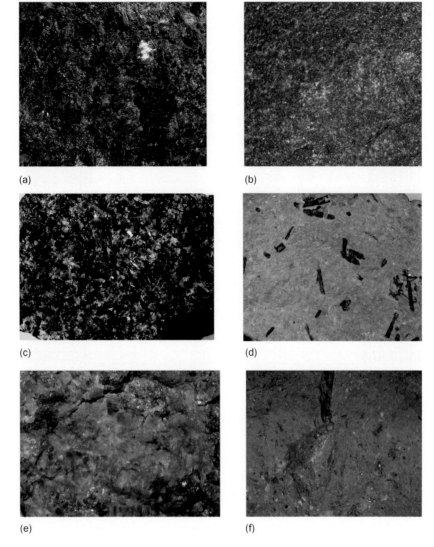

Plate 3.7 Common igneous rocks: (a) gabbro, (b) basalt, (c) diorite, (d) andesite, (e) granite, (f) rhyolite.

Plate 3.9 Examples of common sedimentary rocks: clastic rock (a) conglomerate, (b) arkose, (c) quartz sandstone, (d) shale; chemical sedimentary rock, (e) chert; biological sedimentary rock, (f) coquina.

(a)   (b)   (c)

Plate 3.11 (a) Well-sorted sand versus (b) poorly sorted sand versus (c) an extremely poorly sorted till.

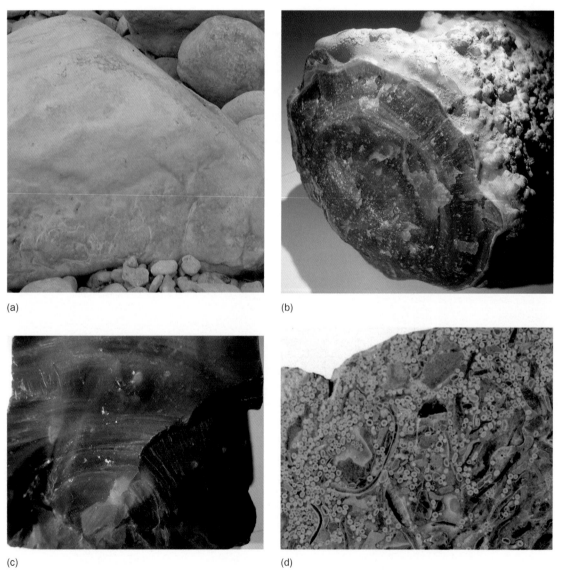

(a)   (b)

(c)   (d)

Plate 3.15 Chemical sedimentary rocks: (a) limestone, (b) travertine limestone, (c) chert, (d) oolitic limestone.

(a)   (b)   (c)

(d)

Plate 3.16 Biological sedimentary rocks: (a) fossiliferous limestone, (b) coquina, (c) coal, (d) two different examples of amber.

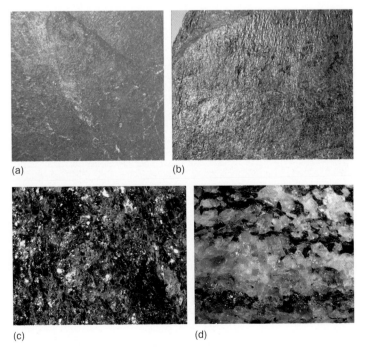

(a)   (b)

(c)   (d)

Plate 3.19 Foliated metamorphic textures: slaty, phyllitic, schistose, gneissic.

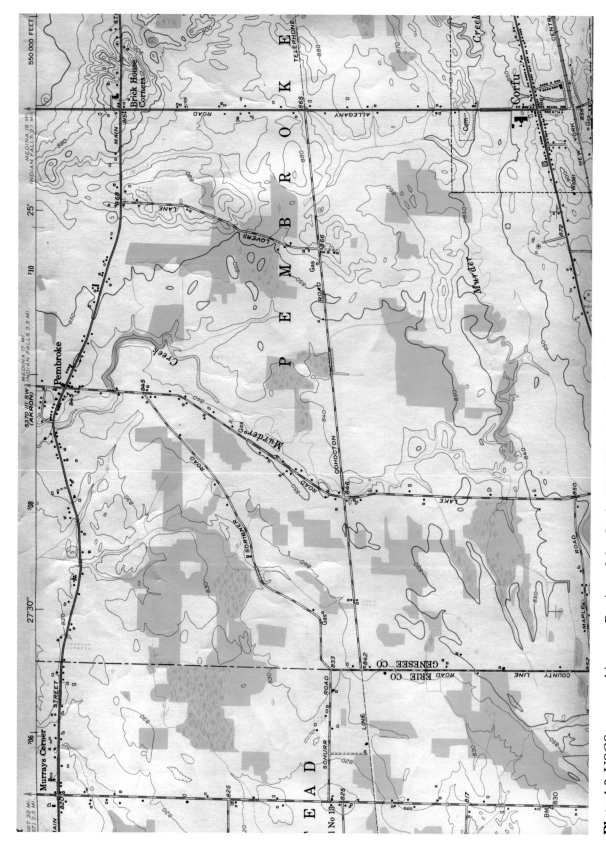

Plate 4.9 USGS topographic map. Portion of the Corfu, NY NW/4 Attica 15′ Quadrangle 1950 edition. *Source:* Courtesy of the United States Geological Survey.

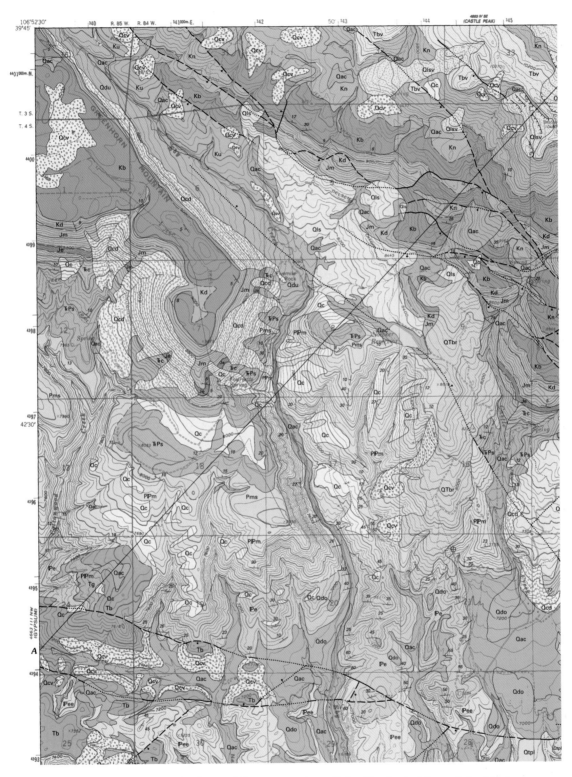

Plate 4.11 Example of a geologic map from a section of the Geologic Map of the Eagle Quadrangle, Eagle County, Colorado by David J. Lidke, 2002.
Source: Courtesy of the United States Geological Survey.

CONTROL DATA AND MONUMENTS

Aerial photograph roll and frame number*

Horizontal control

Third order or better, permanent mark

With third order or better elevation

Checked spot elevation

Coincident with section corner

Unmonumented*

Vertical control

Third order or better, with tablet

Third order or better, recoverable mark

Bench mark at found section corner

Spot elevation

Boundary monument

With tablet

Without tablet

With number and elevation

U.S. mineral or location monument

CONTOURS

Topographic

Intermediate

Index

Supplementary

Depression

Cut; fill

Bathymetric

Intermediate

Index

Primary

Index Primary

Supplementary

BOUNDARIES

National

State or territorial

County or equivalent

Civil township or equivalent

Incorporated city or equivalent

Park, reservation, or monument

Small park

*Provisional Edition maps only
Provisional Edition maps were established to expedite completion of the remaining large scale topographic quadrangles of the conterminous United States. They contain essentially the same level of information as the standard series maps. This series can be easily recognized by the title "Provisional Edition" in the lower right hand corner.

LAND SURVEY SYSTEMS

U.S. Public Land Survey System

Township or range line

Location doubtful

Section line

Location doubtful

Found section corner; found closing corner

Witness corner; meander corner

Other land surveys

Township or range line

Section line

Land grant or mining claim; monument

Fence line

SURFACE FEATURES

Levee

Sand or mud area, dunes, or shifting sand

Intricate surface area

Gravel beach or glacial moraine

Tailings pond

MINES AND CAVES

Quarry or open pit mine

Gravel, sand, clay, or borrow pit

Mine tunnel or cave entrance

Prospect; mine shaft

Mine dump

Tailings

VEGETATION

Woods

Scrub

Orchard

Vineyard

Mangrove

GLACIERS AND PERMANENT SNOWFIELDS

Contours and limits

Form lines

MARINE SHORELINE

Topographic maps

Approximate mean high water

Indefinite or unsurveyed

Topographic-bathymetric maps

Mean high water

Apparent (edge of vegetation)

COASTAL FEATURES

Foreshore flat

Rock or coral reef

Rock bare or awash

Group of rocks bare or awash

Exposed wreck

Depth curve; sounding

Breakwater, pier, jetty, or wharf

Seawall

BATHYMETRIC FEATURES

Area exposed at mean low tide; sounding datum

Channel

Offshore oil or gas; well; platform

Sunken rock

RIVERS, LAKES, AND CANALS

Intermittent stream

Intermittent river

Disappearing stream

Perennial stream

Perennial river

Small falls; small rapids

Large falls; large rapids

Masonry dam

Dam with lock

Dam carrying road

Perennial lake; intermittent lake or pond

Dry lake

Narrow wash

Wide wash

Canal, flume, or aqueduct with lock

Elevated aqueduct, flume, or conduit

Aqueduct tunnel

Well or spring; spring or seep

SUBMERGED AREAS AND BOGS

Marsh or swamp

Submerged marsh or swamp

Wooded marsh or swamp

Submerged wooded marsh or swamp

Rice field

Land subject to inundation

BUILDINGS AND RELATED FEATURES

Building

School; church

Built-up Area

Racetrack

Airport

Landing strip

Well (other than water); windmill

Tanks

Covered reservoir

Gaging station

Landmark object (feature as labeled)

Campground; picnic area

Cemetery: small; large

ROADS AND RELATED FEATURES

Roads on Provisional edition maps are not classified as primary, secondary, or light duty. They are all symbolized as light duty roads.

Primary highway

Secondary highway

Light duty road

Unimproved road

Trail

Dual highway

Dual highway with median strip

Road under construction

Underpass; overpass

Bridge

Drawbridge

Tunnel

RAILROADS AND RELATED FEATURES

Standard gauge single track; station

Standard gauge multiple track

Abandoned

Under construction

Narrow gauge single track

Narrow gauge multiple track

Railroad in street

Juxtaposition

Roundhouse and turntable

TRANSMISSION LINES AND PIPELINES

Power transmission line: pole; tower

Telephone line

Aboveground oil or gas pipeline

Underground oil or gas pipeline

Plate 4.15 Symbols used on topographic maps produced by the USGS. Variations will be found on older maps.

Source: Courtesy of the United States Geological Survey.

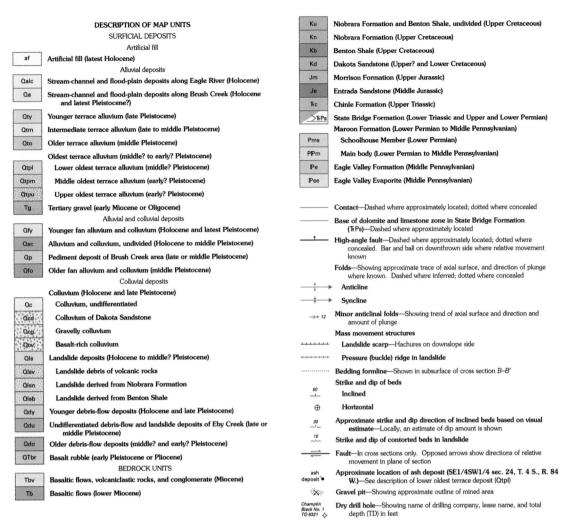

DESCRIPTION OF MAP UNITS
SURFICIAL DEPOSITS
Artificial fill

| af | Artificial fill (latest Holocene) |

Alluvial deposits

Qalc	Stream-channel and flood-plain deposits along Eagle River (Holocene)
Qa	Stream-channel and flood-plain deposits along Brush Creek (Holocene and latest Pleistocene?)
Qty	Younger terrace alluvium (late Pleistocene)
Qtm	Intermediate terrace alluvium (late to middle Pleistocene)
Qto	Older terrace alluvium (middle Pleistocene)

Oldest terrace alluvium (middle? to early? Pleistocene)

Qtpl	Lower oldest terrace alluvium (middle? Pleistocene)
Qtpm	Middle oldest terrace alluvium (early? Pleistocene)
Qtpu	Upper oldest terrace alluvium (early? Pleistocene)
Tg	Tertiary gravel (early Miocene or Oligocene)

Alluvial and colluvial deposits

Qfy	Younger fan alluvium and colluvium (Holocene and latest Pleistocene)
Qac	Alluvium and colluvium, undivided (Holocene to middle Pleistocene)
Qp	Pediment deposit of Brush Creek area (late or middle Pleistocene)
Qfo	Older fan alluvium and colluvium (middle Pleistocene)

Colluvial deposits
Colluvium (Holocene and late Pleistocene)

Qc	Colluvium, undifferentiated
Qcd	Colluvium of Dakota Sandstone
Qcg	Gravelly colluvium
Qcv	Basalt-rich colluvium
Qls	Landslide deposits (Holocene to middle? Pleistocene)
Qlsv	Landslide debris of volcanic rocks
Qlsn	Landslide derived from Niobrara Formation
Qlsb	Landslide derived from Benton Shale
Qdy	Younger debris-flow deposits (Holocene and late Pleistocene)
Qdu	Undifferentiated debris-flow and landslide deposits of Eby Creek (late or middle Pleistocene)
Qdo	Older debris-flow deposits (middle? and early? Pleistocene)
QTbr	Basalt rubble (early Pleistocene or Pliocene)

BEDROCK UNITS

| Tbv | Basaltic flows, volcaniclastic rocks, and conglomerate (Miocene) |
| Tb | Basaltic flows (lower Miocene) |

Ku	Niobrara Formation and Benton Shale, undivided (Upper Cretaceous)
Kn	Niobrara Formation (Upper Cretaceous)
Kb	Benton Shale (Upper Cretaceous)
Kd	Dakota Sandstone (Upper? and Lower Cretaceous)
Jm	Morrison Formation (Upper Jurassic)
Je	Entrada Sandstone (Middle Jurassic)
Ʀc	Chinle Formation (Upper Triassic)
ƦPs	State Bridge Formation (Lower Triassic and Upper and Lower Permian)

Maroon Formation (Lower Permian to Middle Pennsylvanian)

Pms	Schoolhouse Member (Lower Permian)
PPm	Main body (Lower Permian to Middle Pennsylvanian)
Pe	Eagle Valley Formation (Middle Pennsylvanian)
Pee	Eagle Valley Evaporite (Middle Pennsylvanian)

———— Contact—Dashed where approximately located; dotted where concealed

———— Base of dolomite and limestone zone in State Bridge Formation (ƦPs)—Dashed where approximately located

——ᵀ—— High-angle fault—Dashed where approximately located; dotted where concealed. Bar and ball on downthrown side where relative movement known

Folds—Showing approximate trace of axial surface, and direction of plunge where known. Dashed where inferred; dotted where concealed

——→ Anticline

——→ Syncline

→→ 12 Minor anticlinal folds—Showing trend of axial surface and direction and amount of plunge

Mass movement structures

—┴┴┴┴┴— Landslide scarp—Hachures on downslope side

—+++++— Pressure (buckle) ridge in landslide

·········· Bedding formline—Shown in subsurface of cross section B–B′

Strike and dip of beds

60
⊥ Inclined

⊕ Horizontal

30
⊥ Approximate strike and dip direction of inclined beds based on visual estimate—Locally, an estimate of dip amount is shown

15
∿ Strike and dip of contorted beds in landslide

══ Fault—In cross sections only. Opposed arrows show directions of relative movement in plane of section

ash deposit ● Approximate location of ash deposit (SE1/4SW1/4 sec. 24, T. 4 S., R. 84 W.)—See description of lower oldest terrace deposit (Qtpl)

⋉ Gravel pit—Showing approximate outline of mined area

Champlin Black No. 1 TD 6321 ◇ Dry drill hole—Showing name of drilling company, lease name, and total depth (TD) in feet

Plate 4.16 Simplified key for Geologic Map of the Eagle Quadrangle, Eagle County, Colorado by David J. Lidke, 2002.
Source: Courtesy of the United States Geological Survey.

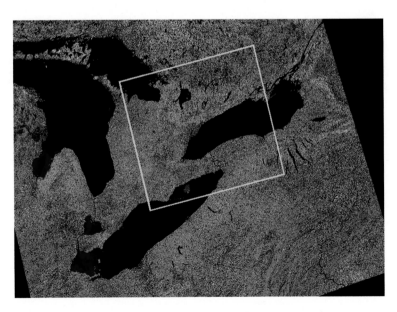

Plate 4.17 Landsat image of the Great Lakes Region of North America ID:LE70170302010125EDC00, taken 5.5.2010.
Source: Courtesy of the United States Geological Survey's Global Visualization Viewer.

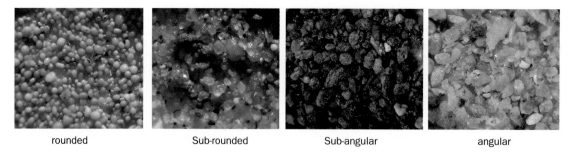

| rounded | Sub-rounded | Sub-angular | angular |

Plate 5.7 Photograph of rounded, sub-rounded, sub-angular, and angular sand grains.

(a) (b) (c)

(d) (e) (f)

Plate 5.11 Images of different types of sediments. The black bar in each image represents 1 mm: (a) river sand, (b) inner shelf sand, (c) beach sand, (d) beach gravel, (e) river gravel with cobbles, (f) beach gravel and cobbles.
Source: Buscombe, Rubin and Warrwick, 2010. Courtesy of the United States Geological Survey.

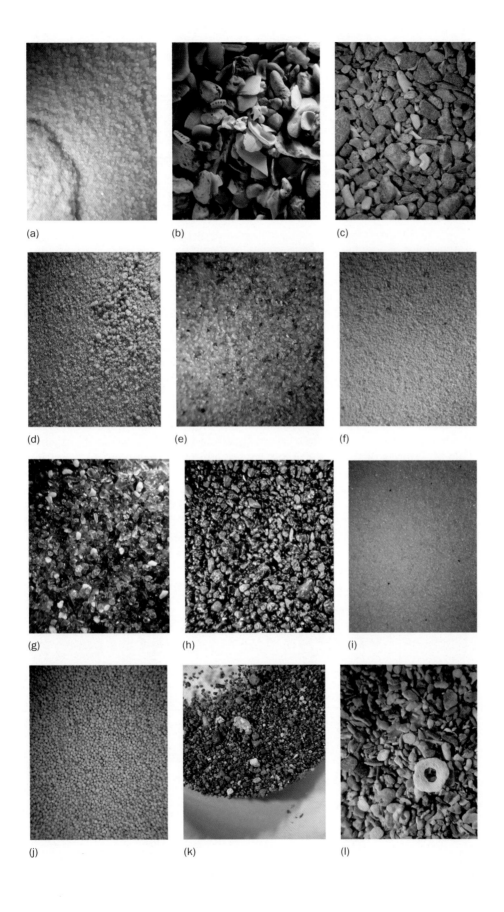

(a)

(b)

(c)

(d)

(e)

(f)

(g)

(h)

(i)

(j)

(k)

(l)

Plate 5.14 Examples of some of the wide variety of distinctive sand types from around the world: (a) White Sands, New Mexico desert sand composed of small soft grains of gypsum; (b) Texas City, Texas beach sand composed of coarse rock fragments and shell pieces with almost no fines; (c) Smith Island, Baja California, Mexico, beach sand composed of shiny metamorphic rock fragments with some organic material; (d) Oak Creek, Nevada desert sand composed of sub-angular grains of quartz with some feldspar; (e) Indiana Dunes, Indiana, dune sand composed of pitted, frosted, rounded quartz grains that will make a squeaking sound when shaken; (f) Half Moon Cay, Bahamas, white oolitic beach sand with small, rounded grains of pink coral; (g) Big Island, Hawaii, green beach sand formed predominantly of olivine with some obsidian and shell fragments; (h) Oahu, Hawaii, black basalt beach sand; (i) Fort Pierce, Florida, mature beach sand composed almost solely of clear, well-polished quartz grains; (j) Antelope Island, Great Salt Lake, Utah, oolitic lake sand; (k) Lake Bratan, Bali, Indonesia, lake sand from a lake that fills a volcanic crater; (l) Vanua Levu, Fiji, carbonate beach sand composed almost entirely of shells, sea urchin spines, and other organics; (m) Coral Pink Sand Dunes State Park, Utah, frosted dune sand; (n) Pismo Beach, California, beach sand; (o) Third Beach, Vancouver, Canada, beach sand; (p) Hoshizuna-no-hama, Iriomote, Okinawa, Japan, star sand; (q) Kalalau, Kauai Island, Hawaii, beach sand; (r) Perissa, Santorini, Greece, black volcanic sand.

Source: Photograph m. and n. by Mark A. Wilson; o. by Bobanny; p. by Geomr; q. by Psammophile; r. by Stan Zurek.

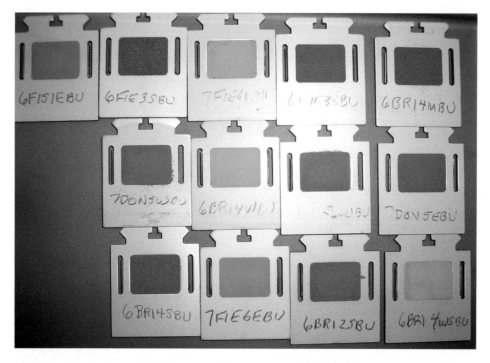

Plate 7.13 A very small example of some of the common soil colors.

Plate 7.15 (b) Soil sample divided into sieve fractions and pan fraction (top right), along with some prepared sample slides.
Source: Photograph courtesy of William Schneck.

Plate 8.3 Pigments for sale on a market stall, Goa, India. Photograph by Dan Brady.
Source: Used courtesy of the Creative Commons Attribution 2.0 Generic license.

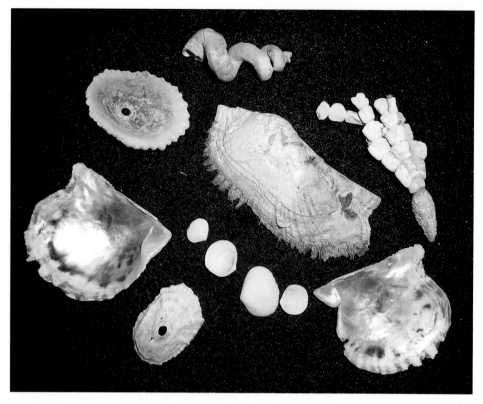

Plate 9.3 Original hard part materials.

Plate 9.4 Fossil from Glass Mountain in Texas, so named because the original material of many of the fossils has been replaced with silica. The fossils are extracted from the limestone by acid dissolution. The resulting silicifed specimens retain incredible detail and much fragile structure that would normally have been lost.

Plate 9.5 Pyritized fossil ammonite.
Source: Photograph by Randolph Femmer courtesy of the National Biological Information Infrastructure (NBII) office.

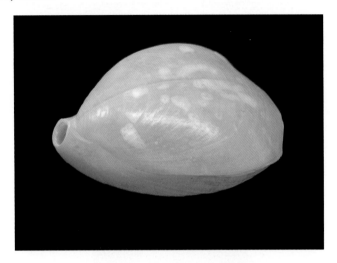

Plate 9.23 Brachiopoda, class Articulata, order Terebratulida. Recent brachiopod from the Philippines, displaying the distinctive shape that causes them to be called *lamp shells*.

Plate 9.34d Mollusca, Ammonoidea.

Plate 9.37 Echinodermata, Crinoidea. A variety of examples of fossil crinoid pieces.

Plate 9.42d Graptolite fossil *Pendeograptus fruticosus*: two overlapping, three-stiped rhabdosomes.
Source: Photograph by Mark A. Wilson.

Plate 9.45 Foraminifera (s) planktonic Globigerina in the Northern Gulf of Mexico.

Plate 9.52 Ostracoda (a) A living ostracod, genus *Spelaeoecia* that is only from marine caves and occurs in Bermuda, the Bahamas, Cuba, Jamaica, and Yucatan, Mexico.

result is often a poor-quality, somewhat porous soil layer. The potentially grey coloration of this layer can lead to misinterpretation as a *gleyed* horizon, a condition that is typically the result of stagnant water and reducing conditions. The key characteristics of the E-horizon are loss of silicate clay, iron and/or aluminum, and a concentration of lightly colored sand and/or silt. If the topsoil gets stripped away, exposing the E-horizon, agriculture is severely affected.

The *B-horizon*, also called the *subsoil*, is characterized by the importation of solutes and fines, a process called *illuviation*. Iron, aluminum, and soluble humic compounds leached out of the overlying horizons will often be deposited here, sometimes giving the layer a distinctive reddish color. Much of the silt and clay carried away from the E-horizon is also typically deposited here. Under some conditions, so many fines accumulate and/or so many crystals precipitate out that the entire layer can become clogged up. If this horizon turns completely into hard clay, it is called *hardpan*. Most B-horizons will have a distinct reddish tinge to them, and well-drained soils will often have brightly colored yellowish-brown to strongly brown B-horizons.

Soil horizons O, A, E, and B together comprise the *solum*, or "true soil." This is where soil-forming processes are active and where living roots, animals, and microorganisms are confined. Below the solum lies the *C-horizon*, a layer composed of partially altered parent material (i.e. rock or glacial till). The material here is weathered sediment that has theoretically not been affected by soil-forming processes and lacks the properties of the O-, A-, E-, or B-horizons. The color is usually that of the underlying unweathered geological material. Shallow soils may lack a C-horizon, instead lying directly on unweathered bedrock. Beneath the C-horizon lies continuous, unaltered bedrock, sometimes referred to as the *R-horizon*.

Now, this was just a generic description of the layers of an idealized soil. When reading the report on a specific soil, additional information, such as the thickness and maturity of each of the horizons, will be provided. Also, the names of each of the master horizons (O, A, E, B, and C) can be appended with lower-case suffixes that provide additional information about their characteristics. For example, Ap indicates a plowed A-horizon, Bc indicates a B-horizon that contains concretions, and f is used to indicate frozen soil. A full list of these suffixes can be found in Schaetzl and Anderson (2005) or a similar soil science textbook.

Soil Origins

Soils form from the natural physical and chemical weathering of rocks. Chemical weathering is responsible for the ultimate breakdown of the original mineral components into minerals that are stable on the Earth's surface, typically quartz and clay minerals. Climate is a dominant factor in determining the types and rates of chemical weathering, thus differing climates result in different soil types.

During chemical weathering, quartz and muscovite mica remain almost unchanged; they just get physically broken down into smaller pieces. Muscovite mica is physically a weak mineral, so it typically gets shredded into tiny clay-sized pieces. If you run your hand through some quartz beach sand, your hand will usually become covered in tiny flecks that sparkle in the sunlight. Those flecks are often all that is left of the large muscovite mica plates commonly found in granite. Mica is somewhat susceptible to chemical weathering, so in certain climates it will be altered into a new mineral. Quartz is highly chemically resistant, and is physically much stronger than mica, so it is typically broken down into sand-sized pieces. The other common silicate minerals are much more susceptible to chemical weathering and are altered into new minerals.

Potassium feldspar ($KAlSi_3O_8$) breaks down into soluble potassium (K^+), bicarbonate ions (HCO^{-3}), and dissolved silica (SiO_2), leaving behind the newly formed residual mineral named kaolinite [$Al_2Si_2O_5(OH)_4$], a hydrated aluminum silicate clay. *Residual minerals* are those that remain following weathering and are stable at the Earth's surface. Plagioclase feldspar, in a similar process, breaks down into soluble calcium bicarbonate, soluble sodium bicarbonate, soluble magnesium bicarbonate, and hydrated aluminum silicate clay minerals. Ferromagnesian minerals like biotite, amphibole, and pyroxene break down into soluble potassium bicarbonate, soluble magnesium bicarbonate, hydrated aluminum silicate clay, hydrated iron oxides like hematite (usually resulting in a reddish-rust color), and dissolved silica. Thus, the composition of the parent material determines what kinds of clay minerals will form the bulk of the mineral fraction of a soil (Table 7.1).

Table 7.1 Soils produced from common parent materials.

Parent Rock	Primary Minerals	Residual Minerals	Soil Properties
Granite, Tonalite, Quartz Monzonite, Granodiorite	Feldspars Micas Biotite Quartz	Clay minerals Clay minerals Vermiculite Quartz	Sandy, gravelly residuum, typically nutrient poor Acid, light-colored or reddish soils also common
	Fe-Mg Minerals	Clay minerals + Iron oxides	
Gneiss and Schist	Similar to above	Similar to above	Silty, sand residuum, typically nutrient poor
Basalt, Gabbro, Diorite	Feldspars Pyroxene/Olivine Magnetite	Clay minerals Clay minerals Iron oxides	Dark, clay-rich soils Low quartz content results in little sand Kaolinite, smectite, vermiculite and halloysite clays common Sometimes dark red
Peridotite, Dunite	Olivine, Serpentine	Clay minerals	Tend to be shallow, nutrient poor and fine textured
Sandstone	Quartz Feldspars	Quartz Clay minerals	Sandy soils, acidic and deep Arkose produces more clay-rich soil. Graywackes often create loamy soil
Quartzite	Quartz	Quartz	Very shallow
Limestone/Marble	Calcite	None	Produces little residuum, just insoluble residue such as chert

Climate determines the intensity of chemical weathering, driving the various chemical reactions resulting in different products. Clay minerals, such as kaolinite, and *sesquioxide minerals* (oxide minerals containing three atoms of oxygen with two atoms of some other element), such as Al_2O_3 and Fe_2O_3, are quite stable at the Earth's surface. Minerals formed from the precipitation of dissolved silica, such as chert and agate, are also quite stable since they are basically quartz.

It should be noted that the gross bulk chemistry of most soils is fairly straightforward. If you remember, just eight elements (O, Si, Al, Fe, Ca, Na, K, and Mg) form approximately 98.59% of the Earth's surface. Therefore, unless you happen to be working in an atypical area of the Earth, simple bulk elemental analysis will not be especially helpful for forensic characterization. An example of this can be seen in Daugherty (1997), where soil samples from San Diego, California; Ruidoso, New Mexico; and Sand Springs, Montana were compared using energy dispersive X-ray fluorescence (EDX) spectroscopy, a common approach for determining elemental composition.

All three soils were described as *mollisols* (see USDA Soil Classifications below) and all of the samples collected were found to be composed of – surprise, surprise – oxygen, silicon, aluminum, magnesium, calcium, potassium, and iron. There was also a significant component of carbon, which again is unsurprising given that these were fertile soils. No diagnostic pattern was found that could be used to distinguish the samples from the three sites, despite their widely separated points of origin, and the amounts of iron and potassium appeared to be virtually identical. All of which is pretty much what you would expect from soils with a similar continental origin. But, while bulk chemistry may not usually be particularly illuminating, there is promise in using trace (less common) elements to distinguish soil samples.

As you no doubt have already noticed, clay minerals and oxide minerals (including quartz) are the most common byproducts of chemical weathering. Thus, clay minerals and quartz are the most abundant contributors to clastic sediment and soil. Since the minerals in a typical soil are too small to see with the unaided eye, and too small to identify using a microscope, alternate methods of identification must be employed if their mineralogy becomes important. First, however, it is important to understand more about clay minerals, also called *phyllosilicates*. *Clay* is a material that is plastic when wet, holds its shape when dry, and becomes hard when fired. *Clay minerals* have all of the above characteristics plus are structurally phyllosilicates, or sheet silicate minerals.

Phyllosilicates (Sheet Silicates)

Technically speaking, clay minerals are hydrous aluminum silicates and are classified as phyllosilicates or sheet silicates. This refers to their distinctive layered structure, which you would be most familiar with through the properties of mica. There is quite a bit of chemical variability within this family of minerals but they all have platy morphology (i.e. are layered) and perfect 001 cleavage (i.e. break cleanly into flat sheets; again, think mica). The phyllosilicate family includes chlorite, serpentine, and talc, in addition to the clay minerals, which include the micas, kaolinite, and montmorillonite, as well as several other less common minerals. This topic is introduced in detail because it is very important to understand the degree of mineralogical variation that can exist in simple "dirt." Just because most of these minerals are almost indistinguishable to the naked eye, they are no less valuable in terms of forensic characterization than the minerals discussed previously.

The basic structure of phyllosilicates is based on two different types of layers, a layer of corner-linked tetrahedra (three-sided pyramids) and a layer of edge-linked octahedra (two four-sided pyramids connected at the base). The tetrahedral layer (T), properly referred to as a *sheet*, is composed of silicon–oxygen tetrahedron, each resting on a triangular face and sharing all three of the oxygen atoms on that face with other surrounding tetrahedra (Figure 7.2). The forth oxygen, the one on the peak of the tetrahedron, is not shared. The resulting structure is an infinite sheet of interconnected six-member rings with a basic formula of Si_2O_5, though aluminum atoms can substitute for up to half of the silicon atoms (Figure 7.3).

The octahedral sheet (O) is composed of two planes of rhombohedrally, or closest-packed, oxygen ions with cations occupying the spaces between the two planes (Figure 7.4). When you connect the centers of the six oxygen ions packed around a cation, the resulting shape is an octahedron. The oxygen ions on the edges of each of the octahedra are shared, also resulting in an infinite sheet. The

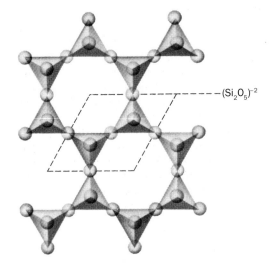

$-(Si_2O_5)^{-2}$

Figure 7.2 Tetrahedral sheet. Silicon–oxygen tetrahedra arranged so that the tips all point in the same direction and the bases are all in the same plane.
Source: Klein and Hurlbut, 1985.

Figure 7.3 Hexagonal ring of six interconnected tetrahedra with a basic formula of Si_2O_5. In an ideal structure, the dashed triangles (outlined by shading) are roughly the same size as the triangular face of an octahedron.
Source: Klein and Hurlbut, 1985.

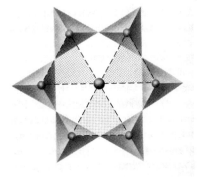

● Apical oxygens
● -OH group

most common cations are Al^{+3}, Mg^{+2}, Fe^{+2}, or Fe^{+3}, though lots of other elements have been found.

Imagine looking down on the octahedral sheet. In the bottom layer each oxygen ion has six nearest neighbors surrounding it and there are six spaces, or dimples, between each of the oxygen ions in the layer (Figure 7.5a). With closest-packing, the upper layer of oxygens will only occupy three of these dimples, leaving three other spaces available for occupation by cations (Figure 7.5b). Since there is a total charge imbalance of −6, this allows for two possible resulting structures. If all of the cations are +2 (e.g. Mg^{+2}, Fe^{+2}) then all three of the spaces will be filled, resulting in a brucite-like structure called *trioctahedral*. If the cations are +3 (e.g. Al^{+3}, Fe^{+3}) only two out of every three spaces will be filled, resulting in a gibbsite-like structure called *dioctahedral*.

The last step to building a clay mineral is to link the tetrahedral and octahedral sheets together. The oxygen-to-oxygen ionic dimensions of both sheets are

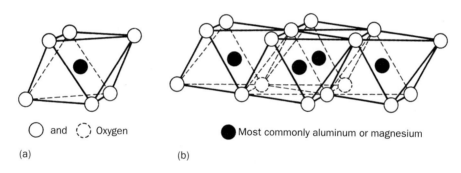

○ and ◌ Oxygen   ● Most commonly aluminum or magnesium

(a)   (b)

Figure 7.4 (a) A single octahedron, (b) octahedral sheet composed of two planes of rhombohedrally, or closest-packed, oxygen ions with cations occupying the spaces between the two planes. All octahedra lie on triangular faces.
Source: Mitchell, 1993.

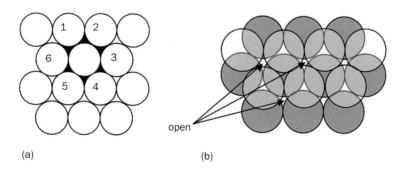

(a)   (b)

Figure 7.5 (a) Top-down view of a plane of oxygen atoms (or hydroxyl ions) in an octahedral sheet showing the closest-packed arrangement, where each anion has six nearest neighbors and is surrounded by a series of dimples, or spaces, (colored black), (b) the upper layer of anions rests in three of the six dimples, leaving three vacant dimples for cations (labeled "open").

approximately the same, so the apical (top of the tetrahedron) oxygen atoms in the tetrahedral sheet can be seen as replacing two out of three of the ions in the plane of the octahedral sheet (Figure 7.6). Structures that have one tetrahedral sheet and one octahedral sheet (like Figure 7.6) are called *1:1 layer silicates*, while structures that have two tetrahedral sheets and one octahedral sheet (like Figure 7.7) are called *2:1 layer silicates*. The bonds between these layers are van der Waals bonds (see Chapter 2 for a discussion on bonding), which are very weak, compared to ionic or covalent bonds, thus explaining why these minerals have perfect cleavage along one plane.

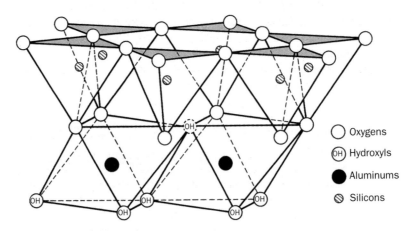

Figure 7.6 Figure 7.6 Diagram of the dioctahedral structure of kaolinite, a 1:1 layer silicate (or a T–O silicate) with a tetrahedral sheet bounded on one side by an octahedral sheet. *Source:* After Mitchell, 1993.

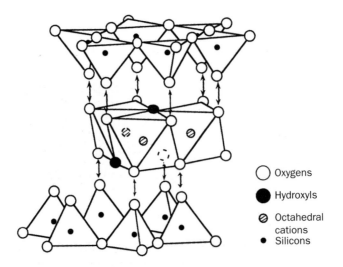

Figure 7.7 Diagram of a 2:1 layer silicate (or T–O–T silicate) with a central octahedral sheet bounded on the top and bottom by tetrahedral sheets.

Figure 7.8 Scanning electron microscope (SEM) image of kaolinite showing distinctive platy structure of a phyllosilicate. The arrow indicates the partings within a single crystal or "booklet" of kaolinite.
Source: Courtesy of the United States Geological Survey.

Some Important Clay Minerals

1:1 Layer Minerals

Kaolin group – This group of dioctahedral clay minerals has three members (kaolinite, dickite, and nacrite) and the basic formula $Al_4Si_4O_{10}(OH)_8$. The different minerals are *polymorphs*, meaning that they have the same chemistry but different structures (polymorph = many forms). Kaolinite group minerals are composed of silicate sheets (Si_4O_{10}) bonded to dioctahedral aluminum oxide/hydroxide layers ($Al_4(OH)_8$) (Figure 7.6). The interior layers are tightly bonded together, while only weak bonding exists between the paired layers (Figure 7.8).

Kaolinite is very common in soils formed by chemical weathering, or during hydrothermal alteration, of alumino-silicate minerals, such as the feldspars. It commonly forms in slightly acidic, well-drained, highly weathered soils in which the base ions like Na^+, K^+, Ca^{2+}, Mg^{2+}, and $Fe^{2+/3+}$ have been leached away. Granitic rocks, because they are rich in feldspar and tend to form acidic soils, are a common source of kaolinite. Dickite and nacrite form in hydrothermal environments and are not usually found in soils. Kaolinite does not swell in water, making it fairly structurally stable. It is commonly used in ceramics, and as a filler in paint, rubber, and plastic. The largest use of kaolinite is in the paper industry, where it is used as a coating to produce glossy paper, like you find in most magazines. If you have ever wondered why a stack of *National Geographic* magazines is so heavy, it is because of the amount of kaolinite used in the paper.

Halloysite has a structure similar to kaolinite; however, it has water molecules attached between the tetrahedral and octahedral sheets that make the structure curl up to form tubes or even spheroids. Halloysite has the general chemical formula $Al_4Si_4O_{10}(OH)_8 \cdot 8H_2O$. It commonly forms through the weathering of acidic, volcanic sediments.

Serpentine group – The serpentine group, which includes chrysotile, antigorite, and lizardite, is composed of trioctahedral minerals with the general formula $Mg_6Si_4O_{10}(OH)_4$. In the latter two members of the group, Fe^{2+} substitutes for some of the Mg^{2+}. Serpentine minerals form in soils derived from ultramafic rocks by the hydrothermal alteration of olivine and pyroxenes. Because they weather easily, they are not commonly found in the clay fraction of soils unless the soil is particularly young and unweathered. Where found, they are often associated with smectite, talc, and chlorite.

2:1 Layer Minerals

This group is composed of minerals with two silicate tetrahedral layers sandwiching either a di- or trioctahedral layer, in a T–O–T stacking sequence.

Talc is a trioctahedral mineral with the general formula $Mg_3Si_4O_{10}(OH)_2$ (Figure 7.9). Talc forms in hydrothermal and metamorphic environments from ultramafic

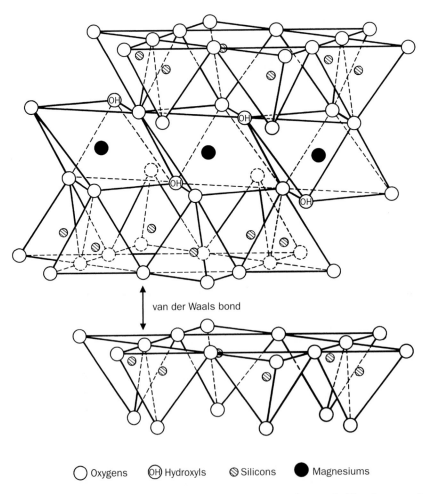

○ Oxygens   ⊕ Hydroxyls   ⊗ Silicons   ● Magnesiums

Figure 7.9 Trioctahedral 2:1 structure of talc. The structure of pyrophyllite is very similar, except that two of the magnesium cations would be substituted with aluminums and the third magnesium would be replaced with a vacant octahedral site, thus pyrophyllite is dioctahedral.

parent material. It can also form from the chemical weathering of pyroxenes and amphiboles. Due to its slippery or greasy feel, the result of weak interlayer bonding, talc is the major ingredient of talcum powder and is a very important industrial mineral. It is very resistant to heat and acid, and is a good insulator. Talc is also used as an opaque filler for paint, soap, and rubber, and is a common anti-caking agent. The related mineral dioctahedral mineral is *pyrophyllite*, which has the formula $Al_2Si_4O_{10}(OH)_2$. Pyrophyllite is very rarely found in soils.

Smectite group – This group contains both di- and trioctahedral members and is composed of several minerals, including saponite, hectorite, nontronite, beidellite, and montmorillonite, the last of which is the most common. The general formula for this group is $(Ca, Na, H)(Al, Mg, Fe, Zn)_2(Si, Al)_4O_{10}(OH)_2 \cdot n(H_2O)$, where n represents a variable amount of water. The smectite group of clays has a T–O–T structure that is similar to that of talc, but absorbed water molecules are positioned between the T–O–T stacks. A key characteristic of the smectites is their ability to absorb water molecules into their structure, causing the minerals to increase in volume when they come into contact with moisture. Thus, the smectites are *expanding* or *swelling clays*. They swell when wet and shrink when dried. In some soil survey documents, soils that contain smectites are identified as *hydric* soils and they can pose several construction difficulties. Smectite-rich soils weaken cohesion and contribute to soil creep and landslides.

The most common smectite is *montmorillonite*, a dioctahedral clay mineral with a general formula of $(Na, Ca)_{0.33}(Al, Mg)_2Si_4O_{10}(OH)_2 \cdot n(H_2O)$ (Figure 7.10). When montmorillonite comes into contact with water, it can expand by several times its original volume, greatly slowing the flow rate, thus it is commonly used in landfill liners, dams, or in other geotechnical engineering projects where the goal is to plug leaks or impede the flow of water. Montmorillonite is the main constituent in a volcanic ash called *bentonite*, which is used in drilling muds. The addition of bentonite gives the drilling slurry greater viscosity (resistance to flow) facilitating the removal of rock and dirt from within a drill hole, and it helps keep the drill head cool during drilling.

While montmorillonite can be very useful in certain contexts, it can also prove very hazardous if it forms a significant component of any soils that intersect tunnels, road cuts, or bridges. Its expandable nature means that changes in soil moisture levels can quickly lead to slope or wall failures. It can also cause serious cracks to form in a building's foundation. As long as the soil remains relatively moist, it will shore up buildings, but if the area falls prey to a drought, the montmorillonite will collapse creating large cracks in the soil, and potentially in all of the concrete structures and pipes located therein.

Smectites tend to form in poorly drained environments with long dry seasons, where soils are rich in silica, Mg^{2+} and Ca^{2+}. They are most abundant in the very fine clay fraction because they are formed through some fairly complex partial weathering or dissolution/precipitation processes. Montmorillonite is common in poorly drained or dry soils that formed from magnesium-rich intermediate or mafic rocks.

Vermiculite group – This group has dioctahedral and trioctahedral end members and a general formula of $Mg_{0.5}(Fe^{+3}, Mg)_3(Si_{4-x}, Al_x)O_{10}(OH)_2 \cdot n(H_2O)$. They also swell in water, though not as much as do the smectites. Vermiculites are thought to be unstable intermediaries of the mica weathering process, though trioctahedral vermiculite can form from the chemical weathering of biotite. While not a major component of soil, vermiculites are found in variable amounts in all major soil

x

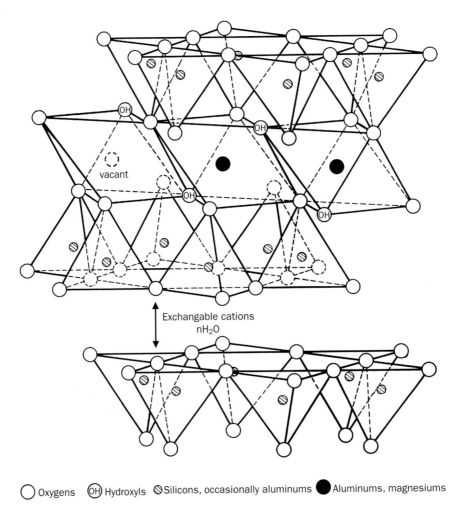

○ Oxygens   ⊕ Hydroxyls   ⊘ Silicons, occasionally aluminums   ● Aluminums, magnesiums

Figure 7.10 Dioctohedral structure of montmorillonite.
Source: Modified from Mitchell, 1993.

groups and are most common in the soils formed in temperate and subtropical climates.

Illite group – Also called hydrous micas, this group of dioctahedral clay minerals has the general formula of $K(Fe, Mg, Al)_2(Si_{4-x}, Al_x)O_{10}(OH)_2 \cdot n(H_2O)$. Illite-type clays are formed from weathering of potassium- and aluminum-rich rocks under basic, or high, pH conditions. Thus, they are formed by the alteration of minerals like muscovite and feldspar. Illite clays are the main constituent of ancient mudrocks and shale. Some researchers use the term illite to refer solely to diagenetic micas.

True micas – This group includes several di- and trioctahedral minerals. The most common trioctahedral forms are *biotite* with a formula of $K(Mg, Fe^{2+})_3Si_4O_{10}(OH)_2 \cdot n(H_2O)$, and *phlogopite*, which has a basic formula of $KMg_3Si_4O_{10}(OH)_2 \cdot n(H_2O)$. By far the most common true mica is the dioctahedral mineral *muscovite*, $KAl_2(Si_3 Al_1)O_{10}(OH)_2 \cdot n(H_2O)$ (Figure 7.11). Less common micas include *celadonite*, *glauconite*, and *lepidolite*. The true micas are the most familiar of all the clay minerals. The only difference between the hand samples that

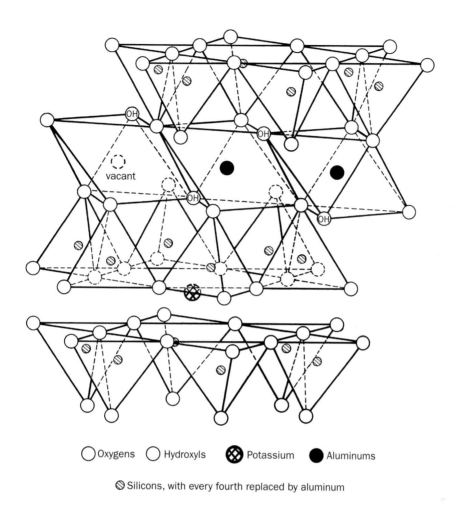

○ Oxygens ○ Hydroxyls ⊗ Potassium ● Aluminums

⊘ Silicons, with every fourth replaced by aluminum

Figure 7.11 Diagram of an idealized muscovite mica structure.
Source: Modified from Mitchell, 1993.

everyone will probably have seen at some point and the clay version of these minerals is size.

Unlike most of the other clay minerals discussed, micas are not the product of a chemical process but are primary minerals inherited from the physical weathering of a parent material. They are found in many different soil environments, though only muscovite and biotite are common. Muscovite is found in granitic igneous rock and in metamorphic rock, while biotite is found in rocks with a higher mafic content. Biotite weathers up to a hundred times faster than muscovite, so it is less common in soils. The other micas are uncommon except in soils where they are found in abundance in the parent rock.

2:1:1 Layer Minerals

Chlorite group – The chlorites are not always considered part of the clay minerals and are instead sometimes treated as a separate group within the phyllosilicates. It is a relatively large and common group. The general formula is $X_{4-6}Y_4O_{10}(OH, O)_8$, where X represents either aluminum, iron, lithium, magnesium, manganese, nickel, zinc, or, rarely, chromium. The Y represents aluminum, silicon, boron, or iron, but most often aluminum and silicon.

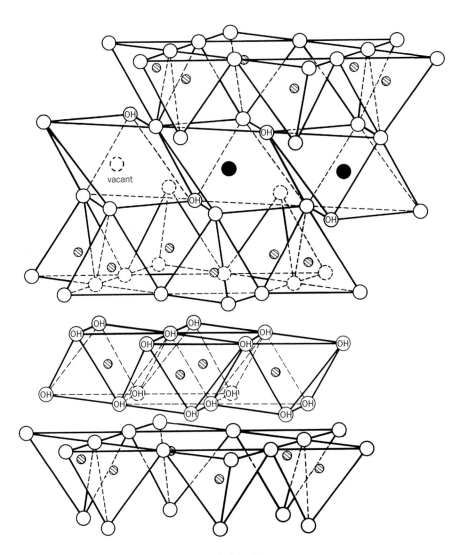

vacant

Figure 7.12 Diagram of the 2:1:1 structure of chlorite.
Source: Based on Grim, 1968.

All of the members of this group have a basic 2:1 layer structure, but they differ from the other 2:1 layer minerals in one unique respect: they contain an additional stable, positively charged octahedral sheet rather than a layer of adsorbed cations, or water, in the interlayer space (Figure 7.12). This additional octahedral sheet consists of two layers of OH^- ions that enclose central Mg^{2+}, Fe^{2+}, or Al^{3+} cations, leading to an overall positive sheet charge. By virtue of this positive charge, the interlayer octahedral sheet neutralizes the negative charge of the 2:1 sheets. Thus, because chlorite has two octahedral sheets, it is called a 2:1:1 layer mineral. Like the micas, chlorites are primary minerals that are inherited through the physical weathering of chlorite-bearing mafic igneous and metamorphic parent rocks. Chlorites are unstable minerals, so it is usually found only in weakly developed soils or in cold climates where chemical weathering is retarded. There are no common industrial uses for chlorites.

Soil Classification

The type and amount of clay minerals found in a soil can give it a variety of different physical properties, but due to the difficulties of identification it is generally not clay mineral content that is used to classify soils. Vasili Vasilievich Dokuchaev (1846–1903) is commonly credited with founding the discipline of soil science. He and his students produced the first scientific classification of soils and developed the first soil mapping methods. But while Dokuchaev, and others, laid the foundations of the modern zonal concept of soil that is applied internationally, soil classification is a highly contentious subject, thus there are a wide number of systems in use around the world. The Food and Agriculture Organization (FAO) of the United Nations, in collaboration with the International Society for Soil Science (ISSS) and the International Soil Reference and Information Centre (ISRIC), have developed a World Soil Classification system, which is useful in general terms but is not intended for use in detailed mapping, thus it is not particularly useful for forensic work (http://www.fao.org/ag/agll/prtsoil.stm).

In the United States there are basically three major classification systems in use: the United States Department of Agriculture (USDA) Soil Taxonomy System, which has the longest history, the Unified Soil Classification System (USCS) from the American Society for Testing and Materials (ASTM), and the American Association of State Highway and Transportation Officials (AASHTO) System (M 145-87). Each of these systems has a different emphasis based on the properties of concern. Engineers are most interested in soil stability and a soil's physical properties because they are trying to determine whether the soil will support a structure such as a highway, runway, or bridge. The USDA is primarily interested in a soil's ability to support crops. Both the USDA and USCS approaches are introduced because both of them may, at one time or another, be of some use to the forensic worker, depending on what types of maps are available. By the way, various introductory geology books, for some odd reason, still use a three "soil types" system (pedalfer, pedocal, and laterite) from the 1930s, which is extremely outdated and should be ignored.

Soil Color

One characteristic that is employed to some degree by all of the different classification systems is soil color (Figure 7.13). Different colors can be reflective of important soil properties. For example, very dark-black or brown soils are usually associated with a very high organic content, though manganese concretions are also often black. Reds and oranges typically indicate the presence of iron oxides such as hematite or lepidocrocite. Yellowish-brown soils often contain the iron oxides goethite or jarosite. Wet soils that are rich in organic matter and have been under prolonged anaerobic conditions (low or no oxygen), as is common in swamps, often develop a muted gray, or washed-out, color (gley soils). Soils that are contaminated with petroleum products frequently develop a greyish-green tint. There are even some soils that appear bluish or purplish. Localized soil areas can contain a wide variety of colors, while others might be highly uniform. When collecting comparison samples, it is important to look for these kinds of variations to ensure that they are fully represented.

Because color is subjective, it should be determined by comparison with a color reference chart. The most common charts in use are the Geotechnical Gauge, which only has chips for six colors but is used by many engineers in the field, the

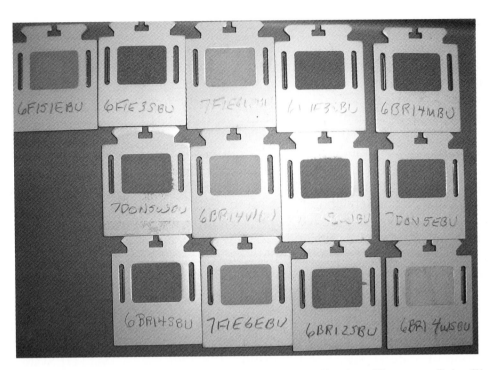

Figure 7.13 A very small example of some of the common soil colors. Please see Color Plate section.

Munsell Rock Color chart, and most commonly, the Munsell Soil Color chart (Munsell Color, 2000). The Munsell system characterizes colors using three attributes: hue, value, and chroma (Figure 7.14). *Hue* technically refers to the wavelength of light that emanates from an object, though it can be thought of as the actual spectral color, like red or yellow. There are five principal hues that correspond to spectral colors: red (R), yellow (Y), green (G), blue (B), and purple (P), plus five intermediate hues that lie halfway between adjacent principal hues: YR, GY, BG, PB, and RP. Each chart in a Munsell book contains several color chips of the same hue. Hue is described using a notation composed of letters (R, YR, etc.) or gley for the greenish-bluish-gray colors common to gley soils, and a number that ranges from 0 (which is not included in the charts) to 10. Five represents the middle of each hue range, while the zero value corresponds to the 10 value of the next hue (which is why they do not include the zero chips). For example, 10R is equivalent to zero YR, and 10YR is equivalent to zero Y. Hue designations are located on the tabs to each of the charts and at the top of each chart.

Value refers to the lightness/intensity of a color, or technically the amount of white or black that is added to a color. It ranges from 0, which is black, to 10, which is pure white. Low numbers correspond to darker colors, while high numbers correspond to very light colors that appear faded. Value is displayed along the vertical axis of each chart. *Chroma* describes the purity of a color or how much gray is in the color. It also ranges from 0, which corresponds to a neutral gray, designated N0, that is a common endpoint for all colors in the Munsell scale, to a theoretical maximum of 20, though soil charts top out at 8. The higher the value, the purer the color; the lower the value, the more washed out or pastel the color. Chroma is listed along the horizontal axis of each chart. The Munsell Soil Color Chart is actually a truncated version of the complete Munsell Color System and

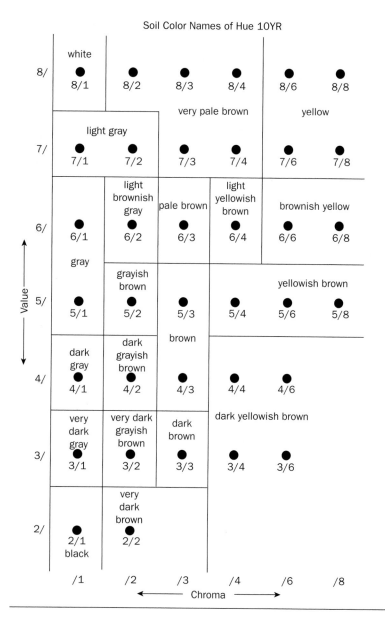

Figure 7.14 Example of the Munsell Soil Color Chart and associated names for hue 10YR soil colors.
Source: Soil Survey Division Staff, 1993.

includes only the colors most applicable to soils, or about one-fifth of the full color range.

Each Munsell chart has a series of color chips that are surrounded by holes in the page. The idea is to look at your samples through the holes in order to find the best color match. When referring to a soil color, the hue is listed first, followed by the value, a slash mark, and finally the chroma. For example, 5YR 5/3 is a reddish-brown color, while 2.5Y 6/8 is an olive yellow color. Each color chip has also been assigned a color name, but these names are often confusing and should not be used in isolation. Soil colors should be determined in bright, uniform midday sunlight and in the absence of sunglasses or other tinted eyewear. In general, color should not be determined under fluorescent lights unless the same location is consistently used for in-house examinations.

Even with the use of soil reference charts, color is still quite subjective. Post *et al.* (1993) found that soil scientists who were asked to determine the dry and moist colors of provided samples agreed on the same Munsell color chip only 71% of the time. To address this, there are some more precise methods of determining soil color. The most common requires the use of a computer-controlled spectrophotometer system (Croft and Pye, 2004).

In forensic work, the dry soil color is usually the most important to determine, because even slight variations in soil moisture levels can cause significant color change. Additional information can be gained by using a series of treatments to further isolate the colors of different components in a soil. For example, Sugita and Marumo (1996) describe a color examination method that employs air-drying, moistening, organic matter decomposition, and iron oxide removal by which they were able to differentiation 97% of their test samples. Because bulk soil color is often determined by the color of its clay fraction, Janssen, Ruhf, and Prichard (1983) developed a procedure to remove a homogeneous sample of the clay fraction to create small clay disks for color comparison. Since soil color can vary horizontally, vertically, and by moisture content, it is important to ensure that the samples being compared are reflective of the same conditions and are from the same spatial location.

As soil color is so variable, even over short distances, it should generally not be used as the sole indicator of whether specimens could have the same source. Petraco, Kubic, and Petraco (2008) discuss a case in which a female runner was found nearly dead in an isolated area of Central Park in New York City. Soil samples were collected from the victim's clothing, from the clothing of several suspects, and from the crime scene, which ranged over several hundred feet. Examiners were able to show several consistencies between the questioned samples (from the suspects' clothing), the control samples (from the crime scene), and the samples taken from the victim's clothing. The authors note, however, that two of the control samples, taken only 15 inches apart, had the same mineral composition, but very different Munsell colors. One sample was identified as coming from a bog (10YR 2/2), while the other sample was taken from an area identified as "the high point area" (10YR 5/6). The difference in elevation created circumstances by which samples taken from approximately the same area had a significantly different organic content, producing very different colors.

Use of Soil Evidence in a Heartbreaking Crime

Tahisha was just six years old when she disappeared on April 23, 1993. She lived in a large apartment complex surrounded on three sides by a solid block wall over five feet high, while the front of the complex had an iron fence six feet tall. An access code was required for entry at the gate. She had been playing catch with her nine-year-old brother, Stefan, when at some point during the game the ball rolled down a hill and toward two satellite dishes. Tahisha had just gone to retrieve the ball when Stefan heard their mother, Marianne, call them to come inside. He went in and told his mother that Tahisha had gone to get the ball.

Ten minutes later, at around 7 p.m., when Tahisha still had not returned, Stefan went back outside to look for her. Marianne started checking with the neighbors and at the apartments of Tahisha's friends, but they still did not find her. By now, it was starting to get dark. Marianne and other adults walked around the entire apartment complex yelling Tahisha's name. Another child found Tahisha's ball inside a fence surrounding the satellite dishes, but she was nowhere to be found and the family called the police.

Officers soon arrived and organized a coordinated effort to check all of the apartments and other areas in the large complex. Within a couple of hours, someone had been contacted in all but five of the apartments. Three of the apartments turned out to be vacant and one belonged to someone known to be out of town. The last one belonged to Terrance Page and was located directly across from the satellite dishes.

Detectives first visited Page at 2:35 a.m. the next morning and asked him the same questions that they had been asking all of the other tenants. Page stated that he had been in his apartment all evening, that he had not seen the girl, and that he did recognize her photograph. Later that day, officers went to question Page again because they learned that he had been seen leaving his apartment at approximately 9:30 p.m. the previous evening, while everyone else was still searching for the missing child. A detective spotted Page in the apartment complex's laundry room and officers followed him back to his apartment, where they asked if they could search his apartment and vehicles. Page agreed, also giving them access to locked areas of his garage. It is unclear, but it seems that the appearance of the officers kept Page from doing his laundry.

That afternoon, while investigators were still working at the apartment complex, a man who had just completed his work shift got a ride home with a friend. During the ride, the man said that he needed to relieve himself, so his friend pulled the car over. The man got out of the car and walked over a hill, out of sight of the road. When he crested the hill, he saw a body at the bottom of the slope, about 15 feet away. The man quickly reported his discovery to the police. Officers determined that the body was Tahisha's. It appeared that she had been thrown from the hill on the side of the road and had rolled into the pit of a mine excavation site.

The next day, an officer was dispatched to Page's apartment in response to a report from the suicide hotline. Page, who had taken an overdose of acetaminophen, was taken to a hospital where he stated that he suffered from stress at work and was upset because the police kept questioning him about the missing girl. While Page was in the hospital, investigators obtained a search warrant for his apartment. Along with several other important items of evidence, they found a shirt with two small spots of blood inside a clothesbasket. Next to the shirt, they found a pair of pants that had a white dust-like material on the legs. Inside a closet, investigators found a pair of dress boots that also had white dust-like material on them.

Forensic geologist Sergeant Erich Junger examined the white dust-like substance, which was also found on the floor mats of Page's vehicle. He determined that the material was a fluffy, white bentonite containing bits of other minerals. He also found that the bentonite was relatively pure, and that the boots and pants did not have any sand or other stray soil that might have been picked up if the wearer had walked around other places in addition to walking in the bentonite deposit. Further, the bentonite on Page's belongings was of the same color and type as the material from the open pit mine where the body was found. It had the same white color and the same general ratio of other minerals mixed in.

For comparison, Junger also examined 27 soil samples collected from the region outside the apartment complex, three samples from inside the complex, five samples from locations around the mine where the body was found, and 19 samples taken from areas around Page's place of work. The two samples taken from inside the quarry contained white bentonite, while the remainder of the samples did not. Samples from six other bentonite mines located within a 50-mile radius of the crime scene were also examined, but only the site where Tahisha was found had white bentonite. At the trial, Junger concluded that the bentonite on Page's belongings was consistent with the soil at the site where Tahisha's body was found. Because the addition of even a minimal amount of impurities can change soil color drastically, "white" soil is relatively rare.

Based on witness testimony and physical evidence including genetic markers and a stick-on earring, along with the white bentonite, Terrance Page was convicted of first-degree murder and the commission of a lewd act on a child under the age of 14 years. During the penalty phase of his trial, Page was sentenced to death. In December 2008, Page committed suicide while on death row at San Quentin.

Soil Moisture

Because soil moisture, the water contained within the pore spaces of a soil, is a highly variable characteristic determined by the height of the local water table, any recent rainfall events, the general humidity of the region, and the storage characteristics of the soil, it can be useful to determine total soil moisture in fresh or *in situ* (in place) samples. This would be most common in environmental or engineering assessments. Some types of soil are able to store much more water than others are, depending on the grain size, chemical composition, percentage of organic material, and degree of compaction. Moisture is general reported as a percentage:

Moisture content = (mass of water in soil/mass of wet soil) × 100%

One of the most common methods used to determine soil moisture is the *gravimetric method* (ASTM D2216 or ASTM D4643). The conventional method is as follows:

1. Collect a soil sample in the field, place it in a waterproof container, and seal the container. Label the container and store it in a cool place for transport back to the laboratory (most people use coolers for this).
2. Label a clean dry container (these are usually aluminum canisters but can be as simple as baby food jars) using a pencil or scribe (inks desiccate and lose weight). Weigh the empty container and record the weight.

3. Fill the container with approximately 25 g of moist soil taken directly from the sealed sample container and immediately obtain the *wet weight* of the soil.
4. Place the sample in a drying oven that has been preheated to 104.5 °C and allow the soil to dry for a minimum of 24 hours.
5. Pull the sample out of the oven and place it into a desiccating chamber (to keep it from absorbing moisture from the air) and allow the container to cool.
6. Weigh the container to obtain the *dry weight* of the soil.
7. Calculate the moisture content as follows:

$$\frac{\text{Moisture}}{\text{content}} = \frac{(\text{mass of wet soil} + \text{container}) - (\text{mass of dry soil} + \text{container})}{(\text{mass of wet soil} + \text{container}) - (\text{mass of container})} \times 100\%$$

A quick alternative supported by the ASTM uses a microwave oven. Follow the same steps as described above, except use a paper cup or microwave safe container to hold the soil, and place the cup into a microwave oven for two minutes at full power. Warning: the soil and the container will be hot. Weigh the sample, record the weight, and place the sample back into the oven for another two minutes. Repeat until the soil achieves a stable weight. Do this in a laboratory microwave, not one that is used for food, and be aware that it can smell pretty awful.

There are also several equipment-based field methods to determine soil moisture, such as the use of a tensiometer, measuring the soil's electrical resistance, or using a neutron probe, but the gravimetric method is still considered the standard. Soil moisture included in most reports will have been determined using the gravimetric method.

Particle Size

Another theme that is common to all soil classification schemes is the categorization of soils/sediments on the basis of particle size. Most often, particles are divided according to their sizes by the use of *sieves*, which are precisely engineered screens attached to metal frames (Figure 7.15a). If a particle does not pass through a screen with a particular size opening, it is said to be *retained* by that sieve. Pouring a soil mixture through a stack of several different size sieves will divide the sample into its particle size fractions as defined by the sieves used (Figure 7.15b).

As you have already seen, there are many different grain-size scales in use. For example, sedimentologists define the division between silt and clay as 0.0039 mm, while civil engineers consider any particles below 0.005 mm clay-sized (Figure 7.16). For reference, a particle 0.07 mm in diameter is about the smallest that can be detected by the unaided eye. When you are interested solely in a soil's engineering properties, once you get below 0.074 mm (the size of the openings in a Number 200 sieve), size is relatively unimportant in comparison to other properties. Particles below 0.002 mm (or 0.001 mm in some grain-size scales) are frequently designated as *soil colloids*. Whenever you read a soil report, it is vital that you understand which referencing system is being used; otherwise, significant confusion can result.

Several methods may be used to determine the size of the particles contained in a soil sample and the distribution of the particle sizes. The most common is probably a *dry sieve analysis*. Sieves are stacked according to size: the sieve with the largest mesh openings is on the top, followed by sieves with progressively smaller

(a)   (b)

Figure 7.15 (a) From top left to bottom right: No. 35, No. 60, and No. 200, three-inch sieves with stainless steel pan and lid, (b) soil sample divided into sieve fractions and pan fraction (top right), along with some prepared sample slides. Please see Color Plate section.
Source: Photograph (b) courtesy of William Schneck.

	FINE EARTH					ROCK FRAGMENTS	150	380	600 mm
						channers	flagst.	stones	boulders
USDA	Clay (fine, co.)	Silt (fine, co.)	Sand (v.fi., fi., med., co., v.co.)	Gravel (fine, medium, coarse)	Cob-bles	Stones	Boulders		
millimeters:	0.0002 .002 mm	.02	.05 .1 .25 .5 1	2 mm 5 20	76	250	600 mm		
U.S. Standard Sieve No. (opening):			300 3 140 60 35 18 10	4 (3/4")	(3")	(10")	(25")		

			Sand		Gravel	Stones	
Inter-national	Clay	Silt	fine	coarse			
millimeters:	.002 mm	.02	.20		2 mm	20 mm	
U.S. Standard Sieve No. (opening):					10	(3/4")	

		Sand			Gravel		Cobbles	Boulders
Unified	Silt or Clay	fine	medium	co.	fine	coarse		
millimeters:		0.74	.42	2 mm 4.8	1.9		76	300 mm
U.S. Standard Sieve No. (opening):		200	40	10 4	(3/4")		(3")	

			Sand		Gravel or Stones			Broken Rock (angular), or Boulders (rounded)
AASHTO	Clay	Silt	fine	coarse	fine	med.	co.	
millimeters:	.005 mm		.074	.42	2 mm 9.5	25		75 mm
U.S. Standard Sieve No.:			200	40	10 (3/8")	(1")		(3")

phi #:	12	10	9	8	7	6	5	4	3	2	1	0	−1	−2	−3	−4	−5	−6	−7	−8	−9	−10	−12
Modified Wentworth	←clay→				← silt →				← sand →				← pebbles →					cobbles	← boulders →				
millimeters:		.002	.004 .008 .016 .031			.062 .125 .25			.5 1		2 mm	8	16	32	64			256				4092 mm	
U.S. Standard Sieve No.:						230 120 60			35 18 10		5												

Figure 7.16 Comparison of five of the different sediment scales in common use.
Source: Schoeneberger, 2002.

mesh openings. The sieve with the smallest mesh size is located at the bottom and is underlain by a catch pan. Most sieves are designated by a number that corresponds to the number of openings per lineal inch (*numbered sieves*). For very large mesh sizes, sieves are designated by the actual size of the opening (*dimensioned sieves*). For example, the Number 4 standard sieve has four openings per lineal inch (or 16 openings per square inch), whereas the quarter-inch sieve has a sieve opening of a quarter of an inch. Table 7.2 lists sieve numbers and the corresponding mesh size in millimeters.

To perform a dry sieve analysis, obtain an oven-dry *representative* sample of soil. This means that the sample you are analyzing must have basically the same properties as the bulk material from which it was taken. You typically do not want to be basing your analysis on a sample that has strange or anomalous properties that are at variance with the bulk of the surrounding materials unless, of course, those anomalous properties are somehow related to the investigation at hand.

Sample Collection

Depending on the nature of the analysis to be performed and the environment involved, determining whether a sample is representative can sometimes be difficult,

Table 7.2 Sieve numbers and mesh sizes in SI units.

Sieve Number	Opening Size (mm)
4	4.750
5	4.000
6	3.350
8	2.360
10	2.000
12	1.680
16	1.180
20	0.850
30	0.600
35	0.500
40	0.425
50	0.300
60	0.250
80	0.180
100	0.150
120	0.125
140	0.106
200	0.075
230	0.063
270	0.053

but in most cases for control and alibi samples it is sufficient to take a sample approximately $10\,cm \times 10\,cm \times 1\,cm$ if surface properties are of the most interest (for comparison to mud on a shoe, for example) or samples approximately $5\,cm \times 5\,cm \times 5\,cm$, when bulk properties are of the most interest.

The most important aspect of collecting soil samples is to obtain them from the appropriate location. Because soil properties vary with depth was well as horizontal distance, it is particularly important to ensure that the samples collected for comparison not only are from the correct location spatially but also are from the same location in the soil profile. If the mud collected in the wheel wells of a suspect's car comes from a muddy ditch where the car's wheels dug in, comparison samples taken simply from the upper couple of centimeters of the soil profile will not be an appropriate comparison. If you are looking at the soil on a shovel or other tool used for excavation, samples of entire soil profiles must be collected for comparison. It is also extremely important that soil samples not be mixed, even if they come from close proximity, and the original texture of the soil should be retained as well as possible. Pye (2007) has a very nice discussion of approaches for soil sample collection from the scene.

The representative size of a sample is dependent on the approximate diameter of the largest particle in the bulk soil. The larger the particles in the soil, the larger the sample must be to accurately capture its bulk properties. For reference, the minimum sample size recommended by the USCS approach is 200 g for soils where the diameter of the largest particle is 2.000 mm (No. 10 sieve), 500 g for soils where the largest particle is 4.750 mm (No. 4 sieve), and 1500 g for soils where the largest particle is 18.85 mm (three-quarter-inch sieve), though in forensic work you must find a way to work with whatever size of sample you have been provided.

Simplified Manual Dry Sieve Method for Particle Size Analysis

1. First, it is important for the sample to be completely dry. Moisture causes soil particles to stick to each other and to the sieve mesh, clogging the openings and affecting your analysis. Most laboratories dry their soil samples in a low-temperature (<105 °C) drying oven for several minutes or hours depending on the initial condition of the soil. Care should be taken to ensure that any materials that could be altered by exposure to heat, such as thin plastics, are removed.
2. Transfer the sample to a tared weigh boat and record the mass of the sample in your laboratory notebook.
3. Assemble the sieves necessary for your analysis. A fairly standard set would include the following: Lid, 5, 10, 35, 60, 120, 200, and catch pan. The USGS uses: Lid, 18, 20, 35, 40, 60, 100, 120, 230, and catch pan. If your sample contains very large clasts, you will need to add larger-sized sieves.
4. Obtain the empty weight of each of the sieves and the catch pan.
5. Assemble the sieves in the above-listed order top to bottom, leaving the lid off. Carefully pour the dry sample through the stack of nested sieves and place the lid on as soon as you have finished pouring. You must pour slowly and gently to minimize the loss of clay particles, which can easily float away in a cloud of dust if you are not careful.

6. If an automated shaker is not available, shake the sieve stack by hand *carefully* for 5–10 minutes. Do not shake in a specific pattern but instead alternate your direction of motion so that the grains are moved randomly about. Occasionally, tap the stack against a counter to help distribute the grains.

7. When you have finished shaking the stack, let the sieves stand for several minutes (at least two, preferably more) to allow the finer grains to settle. If a large cloud of clay starts to float out as soon as you remove the lid, you did not wait long enough.

8. Separate each sieve and set it on a clean paper towel or square of butcher paper. If you did not shake the sample long enough, particles may fall onto the paper. If that happens, pour the contents of the paper into the next lower sieve. If a large amount of sediment falls out from your sieves, you need to re-stack the sieves, pour all of the loose sediment back in, and shake for a longer period.

9. From the top down, weigh each sieve, and record the data into a table similar to the one shown in Figure 7.17.

10. Determine the mass retained in each of the sieves by subtracting the mass of the sieve from the total mass of the sieve plus the sample. Record these values under the heading *Mass Retained*. Sum the mass retained column, including the mass of the sample in the catch pan, and compare this value with the original mass of your sample. A small loss of clay is not unusual, but a loss or gain of more than 2% is considered unsatisfactory.

11. Calculate the *percentage retained* in each sieve (and the pan) by dividing the weight retained by the weight of the *original sample*. This is based on the

Sieve No.	Minimum grain size (mm)	Total mass of sample & sieve (g)	Tare weight of sieve (g)	Mass retained (g) (total − tare)	% retained	% finer
5	4.00					
10	2.00					
35	0.500					
60	0.250					
120	0.125					
200	0.075					
Pan	<0.075					

0%   100%

Total mass of sample = _____ g (Sum of column 5)

Initial mass of sample = _____ g

Figure 7.17 Sample table for collecting grain size analysis data.

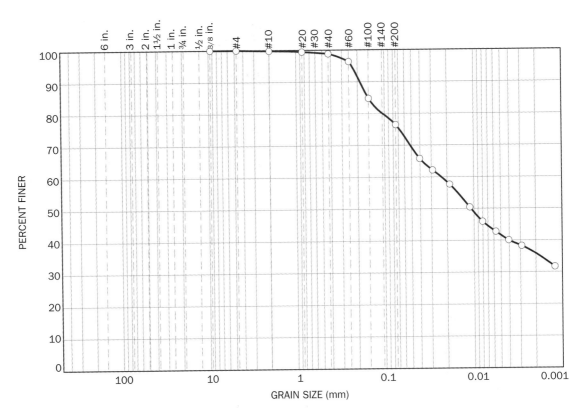

Figure 7.18 An example of a semi-log plot of grain size distribution data.

assumption that the lost material is most likely clay particles that would have passed through all of the sieves.

12. Compute *percent finer* by starting with 100% and subtracting the percent retained on each sieve as a cumulative procedure.

13. Make a semi-logarithmic plot of your data using particle size (log scale) versus percent finer (normal scale), as discussed in Chapter 5 (Figure 5.3; Figure 7.18). The particle size distribution information can be used to determine statistical parameters for your sample, as discussed in Chapter 5.

14. The fraction retained in the No. 120 sieve is commonly used for development of a mineralogical profile. Those grains would often be washed in a solution of 0.1% sodium hexametaphosphate in distilled water to clean off any surface coatings of clay. A typical procedure would include hand shaking or placement in an ultrasonic cleaner for several minutes. The *supernatant* (surrounding liquid) is pipetted or siphoned off and retained in a separate container. The process is repeated until the water remains clear. For analysis of the finest particles by X-ray diffraction, or another analytical method, many researchers combine the sample fraction from the catch pan and the No. 230 sieve (i.e. the two finest fractions).

In general, *wet sieving methods* for particle size analysis should be avoided unless there is a significant amount of sample available for alternative forms of examination, or the possibilities afforded by dry sieving have been exhausted and additional information is still needed. The entire above procedure can be repeated, but instead of shaking, the sample is flushed through the sieves using *ethanol* (ethyl alcohol). The alcohol is allowed to evaporate overnight, or the sieves can be gently heated to accelerate evaporation. Because of it characteristics as a solvent and the potential reaction of any swelling clays, water should generally not be used for this process.

Finally, under some circumstances a more detailed analysis of distribution of soil fines might be necessary. There are a few different approaches, including the hydrometer and decantation methods, as well as some that require more sophisticated equipment, but the *pipette method for silt-clay particle size analysis* is probably the most commonly used. A detailed description of one version of the pipette method (there are several variants) can be found on the companion website (including Tables 7.3 and 7.4; www.wiley.com/go/bergslien/forensicgeoscience).

Soil Classification Schemes

United States Department of Agriculture (USDA) Soil Taxonomy System

In 1949, the Soil Conservation Service System (SCS) began the development of a new method of soil classification that was based on measurable soil characteristics. The process took nearly a decade to complete, and in 1960 the Seventh Approximation Soil Classification System was introduced. This soil classification system, which has undergone numerous minor modifications and is now simply known as Soil Taxonomy, or *the green book*, is still in use today (Soil Survey Staff, 1999). The name of the SCS was changed to the Natural Resources Conservation Service (NRCS) in 1994 when the USDA was reorganized. The most current version of the NRCS soil taxonomy book is available online at the NRCS website (http://soils.usda.gov/technical/classification/taxonomy).

The current version of the NRCS Soil Taxonomy classification system has six levels of classification in its hierarchical structure. The major divisions in this classification system, from general to specific, are orders, suborders, great groups, subgroups, families, and series. There are 12 formal soil orders, 64 recognized suborders, around 300 great groups, and more than 2400 subgroups. At its lowest level of organization, the National Cooperative Soil Survey currently recognizes more than 19,000 different soil series. Taxa are defined based on the presence or absence of defined diagnostic soil horizons.

The NRCS system is widely used by US law enforcement agencies because the survey maps are readily available online at the Web Soil Survey site (http://websoilsurvey.nrcs.usda.gov/app) and there are related NRCS sites that create downloadable files for use with Geographic Information Systems software. Educational materials are also available (http://soils.usda.gov/education). At the most general level, the NRCS Soil Classification System recognizes 12 distinct *soil orders*: alfisols, andisols, aridisols, entisols, gelisols, histosols, inceptisols, mollisols, oxisols, spodosols, ultisols, and vertisols. Short descriptions of each of the soil orders follow, with more information available on the companion website.

Alfisols are well-developed soils that form under forest vegetation and are mostly found in temperate humid and sub-humid regions of the world. Weathering processes leach the surface layers so that there is large-scale illuviation (significant deposition of humus, chemical substances, and fine mineral particles of clay in the B-horizon). Alfisols have relatively high native fertility, light-colored surface horizons, and tend to be very productive soils for agriculture and *silviculture* (the cultivation of forests).

Andisols are soils that develop from pyroclastic volcanic parent materials, including ash and cinders, which weather to form amorphous substances such as allophane, imogolite, and ferrihydrite. These hydrated alumino-silicate substances have high moisture capacities and can fix large quantities of nutrients, making them unavailable to plants. Significant content of volcanic glass is one of the defining characteristics of this soil.

Aridisols are soils that develop in very dry environments. Lack of moisture limits weathering, resulting in poor, thin soil horizon development. Aridisols also often contain clays, calcium carbonate, gypsum, and salts that would have been leached from soils in more humid climates. Limited vegetation growth in arid regions produces little humus, thus aridisols tend to be light-colored.

Entisols are immature soils of recent origin that are characterized by their lack of well-formed horizons. These soils are often found in areas with recently deposited sediments or in settings where deposition outpaces soil development processes, such as steep slopes or dunes. Many are sandy or very shallow, and poor in humus. Given more time, these soils will usually develop into another soil type.

Gelisols are soils that have permafrost within one meter of the soil surface or have gelic materials within one meter of the soil surface and have permafrost within two meters. *Gelic* materials are mineral or organic soil materials that show evidence of cryoturbation (frost churning) and/or ice segregation in the seasonal thaw layer. Low temperatures greatly slow soil development processes, so there is poor horizon development, and hinder the decomposition of organic material, resulting in significant carbon sequestration.

Histosols are organic soils that form in areas of poor drainage with no permafrost, such as wetlands, mucks, or bogs. By definition, they contain at least 20–30% organic matter in the upper horizons. When plant litter rapidly accumulates in water, decay is dramatically slowed by the lack of available oxygen, thus a typical soil profile consists of thick (>80 cm) accumulations of organic matter at various stages of decomposition.

Inceptisols are immature soils that show more horizon development than entisols do, but lack the defining characteristics of the other orders. These soils are widely distributed and found in humid and sub-humid regions, arctic tundra environments, glacial deposits, and in relatively recent deposits of stream alluvium. Inceptisols show limited eluviation in the A-horizon and illuviation in the B-horizon. Given enough time, inceptisols may develop into another soil type.

Mollisols are soils common to grassland or prairie regions. Mollisols have thick, very dark-brown to black surface horizons and are quite fertile. The dark color of the A-horizon is the result of humus enrichment from the decomposition of plant litter. Mollisols are among some of the most important and productive agricultural soils in the world.

Oxisols develop in tropical and subtropical regions where high precipitation and high temperatures drive intense weathering. Oxisols are depleted in nutrients and contain mostly resistant minerals such as quartz, kaolinite clay, and iron and aluminum oxides, which tend to give them a reddish or orange hue. Decomposing plant material is usually the only organic matter. For the most part, they have a nearly featureless soil profile without clearly marked horizons.

Spodosols are soils that develop under coniferous vegetation in cool, wet environments from parent materials that tend to be coarse-textured and rich in sand. Spodosols are formed by *podsolization*, a process by which organic material, aluminum, and sometimes iron, is leached out of the A- and E-horizons to accumulate in the B-horizon. The E-horizon of these soils is normally light-colored, often looking ashen-gray, while the B-horizon is always darker and usually has a distinctive iron hardpan layer. Spodosols have very low native fertility.

Ultisols are soils that form in areas where warm temperatures and the abundant availability of moisture enhance weathering processes, increasing the rates of leaching and clay formation. These soils contain no calcareous material, and resistant minerals such as quartz, kaolinite, and iron oxides dominate. The A-horizon is often stained red or yellow by a high concentrations of iron oxides. These soils are acidic and have low native fertility. Ultisols may lack a distinct E-horizon, but they will always contain a zone where clay accumulates.

Vertisols contain a high percentage of clay minerals that expand and contract in response to the presence or absence of moisture. When wet, soil volume can increase significantly; when dry, shrinkage can cause large cracks to form. This swelling means that water is transmitted very slowly, limiting leaching so that vertisols usually lack well-developed horizons, but they typically have good native fertility. Vertisols are common in areas with shale parent material and heavy precipitation.

Soil Survey Maps

Each of the 12 soil orders listed above is subdivided into a number of suborders, which are further subdivided until you reach the smallest unit, soil series. As a testament to the amount of variation possible, there are more than 19,000 soil series currently recognized in the United States alone. The soil series classification system is used to relate conceptual taxa to physical map units, though in reality there is a disconnect between the two systems. It must be remembered that in reality soil is a continuum with properties that can vary enormously both with horizontal distance and with depth. A soil survey map is an attempt to break this continuum up into discrete units that show similar properties. Mapped units are given a series name and a taxonomic designation for all of the higher levels.

It is important to have a basic understanding of how to read soil survey maps because they are frequently used by law enforcement, are freely available online, and provide a good overview of the variety of soils in a region. Soil maps in the United States were originally released as large paper volumes, but today they are available online at the Web Soil Survey website (http://websoilsurvey.nrcs.usda.gov) or through the Soil Data Mart, a download site (http://soildatamart.nrcs.usda.gov). The simplest way to generate a soil map is at the Web Soil Survey site. Start by clicking on the "Start WSS" button. A new window will open showing a map of the continental United States and with the Area of Interest (AOI) tab active. Other geographic areas of the United States can be selected by clicking on the "View Extent" pull-down menu.

The first step is to select your AOI. There are several methods available for creating an AOI: the navigation window allows for address, state, and county, longitude and latitude, PLSS (Public Land Survey System), and a number of other inputs. You can also simply drag and click in the map window to zoom in on a particular area. To set the map scale, click on the "Scale" button. All soil maps are also overlain upon aerial photos, where available. A full soil survey includes soil maps, map unit descriptions, soil series descriptions, and taxonomic information.

Soil surveys are supposed to be at least 85% pure, although in reality field checks suggest that this figure is rather lower.

Soil maps and additional information for the United Kingdom are available at the National Soils Resources Institute at Cranfield University (http://www.landis. org.uk/gateway/ooi/welcome.cfm). An introduction to the Canadian Soil Classification System is available online from the University of Alberta Environment Studies program (http://www.environment.ualberta.ca/SoilsERM/ class.html). The complete Canadian System of Soil Classification, third edition, (http://sis.agr.gc.ca/cansis/references/1998sc_a.html) and the Canadian National Soil Database are online at the Agriculture and Agri-Food Canada website (http:// sis.agr.gc.ca/cansis/nsdb/intro.html). Many other countries also have soil information available online as well.

USDA Textural Classification

Soil survey maps use the USDA Textural Classification system, which is based solely on the grain size distribution of a sample (Table 7.5). It is very important to note that the size divisions used here are slightly different than the Udden–Wentworth Grain Size Scale used by sedimentologists introduced as Table 5.2 in Chapter 5.

This system employs a tertiary diagram to determine texture classification (Figure 7.19) and is based on the relative percentages of sand, silt, and clay present in a sample. For example, if you determine that your soil sample contains 40% clay, 30% silt, and 30% sand, you would plot these percentages on the diagram to identify the soil's texture as "clay loam." Using this system, any gravel or larger material is ignored for the plotting. For example, if you have a sample that has 10% gravel, 65% sand, 20% silt, and 5% clay, the 10% gravel is discounted. The percentages of sand, silt, and clay are corrected to add up to 100%, as follows: each component's percentage is divided by the total percentage of sand, silt, and clay (in this case 90%) or 0.9:

Sand 65/0.9 = 72.2%   Silt 20/0.9 = 22.2%   Clay 5/0.9 = 5.6%

You plot the new percentages on the diagram to determine the classification of the sample. This sample fits into the category "sandy loam" on the diagram.

Table 7.5 USDA and USCS grain size scales.

USDA Grain Size	Diameter Range (mm)	USCS Grain Size	Diameter Range (mm)
Cobbles and Boulders	>75	Boulders	>300
		Cobbles	75–300
Gravels	2–75	Coarse Gravel	19–75
		Fine Gravel	4.8–19
		Coarse Sand	2.0–4.8
Sands	0.05–2	Medium Sand	0.43–2.0
		Fine Sand	0.08–0.43
Silts	0.002–0.05	Silts	<0.08
		Clay	<0.0075
Clay	<0.002		

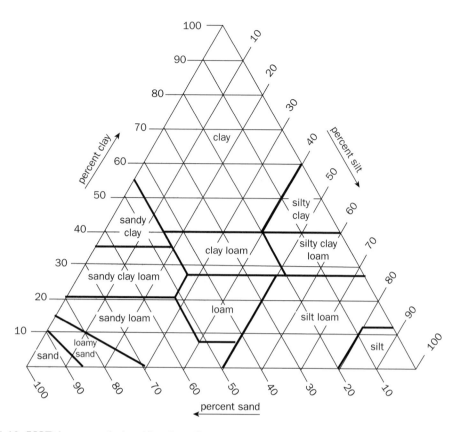

Figure 7.19 USDA textural classification diagram.

Because of the 10% gravel in the sample, you would classify the sample as "sandy loam with gravel." If the percentage gravel were very high, you would classify the sample as "gravelly sandy loam."

The ASTM Unified Soil Classification System (USCS): D-2487

A different system, the USCS classification of soils, is commonly used in the environmental industry in the United States and is based on a combination of visual observation and laboratory tests. This system was co-developed by the United States Bureau of Reclamation and the Army Corp of Engineers in 1942, based on the airfield classification system developed by Arthur Casagrande in 1932. This is the most widely used soil classification system in geotechnical engineering for non-highway and non-airway projects. (For roads, the AASHTO system is used.)

The USCS uses slightly different particle size ranges than the systems discussed previously (Table 7.5). In this system, particle mineralogy is not considered and soils are divided into two broad categories. If more than 50% of the grains are larger than No. 200 sieve mesh (0.075 mm) the soil is considered to be a *coarse-grained soil*. If more than 50% of the grains in a sample are smaller than No. 200 sieve mesh, the soil is called a *fine-grained soil*. In the absence of sieves, reference charts such as Figure 7.20 can be used to help estimate the relative percentages of

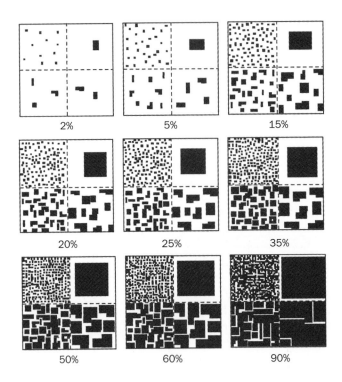

2%   5%   15%

20%   25%   35%

50%   60%   90%

Figure 7.20 Chart for determining approximate percentages of minerals/objects in a sample. Each quadrant of a given box contains the same total area covered. They just use different-sized objects.
Source: Soil Survey Staff, 1999.

materials. Soils are further subdivided into classes on the basis of sorting, called *grading* by engineers, and *plasticity* (how much the soil behaves like modeling clay). There are actually two versions of the USCS method for describing soils: the Laboratory Method and the commonly used field-based Visual–Manual Method.

In the USCS system, the distribution of particle sizes is called *gradation* rather than sorting. A *well-graded* soil is defined as having a good representation of all particle sizes from the largest to the smallest and the shape of the grain size distribution curve is considered "smooth." This is the equivalent of poorly sorted soil. There are a couple of different terms used to describe *poorly graded* soils. If the soil is well sorted, i.e. all of the particles are approximately the same size, the soil would be referred to as *uniformly graded*. If the soil consists of two or more distinct particle sizes with an absence of some intermediate particle sizes, the soil is called *gap-graded*.

Particle size distribution graphs are used to determine a soil's grading numerically via *uniformity coefficient* (Cu) and *coefficient of curvature* (Cc). Reading off of the graph, Cu is defined as the ratio of the grain diameter (in millimeters) corresponding to 60% finer by weight (d60) divided by the diameter corresponding to 10% finer by weight (d10). Well-graded (i.e. poorly sorted) gravels have a Cu value of >4, well-graded (i.e. poorly sorted) sands have a Cu value of >6. The Cc is also used to judge the gradation of a soil. It is calculated as follows: $Cc = (d_{30})^2 / (d_{60} \times d_{10})$. Cc values equal to 1 through 3 are considered well graded. Technically, gravels that have both a Cu of >4 and a Cc between 1 and 3 are classified as well-graded gravel (GW) and sands that have both a Cu of >6 and a Cc of between 1 and 3 are classified as well-graded sands (SW). Sands and gravels not meeting these conditions are termed poorly graded (SP and GP respectively).

A brief description of the steps used to classify a soil using the USCS Visual–Manual Classification Method (including Tables 7.6 and 7.7) and some information on the liquid and plastic limit laboratory methods are provided on the companion website to assist readers in interpreting engineering and environmental soil reports.

Rare Minerals Linked to Crime

One rainy night, a local resident of Front Royal, Virginia called the authorities after spotting a body lying in the mud on a boat landing in the Shenandoah River. Police arrived at the scene to find one man dead and another man shot in the chest but still alive. As they processed the scene, investigators found deep tire marks created by a vehicle leaving the scene at high speed. That night and into the next day, Lieutenant James Cornett painstakingly collected soil samples from all of the tire tracks at the scene, as well as from surrounding sites and from locations a significant distance from the scene. He photographed and mapped all of the places from which samples were collected.

Based on testimony from the surviving victim, investigators were able to link a known drug dealer to the crime, but did not initially have enough evidence for an arrest. Seven days later, after they had obtained search warrants, police seized the suspect's Jeep just as he was going to wash it, and placed the suspect under arrest. The Jeep did not have any blood spatter, but investigators did find gunshot residue and a lot of fresh mud in a vehicle that was otherwise in immaculate condition. On most vehicles, you would expect to find a series of soil layers created over time as the vehicle is driven from one location to another. In contrast, the Jeep, normally driven only in an urban environment, had just one layer of mud.

With the Jeep in custody, Cornett began to collect evidence. First, he scraped all of the big chucks of mud into jars intact in order to preserve any textural clues or pieces of gravel. Next, he used wet oversized cotton balls to collect soil off of the wheel wells, inside of the bumpers, and from locations on the underside of the vehicle. He also collected samples from where mud had splashed inside the Jeep. All of the evidence collected from the scene and from the vehicle was then sent to forensic geologist Sergeant Erich Junger for analysis.

Junger examined all of the samples under a microscope. The samples from the crime scene and the Jeep both showed significant rounding and sorting, and clumps showed cross-bedding, all consistent with an alluvial environment. Determining the mineralogy of the samples, Dr Junger found quartz, spodumene, tourmaline, and feldspar, which were common to the area, but he also found bright-blue grains of azurite and bright-green grains of malachite. Both are rare and, because they are copper carbonates, very easily destroyed by chemical weathering. Neither mineral could have traveled far from its point of origin, nor survived for very long in the water. Junger asked Cornett if there was a copper quarry anywhere near the crime scene. Indeed, there was, about 0.8 km upstream. Junger visited the quarry and found a teal blue lake brightly colored by the dissolution of copper carbonate minerals. The rocks there were rich in azurite and malachite. He then moved to the next landing downstream from the quarry, which was also the crime scene, where azurite and malachite grains were found in the soil. Junger then moved to the next landing downstream, where he sampled the soil and found no trace of either mineral. This effectively delineated the distribution of azurite and malachite in the area.

Based on his findings, Junger confidently stated that the suspect's Jeep had visited the crime scene. This evidence, however, did not place the suspect at the scene, just his vehicle. However, following other lines of inquiry, investigators were able to demonstrate that calls were made from the suspect's cell phone to one of the victim's cell phones the night of the shooting and that the calls had originated in the same county where the shooting occurred. Three months later, on the advice of his attorney, the suspect pled guilty to all of the charges against him and was sentenced to life.

Scene Examination

Given the widely variable characteristics of soils, the least useful method of collection is to simply grab a shovel, dig up several centimeters of soil, and toss it into a bag. This homogenizes the sample, potentially losing significant information. For example, if there is a thin layer of fine sand or reddish mud overlying a soil that is generally a cohesive, brown clay, the incongruent material should be sampled separately. To maximize the information that can be obtained, a standard method that involves the collection of intact soil profiles or of multiple small samples of the entire soil profiles should be used. Using a clean tool, soil sections should be cut out in the shape and size of an airtight collection container. Once the sample is inserted into the container, its original orientation should be marked. If a vertical profile, such as a grave, is being examined, sequential unmixed samples along the entire exposed column should be collected, as well as samples of the mixed materials on the surface and at the base. For a detailed mineralogical analysis, Munroe (1995) recommends collecting soil horizon information by softening the soil with distilled water (or acetone if the soil is coarse-grained) and then pressing down a non-adhesive sheet of acetate.

The goal of such sampling is to characterize an area's soil as thoroughly as possible. Take copious photographs and detailed field notes describing the setting, visible mineralogy, composition, texture, bedding, soil structure, organic content, and the lateral extent of the various components. In the previous case study, it was Lieutenant Cornett's quick and comprehensive work that allowed forensic geologist Junger to link the mud on the Jeep to the crime scene. Remember, the crime occurred on a rainy night. The extra water caused larger-than-normal grains of fresh material to be carried downstream from the mine to the crime scene. If investigators had waited for days or weeks to sample the soil, the vital azurite and malachite grains could have broken down.

Visual Examination of Soil Evidence

One of the most common investigations conducted by a forensic geologist is the examination and comparison of soils samples. Because soil is pervasive on the surface of the Earth, it is commonly found adhering to almost everything, including people, clothing, and physical evidence items. Samples from a crime scene, a suspect's property or effects, or other known sources often must be compared in order to determine if there is a relationship between the samples. The question the examiner must answer is whether the samples could have come from the same place. It is important to recognize that when comparing soil samples the examiner is not looking for a *match*, but instead is looking for characteristics that would *exclude* the possibility of a common origin.

There is a wide range of simple physical properties that distinguish *in situ* soils, but many of these properties can be altered when that soil is removed from its place of origin. Many properties can also vary widely, depending on the season or over a very small area. The properties that are of interest to a forensic geologist are those that remain relatively uniform over a small area but vary considerably over wider distances, and are constant properties of the soil that do not change upon deposition or removal. None of these properties is ideal but each of the following are commonly used to identify soil samples: *color* (which as previously discussed, can be problematic), *mineralogy* (usually via optical mineralogy, but also by X-ray diffraction etc.), *grain size distribution* (discussed in more detail below), and *density separation* by means of density gradient columns (discussed in Chapter 5, this method is more applicable to soils but still has some significant liabilities; though it can be very important under some circumstances, it will not be discussed further here). In addition to classifying soils, the forensic examiner is often expected to identify and compare substances contained in the soil, such as glass, paint chips, brick chips, fertilizer, cinders, and materials that would not commonly be considered soil at all, such as kitty litter. Different potting soils, which are manufactured for use in gardening and commonly contain significant quantities of gypsum, perlite, and sphagnum, are also often distinguishable from each other using these techniques.

Examination Procedures for Soil Samples

One approach to a detailed geologic analysis of soils, originally developed by Skip Palenik, can be found in Murray and Solebello (2002), Murray (2004), Ruffell and McKinley (2008), and Fitzpatrick, Raven, and Forrester (2009). A slightly different approach is found in Sugita and Marumo (1996). The following is an overview of a very simplified procedure *intended for student training purposes*. A more detailed version of the procedure is available on the companion website. Remember, the point of this examination is to see whether you can exclude samples from sharing the same point of origin, which means that you can stop the procedure at any point where the two samples are clearly, demonstrably different:

1. Only work with one sample at a time and unpack the sample while holding it over a large sheet of white butcher paper labeled with the sample number.
2. Conduct a brief visual survey of the sample, noting its general characteristics.
3. *Debride* the sample by removing visible leaves, roots, shells, fibers, paint chips, and other detritus and placing them to the side. These items should then be placed in separate, appropriately labeled containers.
4. If the soil is still moist, determine the color of the soil and then allow the sample to air-dry for a minimum of 12 hours. If soil moisture content is important, place the sample into a 104.5 °C drying oven and use the

gravimetric method described previously. Soil scientists often look at the damp color of soils, but for forensic applications it is often better to use only the color of dry samples for comparison.

5. With a portion of the dry soil sample placed on the labeled white paper, note the relative color of the sample using a Munsell Soil Color Chart. Also, make sure that you note the sample's condition (e.g. 10YR 5/3 dry). If you are examining a soil profile section, there might be several color changes along the sample, and there might be *mottling* (spots or blotches of color), in which case all of the color information should be recorded.

6. Section the sample into two equal portions, if possible, and keep one portion in reserve in an appropriately labeled container.

7. Perform a *grain size distribution analysis* on the remaining portion of the sample using a sub-sample of at least 5 g, preferably 15–20 g, and create a semi-logarithmic plot of the *grain size distribution* using particle size (log scale) versus percent finer (normal scale) (Figure 7.21).

8. Perform an *analysis of the clean color of the largest grain fraction*: for large samples, or where there are clear points of distinction, perform this procedure individually on the fractions from each pan for the largest size fractions (e.g. Nos. 5, 10, 35, and 60). For small samples, you can combine the contents of those sieves. Place each sample in a centrifuge tube with 0.1% sodium hexametaphosphate in distilled water.

9. Shake or ultrasonicate for at least five minutes, or longer if you continue to see a color change. Light materials such as seeds, leaf fragments, and anthropogenic debris, like plastic wrappers or paint chips, will usually float and should be separated off and placed into a sample container.

10. Once you no longer see a color change, centrifuge the tube to spin the solids off from the water. Carefully pour off the excess liquid into a labeled sample container. Dump out the wet solid sample into a metal or glass container, place it in an oven set between 95 and 105 °C and allow it to dry.

11. Determine the Munsell color of the largest fraction and re-examine it under a low-powered microscope, as described previously in Chapter 5.

12. If merited, perform a *heavy mineral separation* (see companion website for details). In general terms, heavy minerals are defined as having a specific gravity (SG) greater than $2.9 \, \text{g/cm}^2$, and light minerals as having an SG less than $2.9 \, \text{g/cm}^2$. Many of the liquids traditionally used for this task are highly toxic and their use should be avoided; however, sodium polytungstate (SPT) has an SG of 2.89 and is non-toxic, so it can be used without a fume hood, though gloves should always be worn. It also evaporates very quickly, so bottles should never be left uncovered and work must be done quickly and efficiently. All labware used for this procedure much be washed immediately after use, as SPT will dry into a concrete-like solid.

13. Examine the cleaned and/or separated mineral fractions under the microscope. Slides may be prepared for detailed mineralogical assessment by using Cargille Meltmount. The procedures described in Chapter 6 for using a petrographic microscope can be employed to identify most of the minerals you will usually find. See Table 7.8 for the optical properties of most common minerals (a longer version is available on the companion website). You will also want to refer to Table 2.4 for habit and cleavage information.

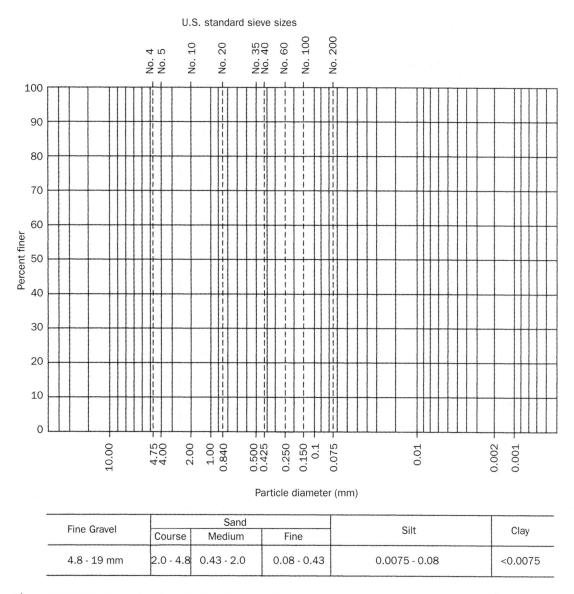

Figure 7.21 U.S. standard sieve sizes

Fine Gravel	Sand			Silt	Clay
	Course	Medium	Fine		
4.8 - 19 mm	2.0 - 4.8	0.43 - 2.0	0.08 - 0.43	0.0075 - 0.08	<0.0075

Figure 7.21 Blank grain size distribution graph.

14. To perform an *analysis of the finest grain fraction* re-examine the color of the fractions from the Nos. 120 and 230 sieves and the catch pan. Compare the color of the fine-grained fraction with the color of the material that was coating the coarse-grains. If the color is different, additional analysis of the grain coating, reserved in a sample container in step 10, might be useful.

15. Examine the material from the #230 pan for plant opals, diatoms, or other microfossil or biological materials.

16. Take the finest fraction from the catch pan (include the material from the #230 pan as well if there is too little sample), and prepare a sample for X-ray diffraction spectrometry (XRD) analysis. This last step would only be taken if necessary and if at all other points the samples being compared cannot be differentiated.

Table 7.8 Optical properties of common or important minerals, based on standard 0.03 mm thin-section preparation.

Mineral	Color in Thin Section	Crystal System	Optical System	Optic Sign	2V°	Mean RI	Mean Birefringence
Actinolite (Amphibole)	Colorless, pale to deep green; x-plr up to mid-2nd order	Monoclinic	Biaxial	–	74–88	1.61–1.63	0.016–0.030
Albite – Anorthite (Plagioclase)	Colorless; x-plr shows twinning (dark and light banding)	Triclinic	Biaxial	+ or –	75–90	1.53–1.59	0.007–0.013
Anatase	Pale brown, reddish brown, deep brown; x-plr creamy upper order colors	Tetragonal	Uniaxial	–		2.49–2.56	0.073 very high
Anhydrite	Colorless; x-plr vivid 2nd order to low 3rd order	Orthorhombic	Biaxial	+	42	1.57–1.61	0.044
Anorthoclase (Potassium/ Alkali Feldspar)	Colorless, sometimes cloudy; x-plr complex twinning	Triclinic	Biaxial	–	0–55	1.52–1.54	0.005–0.008
Apatite	Colorless or pale; x-plr 1st order gray up to lower 2nd	Hexagonal	Uniaxial	–		1.63	0.003
Aragonite	Colorless; x-plr. creamy high order	Orthorhombic	Biaxial	–	18	1.53–1.68	0.155 extreme
Augite (Clinopyroxene)	Colorless, gray, pale green, pale brown; x-plr up to middle 2nd order	Monoclinic	Biaxial	+	25–60	1.68–1.75	0.024–0.029
Azurite	Blue, lt. purple, pleochroic; x-plr 4th order and above	Monoclinic	Biaxial	+	64–68	1.73–1.83	0.108 extreme
Beryl	Colorless to pale; x-plr 1st order gray or white	Hexagonal	Uniaxial	–		1.56	0.004–0.008

Table 7.8 *continued*

Mineral	Color in Thin Section	Crystal System	Optical System	Optic Sign	2V°	Mean RI	Mean Birefringence
Biotite Mica	Brown, brownish green; x-plr up to 3rd or even 4th order	Monoclinic	Biaxial	–	0–33	1.57–1.61	0.028–0.081
Calcite	Colorless w/ rhomb. cleavage; x-plr high order pastel creamy	Hexagonal	Uniaxial	–		1.48–1.65	0.172 extreme
Chalcedony (Quartz)	Colorless, pale-brown, pale, feathery or fibrous, conchoidal fracture	Hexagonal	Uniaxial	+		1.54	0.009
Chalcopyrite	Opaque	Tetragonal					
Chlorite	Light to medium green, pleochroic; x-plr 1st order white or yellow, anom., colors common	Monoclinic	Biaxial	+ or –	0–50	1.56–1.61	0.006–0.020
Corundum	Usually colorless or pale	Hexagonal	Uniaxial	–		1.76	0.005–0.009
Diopside	Colorless or pleochroic green; x-plr upper 2nd order	Monoclinic	Biaxial	+	56–63	1.66–1.75	0.031
Dolomite	Colorless; x-plr. creamy pastels	Hexagonal	Uniaxial	–		1.50–1.67	0.177–0.185 extreme
Enstatite (Orthopyroxene)	Pale pinkish, greenish, yellowish, brownish; x-plr up to 1st order yellow	Orthorhombic	Biaxial	+ or –	58–90	1.65–1.72	0.008–0.016
Fluorite	Usually colorless or very pale w/ oct. cleavage	Cubic	Isotropic			1.434	
Galena	Opaque						

continued

Table 7.8 *continued*

Mineral	Color in Thin Section	Crystal System	Optical System	Optic Sign	2V°	Mean RI	Mean Birefringence
Garnet	Colorless to pale pink dodecahedron	Cubic	Isotropic			1.71–1.87	
Graphite	Opaque	Hexagonal					
Gypsum	Colorless w/ cleavage; x-plr 1st order gray & white	Monoclinic	Biaxial	+	58	1.52	0.010
Halite	Colorless w/ cubic cleavage	Cubic	Isotropic			1.54	
Hematite	Deep red-brown to opaque	Hexagonal	Uniaxial	–		3.00–3.20	0.210–0.280 extreme
Hornblende	Green, yellow-green, blue-green, brown pleochroic; x-plr lower 2nd order	Monoclinic	Biaxial	–	52–85	1.65–1.67	0.020
Hypersthene (Pyroxene)	Pale colored	Monoclinic	Biaxial	+	25–60	1.68–1.75	0.024–0.029
Ilmenite	Opaque	Hexagonal					
Limonite (Goethite)	Dark red, yellow, orange-red, red brown	Amorphous (Orthorhombic)	(Biaxial)	(–)	(0–27)	2.0–2.4 (2.26–2.52)	(0.150)
Magnetite	Opaque	Cubic					
Malachite	Nearly colorless to deep green, pleochroic; x-plr 3rd, 4th, even 5th order	Monoclinic	Biaxial	–	43	1.65–1.90	0.254 extreme
Microcline (Potassium /Alkali Feldspar)	Colorless, clouding; x-plr tartan/cross-hatched twining	Triclinic	Biaxial	–	65–88	1.51–1.53	0.005–0.008
Muscovite Mica	Colorless; x-plr vivid 2nd order up to 3rd order	Monoclinic	Biaxial	–	35–50	1.56–1.60	0.036–0.054
Oligoclase/ Labradorite/ Anorthite (Plagioclase Feldspars)	Colorless; x-plr shows twinning (dark and light banding)	Triclinic	Biaxial	+ or –	75–90	1.53–1.59	0.007–0.013

Table 7.8 *continued*

Mineral	Color in Thin Section	Crystal System	Optical System	Optic Sign	2V°	Mean RI	Mean Birefringence
Olivine	Pale yellowish-green, high relief; x-plr 1st, 2nd and vivid 3rd order interference colors	Orthorhombic	Biaxial	+ or −	47–90	1.63–1.88	0.035–0.052
Opal	Colorless, brown, gray, brown, or other	Amorphous	Usually Isotropic			1.43–1.46	Anom. due to strain
Orpiment/Realgar	Straw yellow color/ yellowish-orange; x-plr 3rd order bright pink, green/ bright blues, greens colors	Monoclinic	Biaxial	+	30–76	2.40–3.00	0.620 extreme
Orthoclase (Potassium / Alkali Feldspar)	Colorless, clouding; x-plr 1st order white	Monoclinic	Biaxial	−	40–70	1.51–1.53	0.005–0.008
Phlogopite Mica	Colorless – pale brown, pleochroic; x-plr 3rd order colors, up to 4th	Monoclinic	Biaxial	−	0–33	1.57–1.61	0.028–0.081
Plagioclase Feldspars	Colorless, sometimes cloudy; x-plr twinning (black and white)	Triclinic	Biaxial	+ or −	75–90	1.53–1.59	0.007–0.013
Pyrite	Opaque	Cubic					
Quartz	Colorless, may be cloudy; x-plr 1st order whitish-gray, tinge of yellow	Hexagonal	Uniaxial	+		1.54	0.009
Rutile	Reddish-brown, pale brown to almost opaque; x-plr high order white	Tetragonal	Uniaxial (may be anom. Biaxial)	+	(Anom. 10)	2.61–2.90	0.286–0.287 extreme
Sanidine (Potassium/ Alkali Feldspar)	Colorless; x-plr 1st order white may show twinning	Monoclinic	Biaxial	−	Varies (0–47)	1.51–1.53	0.005–0.008

continued

Table 7.8 *continued*

Mineral	Color in Thin Section	Crystal System	Optical System	Optic Sign	2V°	Mean RI	Mean Birefringence
Serpentine	Colorless to pale green; x-plr 1st order white gray	Monoclinic (also Triclinic or Hexagonal)	Biaxial	−	Highly variable	1.53–1.6	0.001–0.010
Spinel	Colorless, green, blue, red	Cubic	Isotropic			1.72–1.74	
Sulfur	Pale yellow or yellowish-gray; high order white	Orthorhombic	Biaxial	+	69	1.95–2.24	0.287 extreme
Talc	Colorless; x-plr 3rd order colors	Monoclinic	Biaxial	−	0–30	1.54–1.58	0.046–0.050
Topaz	Colorless to pale; x-plr 1st order white to pale yellow	Orthorhombic	Biaxial	+	44–66	1.61	0.008–0.010
Tourmaline	Highly variable, strongly pleochroic; x-plr up to high 2nd order	Hexagonal	Uniaxial	−		1.62–1.67	0.021–0.029
Tremolite (Amphibole)	Colorless, pale to deep green; x-plr up to mid-2nd order	Monoclinic	Biaxial	−	74–85	1.61–1.63	0.016–0.030

x-plr = cross-polarized light and is description of interference colors or other important information.
anom. = anomalous.

An Introduction to X-ray Diffraction Spectrometry (XRD)

The procedures discussed so far work well with soils that contain sand-sized particles, but what if your soil is composed predominantly of mud or you have reached step 16 in the analysis procedure and still need more information? Optical microscopy is no longer particularly useful for determining mineralogy, as the particles are just too small. Now we need to introduce another new tool. X-ray diffraction spectrometry, usually called XRD, is one of the most powerful analytical tools available for the identification of crystalline materials such as minerals. XRD uses X-rays, a form of electromagnetic radiation with a wavelength between 10 to 0.01 nm (nanometers) and typical photon energies between 100 eV and 100 keV, to interact with crystalline materials in order to get information about their atomic structure (the arrangement of atoms in a crystalline structure).

Like light and all other forms of electromagnetic radiation, X-rays have properties of both waves and particles. Since X-rays are of the same size range as atoms, they can be used to investigate the structural arrangements of molecules in a wide range of materials (Figure 7.22). In a simplified way, you can think of this as tossing ping-pong balls at a structure in a dark room. By throwing the balls from

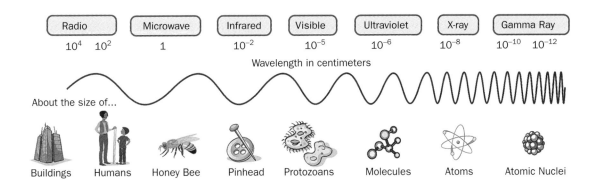

Figure 7.22 Electromagnetic spectrum.
Source: Courtesy of NASA.

various points in the room and carefully determining which balls ricochet off the structure, it is possible to infer the form of the hidden structure. This is the basic idea behind XRD.

X-rays are generally produced using an X-ray tube, a vacuum tube that contains a cathode, which loses electrons, and an anode that collects the electrons. In most X-ray tubes, electrons are emitted from a tungsten filament (cathode) and accelerated across a vacuum by a high-voltage field (15 to 60 kV, though most commonly 40–45 kV) to strike a solid metal target (anode). The most common anodes used in XRD applications are copper (Cu), molybdenum (Mo), iron (Fe), and chromium (Cr). Most geology laboratories use copper tubes. When electrons collide with atoms in the metal target, two different major forms of radiation are generated. The first is called *white* or *continuous radiation* (or sometimes Bremsstrahlung, which means breaking radiation). It is composed of a continuous spectrum of X-rays, the lower limit of which is a function of the voltage across the X-ray tube. Basically, the electrons bombarding the anode as a continuous stream all lose energy in a series of steps, producing a continuous range of wavelengths for the X-rays generated. The wavelength distribution of white radiation is independent of the target material (i.e. all X-ray tubes produce the same white radiation) (Figure 7.23).

The second form of X-rays generated is directly dependent on the target material. The high-energy electrons that strike the anode metal will also occasionally interact with one of the electrons in the inner orbitals of an atom transferring enough energy for the electron to escape. This leaves a vacancy that is immediately filled by an electron that drops in from an orbital further from the nucleus. The energy lost in the drop generates electromagnetic energy (X-rays) of a particular frequency (ν) that depends on the "size" of the jump. The energy difference between orbitals is a function of the number of protons in the nucleus and is therefore different for every element (which is the principle underlying an elemental analysis technique called X-ray fluorescence or XRF). More importantly for XRD, all atoms of the same kind emit exactly the same wavelength of X-rays, thus the X-rays generated are known as *characteristic radiation* (Figure 7.23).

For example, Cu tubes emit 8 keV X-rays and Mo tubes emit 14 keV X-rays, with corresponding wavelengths of 1.54 Å and 0.8 Å, respectively. The energy of a photon is related to its wavelength by the equation $E = hc/\lambda$, where h is Planck's

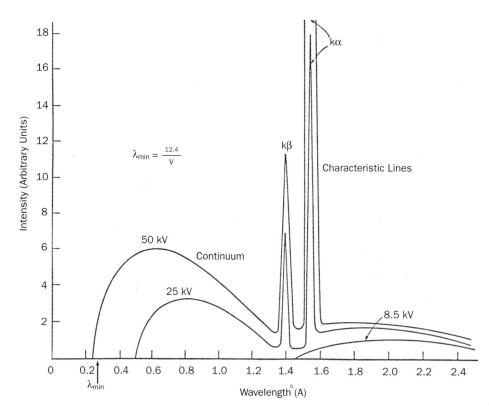

Figure 7.23 Continuous and characteristic radiation for copper. Kα1 is off the scale. *Source:* Jenkins and Snyder, 1996.

constant, c the speed of light, and λ is wavelength. In simple terms energy and wavelength are inversely proportional, thus the shorter the wavelength, the greater the energy. That is why exposure to radio waves (long wavelength radiation) is not a problem, but you have to limit exposure to X-rays (short wavelength radiation), which can be used to "see" inside the human body, not to mention all of your luggage.

When an incoming electron helps pop an electron from the K shell (innermost orbital), the most probable candidates to drop in and fill the resultant vacancy are the closest electrons, so from either the L- or the M-shell (Figure 7.24). The electron that makes this jump most commonly is from L3, resulting in an emission known as K-alpha1 (Kα1) radiation. About half as often, the electron from L2 fills the vacancy to form Kα2 radiation. For reasons beyond this discussion, the electron L1 never contributes. Electrons from the M- (or N-) shell can also sometimes make the jump, forming what is called Kβ radiation. The ratio of intensities for Kα1:Kα2 is always about 2:1 for all of the elements, while Kβ usually ranges from 15 to 30% of Kα (Figure 7.23).

The beam of X-rays that is directed at a crystalline sample in order to determine its structure should be as near to single wavelength (Kα) as possible. Imagine that instead of just throwing ping-pong balls at a structure, you start throwing tennis balls, basketballs, beach balls, and marbles. That would make it much harder to decipher the actual configuration of the hidden structure. In order to narrow the incident radiation down as closely as possible to a single wavelength, a metal filter is used to absorb Kβ radiation. Metal filters are used in a similar way to glass

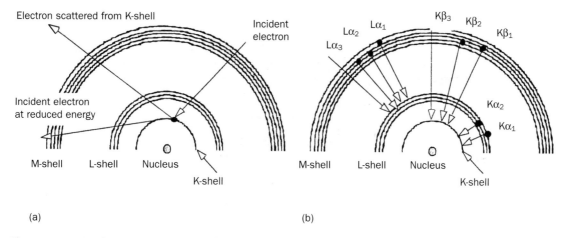

Figure 7.24 (a) Incoming incident electron transfers enough energy that an electron from the K-shell ejects, leaving a vacancy, (b) shows the many possible electrons that can make the jump to fill the vacancy. Only one of these "drops" will occur. The electron that fills the vacancy will emit an X-ray photon.
Source: Modified from Moore and Reynolds, 1997.

Figure 7.25 The incident X-ray beam causes the atoms to oscillate about their lattice points in the crystal structure, scattering radiation spherically. Constructive interference is occurring at the intersections of the secondary waves, for example at the points labeled "*a*" and "*b*".
Source: Courtesy of Christophe Dang Ngoc Chan.

optics, in order to shape, orient, and control a beam of X-rays. The general rule of thumb is that the element one or two atomic numbers less than the anode will have the right characteristics for eliminating $K\beta$ radiation.

The resultant $K\alpha$ radiation is then directed at a crystalline sample (incident beam) and interacts with each atom in the crystal, exciting the electrons in the sample. An electron in the path of the X-ray beam vibrates with the frequency of the incoming radiation becoming a secondary point source of X-rays with the same energy as the incident X-rays, a process called *scattering* (Figure 7.25). Technically, the atomic nucleus is not actually the point source of X-rays, but the electrons surrounding the atom scatter X-rays so that they appear to emanate from the center of the atom. These *secondary X-rays* form interference patterns, much like the interference patterns formed by dropping two rocks into water.

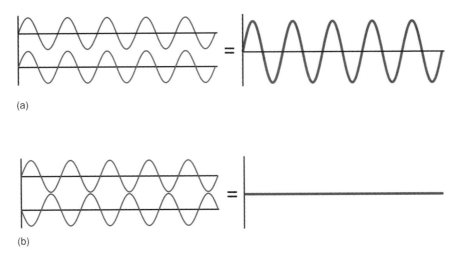

(a)

(b)

Figure 7.26 (a) Constructive interference occurs when waves are in-phase: the peaks and troughs are exactly aligned. The resulting wave has the same wavelength, but the amplitude is doubled; (b) when waves are not in-phase, the peaks and troughs are not aligned: destructive interference occurs. The resultant wave has a lower amplitude and differing wavelength. When the waves are exactly out of phase, they essentially cancel each other out.

A crystal is a complex but orderly repeating arrangement of atoms in which all the atoms in the path of the incident X-ray beam will scatter X-rays simultaneously. Most of the scattered X-rays will destructively interfere, essentially canceling each other out (Figure 7.26b). However, in certain specific directions the scattered X-rays will be "in-phase" and will merge to form a new wave (Figure 7.26a; Figure 7.25). This process of constructive interference is called *diffraction*. Whether the interference is constructive depends on the inter-atomic distances of the crystal and the X-ray's angle of incidence, and is therefore reflective of the sample's crystal structure. Since each crystal is unique, the angles of constructive interference will form unique diffraction patterns.

Part of the definition of the term *crystal* is that the arrangement of the atoms is well constrained and that the entire crystal can be constructed by repeating a specific pattern of atoms called a *unit cell* (similar to the crystal systems discussed in Chapter 6). The directions of possible diffractions depend only on the size and shape of a crystal's unit cell. The intensities of the diffracted waves also depend on the kind and arrangement of atoms in the crystalline structure. The geometry of X-ray diffraction from a crystal is used to determine the unit cell dimensions, and the intensities of the diffracted rays are used to establish the arrangement of atoms.

This might make more sense if you visualize the simplest case, where the unit cell is cubic and composed of only one type of atom (Figure 7.27). In this case, all of the edges of the cube are of equal length, usually designated by the symbol "a" and the corners form right angles. Now, imagine the nucleus of each atom in your cubic structure is bisected by an imaginary plane (Figure 7.27). Each face of the cube is one such plane. The electrons of each of these planes will scatter incident X-rays, acting like a set of parallel mirrors. Each plane will partially "reflect" the incoming X-rays. The scattered secondary X-rays originating from the electrons radiate out in all directions, and are therefore mostly out of phase and interfering destructively (Figure 7.28b). However, at certain orientations, waves that are

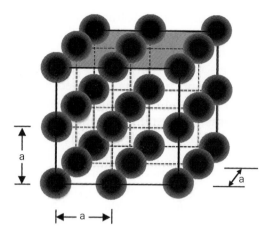

Figure 7.27 Cubic crystalline lattice composed of like atoms. The gray box represents one of the imaginary planes.

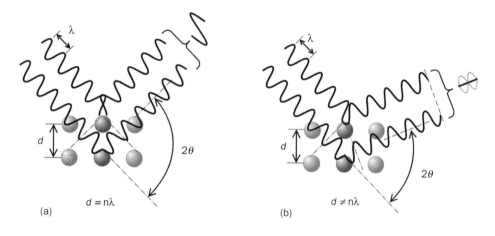

(a) $d = n\lambda$ (b) $d \neq n\lambda$

Figure 7.28 (a) Constructive interference only occurs when the path length difference between reflected waves is an integer multiple of the wavelength; (b) otherwise, the waves will experience destructive interference and cancel out.

Source: Courtesy of Christophe Dang Ngoc Chan.

reflected by adjacent planes can interfere constructively when the wave hitting the second (or third or fourth) plane travels a distance that is equivalent to a whole wavelength further than the wave that is reflected by the first plane. This means that for constructive interference the path length difference between reflected waves must be an integer multiple of the wavelength (i.e. $1 \times \lambda$, $2 \times \lambda$, . . .), or more simply put, the waves must be at exactly the same place in their cycling (Figure 7.28a).

This is called the *Bragg's condition for constructive interference* and is summarized by Bragg's Law:

$$n\lambda = 2d \sin \theta,$$

where λ is the wavelength, d is the inter-atomic separation (d-spacing), and θ is the angle at which the incident X-ray hits the plane of the sample. The n defines the order of the reflection, with n = 1 giving first-order, n = 2 giving second-order, etc. Simple geometry demonstrates that for two waves reflected by adjacent planes, the wave that is reflected by the second plane has traveled a distance of ($2d\sin\theta$) further than the wave that is reflected by the first plane. It is possible to determine the distance between the atoms in crystalline solids if a monochromatic X-ray source is used (only one λ) and the intensity of diffracted X-rays is measured as a function of θ.

By using Bragg's Law, the constructive interference that occurs at specific values of θ can be used to map out the inter-atomic spacing of a solid. The total intensity of the diffracted beam is also dependent on the number of electrons, which in turn is determined by the chemical identity of the atoms. So far, we have only looked at one orientation of our cubic crystal, but as you rotate the cube it turns out that repeating planes of atoms can be constructed in several different ways (Figure 7.29). To keep track of all of the possible repeats in a structure, imaginary planes are classified using *Miller indices*.

The Miller index is written as three numbers, where each number corresponds to a position (h, k, l) orthogonal to the coordinate system, one for each coordinate in three-dimensional space (just to confuse matters a, b, c is used for crystal coordinates rather than the more familiar x, y, z) (Figure 7.30). To determine the Miller indices for a set of planes, start at the origin of the unit cell (the lower left corner) and then determine where the planes intersect each of the three axes (a, b, c) of the cell. Divide the cell dimension along each axis (a, b, c) by the intercept, the resulting number is the index (h, k, l) respectively. For example, in Figure 7.31, in the first image, the plane intercepts the first atom to the right of the origin, and each successive atom past that, so h = (a/a) = 1, and the plane does not intercept any atoms in the b or c directions, giving a Miller index of (100). In the second figure, the plane intersects the first atom in the a direction, and each successive

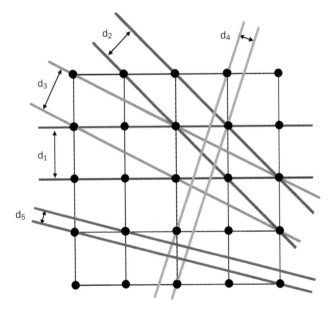

Figure 7.29 Multiple orientations of repeating planes in a cubic structure.

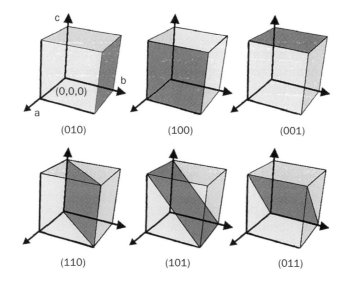

Figure 7.30 Miller indices as applied to a cubic structure.

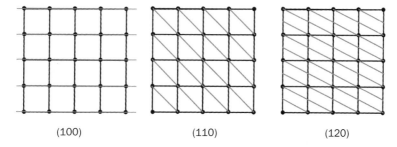

Figure 7.31 Miller indices in a flat plane of atoms.

atom after that (h = 1) and the first atom in the b direction, and each successive atom after than (k = 1), so the index is (110). Only whole numbers are used; so if the intersection occurs halfway along b (as in the last image in Figure 7.31) then k = b/(b/2) = 2.

The important point here is that each set of imaginary planes can satisfy the Bragg's condition and give a constructive interference diffraction peak. To account for all possible planes, the Bragg's Law for a cubic system can be rewritten as:

$$\sin^2 \theta = \left(\frac{\lambda^2}{4a^2} \right)\left(h^2 + k^2 + l^2 \right),$$

where a is the unit cell dimension, λ is the wavelength of the radiation, (h, k, l) are the Miller indices, and θ is the Bragg angle. Obviously, mathematical expressions can get even more complex for other types of unit cells, but the key point here is that you do not just get one or one set of diffractions from a crystalline solid but a whole bunch of diffractions of differing intensities that are reflective of all of the possible repetitions of the atomic structure. This means that all crystalline materials with different chemical formulas and/or different structures (remember the definition of a mineral?) will have unique patterns of diffraction, making identification possible. It also means that to fully describe a crystal, and account for all of the possible repeating sets of planes, it needs to be exposed to incident X-rays from all directions, in order to get all possible diffractions.

The easiest way to do this is to examine a loose powder (such as the fines from a soil sample) that is composed of a very large number of individual crystals and where all possible orientations of each of these crystals will simultaneously be present. This technique, called powder XRD, is used to ensure that the Bragg's condition for constructive interference will be fulfilled for all possible planes generating a diffraction pattern. For geologic work, typical sample preparation often consists basically of grinding the sample in a mortar and pestle if necessary, sieving it, and packing it into a sample holder. The sample holder is then positioned in the XRD and a diffraction pattern is obtained by exposing the sample to a beam of incident X-rays at a series of angles (i.e. either the X-ray tube or the sample slowly rotates, incrementally changing the incident angle θ). A detector is synchronized to rotate in the opposite direction, also at θ to the sample, recording the pattern of diffraction.

The resulting diffraction pattern can be analyzed to determine the structure and the unit cell dimensions of a sample. For the simple cubic unit cells, all possible combinations of integer Miller indices are possible, but for more complex crystalline structures there can be certain combinations of Miller indices that are eliminated by destructive interference. Once a diffraction pattern has been generated, Bragg's Law is used to convert the angles at which diffraction occurs (θ) into atomic spacings (d). From there it is possible to determine unit cell dimensions, and ultimately ionic radii.

The good news, if this discussion has gotten a bit opaque, is that for forensic purposes we are not interested in fully determining the inter-atomic geometry of our samples. Instead, the diffraction patterns can be used somewhat like fingerprints, and by using computer software the identity of the crystalline components of a sample can be determined. Another great strength of XRD is that the technique is able to identify individual crystalline components of mixtures, like soils. The diffraction pattern of a mixture is composed basically of the overlain patterns of the individual components. Pattern-matching computer software can be used not only to identify individual crystalline (or mineral) components but also to give a rough estimate of the proportions. A much more complete discussion of XRD diffraction analysis and its use in the identification of clay minerals can be found in Moore and Reynolds (1997).

A description of *sample preparation for XRD analysis*, which includes methods for creating *random* air-dried samples (in back-loading mounts) and three types of *oriented* samples (air-dried, glycolated, and heated) is available on the companion website.

Interpreting a Diffraction Pattern

For each sample analyzed, a *diffraction pattern*, or *diffractogram*, will be produced. Simple patterns will have only a few clearly defined peaks, while complex patterns could have dozens of peaks (Figure 7.32). In modern analytical laboratories, diffractograms are interpreted using pattern-matching data-processing software such as MDI's Jade or PANalytical's HighScore, but you can also determine what

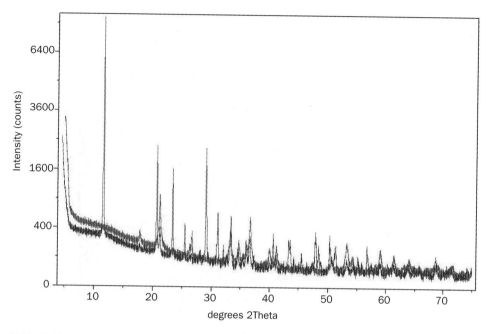

Figure 7.32 Diffraction patterns for two different brands of golden ochre pigment.

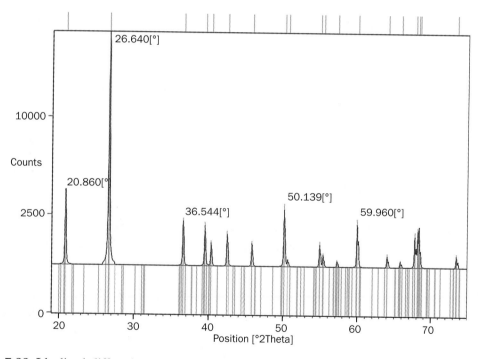

Figure 7.33 Idealized diffraction pattern for quartz.

minerals are present in a sample by hand. This is good practice for understanding how well or how poorly a software package is working. The following example uses Figure 7.33 but will work on any simple diffraction pattern.

To determine what minerals are in the sample you must start by identifying each of the peaks in the diffraction pattern. If there are several peaks, you want to concentrate on the highest peaks in the pattern. Convert each peak into a d-spacing by applying Bragg's Law ($2d \sin\theta = n\lambda$), which converts to $d = (n\lambda / 2 \sin\theta)$. Remember that you have to divide 2θ by two before taking the sine. This conversion can be done using a calculator, or you can create an Excel program, but remember Excel works in radians, so you will need to multiply each θ by pi (3.1415 . . .) and divide it by 180 to convert it from degrees to radians. For this example, $\lambda = 1.54$ Ångstroms, the constant for CuKα radiation and this is a one-dimensional analysis, which means that you can set $n = 1$.

For example, if you look at the peaks in Figure 7.33 you will see that the largest peak occurs at a 2θ angle of 26.6°, the second-largest peak occurs at a 2θ angle of 20.9°, and the third-largest peak occurs at a 2θ angle of 50.1°. It is very important to note that most diffraction patterns do not actually start at zero, because you would never set your instrument to fire X-rays directly into the detector. Instead, most patterns will usually start at a low angle between two and eight degrees 2θ. Make sure that you determine the actual value for the origin of each diffraction pattern you are interpreting. Applying Bragg's Law to the peaks from Figure 7.33 allows us to translate into d-spacings of 3.34Å, 4.25Å, and 1.82Å respectively. Once you know the d-spacings, you try to match them to specific minerals (Table 7.9).

The table lists values for the three largest (most intense) peaks in the form of interplanar spacings (d-spacings), the relative intensities ($I/I_o \times 100$), and the mineral/compound name with a chemical formula. Relative intensity works like this: the largest peak is designated I_o and all of the other peaks are listed as some percentage of that height ($I/I_o \times 100$). For example, a peak that is half as high (has half the intensity) will be listed as (50). Minerals are listed in decreasing order of the d-spacing of the largest peak (i.e. by smallest 2θ value for most intense peak). Minerals listed with the asterisks (*) are the most common minerals, and those designated with a (†) have highly variable structures.

So, for our example in Figure 7.33 we are looking for a mineral that has a d-spacing of 3.34Å for its largest peak, 4.25Å for its second-largest peak, and 1.82Å for its third-largest peak. Looking in Table 7.9, there are three minerals that have peaks around 3.34Å: cinnabar, quartz, and microcline. By looking at the second- and third-largest peaks, it becomes clear that the mineral in Figure 7.33 cannot be either cinnabar or microcline because their respective d-spacings are quite different from those in our figure. However, there is a clear complement with quartz. If we still had several possible candidates, the next step would be to measure the relative intensities of the second and third peaks by placing a ruler on the diffractogram and measuring the height of all three peaks. The height of the tallest peak would be set to I_o and the relative intensities of the second and third peaks would be calculated using $I/I_o \times 100$.

For practice look at Figures 7.34 and 7.35 and see whether you can identify minerals 00-016-0613 and 01-084-1304. The numbers cited are International Centre for Diffraction Data (ICDD) reference codes. The IDCC sells powder diffraction databases that contain information on thousands of materials that can be used to identify an enormous variety of organic and inorganic crystalline material (http://www.icdd.com). It is a fee-based service and quite expensive, but serves as the reference standard for XRD analysis. As an alternative, the online Mineralogy Database has a large X-ray diffraction peak library that lists minerals by largest peak d-spacing (http://webmineral.com/MySQL/xray.php) and allows users

Table 7.9 Inter-atomic spacing of three largest peaks for several common minerals. (A more complete version of this table is available on the companion website.)

D-spacing [Å] (I/I$_o$ %)			
Largest peak	2nd largest peak	3rd largest peak	Mineral/Compound
21.5(100)	4.45(55)	2.56(35)	⋆ Montmorillonite Na$_{0.33}$ (Al$_{1.67}$Mg$_{0.33}$) Si$_4$O$_{10}$(OH)$_2$·(8H$_2$O)
15.0(100)	4.5(80)	3.02(60)	⋆ Montmorillonite Ca$_{0.2}$ (Al, Mg) $_2$Si$_2$O$_{10}$(OH)$_2$·4 (H$_2$O)
14.3(100)	1.53(70)	4.57(60)	† Vermiculite Mg$_{0.6}$(Mg,Al,F^{++})$_3$(Si,Al)$_3$(OH)$_2$·4(H$_2$O)
10.1(100)	2.6(50–41)	3.41 (38)	⋆ Biotite K(Mg,Fe^{++})$_3$AlSi$_3$O$_{10}$(OH)$_2$
10(100)	2.59(100)	4.53(80)	Glauconite (K,Na) (Fe^{+++},Al,Mg)$_2$(Si,Al)$_4$O$_{10}$(OH)$_2$
9.3(100)	4.5(60–40)	1.52 (55)	⋆ Talc Mg$_3$Si$_4$O$_{10}$(OH)$_2$ [May have peak 3.21(90)]
8.52(100)	2.73(46+)	3.16(29+)	Ferrohornblende Ca$_2$[Fe$_{++4}$(Al,Fe$_{+++}$)] Si$_7$AlO$_{22}$(OH)$_2$
8.43(100)	2.72(75+)	3.13(54)	Magnesiohornblende Ca$_2$[Mg$_4$(Al,Fe^{+++})] Si$_7$AlO$_{22}$(OH)$_2$
7.63(100)	4.28(100)	3.06(75)	Gypsum CaSO$_4$·2(H$_2$O)
7.16(100)	1.49(90)	3.58(80–60)	⋆ † Kaolinite Al$_2$Si$_2$O$_5$(OH)$_4$ (broad peaks) [4.36(60–50)]
4.85(100)	4.83(35)	4.33(18)	Gibbsite Al(OH)$_3$
4.42(100)	3.62(90–60)	7.3-.5(90–60)	⋆ † Halloysite, dehydrated Al$_2$Si$_2$O$_5$(OH)$_4$ (broad peaks)
4.4(100–67)	10–10.7 (100–37)	3.34 (100–64) 2.58 (100–53)	⋆ † Illite (K,H$_3$O)(Al,Mg,Fe)$_2$(Si,Al)$_4$O$_{10}$[(OH)$_2$, (H$_2$O)] [may have peak 5.0(35 – 50)]
4.183(100)	2.45(69–25)	2.69(40–30)	Goethite Fe^{+++}O(OH)
3.85(100)	3.21(60–36)	3.44(40)	Sulfur S$_8$
3.45(100)	1.9(22)	2.4(18)	Anatase TiO$_2$
3.5(100)	2.85(35)	2.21(23)	Anhydrite CaSO$_4$
3.39(100)	1.98(63)	3.27(56)	Aragonite CaCO$_3$
3.35(100)	2.863(95–90)	1.98(35–28)	Cinnabar HgS
3.342(100)	4.257(22)	1.82(14)	Quartz SiO$_2$
3.3(100)	3.2(96–83)	4.22(74–70)	† Microcline KAlSi$_3$O$_8$

continued

Table 7.9 *continued*

D-spacing [Å] (I/I$_o$ %)			
Largest peak	**2nd largest peak**	**3rd largest peak**	**Mineral/Compound**
3.26(100)	3.22(90)	3.76(75) 3.23(75)	Sanidine (K,Na)(Si,Al)$_4$O$_8$
3.2(100)	3.18(78)	4.04(60–52)	Anorthite CaAl$_2$Si$_2$O$_8$
3.32(100)	3.23(80+)	3.77(76)	Orthoclase KAlSi$_3$O$_8$
3.18(100)	3.2(90–30)	4.02(81)	† Albite NaAlSi$_3$O$_8$
3.04(100)	1.87(20)	1.91(19) 2.10(16)	Calcite CaCO$_3$
2.88(100)	2.19(22+)	1.79(18+)	Dolomite CaMg(CO$_3$)$_2$
2.82(100)	1.99(57)	1.63(15)	Halite NaCl
2.8(100)	2.7(60–44)	2.77(55)	† Apatite Ca$_5$(PO$_4$,CO$_3$)$_3$(OH,F,Cl)
2.79(100)	1.73(35–27)	3.59(36–25)	Siderite Fe^{++}CO$_3$
2.74(100)	2.1(47)	1.7(38)	Magnesite MgCO$_3$
2.74(100)	2.54(69)	1.7(44)	Ilmenite Fe^{++}TiO$_3$
2.7(100–92)	1.6(100–69)	2.4(79–53)	Pyrite FeS$_2$
2.7(100)	2.5(77–70)	1.7(45–41)	Hematite Fe$_2$O$_3$
2.65(100)	2.96(40–35)	1.58(39–30)	Grossular Ca$_3$Al$_2$(SiO$_4$)$_3$
2.57(100)	1.54(50–30)	2.88(40–33)	Almandine Fe^{++}3Al$_2$(SiO$_4$)$_3$
2.56(100)	10(95–65)	4.46(85) 3.5(65+)	★ † Muscovite KAl$_2$(Si$_3$Al)O$_{10}$(OH,F)$_2$
2.40(100)	1.70(50)	2.77(46)	Lime CaO
1.53(1)	4.58(1)	15.8(0.8)	Hectorite Na$_{0.3}$(Mg,Li)$_3$Si$_4$O$_{10}$(OH)$_2$ (questionable pattern)

to search for minerals by d-spacing. The downside of this service is that there are some errors and some of the data are rather outdated.

It is important to note that the normal variations in the elemental composition of most minerals will cause variations in the resultant diffraction patterns, such as slightly shifting peak location or altering relative intensity. Whenever sodium substitutes for calcium, or magnesium substitutes for iron, the inter-atomic spacing of the crystalline lattice will be altered. This means that mineral structures that support lots of substitutions (like feldspars or apatite) can have a variety of subtly (or not so subtly) different diffraction patterns that are all equally valid. Other minerals, like diamond or halite, which have simple chemical formulas and do not

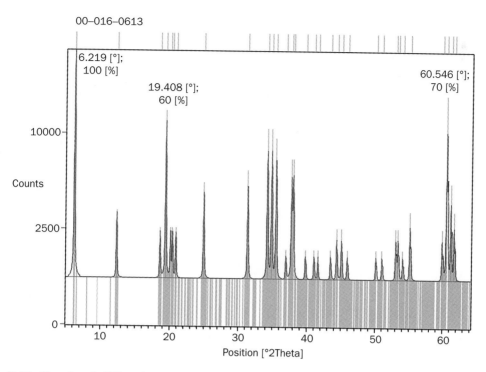

Figure 7.34 Simulated diffraction pattern for mineral 00-016-0613 (ICDD code).

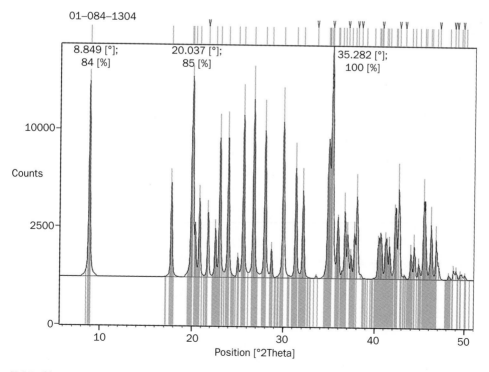

Figure 7.35 Simulated diffraction pattern for mineral 01-084-1304 (ICDD code).

allow much in the way of substitution, will have a more definitive diffraction pattern.

Minerals with poorly formed crystals, like many clay minerals, typically have very broad peaks, while well-formed crystals form very sharp diffraction peaks. Amorphous samples will create diffraction patterns that resemble broad shallow lumps with no clear peaks at all. Variations in moisture can also have a significant impact on inter-atomic spacing, causing shifts in the location of diffraction peaks. This is why it is sometimes necessary to heat treat samples or to glycolate them in order to elucidate the sample's true crystalline structure. It is awfully unlikely that any given sample will match a pattern in the ICDD diffraction library exactly. There will usually be some variation due to peak shift and possible discrepancies in percentage intensity, but no significant peaks should ever be missing entirely.

Using the Mineralogy of Crime

On September 18, 2000, Mr Holding arrived home from work at around 5 p.m. to find signs of a disturbance and that his wife's car was gone. He called the police and reported that his wife and his mother-in-law, who was visiting and due to return to Hong Kong in two days, were missing. Authorities found a broken bottle and some blood in the home and quickly organized a search. The next morning police found the car, which had broken down, and Matthew Holding, the missing woman's son, on the Moonta to Kadina road in Yorke Peninsula, South Australia.

The car contained a bloodstained shovel with soil-like material on the blade. The police searched the surrounding bush land for the women but no signs were found and Matthew was not talking. In the following days, Yorke Peninsula was intensively searched but still no graves or bodies were found. Meanwhile, soil scientists at the CSIRO Land & Water Division/Center for Australian Forensic Soil Science examined the soil recovered from the shovel and the boots worn by Matthew Holding.

Along with determining other soil characteristics, scientists used optical microscopy to discover angular quartz grains and clay in the samples. They next used XRD to determine the samples' mineralogy, identifying crystalline kaolinite, mica, and talc. This effectively ruled out Yorke Peninsula as a possible source of the soil. Using geologic and soil maps, researchers determined that the most likely point of origin for the soil was a gravel quarry in the Adelaide Hills, nearer to the Holding's home. Police and scientists investigated the quarry, which was partly flooded, and collected comparison samples. X-ray diffraction patterns of the samples from the quarry were virtually identical to that of the soil from the shovel, strongly suggesting that the women were buried somewhere at the site.

Based on this evidence, detectives repeatedly visited the quarry and three weeks later fox activity finally led investigators to one of the bodies. The second body was then located 50 meters away in a shallow grave similar to that of the first. Police had searched the quarry previously, shortly after the women disappeared, but the area where the bodies were found was under water at the time. The soil evidence convinced them to go back. Matthew Holding, who apparently suffers from severe schizophrenia, was sentenced to a minimum of 18 years.

Summary

For a variety of reasons, soil is the most commonly encountered form of geologic evidence. Soil is ubiquitous, usually transfers easily and can potentially provide a wealth of information. This chapter was just a brief introduction to a complex and wide-ranging subject area. There are entire books, like Pye's (2007) *Geological and Soil Evidence*, dedicated to this topic, not to mention all of the books that discuss soils from an engineering or agricultural perspective. Because soils are so complex, no single, one-size-fits-all procedure for their analysis and characterization could ever be appropriate for all possible situations. This chapter gave an outline of some of the more commonly applied approaches and should have provided the reader with a feel for the types of information that soil evidence can provide.

Further Reading

For additional information on mineral properties, the Mineralogical Database at http://mindat.org is an extremely useful resource. It lists physical properties, optical properties, and even X-ray diffraction information.

MacKenzie, W. S. and Guilford, C. (1980) *Atlas of Rock-forming Minerals in Thin Section.* Halsted Press Book, John Wiley & Sons Ltd, New York.

Nesse, W. D. (2003) *Introduction to Optical Mineralogy*, 3rd edn. Oxford University Press, New York.

References

Croft, D. J. and Pye, K. (2004) Colour theory and the evaluation of an instrumental method of measurement using geological samples for forensic applications. In: K. Pye and D. J. Croft (eds), *Forensic Geoscience: Principles, techniques and applications.* Geological Society of London, Special Publications **232**: 49–62.

Daniels, F. J. (2006) *Dead Center: The shocking true story of a murder on Snipe Mountain.* Penguin, New York.

Daugherty, L. A. (1997) Soil Science contributes to an airplane crash investigation, Ruidoso, New Mexico. *Journal of Forensic Sciences* **42**(3): 401–5.

Fitzpatrick, R. W., Raven, M. D., and Forrester, S. T. (2009) A systematic approach to soil forensics: Criminal case studies involving transference from crime scene to forensic evidence. In: Ritz, K., Dawson, L. A., and Miller, D. (eds), *Criminal and Environmental Soil Forensics.* Springer Publishing, Dordrecht, The Netherlands, pp. 105–28.

Grim, R. E. (1968) *Clay Mineralogy*, 2nd edn. McGraw-Hill, New York.

Janssen, D. W., Ruhf, W. A., and Prichard, W. W. (1983) The use of clay for soil color comparisons. *Journal of Forensic Sciences* **28**(3): 773–6.

Jenkins, R. and Snyder, R. L. (1996) *Introduction to X-ray Powder Diffractometry.* John Wiley & Sons Ltd, New York.

Klein, C. and Hurlbut, Jr., C. S. (1985) *Manual of Mineralogy*, 20th edn. John Wiley & Sons, Inc., New York.

Mitchell, J. K. (1993) *Fundamentals of Soil Behavior.* John-Wiley & Sons Ltd, New York.

Moore, D. M. and Reynolds Jr., R. C. (1997) *X-ray Diffraction and the Identification and Analysis of Clay Minerals*, 2nd edn. Oxford University Press, New York.

Munroe, R. (1995) Forensic geology. *Royal Canadian Mounted Police Gazette* **57**(3): 10–17.

Munsell Color (2000) *Munsell Soil Color Charts*. Revised washable edition, 10pp + charts. GretagMacbeth, New Windsor, NY.

Murray, R. C. (2004) *Evidence from the Earth*. Mountain Press Publishing Co., Missoula, MT.

Murray R. C. and Solebello, L. P. (2002) Forensic examination of soil. In: R. Saferstein (ed.), *Forensic Science Handbook: Volume 1*, 2nd edn. Prentice Hall, Upper Saddle, NJ, pp. 615–33.

Petraco, N., Kubic, T. A. and Petraco, N. D. K. (2008) Case studies in forensic soil examinations. *Forensic Science International* **178**: 23–7.

Post, D. R., Bryant, R. B., Batchily, A. K. *et al.* (1993) Correlations between field and laboratory measurements of soil color. In: J. M. Bigham and E. J. Ciolkhoz (eds), *Soil Color*. Soil Science Society of America Special Publication **31**: 35–49.

Pye, K. (2007) *Geological and Soil Evidence: Forensic applications*. CRC Press, Boca Raton, FL.

Ruffell, A. and McKinley, J. (2008) *Geoforensics*. John Wiley & Sons Ltd, Chichester.

Schaetzl, R. and Anderson, S. (2005) *Soils: Genesis and geomorphology*. Cambridge University Press, Cambridge.

Schoeneberger, P. J., Wysocki, D. A., Benham, E. C., and Broderson, W. D. (eds) (2002) *Field book for describing and sampling soils: Version 2.0*. Natural Resources Conservation Service, National Soil Survey Center, Lincoln, NE, available online http://soils.usda.gov/technical/fieldbook/.

Soil Survey Division Staff (1993) *Soil Survey Manual*. Soil Conservation Service. U.S. Department of Agriculture Handbook 18, U.S. Gov. Print. Office, Washington, DC, available online http://soils.usda.gov/technical/manual/.

Soil Survey Staff (1999) *Soil Taxonomy: A basic system of soil classification for making and interpreting soil surveys*, 2nd edn. USDA Natural Resources Conservation Service, Washington.

Sugita, R. and Marumo, Y. (1996) Validity of color examination for forensic soil identification. *Forensic Science International* **83**: 201–10.

Chapter 8
The Geology of Art

Geological materials have been used as artistic media for thousands of years. Minerals are used as pigments for paint, stone is used for sculpture, and clay is used for creating pottery. Thus, the same techniques we have used for mineral, sand, and soil identification can be applied to identifying the materials used in many types of artwork. More to the point, these tools are commonly employed for unmasking fakes and forgeries. However, there are some difficulties with this new type of assessment.

As with gems and gemstones, the application of destructive analytical techniques is typically not allowed, as damaging potentially priceless works of art is frowned upon. *Polarized light microscopy* is still commonly employed, but the samples used are typically very, very small. There are several nondestructive tools that we can employ to aid in the identification process, such as XRD (introduced in Chapter 7), X-ray fluorescence (XRF), and Raman spectroscopy. It is important to keep in mind that, as with any other forensic examination, the researcher is looking for things that would exclude a sample from a set – as the inclusion of a modern synthetic pigment would exclude an item from being a Renaissance painting, for example – and not attempting to affirm the identity of an art object.

A Brief Examination of the Vinland Map

In October 1957, an antiquarian bookseller began offering a 15th-century manuscript called the Tartar Relation for sale. The manuscript contained a previously unknown version of the story of the 1245 to 1247 expedition of Friar John of Plano Carpini, one of the first Europeans to travel through the empire of the Mongols, who were at the time known as Tartars. By itself, the manuscript was a minor historical curiosity, but bound with it was a sheet of *parchment* (i.e.

An Introduction to Forensic Geoscience, First Edition. Elisa Bergslien.
© 2012 Elisa Bergslien. Published 2012 by Blackwell Publishing Ltd.

made from animal skin and not paper) measuring approximately 27.8 cm by 40 cm containing a hand-drawn map (Figure 8.1). The map included outlines of Greenland, Iceland, and an area labeled "*Vinlanda Insula a Byarno re pa et leipho scoijs*," which translates as "Island of Vinland, discovered by Bjarni and Leif in company." If authentic, the map, commonly referred to as the Vinland Map, would include the oldest-known representation of North America, predating Columbus's expedition by over 50 years. Clearly, this was a find that could radically change our historical understanding of the discovery of America.

After examining the manuscript, the British Museum declined to make a bid, as the curator of manuscripts, Bertram Schofield, concluded that the map was a fake. Then, through a rather complex series of events involving bookseller Lawrence Witten and the curator of classics at the Yale University Library, Thomas Marston, the manuscript was purchased by an anonymous benefactor and donated to Yale University. The donor was later revealed to be philanthropist Paul Mellon.

While there is now archeological evidence that the Vikings discovered North America before Columbus, at the time of the map's purchase no such evidence had yet been uncovered, and this was a hotly debated topic. More importantly, the map's date and provenance strongly suggested that Europeans, possibly even Columbus himself, knew about the existence of "the New World" and even its rough location. Because of the potentially heated controversy surrounding the Vinland Map, Yale kept the donation quiet and provided only a select few researchers access. Then after seven years of silent, covert study came a stunningly timed Columbus Day announcement in 1965, when Yale University unveiled both the map and the publication of *The Vinland Map and the Tartar Relation* (Skelton, Marston, and Painter, 1965, 1995), a book pronouncing the Vinland Map to be authentic and that the Vikings were the first Europeans to visit the New World. Since that day, the Vinland Map has been the focus of impassioned debate.

Scholarly dissatisfaction with the findings reported in the book started almost immediately and led Yale, in 1972, to commission a team headed by Walter McCrone to examine the map. Using polarized light microscopy and other microanalytical techniques, McCrone's team discovered that the ink lines on the Vinland Map, which appear to be black ink surrounded by a yellow discoloration, were actually double lines composed of a broad yellow stoke with a narrow black line drawn carefully down the center (Figure 8.2). This might sound inconsequential, but it became highly significant.

There were two basic types of black ink used in medieval manuscripts: *carbon-black ink*, made from charcoal or lamp-soot, and *metal gall ink*, most commonly *iron gall ink*, usually made by mixing gallnuts (ball-like structures that form when wasps lay their eggs in the growing bud of a tree) and ferrous sulfate (often iron-rich earth or created by pouring acid over old nails). Carbon black is stable, but iron gall inks, where the black color is the product of a chemical reaction, deteriorate over time, causing the parchment or paper next to the ink to yellow and potentially disintegrate. Thus, what appeared at first glance to be the natural byproduct of iron gall ink aging actually turned out to suggest forgery.

McCrone took 29 ultramicroscopic ink samples from the map for analysis, as well as 25 ink samples from the manuscripts that had originally been bound with the map. Examiners found that the inks used in the *Tartar Relation*, and in the *Speculum Historiale* (a second manuscript that was found to have originally been bound with the map), were iron gall inks, but the black ink used on the Vinland Map was found to be carbon based, and quite dissimilar to any other medieval ink tested. Also, Walter McCrone Associates, Inc. reported in 1974 that the ink on the map contained *anatase*, a crystalline form of the compound titanium dioxide that was not commercially available until 1917 at the earliest.

Titanium dioxide does occur naturally, most commonly in the form of the mineral rutile, but structurally rutile is quite different from anatase. Anatase itself is also a naturally occurring mineral but is usually brown or gray, not pure white. The crystals that McCrone found on the map were smooth, rounded rhombohedral shapes of nearly identical size. Natural pigments are typically angular, poorly sorted, and significantly coarser than synthetically generated pigments. Chemically, the anatase on the map was highly pure, while naturally occurring mineral anatase typically contains significant amounts of iron and manganese. Based on these findings, the Vinland Map was declared to potentially be a forgery.

Then in 1987, Thomas Cahill of the University of California-Davis and his coworkers used particle-induced X-ray emission (PIXE) to examine the Vinland Map and its associated manuscripts (Cahill *et al.*, 1987). PIXE works by directing a beam of photons through a sample, inducing the generation of X-rays as some inner-orbital electrons are popped out and outer-orbital electrons jump in (as described previously in the introduction to XRD in Chapter 7). By measuring the number of X-rays and their energy levels, PIXE can determine the amount of each element present (from silicon through uranium) at concentrations as low as a few parts per million. It is important to note that while PIXE can be used to determine elemental composition, it does not provide information about elemental state, so you could not tell whether the titanium found is in the form of anatase, rutile, or some other compound. Cahill and his co-authors determined that there was no titanium in the *Tartar Relation* or in the *Speculum Historiale*. They also determined that while there was titanium on the Vinland Map it was apparently a thousand times less than the amount reported by McCrone. Based on these results, Cahill concluded that McCrone's analysis was incorrect.

In response, McCrone (1988) published a much more extensive article describing his team's findings, including the results of studies performed using XRD, scanning electron microscopy (SEM), transmission electron microscopy (TEM), electron microprobe analysis (EMA), and ion microprobe analysis (IMA). In brief, he stood behind his original conclusion, that the Vinland Map was most probably a forgery. The discrepancy lies in how you calculate the percentage of titanium found via the different analytical methods.

In 1995, partially on the basis of Cahill's findings, Yale University Press issued a second edition of *The Vinland Map and the Tartar Relation* that included a contribution by Cahill, but not one by McCrone, and stopped just short of calling the map authentic. The new publication also begged the question of why the black lines on the map are surrounded by a yellow discoloration inconsistent with the use of carbon-based ink. In February 1996, Yale held a symposium to discuss the map and did not invite Walter McCrone, who showed up anyhow.

In 2002, Donahue and his coworkers published a paper reporting the radiocarbon age of the parchment on which the Vinland Map is drawn. Using accelerated mass spectrometry (AMS) they found that the parchment dates to AD 1434 +/− 11 years, with a 95% confidence interval that the age range of the parchment is between AD 1411 and 1468. Donahue suggested that the age of the parchment supported the authenticity of the map through the rather tortuous logic that "based on the known history of the map . . . a putative forger . . . could not have had a 'target date' of AD 1434 to aim at because that date was not to be established as the appropriate date for the parchment until after the map was acquired." Basically, he was stating that a forger could not have known that they would need 15th-century parchment because radiocarbon dating was not a widely known technique prior to the acquisition date of the *Tartar Relation* (Donahue, Olin, and Harbottle 2002). Strangely, this quote suggests that a potential forger would not know the appropriate timeframe for their own forgery and greatly underestimates the demonstrated abilities of known forgers. Besides, there is an even simpler explanation available. Since the Vinland Map was bound with two genuine 15th-century manuscripts, and matching wormholes show that all three documents were once bound together, a forger could simply have used a blank page from part of the original document. No extraordinary foresight was needed to end up with parchment of the appropriate age.

Given the contentious state of things, another analytical method was brought to bear on the problem by Katherine Brown and Robin Clark in 2002. They used *Raman microprobe spectroscopy* to analyze the black ink on the Vinland Map. In Raman microscopy, samples are illuminated with monochromatic laser light. Most of the light that scatters from the sample is of the same frequency of the incident laser light, but a very tiny portion of the light experiences a frequency shift. This effect, named after its discoverer, C. V. Raman, can be used to determine chemical composition, molecular state, and even bonding energies. Thus, unlike PIXE, this technique can determine not only elemental composition but also mineralogy as well. Brown and Clark found that the black ink in the *Tartar Relation* gave only a weak carbon spectra, and was intensely fluorescent, a feature common in iron gall ink. However, on the Vinland Map in the black areas, the ink had spectra characteristic of carbon, while the yellow areas were clearly found to contain titanium. Plus, this technique allowed examiners to determine that the titanium was in the form of anatase and not another titanium polymorph, such as rutile. Thus, Brown and Clark supported McCrone's original conclusion, that the presence of anatase in the Vinland Map's ink strongly suggests that the map is a modern forgery.

Since 2000, Jacqueline Olin of the Smithsonian Initiation has authored articles relating possible ways that a medieval ink could legitimately contain anatase, including referencing a 1974 experiment in which her preparation of an iron gallotannate ink contained poorly crystalline anatase (Olin, 2000, 2003). However, the anatase resultant from the process she describes does not have the same properties as the anatase found on the map by McCrone, or Brown and Clark. Olin's arguments do not take into account that the ink on the Vinland Map has clearly been identified a carbon-based ink. She also never explains why a black ink that is clearly not an iron gallotannate ink would have a yellow deterioration halo (Figure 8.2).

This discussion presents just a brief overview of the history of the Vinland Map. There have been even more articles published, and strangely the debate still goes on. Interested parties are directed to the list of articles in the Further Reading section for additional information. For a book that discusses the wider issues involved in the history of the Vinland Map see Seaver (2004). There is also a very nice website, The Vinland Map: Medieval or Modern? (http://www.webexhibits.org/vinland), that allows you to explore the issues yourself, including more details about the history of the Vinland Sagas (Viking stories about the discovery and exploration of North America) and the cartography of the map.

Geologic Media and Art Forgery

The field of art crime is enormous and wide reaching. Not only in terms of forgery and fraud but also in terms of stolen art and its use as currency in the international drug trade. The information included in this chapter examines just the smallest subset of this topic where geological materials are used as artistic media. If you are interested in a better understanding of the broader history of art forgery and the difficulties involved, I would suggest, for their contrasting views, *The Art Forger's Handbook*, a how-to book written by Eric Hebborn (2004), an infamous forger, and *False Impressions: The hunt for big time art fakes* by Thomas Hoving (1997), former director of the Metropolitan Museum of Art.

Technically, for something to be a *forgery* it has to have been created, or altered, with the intent to defraud. For example, a painting in the style of Monet is not a forgery unless you also add Claude Monet's signature or try to sell it as an authentic Monet. This would also include taking an unattributed artwork, or a work by an unknown artist, and then selling it as the work of a famous artist. The most common type of forgery is a *pastiche*, or an object composed of elements from a variety of different authentic materials. These are generally exposed through careful comparison of known works with the suspected work. Forgeries that are constructed purely from the imagination of the forger, without reference to original sources, are considered rare since they would be difficult to get accepted for sale, and in general the whole point of forgery is to make money. However, there is a long history of art objects being created based on some fragmentary historical reference or commonly known legend. This gives their forgeries verisimilitude and often can ensure a successful sale to the sometimes fanatical devotees of such stories. The Vinland Map would probably fall into this category.

Direct copies of known works of art are also fairly rare, though there have been some very famous examples of this type of activity. For instance, Ely Sakhai, a Manhattan art "dealer," would buy authentic mid-level paintings by famous artists like Renoir, Chagall, and Klee, at the major New York auction houses. He would then have copies made, taking pains to ensure that they were exact duplicates, even down to the markings on the back of the canvases. Sakhai would then allow some time to pass, often a couple of years or more, and then privately sell the copies using the *provenance* (ownership history) of the authenticated item. Years later, he would then sell the original, often at the auction house from which he first made

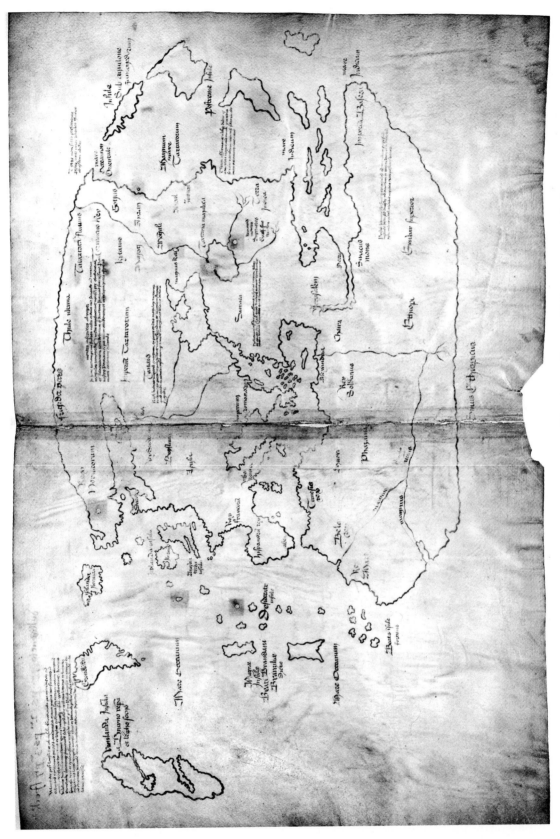

Figure 8.1 The Vinland Map, allegedly a 15th-century *Mappa Mundi*, redrawn from a 13th-century original. *Source:* Owned by Yale University.

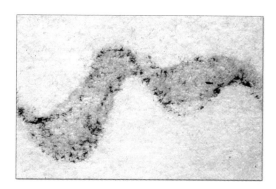

Figure 8.2 Close-up of ink on the Vinland Map showing the purported yellow deterioration halo. *Source:* Courtesy of Yale University.

the purchase. Sakhai was caught when the same painting (Gauguin's *Vase de Fleurs*) turned up for sale at both Sotheby's and Christie's (famous auction houses) at approximately the same time. In 2005, he was sentenced to 41 months in prison and ordered to pay $12.5 million in restitution.

A *fake* is a real object that has been altered in some way, such as adding a few lines to a stamp or altering the mintmark on a coin, turning a common item into something rare and valuable. Another example of a fake would be to take a genuine painting from a particular period and altering it so that it now includes a cat, dog, or coat-of-arms (which would drive up the selling price), or so that the subject appeared to be doing or holding something unusual. For example, portraits in which the subject is smiling are very rare in Renaissance art. The difference between a forgery and a fake is sometimes opaque and the terms are often used interchangeably. The key distinction that elevates both forgeries and fakes into criminal activity is the intent to deceive, usually in order to increase profitability, though not always.

It is also important to note that scientific analysis cannot be used to prove that a work of art is genuine, but it can be used to demonstrate that an item is *not characteristic* of the purported period of its creation. Scientific analysis is used to look for anomalies, but even if every test demonstrates that the materials and techniques used are appropriate, it still cannot be used to say that an item is authentic. Science is employed for the detection of forgery, while the work of *authentication* (the act of establishing that something is authentic or genuine) lies in the hands of specialists, usually art historians, and outside the purview of the forensic investigator.

Mineral Pigments

There are actually many ways in which the world of the geosciences and the world of art intersect, but we will start with what is probably the most common: the use of geological materials as colorants. Technically, a *pigment* is a finely ground, insoluble substance that imparts its color to another material (Figure 8.3). This differentiates pigments from *dyes*, which are soluble and impart color by staining or by being absorbed. Pigments require a *binder*, or an adhesive liquid, in which the color is suspended in order to be secured to a surface, such as canvas, paper, or parchment.

Figure 8.3 Pigments for sale on a market stall, Goa, India. Photograph by Dan Brady. Please see Color Plate section.
Source: Used courtesy of the Creative Commons Attribution 2.0 Generic license.

Pigments can be subdivided into categories depending on their origin as follows:

- Inorganic (mineral)
  - native earths (ochre, azure, etc.)
  - calcined native earths (*calcining* is thermal treatment, typically at temperatures below melting point, or oxidizing by means of heating, e.g. pigments such as burnt umber)
  - inorganic synthetic colors (synthetic pigments);
- Organic
  - vegetable
  - animal
  - synthetic organic pigments;
- Lakes: Lakes are soluble dyes, usually organic in nature, that have been attached to inert, semi-opaque, support material, usually inorganic, to render them insoluble and usable as pigments.

Historically, inorganic pigments were derived from naturally occurring mineral deposits that were collected and finely ground. Usually, these pigments were then mixed with some type of naturally occurring binder before application to a surface. Human use of mineral pigments predates the rise of *Homo sapiens*, with recent discoveries pushing back the first confirmed use to over 164,000 years ago. Cave paintings fashioned over 70,000 years ago were produced using charcoal and *ochre*, a generic term for iron oxide pigments. Through time, more naturally occurring materials were introduced to the palette as new trade routes opened and new substances were discovered.

For some colors, such as pure white, or grass green, there are no good, naturally occurring pigments. Consequently, the alteration of inorganic pigments through exposure to heat, or chemical reaction, in order to improve color or create a new color, goes at least as far back as the civilizations of ancient Egypt and Rome.

Other pigments, though beautifully colored, are highly toxic, or unstable and liable to react catastrophically. To counter these difficulties, pigments with better properties were constantly being sought and, as the science of chemistry evolved, new methods of creating pigments were discovered. Reaching into the modern era, safer, more stable, and more intensely colored synthetic replacements began to replace many of the less acceptable pigments. Thus, over time, many different pigments have fallen in and out of favor. As demonstrated with the Vinland Map, one of the most commonly used tools for assessing authenticity is to examine all of the pigments used in a painting, or other art or historical object, to determine whether they are appropriate for the known time of the object's production.

A Summary of the Properties of Some Common Pigments

The following is a list of some of the most common pigments used throughout history, with an emphasis on mineral pigments. They are listed by color and then roughly in order of their importance. It is by no means exhaustive. For interested parties, there are several reference books that discuss the history and/or physical properties of pigments much more extensively. A couple of suggestions are *Pigment Compendium: Optical microscopy of historical pigments* (Eastaugh *et al.*, 2004) and the *Artists' Pigments: A handbook of their history and characteristics* series produced by the National Gallery of Art (various dates). For a less technical introduction to pigments, and a fascinating look at some related geology, see Victoria Finlay's (2004) *Color: A Natural History of the Palette*. There is also a very useful website called Pigments through the Ages (http://www.webexhibits.org/pigments).

Black Pigments

Bone black and ivory black are both created by burning ivory or bone inside of sealed containers (anoxic environments). *Bone black* is bluish-black or brownish-black in color and fairly smooth in texture. *Ivory black*, which tends to be a deep bluish-black, is a term that used to refer just to pigments made by charring ivory waste from the comb-making industry, though any ivory could be used to make this pigment. It was considered the most intense of all the black pigments. Over time, the term became corrupted, and these days ivory black is often used to refer to any black produced from animal bones. Ivory and bone black are denser than carbon black or lampblack, and produce smooth, slow-drying paints with good hiding power. They are also lightfast and non-reactive. Bone black contains approximately 10% carbon, 84% calcium phosphate [$Ca_3(PO_4)_2$], and 6% calcium carbonate ($CaCO_3$). Microscopically, it is composed of medium to fine, irregularly shaped particles between 1 and 20 μm, mostly opaque, though some may be transparent and some brown.

Lampblack is nearly pure (99%), amorphous carbon that has been collected from the dense smoke produced by burning mineral oil, tar, pitch, or resin (petroleum products) in an anoxic environment. It is a soft brownish- or bluish-black pigment that is very stable and unaffected by light, acids, and alkalis. Since it is not a true black, it is very good for making shades of gray. Microscopically, it is very fine (<1 μm), opaque, uniform, and homogeneous. Often under the microscope, it is only possible to make out aggregates. Since it is produced from petroleum products, it does not wet well with water. Lampblack was made commercially at least as far back as the 1700s.

Carbon black or *charcoal black* (also called *chars*) is a class of pigments derived from the partial burning or carbonizing of wood, fruit pits, or other organic materials in an anoxic environment. In this case, the resulting product is not pure carbon, instead containing many different impurities. Carbon blacks are a deep brownish-black in color and more significantly more granular than lampblack (1–100 µm). Microscopically, some cellular structure might still be visible and curved fractures can sometimes be seen. A good carbon black wets well in water and is very stable, unaffected by light, acids, and alkalis.

Vine black is a special form of charcoal black that is made from young shoots of grapevines and was referred to in medieval times as the best of blacks, though it really is more of a bluish-black. The vine sprigs were carbonized by packing them tightly in little bundles that were placed in casseroles (a type of deep pot or dish), which were then covered, sealed, and baked slowly in an oven. The vines must not be burnt in the air or they would reduce to gray/white ashes instead of to carbon. The resulting charcoal can be used in sticks for drawing or powdered for use as a paint pigment.

Coal is a rock formed of carbon, hydrogen, and oxygen formed by the *digenesis* (compaction and chemical alteration) of plant material. Coal was occasionally used as a black pigment, though it does not produce deep black and was generally considered unsatisfactory. Under magnification, coal consists of fragments of land plant material, including wood, cuticle (the waxy surface found on some leaves), sap, resin (amber), spores, and pollen. Each of these can be present in varying degrees of degradation due to decay near the surface and "cooking" due to burial in thick sediments.

Graphite is a crystalline form of carbon widely distributed as a mineral that has long been used as a writing material (the "lead" in pencils is actually graphite). It gets its name from γραφευ the Greek for "to write." Graphite has a greasy feel, a submetallic to dull luster and perfect cleavage. It was not used as a pigment very often, due to its dull gray color. Under the microscope, graphite particles range from irregular flakes to hexagonal plates that are opaque to weakly translucent. The particles vary from fine to very coarse in size and often display fractures.

Iron gall ink, while not technically a pigment, is also a very important black. It is made from tannin or gallotannic acid from oak galls (galls are irregular plant growths which are stimulated by the reaction between plant hormones and growth-regulating chemicals produced by some insects, particularly wasp eggs.) When the tannin from the oak gall is combined with ferrous sulfate, it forms ferrous gallotannate, which turns black on exposure to air as the iron oxidizes to ferric gallotannate. This form of ink has been in use since the very early medieval times.

White Pigments

Bone white or *calcined white* is composed of 85–90% calcium phosphate $Ca_3(PO_4)_2$ with some minor calcium carbonate ($CaCO_3$). It is created when bone is burned in an oxygen atmosphere, reducing it to ash (unlike bone black, which is created by incomplete combustion in an anoxic environment). Bone white is a grayish-white and slightly gritty, with particles that are usually <100 µm. Because the white color is not very pure, it was more often used to give texture or grit to paper or parchment than as a paint pigment.

Chalk (also called *whiting* or *lime white*) is one of the many natural forms of calcium carbonate ($CaCO_3$). It occurs in deposits all over the world, most famously the White Cliffs of Dover, England. Natural chalk is a soft, white, grayish-white, or yellowish-white, very fine-grained, extremely pure limestone, composed mainly of coccolith biomicrites (the skeletal elements of submicroscopic planktonic green algae) associated with varying proportions of larger microscopic fragments of bivalves, foraminifera, and ostracods. The upper section of the ocean, where sunlight penetrates, is home to billions upon billions of planktonic (floating) species (like coccoliths and foraminifera) that form calcium carbonate shells. When they die, their skeletons sink to the bottom of the ocean, mix with the hard parts of other organisms, and are compressed through time to form sedimentary rock, such as chalk and limestone.

The color and quality of chalk varies widely from place to place, with deposits from England, France, Belgium, and Denmark considered the highest in quality, while the deposits in the United States are considered too poor to be used as a pigment. Chalk is composed of fine particles (1–10 μm) that are fairly homogeneous microscopically. At very high magnification, fossilized shell materials and foraminifera can be seen, though it should be noted that coccoliths can only be seen using an SEM. Chalk has poor covering power and discolors in oil, so it has limited usefulness as a pure white. It is relatively stable but will decompose upon heating or exposure to acids. In fact, like limestone, chalk can be identified by its vigorous effervescence when exposed to dilute hydrochloric acid (HCl). The major reason chalk was used as a pigment is that, unlike lead white, it is chemically compatible with orpiment and most other pigments, and it does not darken on exposure to air when used in watercolors. It is most often found in the *grounds* (preparatory background) and support structures of paintings, and is often used as a filler or to increase opacity. Chalk is also a major component of whitewash.

Shell white is made from eggshells or seashells and is chiefly composed of calcium carbonate ($CaCO_3$). See also Chalk.

Gypsum is a naturally occurring calcium sulfate dihydrate ($CaSO_4 \cdot 2H_2O$), which occurs worldwide in several different crystalline forms. Raw gypsum is of little use except as an inert filler material. The most important use of gypsum is in the preparation of plaster of Paris and frescos. Particles are typically 5–50 μm in size, monoclinic (with oblique extinction under cross-polarized light), with three perfect cleavages that often form tabular rhombs when ground. Gypsum is mostly stable, though soluble in strong acids and subject to efflorescence (crystal growth) in damp environments.

Lead white (flake white, white lead, Cremnitz white, Flemish white) is a lead carbonate [$2PbCO_3 \cdot Pb(OH)_2$], normally composed of about 70% lead carbonate and 30% lead hydrate. It occurs naturally as the mineral cerussite, but the mineral has never been an important white pigment. Instead, lead white is one of the oldest-known synthetic pigments and was used in ancient Greece and Rome. It was used in China as early as 300 BCE and was probably the most important white pigment in the history of Western painting. Lead white was the only white commonly used in European paintings until the 19th century, and was used in makeup for years. It was partially replaced by zinc white in the 19th century and almost fully replaced in the 20th century by titanium white.

Lead white is permanent and unaffected by light, but turns black in the presence of sulfides (in the air or in neighboring pigments), which severely limited its use in watercolor. However, as long as lead white is sealed in a binder, as with varnished oil paintings, this reaction is much less likely to occur. As a pigment, it has some fine qualities, such as low oil absorption, good flow, and rapid drying, though clearly one of the major disadvantages is the high toxicity of lead. Microscopically, lead white particles are rounded hexagons with parallel extinction that range in size from 1 to 50 μm. Cerussite is orthorhombic, forming angular, shard-like transparent particles when crushed. Under cross-polarized light, cerussite has very high birefringence, as does lead white.

Zinc white (Chinese white) is a pigment composed of zinc oxide (ZnO). It was manufactured as a replacement for lead white, a use first suggested in a 1782 report by Guyton de Morveau. Zinc white actually appeared for use in France and England in the late 18th century, but artists found the covering power insufficient, and it did not catch on for mainstream use. Then, in 1834, Messrs Winsor and Newton Ltd of London introduced a watercolor under the trade name *Chinese white* that contained a denser form of zinc oxide. The resultant white was very bright and had good covering power. Some artists began to use the new white, but there were still many people who considered lead white to be superior.

Initially, use of zinc white was limited to watercolors, due to its poor drying qualities in linseed oil, the base for oil paints. However, in 1844 LeClaire in France discovered a method to overcome this difficultly by using a different oil formulation. By 1850, zinc oxide was commonly being used in oil paints as well and zinc white paints were being manufactured throughout Europe. Zinc white eventually replaced lead white as the most important white pigment for a brief period and is still in use today.

Zinc white is a pure, cold white that is less dense than lead white but considered to be more brilliant and better for use in highlights and for underpainting. However, zinc white has a tendency to become brittle when dry, potentially creating a fine network of cracks where large amounts are used. It also has a slow drying time, absorbs oil, and has poor flow. Zinc oxide is derived from smoke fumes so the particles are very fine (<2 μm) and can be difficult to observe except at very high magnification. The particles are typically transparent to translucent, sometimes showing a pale yellow color and appear rounded. For oil painters, a mixture of lead white and zinc white often provided the best properties, so finding both in the same painting is not unusual. Similar to lead oxide, zinc oxide reacts with sulfides to create the compound ZnS, but unlike lead, which is blackened by this reaction, zinc sulfide is white.

Barium white (blanc fixe, permanent white, and constant white) was also developed as a replacement for lead white. Barium white is composed of barium sulfate ($BaSO_4$), which can be obtained from the mineral barite or produced synthetically. It is extremely inert and is unaffected by heat, light, or strong chemicals. However, it does not have much hiding power, as the particles are fairly transparent. Both blanc fixe and natural barium sulfate are too transparent to be usable as an oil white. This transparency, however, makes them good extenders for oil colors and good inert substrates for the creation of lake pigments. Microscopically, barium white particles are colorless and extremely fine-grained (<2 μm) with low birefringence.

Lithopone was also developed in the 19th century, introduced after 1874. It was a combination of zinc sulfide (30% ZnS) and artificial barium sulfate (70% $BaSO_4$). It cannot be differentiated from barium sulfate using optical microscopy.

Titanium white is newest white pigment and the most important white pigment of the 20th century. Titanium dioxide (TiO_2), the primary component of titanium white, was first discovered in 1821, but it was not until 1916 that difficulties in purification were overcome so that it could be mass-produced. The titanium used in manufacturing usually comes from ilmenite ore, or deposits of the mineral rutile. Titanium dioxide was first used for painting in around 1920. Microscopically, titanium dioxide particles are translucent, yellowish-brown, though they appear white in reflected light. The particles are very fine (<1 μm), typically well rounded, and show high birefringence under cross-polarized light.

Titanium white has the greatest hiding power of all whites, is extremely stable, is unaffected by heat, light, or dilute chemicals, and is non-toxic. These properties mean that this "whitest of the white" pigments is used for a range of applications, from coating paper to whitening foodstuffs, and is found in products such as cosmetics, sunscreen, and toothpaste. It is almost ubiquitous in the modern world and its presence has exposed several forgeries.

Titanium dioxide also commonly occurs in the mineral forms of rutile and anatase, but neither mineral has ever been widely used as a paint pigment. Rutile particles are tetragonal, transparent to translucent yellowish-brown, and usually angular shards that range in size from fine to very coarse. Anatase particles are also typically angular shards that range from fine to coarse, but differ from rutile because they are translucent and pleochroic, changing from bluish-gray to olive green in color, depending on the angle of incident light.

The L'Infante (Brainerd, 1988)

Many years ago, a collector/dealer named Andrew Brainerd purchased a small oil painting at a gallery in Amsterdam. It was the portrait of a young girl, with her wavy blonde hair parted on the side, wearing an ornate gown trimmed with black lace. The painting was clearly a stylized copy of Diego Velázquez's 1664 (or 5) portrait of L'Infante Marie-Marguerite, which has been hanging in the Musée du Louvre in Paris since 1816. Brainerd fell in love with it and purchased the painting for around $8300.

After spending several years investigating the history of the painting, Brainerd contacted Walter McCrone in 1983, and asked the famous microscopist to analyze the pigments on the painting. Using fine-tipped needles, Dr McCrone removed microscopic samples from 15 different areas of the painting and analyzed them using a polarized light microscope. The pigments all turned out to be typical of a 19th-century artist. Given this encouragement, Brainerd told McCrone that he suspected the painting was by the French painter Édouard Manet. Manet had been a great admirer of Velázquez and was known to have made a copy of L'Infante sometime around 1859. The fate of that oil painting had always been a mystery, coming to be known as the "lost Manet."

Brainerd asked McCrone if there was anything they could do to help pin down whether his painting could be the lost Manet. Thinking about it, McCrone recognized that three of the pigments he found, the lead white, cobalt blue, and vermilion, had atypical characteristics. The lead white was a rare form of lead carbonate composed of hexagonally shaped columns or prisms, rather than the more typical flat hexagonal plates. The cobalt blue pigment had an unusually low refractive index (RI; less than 1.66), so low that he had originally assumed that the blue pigment was smalt until elemental analysis confirmed otherwise. Finally, the vermilion was unusually pure, showing only a trace of silicon. These unusual characteristics could be used as a basis for comparison.

McCrone needed access to other known Manet paintings from the same period to see whether these same unusual characteristics showed up. One such painting was *Le ballet Espagnol*, which McCrone was given permission to sample. Once again, using polarized light microscopy, he found that the pigments in *Le ballet Espagnol* were generally similar to those found in Brainerd's *L'Infante*. Moreover, the unusual white lead prisms, the low RI cobalt blue, and the pure vermilion were also all present. Examining his records, McCrone found that in all of his examinations of lead white pigment, nearly 50 from all over the world, only *Le ballet Espagnol* and Brainerd's *L'Infante* had white lead with that unique prism shape.

Still, this could all just have been a remarkable coincidence, so McCrone decided to check the particle size distribution of the lead white particles in each of the paintings. He carefully measured the longest dimension of each of the prisms visible under the microscope in one sample from each of the paintings. He found very similar size distributions and decided to do one further test: a trace element analysis using an instrument called an electron microprobe. As discussed in previous chapters, the Earth's crust is primarily composed of just eight elements, which are found just about everywhere, so bulk elemental composition is generally not a good tool for forensic comparison. However, the distribution of the remaining elements varies widely, potentially making them useful tools for a characterization of different samples. McCrone and his team carefully isolated four lead white prisms, then washed and mounted them on a labeled beryllium plate. The results of the analysis can be seen in Table 8.1.

The agreement between the first three crystals is remarkable, and the Ballet #2 crystal, which was the smallest of the four, still shows a high level of agreement for the less common elements (Sb, Cu, Fe, Ag, Sn, and Zn). Four crystals is not a large enough sampling to statistically analyze, but based on these results McCrone believed that there was a very high probability that both of the white lead pigments were made from the same raw materials, using the same equipment, and most likely were from the same production lot, possibly even from the same tube.

Additional samples were acquired from a third Manet painting, *La Femme á la Cruche*, and analyzed using polarized light microscopy. Once again, lead white prisms and low RI cobalt blue appeared. There was substantial visible restoration on the painting so additional analysis was not possible. Still, on the basis of the data gathered, McCrone (1988) was able to state that if "we consider all of the evidence together we are as close to proof of common origin as one could come, short of eyewitness evidence." However, even with this level of assurance, there is still not conclusive and final proof that Brainerd's *L'Infante* is the lost Manet. This rare example is as close as one is ever likely to get to scientifically confirming that a painting might be authentic.

Table 8.1 Composition of four lead white pigment particles using an electron microprobe.

Source Painting	Element*								
	Al	Sb	Cu	Fe	Mg	Si	Ag	Sn	Zn
Infante #1	0.21	0.022	0.057	0.032	0.22	0.23	0.008	0.018	0.071
Infante #2	0.18	0.017	0.051	0.029	0.22	0.20	0.007	0.015	0.067
Ballet #1	0.22	0.022	0.058	0.035	0.22	0.27	0.010	0.019	0.076
Ballet #2	0.42	0.032	0.064	0.042	0.40	0.62	0.014	0.024	0.078

*The numbers listed are the ratios of X-ray counts for each element relative to the X-ray counts for lead for a counting time of 4000 seconds. Thus, this information is not qualitative, like part per million data, but relational.
Source: From Brainerd, 1988, Table 5, page 169.

Earth Colors: Red, Yellow, Orange, and Brown Pigments

Ochre, a generic term for natural earth materials containing iron oxides, ranges in color from red and orange to dull yellow and brown. There are even some rare green varieties. Ochre has been used as a pigment from prehistoric times and is still in common use today, not only in paints but in cosmetics as well. *Red ochre*, sometimes called *bole*, is colored by anhydrous iron oxide, Fe_2O_3, usually *hematite*, the most common red iron oxide mineral. Under plain polarized light, hematite appears as translucent to opaque red–orange–brown particles that range widely in size and shape and will often include a variety of impurities. Synthetic iron oxide particles are well sorted, very fine, and translucent to nearly opaque red or reddish-orange. *Yellow ochre* is chiefly colored by the presence of hydrated iron oxides, most commonly the mineral *goethite*, $Fe_2O_3 \cdot H_2O$. Microscopically, goethite appears as translucent yellow crystals that may have a fibrous or acicular appearance. Particles range widely in size from very fine to coarse, and are highly birefringent. Yellow ochre turns red when exposed to high heat, thus roasting is a common treatment to darken the color. Such colors are indicated by adding the term *burnt* to their name. *Brown ochre* is usually almost pure *limonite*, though it might also be a mixture of hematite, goethite, and magnetite.

Colors vary widely around the world and the best ochres are usually named for their points of origin, such as *French ochre, raw sienna*, a variety of yellow ochre from Sienna, Italy with a deeper tint that is considerably more transparent than other varieties, or the famous red ochre known as *sinopia*, also in Italy. *Burnt sienna* is prepared by calcining raw sienna, turning the pigment a warm, reddish-brown. Microscopically, all natural ochre is heterogeneous in particle size, shape, and composition, formed of a mixture of colorless silica particles, clay and semi-opaque, pale-yellow, orange, and brown isotropic particles. Ochre is stable and resistant to dilute acids and alkalis.

Mars red is one of a range of Mars colors (red, yellow, orange, and violet) that are manufactured iron oxides made by precipitating a mixture of soluble iron salts and alum with an alkali, such as lime or potash. The result is a yellow mixture of ferric and aluminum hydroxides (including gypsum if lime is used). Heating of varying durations and intensities produces the other Mars shades. The resultant pigments are very fine and homogeneous, but have no advantage over naturally occurring iron oxides.

Red lead (*minium, orange mineral*) is an orangey-red pigment produced by prolonged heating (at a temperature of about 480 °C) of white lead in an oxygen environment to produce lead tetroxide Pb_3O_4. The story goes that red lead was accidentally created when a shipment of white lead was roasted in a fire during Roman times, but whether that is true, it is one of the oldest synthetic pigments. The mineral *minium* is a naturally occurring variety of red lead that is formed by the chemical weathering of *galena*, a sulfide. It is most commonly found in the portions of an ore deposit that have been oxidized. Thus, minium is considered a secondary mineral (i.e. produced by weathering). Red lead, which actually varies in color from a scarlet to an orange, is a dense, heavy pigment with excellent opacity. Microscopically, it can be crystalline or amorphous, depending on its source, and some of the particles are transparent, while others are orangey-red. Chemically it is very reactive, turning dark when exposed to sulfides, and turning chocolate brown over time when exposed to light. Red lead is commonly found on art objects from antiquity and is still available today.

Massicot and *litharge* are both names that have been used for yellow monoxide of lead (PbO), sometimes synonymously, though there is actually a distinction. Massicot is unfused PbO that is produced by the gentle roasting of white lead at about 300 °C. The resulting color is not a very intense yellow, but it has good hiding power and properties similar to those of white lead. Litharge, or flake litharge, is the fused, crystalline oxide formed by the direct oxidation of molten metallic lead. It is more orange in color than massicot. Microscopically, massicot forms translucent, pale yellow to yellowish-brown crystals, commonly rough prismatic forms with high relief and high birefringence. Litharge is relatively unstable, forming pale-yellow, semi-transparent crystals to opaque reddish-brown aggregates. Neither of these colors sees modern usage.

Vermilion (*cinnabar*) is a brilliant red pigment composed of mercuric sulfide (HgS). It occurs in nature as the mineral cinnabar, which is also the principal ore of mercury and is found worldwide. The crushed mineral has served as a pigment for centuries and was well known to the Greeks and Romans. Cinnabar has also been used in China since prehistoric times. Various methods of combining mercury and sulfur have also been known since very early times and several recipes can be found in alchemic texts. Chemically, and physically, synthetically created vermilion does not differ from the natural form. Vermilion is one of the heaviest pigments and has excellent opacity. It can vary in color from a strong red to a reddish-orange. Microscopically, the particles are translucent, deep orangey-red and have a waxy luster under reflected light. Many of the particulates may also have perfect cleavage, a property that differentiates it from hematite. It is fairly stable, frequently found in Roman wall paintings unchanged, and is chemically non-reactive. However, the pigment is sensitive to light and will darken when exposed to direct sunlight. This pigment is also extremely toxic.

Gold, genuine gold, was occasionally used as a pigment, though it was more often used in thin sheets (gold leaf) or as an additive to ink (gold powder). Gold is metallic, non-reactive, opaque, and is the only thing that looks truly golden in color.

Orpiment (*king's yellow*) is yellow sulfide of arsenic (As_2S_3), which occurs naturally in small quantities worldwide, and has a long history of use. Orpiment was the most unpopular of all traditional colors, not due to its highly poisonous nature but because it created offensive odors, dangerous fumes, and, most damningly, was highly incompatible with colors containing copper or lead.

Orpiment is brilliant yellow when pure, with fair covering power, though it is slow to dry. Because it was such a good yellow, it still saw significant use until synthetic replacements were discovered in the 18th and 19th centuries. Microscopically, orpiment has very high relief and a distinctive straw-yellow color. The mineral has perfect cleavage, frequently forming coarse-grained, angular crystals with a crosshatched appearance. The larger particles glisten in reflected light and have a waxy luster. Crystals are monoclinic, or sometimes fibrous, and strongly birefringent. It is often found in association with orangey-red particles of realgar.

Realgar is a naturally occurring orangey-red sulfide of arsenic (As_2S_2) that is closely related to orpiment. It occurs as widely in nature as orpiment but does not appear to have been used as much as a pigment, probably because there have always been several other, more stable, reds available. Realgar appears as angular, medium to coarse yellowish-orange particles, some of which may exhibit pleochroism (orangey-yellow to orangey-red), with high relief.

Cadmium pigments, including cadmium red, cadmium yellow, and cadmium orange, were introduced in the mid-1800s. Stromeyer discovered metallic cadmium in 1817 but production of the cadmium pigments was delayed because of the scarcity of the metal. Cadmium yellow is a cadmium sulfide (CaS) that is precipitated from an acid solution of a soluble cadmium salt with an alkali sulfide. The resulting pigment, which ranges from bright lemon yellow to deep orange (cadmium orange), is opaque, permanent, and non-poisonous, and was first introduced to the public at the Great Exhibition of 1851, though it was said to have been made first in 1846. Microscopically, the differences in color relate to whether the sulfide is crystalline or amorphous. The particles of amorphous orange cadmium sulfide are also much larger in diameter than the crystalline yellow particles. Cadmium sulfate occurs naturally as the mineral *greenockite*, but there appears to be no recorded use of it as a pigment.

Cadmium red, a cadmium sulfo-selenide [$CdS(Se)$] or cadmium selenide ($CdSe$), was not made until about 1910. It is created by precipitating cadmium sulfide with a mixture of sodium sulfide and selenium. By varying the proportions, it is possible to get a range of colors from vermilion (which it has now replaced) to deep maroon. Microscopically, it appears as tiny red globules less than 1 μm in diameter and without any appearance of crystallinity.

The cadmium pigments have high refractive indices and are relatively lightfast (except for some of the pale colors being produced today), though older varieties have been known to turn brown in outdoor frescos. They are also permanent and compatible with most other pigments, depending on the amount of free sulfur, though cadmiums in general should not be used with lead-based pigments. When in contact with copper colors, such as emerald green, cadmium pigments turn black. The particle sizes of the deeper cadmiums are about 50 times larger than the paler varieties. They are transparent particles that appear in clusters, microscopically. Cadmium pigments can be dissolved in hydrochloric and nitric acids, though they are unchanged by sodium sulfide.

Naples yellow (antimony yellow) is essentially lead antimoniate [$Pb_3(SbO_4)_2$] and varies in color from sulfur-yellow to orangey-yellow. Its early history is obscure, though it is known to have been used as early as 500 BCE and is said to have been found on the tiles of Babylon. Recipes for making this pigment appear in the middle of the 18th century, making it another of the oldest-known synthetic pigments. Naples yellow has a crystalline structure identical to that of the rare mineral *bindheimite* and is very heavy and dense, with exceptional covering power.

It is permanent, totally unaffected by light, and compatible with other colors, used in oil, tempera, and even frescos. It dries quickly, but, as with all lead pigments, is poisonous. Microscopically, it is composed of fine (1–5 μm), rounded to angular, pale-yellow particles.

Umber is a generic term that is applied to iron oxides that range in color from cream to deep brown. *Raw umber* is a brown-earth pigment similar to ochre (sometimes even considered a type of ochre) but contains manganese dioxide as well as hydrous ferric oxide $[Fe_2O_3 \cdot MnO_2 \cdot nH_2O + Si + AlO_3]$. The best umber has a warm, reddish-brown color with a greenish tint. Microscopically, the pigment is heterogeneous in composition and particle size. It contains significant goethite, which appears as fine, dark yellowish-brown semi- or completely opaque, and other orange, yellow, and clear particles. Umber is a durable pigment and stable, unless it has significant humus (organic matter), which will cause it to fade.

Burnt umber, a combination of iron oxide, oxide of manganese and clay, is made by burning raw umber to drive off the liquid content. Heating changes the ferric hydrate to ferric oxide, which is redder and warmer than raw umber. Microscopically, it appears almost the same as raw umber, except it is a bit redder, dominated by hematite rather than goethite, and is more transparent. Umber dissolves partially in acids, leaving a yellow solution; hydrochloric acid gives it an odor of chlorine and in alkalis it discolors a little. Umber is completely lightfast and unaffected by gases.

Bitumen (asphaltum, Judaic bitumen, Antwerp brown) is a brownish-black, naturally occurring mixture of hydrocarbon with oxygen, sulfur, and nitrogen. It is an amorphous solid or semi-solid material that occurs near petroleum deposits. Widely distributed, the bitumen used in European painting probably came from the Middle East. Known in Egypt and Mesopotamia from very early times, this pigment is little used now. The pigment is partially soluble in oil, in which medium it is fairly stable, and creates a semi-transparent reddish-brown. Microscopically, it can appear as little brown flakes without structure. It is soluble in turpentine, naphtha, and organic solvents, while it is insoluble in water. Because of this latter property, it was typically not used in tempera, fresco, or encaustic mediums. Bitumen is one of the least-desirable pigments, as it never truly dries and the color is not permanent. It was very popular in 16th- and 17th-century oil paintings, in which it is often responsible for a characteristic cracking, or "alligatoring," of the surface.

Mummy is a brown, bituminous pigment that was once actually prepared from the bones and bodily remains of Egyptian mummies that had been preserved using bitumen. Some early authorities claimed that the ancient remains made a better pigment than fresh bitumen, and mummy was favored by some artists in the 16th to 18th centuries, though limited availability kept it from being widely used. With properties similar to bitumen, it was more durable and less prone to cracking. For obvious reasons, this pigment is no longer obtainable.

Blue Pigments

Azurite (mountain blue, lapis armenius, azurium citramarinum, blue bice) is a naturally occurring blue pigment derived from the mineral azurite, a basic copper carbonate $[2CuCO_3 \cdot Cu(OH)_2]$. This mineral occurs in the upper oxidized portions of copper ore deposits found in various parts of the world. It is commonly associated with malachite, the far more abundant green basic carbonate of copper. Azurite was used by the ancient Egyptians and, despite ultramarine having received greater acclaim, was the most important blue pigment in European painting throughout the Middle Ages and Renaissance, well into the 17th century. A synthetically created version of this pigment is called *blue verditer*. The invention of Prussian blue in 1704 brought an end to the widespread use of this pigment.

To prepare the pigment, mineral azurite is simply ground into a powder, washed, and sieved. Coarsely ground azurite produces dark blue, while fine grinding produces a pigment that is lighter in color. Very fine particles are rather pale, greenish-sky blue, and not much admired for painting. If it is not ground finely enough, it is too sandy and gritty to be used as a pigment, but too much grinding produces a pale, weak color. To achieve a balance, the medieval system of manufacture included washing it to remove any mud and then separating the different grains by the processes of *levigation* and *elutriation* (wet grinding a substance into powder and suspending it in a solution to separate particles by density and/or size) using water solutions of soap, gum, and lye.

To produce a solid blue, it was necessary to apply several coats of azurite, but the result was quite beautiful. The actual thickness of the crust of blue added to the richness of the effect, and each tiny grain of the powdered crystalline mineral could add a sparkle. Given the relatively coarse particles used to get a saturated color, *glue size*, an adhesive created from the bones, skin, and/or organs of animals, was often used as a binder to hold the pigment grains firmly in place. Glue size is more easily affected by protracted exposure to damp and washing than other binders, and the blues in wall paintings have sometimes perished through the destruction of their binder where other colors have withstood through time. Azurite blue can also be adversely affected by the use of varnishes that surround the particles of blue. As the varnish yellows and darkens, the power of the azurite to reflect blue light is destroyed. A large number of blacks in medieval paintings were originally blues, obscured by the discoloration of the varnish.

Azurite sometimes looks a little like lapis lazuli, and the two were often confused in the Middle Ages. To tell them apart with certainty, the stones were heated red-hot. This treatment turns azurite black while true lapis is not injured. Under normal conditions, azurite is very stable. It does not blacken from the effects of sulfur gases. The color can, however, be ruined by the presence of acids, alkalis, and, obviously, heat.

Microscopically, azurite appears as translucent blue crystals that often display a greenish cast. The larger the particle, the deeper blue the color, and large particles will exhibit pleochroism from blue to greenish-blue. The crystals are typically angular shards usually ranging from 5 to 50 μm in size. Azurite has poor cleavage, but crushed particles often display conchoidal fracture.

Ultramarine (*azzurrum ultramarine, azzurrum transmarinum, azzurro oltramarino, azur d'Acre, pierre d'Azur, lazurstein*) is probably the most famous pigment in history. Genuine ultramarine is a pigment obtained from the very rare, semi-precious rock *lapis lazuli* [$(Na,Ca)_8(Al,Si)_{12}O_{24}(S,SO_4)$], which is a mixture of the blue minerals *lazurite* ($Na_3CaAl_3Si_3O_{12}S$), *huaynite*, *sodalite*, and/or *nosean*, and some amount of the minerals calcite and pyrite. Diopside, augite, mica, and hornblende may also be present in small amounts. Lapis lazuli forms from contact metamorphism of limestone by alkali-rich magma. The most important deposits of lapis lazuli are in Afghanistan, from whence it was traded throughout the Mediterranean and Europe during the Middle Ages. The history of the lapis trade is fascinating in and of itself.

The earliest-known use of ultramarine as a pigment was in sixth- and seventh-century wall paintings in cave temples at Bamiyan, Afghanistan. In its earliest use, lapis lazuli was simply ground into powder, resulting in a dull blue, as seen in Byzantine illuminated manuscripts. Sometime in the 12th or 13th century, new methods for purifying and concentrating the blue minerals in the lapis were developed in the West, though the raw material still came solely from the East. Ultramarine was widely used in medieval Europe, especially in illuminated manuscripts and panel paintings, alongside vermilion and gold. The highest-quality and most intensely blue-colored ultramarine was often reserved for depictions of the robes of Christ and the Virgin. Because of the limited supply and intensive preparation process, it was as expensive or more expensive than gold. To get an idea of the difficulties involved in extracting ultramarine, read Chapter 62 "On the character of ultramarine blue, and how to make it" in *Il Libro dell' Arte* by Cennino D'Andrea Cennini (1960).

Natural ultramarine has a high stability, even in strong light, as demonstrated by paintings 500 years old that have as intense and pure a blue color as freshly extracted pigment. It is also unaffected by heat or by alkalis but is easily decomposed by even dilute acids, with a complete loss of color. Exposure to sulfur dioxide pollution, or significant moisture, can result in a grayish or yellowish-gray mottling, commonly called *ultramarine sickness*. Due to its cost, ultramarine was almost never mixed with other pigments. Because of the difficulties in obtaining the necessary raw material, and the laborious processing required, use of natural ultramarine rapidly declined with the introduction of a synthetic version (*French ultramarine*) in 1828.

Microscopically, the synthetic and natural varieties are easily distinguished. Natural particles are a clear, intense blue, sometimes almost violet blue, but they are not pleochroic. They appear opaque in reflected light but can be translucent in transmitted light. The natural particles are typically angular shards that display conchoidal fracture and include a wide range of sizes. Lazurite is isotopic and will go extinct under cross-polarized light, but the other blue minerals commonly found in lapis are anisotropic, so this property is not diagnostic, though all of the blue minerals have a very low index of refraction. Ultramarine also commonly includes birefringent particles of calcite and small golden particles of pyrite.

Particles of synthetic ultramarine are an evenly colored deep violet blue, very fine-grained (usually in the range of 0.5–5 μm, though older formulations may contain particles as large as 30 μm) and are extremely well sorted. Synthetic ultramarine is cubic, and will not contain impurities like natural ultramarine, so under cross-polarized light, samples should go completely extinct.

Prussian blue (Paris blue, Berlin blue, Antwerp blue, Chinese blue, iron blue), a synthetic blue pigment with a coppery reddish sheen, is often called the first of the modern pigments. Chemically, Prussian blue is a hydrated iron hexacyanoferrate complex or ferric ferrocyanide $\{Fe_4[Fe(CN)_6]_3\}$. It was first synthesized in Berlin around 1704 when the colormaker Diesbach accidentally formed the pigment while experimenting with the oxidation of iron. It became widely available to artists by 1724 and quickly displaced azurite from the European palette. It has been highly popular since its discovery and is still in use today.

Prussian blue is a low-density, transparent isotropic and is composed of particles so fine that they cannot even be seen under optical magnification (0.01–0.02 μm). What can be seen under the microscope are deep-blue colloidal aggregates that vary in opacity. Prussian blue is fairly permanent with respect to light and air, and has high staining and tinting strength. It is resistant to dilute acids but extremely sensitive to alkalis, which turn it brown, and is instantly discolored by potassium hydroxide.

Smalt (*starch blue*) is a pigment composed of ground blue-tinted glass and was the earliest of the *cobalt pigments*. It is an artificial amorphous potash silicate (i.e. glass), strongly colored by the addition of cobalt oxide, which is then moderately finely ground to create a pigment. First described by Borghini in 1584, but in use much earlier, it is unclear when or where smalt was first created. Cobalt ores were used for coloring glass in Egyptian and classical times, though there is scholarly debate as to whether the cobalt was an intentional additive or an impurity. The principal source of cobalt used in Europe during the Middle Ages appears to be the rare mineral *smaltite*. In the 17th and 18th centuries, other associated cobalt minerals, such as *cobaltite* and *erythrite*, were probably used as well.

To make the pigment, cobalt ore is roasted to create cobalt oxide, which is then melted together with quartz and potash, or added to molten glass. The molten mixture is poured into cold water, causing it to disintegrate into shards (somewhat akin to lava pouring into the ocean, instantly solidifying to form obsidian while shattering into fragments). The shards are ground in water mills, or using a mortar and pestle, and *elutriated* (a method of washing that separates the finer particles from the coarser ones). Several grades of smalt were made, sorted by cobalt content and grain size. Color intensity is directly related to particle size, and only coarse grains are acceptable for use as a pigment. Since smalt is glass, the transparent particles have a low RI, giving it little hiding power.

Smalt particles are amorphous and have variable color intensities, which appears to be uneven across particles of variable thickness. The larger particles are purplish-blue by transmitted light, while the smaller particles are clear blue. Microscopically, smalt displays conchoidal fracture, and particles with square or angular corners are common, as are sharp splinters. Tiny air bubbles are also common, as are birefringent inclusions. Smalt was quite common throughout the 16th and 17th centuries, when it was replaced by *cobalt blue* and synthetic *ultramarine*, though it is still available today.

Cobalt blue (*Thénard's blue*) is the most important of the cobalt pigments. Manufactured commercially in late 1803 or 1804, it is a modern replacement for smalt created by calcining a mixture of cobalt oxide and aluminum hydrate to form cobalt aluminate ($CoO \cdot Al_2O_3$). Cobalt blue is unaffected by acids, alkalis, and heat. It is chemically insoluble and unchanged, even in strong hydrochloric acid. Cobalt blue is lightfast and useful in all techniques. It dries quickly in oil but requires a large amount of binder so it is susceptible to cracking and to yellowing. It is also totally stable in watercolor and fresco techniques. Because this pigment is expensive, even today, it is liable to be replaced or augmented by synthetic ultramarine or even blue lakes. Microscopically, the particles are non-crystalline in appearance, moderately fine (1–50 μm), irregular in size, well-rounded, isotropic, and bright blue in transmitted light.

Indigo (*aniline blue*) is a blue vegetable color used both for dying cloth and for painting, an unusual combination. It is acquired from a variety of plants, most commonly of the genus *Indigofera*, and has been used as a dye for centuries. India was the primary supplier of indigo to Europe, beginning sometime in the 1st century or earlier, and a plant of Indian origin, *Indigofera tinctoria*, was probably the chief source of the indigo trade. To create the dye, plants are shredded, packed into large vats, and allowed to ferment. The dark precipitate is strained, pressed, and formed into cakes. To create a pigment, the precipitate is dried and ground into powder. Indigo was grown all over the world until the 1900s, when synthetic versions became available. Chemist Adolf von Baeyer was the first to synthesize indigo in 1878, but commercial manufacturing was not possible until 1897. By 1913, a synthetic version called aniline blue took over, almost entirely replacing the use of indigo as a pigment.

Indigo has fair tinting strength, good lightfastness (light resistance) though it fades in strong sunlight, good to moderate alcohol resistance, and low oil resistance. It is heat resistant to 150 °C and is resistant to air. Alkalis dissolves it to form the sodium salt *indigo white*, which oxidizes into many shades of blue. It is stable when exposed to hydrogen sulfide.

Green Pigments

Malachite (Mountain Green) is a green basic copper carbonate ($CuCO_3 \cdot Cu(OH)_2$), this mineral is found in association with the blue mineral azurite, which contains less bound water, and occurs in various parts of the world in the upper oxidized portions of copper ore deposits. Geologically, azurite is the parent and malachite a weathered form of the original blue deposit. It was used as a pigment in ancient Egypt and is perhaps the oldest-known bright-green pigment. In fact, Egyptians probably used the pigment as eye paint even before the first Egyptian dynasty. In western China, malachite is found in many paintings from the ninth and tenth centuries. Europeans commonly used malachite in medieval times and during the Renaissance. Synthetic green pigments replaced malachite by about 1800.

Malachite is prepared in the same manner as azurite. The natural mineral is crushed, ground to a chalk-like powder, washed, and levigated to sort it into grades. The powder becomes paler the more it is ground. Microscopically, malachite is composed of monoclinic crystalline particles that have a clear, faint bottle-green color by transmitted light, show high relief, strong birefringence, and pleochroism. Since it is a carbonate, it decomposes in even weak acid, and it blackens when heated. It is unaffected by sunlight and has remained unchanged in many paintings for centuries.

Terre Verte (Green Earth, Verona Green) is a pigment that has been in use since before classical times. Terre verte was applied in ancient Roman wall paintings and found in pots as a prepared pigment in Pompeii. The name terre verte, which means "green earth," is applied to several different complex silicate minerals. Green earth ranges in color from a neutral yellowish-green to a pale greenish-gray, with the neutral sage-green color being the most coveted. It occurs widely, though there are only minor deposits of a quality sufficient to use as a pigment. The most important variety found in medieval painting is the light, cold-green mineral *celadonite*, found chiefly in small deposits in rock in and around Verona, Italy. Other important mineral forms are *chrysocolla*, a bluish-green mineral found in Israel, Democratic Republic of Congo, and England, and the yellowish-olive *glauconite* from central Europe. The relatively dull color meant that terre verte was rarely used in landscapes during the Middle Ages, though medieval Italian painters commonly employed it as a foundation for flesh tones.

Some modern manufacturers are selling products labeled as terre verte that are in reality mixtures of chromium oxide, bone black, and ochre, since the natural product is scare. There have also been reports of colormen selling green synthetic ultramarine as terre verte. Green earth has low hiding power and microscopically is composed of coarse, rounded, smoky-green particles and transparent silica particles. The pigment turns reddish-brown when strongly heated; however, it is otherwise quite stable, dissolves partially with a yellowish-green color in hydrochloric acid, but not in alkalis, and should not dissolve in water, alcohol, or ammonia.

Verdigris (*from Vert De Grece or Green of Greece; also Viride Aeris*) is a term applied to a range of bright-green copper acetates ($CuCH_3COO$), which have been produced since Roman times by treating copper sheets with acetic acid (vinegar, wine, or urine) fumes and scraping the resultant corroded crust. A number of different methods for obtaining verdigris have been described over the ages. A related pigment, *salt green*, is created by spreading copper sheets with honey and sprinkling them with salt before they are treated with acid, resulting in a somewhat different shade of green. The copper acetates range in color from pale green to turquoise blue. Verdigris was commonly widely used by European painters from the 13th to the 19th century.

Well-crystallized verdigris particles are hexagonal and rhombic with distinct boundaries, though it can be composed of fine needles instead, while some varieties of copper acetate are amorphous. The particles are typically strongly birefringent and pleochroic, changing from pale green to deep greenish-blue. Verdigris is the most reactive and unstable of the greens, notorious for darkening with age (which actually has more to do with mixing the pigment with resin than any inherent property of the pigment itself) and a synthetic substitute was sought early. It is slightly soluble in water and readily soluble in acids; in the presence of HCl, verdigris forms a green solution. It is incompatible with lead white, which causes it to blacken, turns into a black residue when heated, and unless it is isolated in a binder the color is *fugitive* (i.e. it fades).

Scheele's Green is a copper arsenate compound ($CuHAsO_3$) developed by the Swedish chemist Scheele sometime prior to 1777, when he wrote a letter to another scientist discussing its high toxicity. In 1778, the Stockholm Academy of Sciences published his detailed instructions for making the pigment. First, potash and powdered arsenic sulfide (As_2O_3) were dissolved in water and heated. The alkaline solution that resulted was added, a little at a time (because of foaming), to a warm solution of copper sulfate. When allowed to stand, a green precipitate would settle out. The liquid was poured off and the precipitate washed and then dried on low heat. Its popularity as a pigment was brief, as its yellowish-green color faded rapidly, and sulfur-bearing air and exposure to sulfide caused it to blacken. Scheele's green can be identified under the microscope as small and large irregular-shaped green flakes that are slightly transparent.

Emerald Green (*Paris Green, Veronese Green, Schweinfurt Green*) is a synthetically produced basic copper acetoarsenite [$Cu_3(AsO_4)_2 \cdot 4H_2O$ or $3Cu(AsO_2)_2 \cdot Cu(CH_3 COO)_2$]. It was developed in 1808 as a replacement for Scheele's green, and first produced commercially by the firm of Wilhelm Sattler at Schweinfurt, Germany in 1814. It is a brilliant bluish-green color quite unlike any other green pigment, has fair hiding power, and has been used to make an imitation patina on bronzes. Microscopically, it consists of small, rounded grains, uniform in size. At high magnification, the particles have a radial structure and many have a small dark pit in the center. Emerald green is readily decomposed by acids and warm alkalis and is blackened by heat and sulfur-bearing air or sulfide, though it is fairly stable in oil or under varnish. Its highly toxic nature and incompatibilities limited this pigment's popularity.

Viridian (*Vert Émeraude, Guignet's Green, Permanent Green*) is a transparent, bright-green, hydrous chromic oxide ($Cr_2O_3 \cdot 2H_2O$) that is formed by heating a mixture of alkali chromate and excess boric acid to dull red. Oxygen gas and water are released and the resulting salt is put into vats of cold water and allowed to hydrate. The pigment is then collected, ground, washed with hot water, and dried again. Guignet of Paris patented this process for manufacturing viridian, or transparent oxide of chromium, in 1859. The resulting pigment is a deep, cool green. It has great tinting strength and is very stable and unaffected by acids, alkalis, or light, though heat causes it to dehydrate into anhydrous chromic oxide (Cr_2O_3), which is a dull, opaque-green (army-green) color. Viridian is non-poisonous and replaced in large part both verdigris and emerald green. Microscopically, viridian is distinguished by its slightly rounded, poorly sorted, large, bright-green particles that are transparent, strongly birefringent, and anisotropic in polarized light.

Better than the Real Thing

According to Giorgio Vasari's (2006) *Lives of the Most Eminent Architects, Painter and Sculptors of Italy*, Michelangelo di Lodovico Buonarroti Simoni (1475–1564), he of the famed statue *David* and the paintings on the ceiling of the Sistine Chapel, commonly produced forgeries in his youth. This actually is not as outrageous as it might seem, for it is still common practice for art students to learn their craft by copying the works of masters. Michelangelo, as an apprentice to the painter Domenico Ghirlandaio (1449–1494), did so and "he also copied drawings of the old masters so perfectly that his copies could not be distinguished from the originals, since he smoked and tinted the paper to give it the appearance of age. He was often able to keep the originals and return his copies in their stead." Usually such copies are not really forgery in the criminal sense, and it was a common enough practice that it causes headaches to museum curators and art historians to this day.

The most famous of Michelangelo's forgeries, produced in 1496, was of a life-sized *Sleeping Cupid* done in the style of the ancient Romans. This was very early in his career and it is unclear whether he was really culpable in the matter or not. As part of the process, the sculpture was buried in acidic earth, either by Michelangelo or by someone else, in order to create the impression of age. At some point later, an art dealer named Baldassare del Milanese sold the statue to the Cardinal of San Giorgio for 200 crowns. It appears Baldassare may have sold the Cupid on commission for Michelangelo, to whom he lied about its purchase price, paying Michelangelo only 30 crowns. Thus, it might actually have been Baldassare who committed the fraud, misrepresenting the statue as ancient.

Anyhow, the cardinal eventually discovered that his ancient Cupid had in fact been made recently in Florence and insisted on getting his money back. Baldassare apparently did return the money and regain possession of the Cupid. Vasari's also notes that the cardinal "did not escape blame for not recognizing the merit of the work, for when moderns equal the ancients in perfection it is a mere empty preference of a name to the reality when men prefer the works of the latter to those of the former . . . What matter, they laughed, whether it were modern or not? . . . the whole affair increased Michelangelo's reputation." Michelangelo was actually summoned to Rome and engaged by the Cardinal of San Giorgio, with whom he stayed for nearly a year.

Baldassare later displayed the Cupid in the house of Cardinal Ascanio Maria Sforza and it was eventually purchased by Cesare Borgia and given to Guidobaldo da Montefelre, the Duke of Urbino. The subsequent history of the Cupid is tangled up in the Borgia history of Italy and things rapidly become quite confusing, but the statue eventually arrived in England as part of a collection purchased for Charles I in the 1630s. There are numerous Cupid statues that have at one time or another been put forward as being the Michelangelo, but according to most authorities the true sculpture is now lost, probably destroyed in the fire that swept Whitehall Palace in 1968.

This story raises several of the troubling points about authenticating and valuing works of art. Which would be worth more, an ancient Roman sleeping Cupid or Michelangelo's copy of a Roman sleeping Cupid? If an artwork is beautiful, appreciated, and honored, does it matter whose name is attached? Now contemplate the fact that some forgers have through time become famous enough that their forgeries are now being forged. Thankfully, while these questions are interesting, they are not the concern of the forensic investigator. It does, however, increase our appreciation for the passions and complexities of the art world.

Collecting a Sample for Microscopic Examination (McCrone, 1982)

This brief introduction does not qualify anyone to analyze questioned artwork, but it will give you a head start on the practice. The best way to get practice identifying pigments is to acquire some powdered pigments (i.e. not in a binder) from an artist's supply house (some include Natural Pigments (http://naturalpigments.com/) or Sinopia (http://www.sinopia.com/)). This will give you a chance to look at the pigments without interference. After you have become familiar with their properties, you can try to combine them with different binders (gum arabic for watercolor, rabbit glue, egg tempera, linseed oil, etc.) in order to understand how the binder alters or interferes with some properties.

The object of sampling artwork or artifacts is to take as small a sample as is possible that is still sufficiently large enough for analysis (which, believe it or not, is only around one nanogram for a skilled microscopist). There are basically two ways to sample a painting. One is to take a *cross-section* (a perpendicular slice that includes everything from the varnish on top to the layer in contact with substrate) and the other is to collect a *flake* (a small surface fragment that does not contain all of the layers). The first step is to place the artwork under a stereomicroscope, preferably one mounted on a swing arm. All sampling should be done while viewing the artwork through the microscope, and taking meticulous records (including photographs) of each sampling area and the subsequent location of each collected sample.

Cross-sections should generally only be taken at the edge of a painting or from areas that are damaged, for example along a crack, and are obtained by using a *microscalpel* (homemade microscalpels can be created by carefully breaking the edge of a razor blade into 1–2 mm sections using needle-nosed pliers and then putting one of the shards into a needle-holder.) Only a very tiny amount of material should be collected, preferably only a couple of nanograms. The freed cross-section is mounted in Meltmount (or a similar material; Aroclor is not recommended as that is the trade name of highly toxic PCBs) with the layers oriented perpendicular to and near the flat upper surface. Before the Meltmount cures completely, use a sharp razor blade to form a planer cross-section on the slide, or if the sample is too hard or too fragile allow the Meltmount to cure completely and grind the surface down to create a thin-section.

The best approach for collecting a flake is to use fine-tip tungsten needles (available from McCrone at http://www.mccronemicroscopes.com/), which are fairly expensive, or, as a significantly cheaper alternative, a fine steel dissecting needle (commonly used in biology classes). Again, look for a damaged area, crack, or deteriorated area. If the paint surface is pristine, take only a tiny submicrogram sample invisible to the naked eye. The needle is carefully pressed into the paint to pop out a tiny amount of sample. Care must be taken not to accidentally propel the flake off into space (the *tiddlywink effect*) causing it to disappear forever. Sometimes, the sample will adhere to the sampling needle; sometimes, it will need to be picked up using the needle. In either case, the flake is transferred to a clean slide by gently rolling the needle over the glass. Record the spot where each flake is deposited using a scribe or permanent marker. If the flake will not come off of the needle, apply a very tiny (1 mm^3) droplet of amyl acetate (this substance is highly flammable and toxic, so use it only in very small quantities) to the glass slide using a micropipette or a capillary paintbrush. Touching the particle to the amyl acetate will cause transfer and the liquid will evaporate fairly quickly, securing the flake to the slide.

Flakes can be treated in a similar fashion to cross-sections, if they are big enough, or they can be carefully crushed to create a loose powder. The latter is done by placing the flake on a clean microscope slide and immediately taping a clean cover slip down over the top with strips of transparent tape along two sides. Now, very carefully, roll the side of a dissecting needle back and forth over the coverslip, gradually increasing the pressure. Do not crack the coverslip! Work slowly, observing your progress under a stereomicroscope, until the flake has broken apart. Prepare another labeled glass slide with tiny drops of Meltmount (0.2–0.5 mm diameter sphere) and, working under a microscope using a needle, transfer samples of pigment from your powdered sample. Once the pigment has been dispersed in the Meltmount, cover it with a cover slip. You can create small cover slips by taking a normal-sized slip, dividing it into quarters using a scribe to etch in lines, and then carefully breaking the cover slip along the lines. Take care not to get the cover slips dirty.

In order to become adept at collecting and preparing pigment samples, a significant amount of practice is required. Try to find appropriate paintings at secondhand shops or in flea markets in order to learn how to do this properly.

A summary of the steps used to perform a *microscopic analysis of pigments* is available on the companion website (www.wiley.com/go/bergslien/ forensicgeoscience). In brief, use Köhler illumination to determine the basic properties (color, shape, RI, etc.) of each pigment, and then use cross-polarized light to determine birefringence, extinction, and interference colors. Compare your pigment's properties to Table 8.2 and the written descriptions. In general, natural pigments include a range of particle sizes, tend to be more angular, and are in general coarser than synthetic pigments. Because most synthetic pigments are the product of a chemical reaction rather than mechanical grinding, they tend to be almost homogeneous in size, very fine, and well rounded.

A brief exercise on *microchemical testing of white and blue pigments* is available on the companion website, as is a list of other *pigment chemical reactions* that can aid in the identification of pigments that have lost their original color. There is also a brief discussion of the use of *minerals in cosmetics* for interested parties.

Raman Spectroscopy

Raman spectroscopy is based upon the Raman scattering effect, discovered in 1928 by Sir Chandrasekhara Venkata Raman. When a beam of light strikes an object (gas, liquid, or solid), the incident photons are either *absorbed* (transformed into vibrational motion, or heat, by the electrons in the sample), *transmitted* (pass through the object essentially unchanged, as through a transparent or translucent object), *reflected* (bounce off the surface of the object), or *scattered* (re-radiated in all directions). We perceive the reflected light as color. For example, a red apple absorbs the orange through violet wavelengths and reflects the red wavelengths back to our eyes.

The majority of the scattered photons collide elastically and have exactly the same wavelength as the incident photons (Rayleigh scatter), but a tiny proportion of the incident photons experience inelastic collision and the scattered radiation is shifted to a different wavelength (Raman scatter). The change in wavelength is related to the structure of the impacted atom. Because Raman scatter is a relatively weak effect, and can only be interpreted if the incident light is monochromatic (i.e. of a single wavelength), technical limitations prevented it from becoming a common laboratory tool. However, the advent of single-wavelength lasers, improved optics, and charge-coupled device (CCD) detectors has turned Raman spectroscopy into a low-cost and increasingly common analysis tool.

When the laser light from a Raman instrument hits the sample, the sample generates Rayleigh and Raman scatter in all directions. The resultant *Raman spectrum* is a plot of the intensity of Raman scattered radiation (y- axis) as a function of its frequency difference from the incident radiation (x-axis, reported in cm^{-1}) (Figure 8.4). Different bonds result in different wavelength shifts, thus the spectrum of shifted photons creates a pattern of characteristic peaks that allows materials to be distinguished.

Table 8.2 Characteristics of some common mineral pigments.

Name	Color	Size (μm)	Crystal System	Optical System	RI	Pleo.	Biref.	Extinc.	Earliest Date of Use
A. Black Pigments									
Bone black	Black	1–25	Amorphous/ Hexagonal	Opaque/ Uniaxial(–)	~1.65		0.01	–	4th-cent. BCE
Ivory black	Black	1–25	Amorphous						
Lamp black	Black	<1	Amorphous	Opaque					3000 BCE
Carbon black	Black	1–100	Amorphous	Opaque					Ancient
Vine black	Black	1–100	Amorphous	Opaque					Roman
Coal	Black/ brown	Varies	Amorphous	Opaque					?
Graphite	Black	Varies	Hexagonal	Opaque, RI > 1.662					?
B. White Pigments									
Bone White	White	<100	Amorphous		~1.65	–	0.005	Undulose	Ancient
Chalk	White	1–10	Trigonal	Uniaxial (–)	1.48– 1.66	–	0.154– 0.174	Undulose	Ancient
Shell	White								
Gypsum	White	5–50	Monoclinic	Biaxial (+)	1.52	–	0.010	Oblique	Ancient
Lead White	White	1–50	Orthorhombic	Biaxial (–)	1.80– 2.09	–	0.15– 0.273	Parallel	500
Zinc White	White	<2	Hexagonal	Uniaxial (+)	1.9– 2.1	–	0.016	Too small	1843
Barium White	White	<1	Orthorhombic	Biaxial (+)	1.63– 1.65	–	0.010	Too small	19th c.
Lithopone	White	<2	Mix of Barium White and Zinc White					Too small	1874
Titanium White– Anatase	White	<1	Tetragonal	Uniaxial (–)	2.52– 2.55	–	0.073	Too small	1923
Titanium White– Rutile	White	<1	Tetragonal	Uniaxial (+)	2.71– 2.72	–	0.29	Too small	1947

Table 8.2 *continued*

C. Yellow Pigments

Yellow Ochre	Yellow	5–50	Varies	Varies	>2.2	–	0–0.15	Varies	Ancient
Goethite	Yellow Orange	1–50	Orthorhombic	Biaxial (–)	2.26–2.52	Yes: reds	0.138–0.21	Parallel–Undulose	Ancient
Litharge	Light Yellow	5–100	Tetragonal	Uniaxial (–)	2.53–2.67	–	0.20	Parallel/ some twinkle	Ancient
Orpiment	Yellow	1–30	Monoclinic	Uniaxial (–)	2.40–3.02	Yes: yellows	0.62	Parallel	Ancient
Cadmium Yellow	Yellow	~1	Hexagonal	Uniaxial (–)	~2.5	–	0.023	Hard to see	1845
Naples Yellow	Yellow	1–5	Cubic	Isotropic	2.01–2.28	–	–	Isotropic	500 BCE
Chrome Yellow	Yellow to Amber	Rods 1 μm × 1–10 long	Monoclinic	Biaxial (+)	2.31–2.66	Yes: yellows	0.35–0.37	Inclined	1809
Zinc Yellow	Pale yellow	1–2	Anisotropic	?	>1.66	Yes: yellows	0.10 +	Undulose	Early 19th c.

D. Red Pigments

Red Ochre	Red	1–3+	Varies	Varies	>2.7	–	0.21	Complete	Ancient
Hematite	Red	1–50	Trigonal to disordered	Uniaxial (–)	2.87–3.22	Yes: reds	0.21–0.28	Parallel	Ancient
Mars Red	Red	>1, uniform	Trigonal to disordered	Uniaxial (–)	2.74–3.01	–	0.21–0.23	Parallel	?
Red Lead	Orange	1–50	Tetragonal	Uniaxial (+) or (–)	2.41–2.42	Yellowish-orange	0.01	Parallel to Undulose	Antiquity
Vermilion	Red	1–30	Trigonal	Uniaxial (+)	2.81–3.25	Orangey-red	0.33–0.35	Parallel	Cinnabar ancient/ synthetic 13th c.
Realgar	Yellow to Orange	1–30	Monoclinic	Biaxial (–)	2.53–2.70	Yellowish-orange	0.166	Parallel	Ancient
Cadmium Red	Red	~1	Hexagonal	Uniaxial (–)	2.51–2.53	–	0.00–0.02	Too small	1910
Chrome Red	Orange	Rods 1 μm × 1–10 long	Monoclinic	Biaxial (+)	2.31–2.66	Yes	0.35–0.37	Inclined	Early 19th c.

continued

Table 8.2 *continued*

E. Orange and Brown Pigments

Brown Ochre	Yellow to Brown	1–50	Usually a mixture of goethite, hematite, and magnetite with magnesium oxides						Ancient
Raw Umber	Yellow	1–50	Usually a mixture of magnesium hydroxides, oxides, goethite, and hematite. RI ~2.0, mostly isotropic						Ancient
Burnt Umber	Brown to Red	1–50	Usually a mixture of magnesium hydroxides, oxides, and hematite. R.I. ~2.2, mostly isotropic, aggregate particles						Ancient
Bitumen (Asphaltum)	Red to brown	–	Amorphous, translucent to opaque, conchoidal fracture		1.64 to <1.66	–	–	Acts Isotropic	Ancient
Raw Sienna	Orange	1–50	Usually a mixture of goethite, hematite, and magnetite with magnesium oxides. RI > yellow ochre						Ancient
Burnt Sienna	Reddish–orange	1–30	Goethite transforms to "disordered hematite" gives pigment more uniform color						Ancient
Lepidocrocite (in Limonite)	Yellow Orange Brown	1–50	Orthorhombic	Biaxial (–)	1.94–2.51	Yes	0.57	Parallel	Ancient

F. Blue Pigments

Azurite	Blue (poor cleavage)	1–50	Monoclinic	Biaxial (+)	1.73–1.84	Blue to blue-green	0.108–0.116	Parallel Anomalous Oblique	Ancient
Ultramarine (just the blue Lazurite particles)	Blue	1–50, highly variable	Cubic to Pseudo-cubic	Isotropic	1.5	–	–	Isotropic to weakly anisotropic	Ancient
Synthetic Ultramarine	Blue	Usually 0.5–5	Cubic	Isotropic	1.50	No	–	Isotropic	1828
Prussian Blue	Blue	0.01–0.5	Cubic	Isotropic	1.56	–	–	Isotropic	1704
Smalt	Blue	10–50	Amorphous, may show inclusions, conchoidal fracture		1.46–1.55	–	–	–	~1500
Cobalt Blue	Blue	1–50	Cubic	Isotropic	>1.66–1.72	–	–	Isotropic	1775
Indigo	Indigo/Dark Blue	1–10	? Rounded particles	Anisotropic	>1.66 variable	Blues	Weak	Mostly disappears/Parallel	Antiquity

Table 8.2 *continued*

G. Green Pigments									
Malachite	Green to Bluish-green	1–50	Monoclinic	Biaxial (–)	1.65–1.90	Colorless to green	0.25	Parallel Anomalous Oblique	Ancient
Terre Verte	Olive to Bright Green	1–50	Mixture of minerals glauconite and/or celadonite, with chlorite, quartz and iron oxides		1.59–1.66	–	0.01–0.03	Undulose	Ancient
			Monoclinic	Biaxial (–)					
Verdigris	Green	1–30	Monoclinic	Unknown	1.53–1.56	Lt. blue to lt. green	0.03	Parallel Undulose common	Roman
Scheele's Green	Lime Green	Med. to coarse	Amorphous to Crystalline	Unknown	1.55–1.75	–	None-weak	Does not go completely extinct	1777
Emerald Green	Pale Green	1–10	Monoclinic	Unknown	1.71–1.78	–	0.06–0.07	Undulose	1814
Viridian	Pale Blue Green	1–10	? Rounded	Biaxial (?)	1.62–2.12	–	0.03–0.50	Undulose	1838
Chromium Oxide	Green	0.1–1.0	Trigonal–Hexagonal	Uniaxial (+)	~2.5	–	High	Parallel	Early 19th c.

Key: Pleo. = pleochroism; Biref. = birefringence; Extinc. = extinction; c. = century.

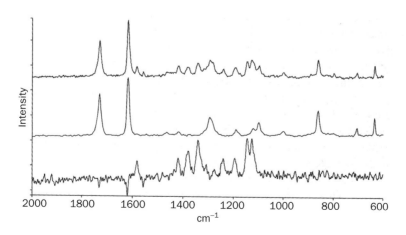

Figure 8.4 Raman spectra of dye in fiber. The top spectrum is from the dyed fiber, the middle one is from the fiber, and the bottom one is the difference of the previous two spectra. *Source:* Smith and Dent, 2005.

Raman spectra can be used in two ways. At the simplest level, Raman spectra can be treated as "fingerprints" of the molecular structure, similar to the way we used XRD spectra (introduced in Chapter 7). By comparing the Raman spectra of an unknown with catalogs of reference spectra, the unknown can be identified. There are several libraries of Raman spectra for minerals and pigments available free online (see, for example, the Caltech RRUFF Project at http://rruff.info/; UCL Raman Spectroscopic Library at http://www.chem.ucl.ac.uk/resources/raman/index.html; RASMIN at http://riodb.ibase.aist.go.jp/rasmin/E_index.htm; and Raman Spectra Database, Dipartimento di Scienze della Terra, at http://www.dst.unisi.it/geofluids/raman/spectrum_frame.htm). There are also reference libraries for such diverse materials as plant fibers, waxes, organic compounds, and narcotics. The Infrared and Raman Users Group website is another valuable resource (http://www.irug.org/ed2k/search.asp).

At a deeper level, it is possible to recognize, at specific positions in the spectrum, bands that can be identified as *characteristic group frequencies* associated with the presence of localized units of molecular structure such as hydroxyls. Raman allows researchers to gather information about the lattice structure of ionic molecules in the crystalline state, the internal covalent structure of complex ions, and the ligand structure of coordination compounds both in the solid state and in solution. This ability to discern information about molecular state is how Brown and Clark (2002) were able to distinguish anatase from rutile, both of which are forms of titanium dioxide, but which have different atomic structures (Figure 8.5). Raman spectrometry is becoming an increasingly popular technique for identifying and studying inorganic solids because no sample preparation is necessary and the technique is nondestructive.

Chromatography

Many of the methods used for the analysis of organic compounds are based on chromatography. In 1906, the Russian botanist Mikhail Tswett (Michael Tsvett) discovered that the green color from plant leaves could be separated into several

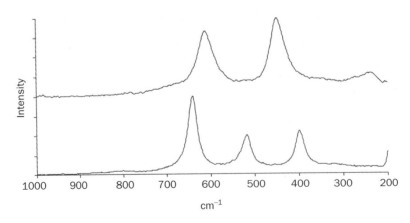

Figure 8.5 NIR FT Raman spectra of TiO_2: rutile at the top and anatase at the bottom. *Source:* Smith and Dent, 2005.

different colored materials by grinding the leaves up in petroleum ether and letting the resultant liquid trickle down a glass tube filled with powdered chalk or alumina. As the mixture seeped downward, different pigments demonstrated varying degrees of readiness to attach to the absorbent, thus separating individual pigments into layers inside the tube. Mikhail Tswett named the process *chromatography* (or, translated from its Greek roots *chroma* and *graphein*, "color writing").

Chromatography is a vital analytical tool in chemistry, biology, and forensic science. The best way to understand this procedure is to look at its simplest form, *paper chromatography*. To perform a separation using paper chromatography, a small drop of liquid about the size of this letter "o" is placed on a strip of filter paper. The filter paper is then placed in a suitable mixture of solvents, which will wick up into the paper via capillary action. As the solvent moves up the paper, it will interact with the spot sample, breaking it apart into several components, determined by their relative rates of dissolution (i.e. how fast each component will dissolve into the solvent). The spot will begin to move, breaking up into several different spots, some of which will move quickly up the paper, and some of which will move very slowly. Some of the spot might not move at all, indicating that those components are not soluble in the selected solvent. The separated components are often a significantly different color than the original spot. As the solvent wicks up the paper, the spots will eventually resolve into several separate spots; their relative rates of progress can be used to identify each component.

All forms of chromatography employ two basic phases: a *stationary phase* and a *mobile phase*. In paper chromatography, paper is the stationary phase, while the liquid solvent is the mobile phase. The two phases basically compete to attract the various components of the sample being analyzed. Because each component of the sample will have a different degree of attraction to the solvent versus the paper, each will be carried a different distance by the mobile phase. This is what causes the sample to separate.

Thin-layer chromatography (TLC) works virtually the same way as paper chromatography, except rather than using filter paper TLC uses plates coated with a thin layer of a solid adsorbent, like silica or alumina. The same basic principles also lie behind much more sophisticated techniques, such as *high-pressure liquid chromatography* (HPLC) or *gas chromatography* (GC), which use gas as the mobile phase while columns containing different inert materials act as the stationary phase.

In paper chromatography, and TLC, once the separation is complete (preferably the point at which each spot is completely separate), the paper/plate is removed from the solvent and dried. The dried paper/plate is called a *chromatogram*. Each mixture of substances will generate a characteristic chromatogram displaying the sample's components separated by some distance. The amount of movement of each chemical component in a mixture can be quantified as R_f *values* where

$$R_f = \frac{\text{distance of component travel from starting point}}{\text{distance of solvent travel from the starting point}}$$

Since no component can travel further than the solvent, all R_f values should be equal to or less than 1.0. R_f values vary depending upon the type of filter paper used, the type of solvent used, the distance of the starting point (origin) from the solution front, and the manner of development of the chromatogram. Under controlled conditions, chromatography can be used to identify unknown substances and for comparisons. The type of chromatogram and the R_f values of the unknown substance are compared with the chromatogram and the R_f values of a standard chemical in order to determine the unknown's identity.

To accurately compare samples using this method, spots of the known and questioned samples to be compared need to be run under the same conditions. Quality-controlled TLC plates are manufactured specifically for this purpose and tables of TLC R_f values for various solvents and substances are available in the literature. A variety of chromatography techniques are commonly used forensically to analyze blood samples, explosive residue, drugs, accelerants, inks, and paint samples.

Inks

Ink is composed of organic or inorganic pigments or dye suspended in a solvent. The first inks were made of fruit or vegetable juices, secretions from squid, cuttlefish, or other cephalopods, and tannin from galls, nuts, or the bark from trees. The first synthetic ink was made thousands of years ago from charcoal mixed in glue. Modern inks can generally be divided into three classes: writing inks (i.e. ballpoint pen), conventional printing inks (used in newspapers and magazines), and non-impact printing inks (ink-jet printers and electrophotographic technologies).

Inks are commonly examined using TLC (ASTM, 2005), though subsequently several nondestructive methods have been put forward as alternatives. TLC is still popular because it is rapid, inexpensive, and minimally destructive. Modern pen companies use exclusive ink formulas that are specific to each brand name of pen and change these formulas periodically. The United States Secret Service, which was originally established to suppress the counterfeiting of US currency, maintains an ink library consisting of more than 8500 thin-layer chromatography plates as standards. As a result, it is often possible to distinguish pen types and relative ages of ink via chromatography. Black inks in particular yield extremely characteristic chromatograms with distinctly colored separation bands. A simple simulated forensic exercise involving the *paper chromatographic separation of inks* is available on the companion website.

Summary

This chapter was, believe it or not, a very short introduction to the use of minerals as pigments for artwork. Much more detailed information on art pigments is available from the resources listed below. Plus, there are many other ways in which geology and art intersect. The information and methodology used here applies anytime minerals are used as colorants or additives, so the forensic applications are myriad.

Further Reading

Vinland Map

Harbottle, G. (2008) The Vinland Map: A critical review of archarometric research on its authenticity. *Archaeometry* **50**(1): 177–89.

McCrone, W. C. (1976) Authenticity of medieval document tested by small particle analysis. *Analytical Chemistry* **48**(8): 676A–9A.

McCrone, W. C. and McCrone, L. B. (1974) The strange case of the Vinland Map. *Geographical Journal* **140**: 212–214.

Towe, K. M. (2004) The Vinland Map ink is NOT mediaeval. *Analytical Chemistry* **76**(3): 863–5.

Towe, K. M., Clark, R. J. H., and Seaver, K. A. (2008) Analyzing the Vinland Map: A critical review of a critical review. *Archaeometry* **50**(5): 887–93.

General resources on pigment information

Artists' Pigments 1780–1880: History and Uses (http://lilinks.com/mara/history.html)

History of Pigments (http://www.winsornewton.com/about-us/our-history/history-of-pigments/)

Odegaard, N., Carrol, S., and Zimmt, W. S. (2000) *Material Characterization Tests for Objects of Art and Archaeology*. Archetype Publications, London.

Pigments through the Ages (http://www.webexhibits.org/pigments/)

Thompson, D. V. (1956) *The Materials and Techniques of Medieval Painting*. Dover Publications, New York.

Art supply stores that sell raw pigments

Kremer Pigments Art Supplies (http://www.kremerpigments.com/)

Natural Pigments Art Supplies (http://www.naturalpigments.com/)

Sinopia Pigments (http://www.sinopia.com/)

References

ASTM (2005) Standard Guide for Test Methods for Forensic Writing Ink Comparison, ASTM International, West Conshohocken, PA, DOI: 10.1520/E1422-05, http://www.astm.org/Standards/E1422.htm.

Brainerd, A. W. (1988) *The Infanta Adventure and the Lost Manet*. Reichl Press Inc., Long Beach, Michigan City, IN.

Brown, K. L. and Clark, R. J. H. (2002) Analysis of pigmentary materials on the Vinland Map and Tartar Relation by Raman microprobe spectroscopy. *Analytical Chemistry* **74**: 3658–61.

Cahill T. A., Schwab, R. N., Kusko, B. H. *et al.* (1987) The Vinland Map, revisited: New compositional evidence on its inks and parchment. *Analytical Chemistry* **59**(6): 829–33.

Cennini, C. d'A. (1960) *The Craftman's Handbook: The Italian "Il Libro Dell' Arte"*, (trans. D. Thompson). Dover Publications, New York.

Donahue, D. J., Olin, J. S., and Harbottle, G. (2002) Determination of the radiocarbon age of parchment of the Vinland Map. *Radiocarbon* **44**(1): 45–52.

Eastaugh, N., Walsh, V., Chaplin, T., and Siddall, R. (2004) *Pigment Compendium: Optical microscopy of historical pigments*. Elsevier Butterworth-Heinemann, London.

Finlay, V. (2004) *Color: A natural history of the palette*. Random House: New York. [Also released as *Colour: Travels through the paintbox*.]

Guyton de Morveau L.B. (1782) Recherches pour perfectionner la préparation des couleurs employées dans la peinture. Nouveau Mémoires de l'Académie de Dijon, **Premier semester**: 23.

Hebborn, E. (2004) *The Art Forger's Handbook*. Overlook Press, New York.

Hoving, T. (1997) *False Impressions: The hunt for big time art fakes*. Touchstone, New York.

McCrone, W. C (1982) The microscopical identification of pigments. *Journal of the International Institute for Conservation: Canadian Group* 7(1 and 2): 11–34.

McCrone, W. C. (1988) The Vinland Map. *Analytical Chemistry* **60**: 1009–18.

McCrone, W. C. and McCrone, L. B. (1974) The strange case of the Vinland Map. *Geographical Journal* **140**: 212–214.

National Gallery of Art (1986–2007) *Artists' Pigments: A handbook of their history and characteristics*. Volume **1** (1986); Volume 2 (1993), Volume 3 (1998), Volume 4 (2007). National Gallery of Art, Washington and Oxford University Press, New York.

Olin J. S. (2000) Without comparative studies of inks, what do we know about the Vinland Map? *Precolumbiana* 2(1): 27–36.

Olin, J. S. (2003) Evidence that the Vinland Map is medieval. *Analytical Chemistry* 75(23): 6745–7.

Skelton, R. A., Marston, T. E., and Painter, G. D. (1965) *The Vinland Map and the Tartar Relation*, 1st edn. Yale University Press, New Haven, CT and London.

Skelton, R. A., Marston, T. E., and Painter, G. D. (1995) *The Vinland Map and the Tartar Relation*. 2nd edn. Yale University Press, New Haven, CT and London.

Seaver, K. (2004) *Maps, Myths, and Men: The story of the Vinland Map*. Stanford University Press, Palo Alto, CA.

Smith, E. and Dent, G. (2005) *Modern Spectroscopy: A practical approach*. John Wiley & Sons Ltd, New York.

Vasari, G. (2006) *Lives of the Most Eminent Architects, Painter and Sculptors of Italy*, (trans. G. C. de Vere). Modern Library, New York.

Chapter 9
Fossils and Microfossils: Traces of Life

Fossils are of interest to the forensic worker for a few different reasons. Just like soil and rocks, fossils can be used to help characterize evidence and as tools for comparison. Because certain types of fossils are only found in certain areas, they can be used to help locate source areas, sometimes very precisely. Plus, fossils themselves can sometimes be worth a substantial amount of money, and there is, unfortunately, a thriving industry in stolen material. The goal of this chapter is to enable readers to recognize common fossils and microfossils that have the potential to be useful in forensic work.

Geologic Time and Index Fossils

Consider the story of Viswa Jit Gupta, a formerly prominent Indian paleontologist at the Panjab University of Chandigarh, with over 400 publications to his name. *Paleontology* is the study of fossils and the history of life on Earth. At first glance, it does not appear to be a research field that would have much potential for criminal activity; however, it is here that one individual apparently has managed to perpetrate "perhaps the biggest geological hoax of all time" (Oliwenstein, 1990). The Himalayas are of great interest to geologists the world over because they are still actively rising as the Indian subcontinent collides with Asia. Dr Gupta's publications form a significant portion of the biogeographical literature available on the Himalayas and have potentially confused the entire geologic history of the area. To understand how this works, it is first necessary to understand the Geologic Time Scale (Figure 9.1).

Based in large part on the distribution of fossils, geologists use a timescale that breaks the Earth's history up into a series of major time segments: *eons, eras, periods, epochs,* and *ages.* An eon is the longest formal division of geologic time, so

An Introduction to Forensic Geoscience, First Edition. Elisa Bergslien.

YOUNGEST

Eon	Era	Period		Epoch	Ma
Phanerozoic	Cenozoic	Quaternary		Holocene	0.01–0
				Pleistocene	2.6–0.01
		Tertiary	Neogene	Pliocene	5.3–2.6
				Miocene	23.0–5.3
			Paleogene	Oligocene	33.9–23.0
				Eocene	55.8–33.9
				Paleocene	65.5–55.8
	Mesozoic	Cretaceous			145.5–65.5
		Jurassic			201.6–145.5
		Triassic			251–201.6
	Paleozoic	Permian			299–251
		Pennsylvanian	Carboniferous		318–299
		Mississippian			359–318
		Devonian			416–359
		Silurian			444–416
		Ordovician			488–444
		Cambrian			542–488
Proterozoic	Precambrian				2500–542
Archean					3850–2500
Hadean					~4600–3850

OLDEST

Figure 9.1 Geologic time scale.
Source: Dates from the Geological Society of America, 2009.

long in fact that there have only been four eons: the *Hadean* (after Hades), for which there is no significant rock record (thus it is sometimes excluded); the *Archean* (meaning "ancient"), which offers limited evidence of life and surface conditions that were not yet modern Earth-like (i.e. little or no oxygen in the atmosphere); the *Proterozoic* ("early life"), in which conditions became modern Earth-like for the most part and in which multicellular life forms appear as fossils; and the *Phanerozoic* ("apparent life"), which has abundant signs of life and in which we live. Collectively, the first three eons are called the *Precambrian*, that stretch of geologic time from the formation of the Earth itself to the beginning of the Cambrian Period at the start of the Phanerozoic Eon. This immensely long stretch of time – some four billion years – saw the formation of the Earth as a planetary body, including the geosphere, atmosphere, and hydrosphere, as well as the appearance of life. There has been a recent proposal to break the history of the Earth up into two informal "*supereons*" with the proposed names of *Pregeozoic* (the supereon before life) and *Geozoic* (the supereon after the appearance of life), though the placement of the boundary between the two is the subject of much debate.

The Phanerozoic Eon, which started ~542 *million years ago* (abbreviated *Ma*), is broken into three eras: the *Paleozoic* ("old life") in which life became increasingly diverse and moved from the oceans onto land and into the air; the *Mesozoic* ("middle life"), commonly called the age of the dinosaurs; and the *Cenozoic* ("recent life"), which could be called the age of mammals. Eras are themselves broken down into periods. The names of the periods are derived from geographic locales where the period was first identified or from Latin roots for characteristic deposits formed within the period. The periods are further broken into epochs, which in turn are broken down into ages (or stages), the smallest division of geologic time. Each of these divisions of time is bracketed by important paleontological events, such as a mass extinction or major observable changes in fossil assemblages.

While it is possible to use radiometric-dating methods to put *absolute ages* (i.e. an actual number) on igneous and some metamorphic rocks, the rocks that form the bulk of the surface of the Earth, and that contain fossils, are sedimentary, and sedimentary rocks generally cannot be so dated. Radiometric dating methods give you information about when rocks solidify from a melt or the time since metamorphism ceased. Because most sedimentary rocks are made from pieces of other rocks, the radiometric ages you get would be for the creation of the individual clasts, not the sedimentary rock formed later. Layers of volcanic ash, for example, can be given absolute dates, but what about the ages of the sedimentary layers above and below the ash bed? What about sedimentary layers with no volcanic material? To create the timescale, geologists also must use *relative ages*, which are based on the sequence of events that must have occurred to create the layers of sediment visible now. It is, in a way, much like a forensic investigation.

There are a series of basic principles, developed by Nicolas Steno in the late 1600s, which geologists use to reconstruct the Earth's history. The *law of original horizontality* states that sediments are initially deposited in layers that are horizontal or nearly so. If you have ever watched snowfall over the winter to form a thicker and thicker pile, that is basically the same idea. This is easier to envision if you think about an environment such as the seafloor or the bottom of a lake, which is where most deposition occurs. Any storm or flood carrying sediment will deposit it in a flat layer because the particles have to settle through the water column under the influence of gravity. So, broadly speaking, layers of sediment all start out horizontal, and if the layers you are looking at are crumpled up or vertical, something happened after deposition to orient them that way.

The *law of lateral continuity* is a related idea, and simply means that sediment layers are also continuous horizontally for large distances. If you imagine driving down a road with rock walls on either side, you can see that the layers on the left side of the road are the same as the layers on the right side of the road. Therefore, before the road was blasted through, the two sides hooked together to form a continuous horizontal expanse of rock. This applies to layers separated by rivers, lakes, and fault lines as well.

Another related principle is the *law of cross-cutting relationships*, which states that any feature that cuts across a rock or layer of sediment must be younger than that

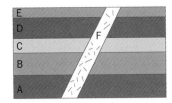

Figure 9.2 The igneous intrusion labeled "F" is younger than the sediment layers ("A–E") through which it cuts. The oldest layer in the picture is A and the sedimentary layers get increasingly young, moving upward.

rock or sediment through which it cuts (Figure 9.2). So the road that you are driving on, while looking at the layers of rock, must be younger than the layers of rock. Again, the same idea applies to other geologic features, like canyons, fractures, and intruded masses of igneous rock.

The *law of superposition* states that in any undisturbed sequence of sedimentary rocks the rocks on the bottom are older than the rocks on top (Figure 9.2). If you imagine newspapers being dropped in a box after they have been read, the paper at the bottom will have the oldest date and the papers will get younger as you work up the pile (referred to as *younging upward*). The same applies to sedimentary rock layers. If you have an ash layer in the middle of an outcrop or *stratigraphic column* (series of rock layers), an absolute age can be determined for the ash, plus you know that the sedimentary layers below that ash bed must be older than that date and the layers above the ash must be younger than that date.

The law that is most important to understanding the importance of fossils as forensic tools, and the effect of Dr Gupta's chicanery, is the *principle of fossil succession*. Basically stated, fossil succession means that each chunk of the Earth's history has a unique collection of fossils representative of that time. As some organisms go extinct and others evolve into new forms, there are successive changes. A *fossil assemblage* is a group of fossils that can be used to determine the relative age of the rock in which it is found. Some fossil species exist for long periods, several millions to tens of millions of years or even more. Finding such a fossil in two different formations allows you to *correlate* (relate in time) those formations only to an accuracy of millions to tens of millions of years, or however long those fossils persisted. *Lingula* (an invertebrate animal that is a form of inarticulate brachiopod) have been around since the Early Cambrian and are still found living in the modern ocean showing essentially no change in external form. Clearly, *Lingula* is of little help in determining the period during which a sedimentary layer was deposited. On the other hand, if you find a *Tyrannosaurus rex* fossil, you have to be looking at rock that is between 85 and 65 million years old. This does narrow things down quite a bit to within a single period, the Cretaceous, but is still of limited help for determining exactly when a sedimentary layer was deposited or relating that layer to other, greatly distant, layers where that fossil is found.

There is a special group of fossils, called *index fossils*, which are used to determine with great certainty the specific age of the rock in which they are found. A good index fossil that is useful for making *biostratigraphic correlations* (i.e. fossils that can be used to relate two separate sequences of rock in geologic time) is generally a species that existed for a short window of time, usually meaning less than a million years, and is thus restricted to a narrow stratigraphic interval (i.e.

the layers of rock where the fossil appears). Other attributes that make for good index fossils are a wide geographic distribution to allow correlations between rocks over a large area, a wide environmental range so that the fossil can be found in a variety of different sedimentary layers, and ease of identification. This makes index fossils of interest to forensic workers as well, since they can be used to help determine the potential source areas of geological material. Regional exposures of rock with a particular type of index fossil or set of fossils are often limited to small, discrete areas.

For example, during the Cretaceous Period, squids with coiled shells, called *ammonites*, were very abundant in the oceans. They evolved very rapidly, producing many distinct, short-lived species, many of which were also very widespread. By using the fossils of ammonites, Cretaceous formations deposited on the Atlantic side of North American can be *chronostratigraphically* (i.e. layers of rock formed at the same time) correlated to formations found on the western side of North America, and around the world, despite the complete lack of any *lithostratigraphic* (layers of rock physically connected or connectable) link between the regions. For example, the ammonoid *Turrilites* has worldwide distribution and is only found in rocks deposited between 99.6 and 93.5 million years ago. Biostratigraphy is by far the most important method for ordering strata in time and is an essential tool for geologists.

Most index fossils are not large, like *T. rex*, but are small invertebrates (i.e. lacking backbones). In fact, many of these fossils are microfossils (tiny fossils), like *conodonts*, which are, loosely speaking, the teeth of a small extinct organism related to the most primitive fish. Conodonts are so small that a group of them can fit easily on the top of pin, but they are also very distinctive. Dr Gupta first reported finding conodonts in the Himalayas in 1967 (Gupta, Rhodes, and Austin, 1967). Conodonts have changed so significantly through time that individual conodont species can often be used to determine the age of a stratigraphic layer to within a million years. So when Gupta published that the same assemblage (group) of conodonts had been found at numerous locations in northeastern India and Nepal, and reported their age to be anywhere from Early to Late Devonian, a span of time that stretches from 416 to 359 Ma, clearly there was a problem.

Imagine a scientist claiming to have found a *T. rex* in a rock that is only 55 million years old (~10 million years after the mass extinction of the dinosaurs occurred). That would cause a major upheaval in geology and would be an international sensation. Dr Gupta was basically claiming to have found the same types of things, only involving fossils, like conodonts, that are not widely known by the general population. Worse, it turns out that the conodonts described in these papers were probably not even from the Himalayas at all, but were from the very distinctive North Evans Limestone at Amsdell Creek in Buffalo, New York.

An Australian scientist eventually exposed Dr Gupta's activities in a series of articles; the most widely distributed was in *Nature* (Talent, 1989a). This initiated a flood of articles (Gupta, 1989, 1990a, 1990b; Lewin, 1989; Talent, 1990, 1995) and published arguments. Most bizarrely, Talent and a few other scientists who had co-authored articles on the fraud began to receive death threats. The technician responsible for preparing and photographing the specimens used in many of Gupta's publications announced that he could reveal information bearing on the case and, two days later, while standing in front of his house, he was killed by a hit-and-run driver (McBirney, 2004). Investigators did not find any conclusive evidence linking Gupta to any of these incidents, though he was eventually found guilty of plagiarism, stealing fossils, recycling fossils into several papers, and making false claims (Jayaraman, 1994). Gupta was censured, barred from any administrative post at Punjab University, and would no longer receive annual salary increases, but he was allowed to keep his job and to keep supervising students. It will take years to straighten out the chronology and biogeography of the Himalayas.

The point of this discussion is that many types of fossils can be very clearly linked to specific points in geologic time and, more importantly for forensic work, to particular locations. Thus, if rocks or debris containing fossils are found associated with a crime scene, the fossils can potentially be used to locate the point of origin of that geological material. They also provide another tool for characterizing and determining the associations of geologic evidence, such as sand or soil samples. Therefore, a general familiarity with common fossil types, and the ability to recognize their potential importance, should be of value to anyone confronted with geologic evidence.

An Introduction to Fossils

When most people think of fossils, the first thing that probably comes to mind is bones, like dinosaur bones. However, the most common types of fossils do not look anything like bones. There are several different types of fossils that vary from easily recognizable material to very odd-looking impressions. Technically, fossils can be either the physical remains of organisms or evidence of the behavior of organisms preserved in the rock record. Virtually all fossils are found in sediment or sedimentary rock, as the temperatures and pressures associated with igneous and most metamorphic processes destroy organic material. Fossils are typically found in depositional environments, reaching highest abundance in marine sediments, though they can be preserved in terrestrial depositional environments as well.

Fossils are divided into two primary categories: *body fossils* that are part of the original organism and *trace fossils* (also called *ichnofossils*), which are the preserved physical marks of the behavior of an organism. *Microfossils* are fossils that are too small to be studied without the aid of a microscope and include both the complete remains of microscopic organisms and microscopic small parts of larger organisms. The scientists who study fossils and the evolution of life on Earth are called *paleontologists*.

Body fossils can be preserved in a variety of ways. In general, only organisms with hard parts such as shells, bones, or teeth are preserved. Hard parts are typically formed from one of the following materials: *calcium carbonate* ($CaCO_3$ or calcite and aragonite), produced by corals, oyster shells, and starfish; *silica* (SiO_2 or cryptocrystalline quartz), produced by diatoms, radiolarians and some sponges; *calcium phosphates* (such as hydroxyapatite), which includes vertebrate bones and teeth; or one of a variety of complex organic materials such as *cellulose*, produced by plants and some algae; *chitin*, most commonly found with arthropods (think insects); and *spongin*, which is produced by types of sponges like the one you might have in your bathtub (a "natural sponge," not a *Luffa*, which is a type of plant). Organisms' hard parts are not solid materials, but are *porous* (full of spaces). This pore space is occupied by organic material in life that upon death begins to rapidly decay, leaving behind empty voids. There are also many organisms that have no hard parts, such as jellyfish, which are only preserved in rare instances.

The most common path to becoming a fossil starts with an organism dying and then somehow escaping scavenging and decomposition. Next, the organism must be buried in sediment, which can occur very quickly after death, as in the event of a volcanic ash fall, or very slowly, as when organisms fall into deep, anoxic water. As you can guess, sometimes the first two steps can occur simultaneously, for example in a giant underwater landslide or mudslide. Such events kill organisms almost instantaneously while burying them at the same time. The buried remains must also rest undisturbed in the sediment until *lithification* (the transformation of the sediment to sedimentary rock). Given these conditions, it is much easier for hard parts such as teeth, bones, and shells to become fossilized than for soft tissues or plant material, which tend to decay and/or *disaggregate* (break or fall apart) much more easily. Thus, the fossil record is heavily biased toward marine animals with robust hard parts. Paleontologists estimate that only around 15% of organisms are preserved in the fossil record.

Types of Preservation

The study of the formation and preservation of fossils is called *taphonomy*. There are several different types of preservation and several different approaches to classification. The following list is a fairly generic version of a classification system for types of preservation. It is important to be able to understand the different ways that fossils can be formed in order to know what they can look like. It is also worth noting that a single fossil may fall into more than one of these categories.

Original preservation, preservation of the original chemical composition, is typically confined to geologically young fossils where the associated sediments have not yet undergone lithification. Examples of this type of preservation include *soft-tissue preservation* and *original hard part preservation*. Soft-tissue preservation, where organic materials such as organs, skin, and hair are preserved, only occurs under exceptional conditions). Forms of this type of preservation include *encasement* in amber and *mummification* via freezing, chemical reaction, lack of oxygen, extremely arid conditions, or dehydration and preservation in oily plant debris (as in the Geiseltal Formation in Germany) or tar (as in the Rancho La Brea tar pits in Los Angeles). This only occurs for geologically young specimens that date back at most a few million or tens of millions of years, beyond which time nothing but chemical residues of organic matter will remain. Most well-known examples of this type of preservation are not technically considered fossils. Ötzi the Iceman (see Chapter 11) is a famous example of soft-tissue preservation. Also in this category are finds of ice age mammals such as woolly mammoths, horses, caribou, and several other species in the tundra of Siberia, Alaska, and the Canadian Yukon. They are sometimes mistaken for modern animals but break roughly into two groups, those aged 50,000–25,000 years old and those aged 15,000–10,000 years old.

Much more common is preservation of original hard parts. Sometimes, organic material can be preserved intact or nearly intact, without any significant alteration. Teeth, bones, and shells may survive for many millions of years almost unchanged if they are incorporated into suitable, low-oxygen sediment (Figure 9.3). The enamel that coats teeth is very resistant, as is calcium carbonate and the calcium phosphate that forms bones. Pollen and some microfossils can also survive for long periods virtually unaltered from their original form. Geologically speaking, fossils in this state of preservation are still quite young. The longer a fossil is subjected to burial, compression, and differing chemical environments, the more likely the original material is to be changed.

Figure 9.3 Original hard part materials. Please see Color Plate section.

The older the fossil, the more likely it is to be in a state of *altered preservation*. As sediments are buried deeper in the Earth and gradually lithify into rock, the associated fossils also undergo alteration to a greater or lesser extent. Examples of this type of preservation include permineralization, recrystallization, replacement, formation of casts and/or molds, and carbonization. *Permineralization*, the most common form of alteration, is a process in which porous organic structures, such as wood and bone, have their microscopic pore spaces, left vacant by decay, filled by minerals precipitated from groundwater. The original hard parts remain but are encased in extra material that fills in the pores. The resulting fossil is heavier and denser than the original material. With this type of fossilization, the fine details of microscopic structure are generally preserved, occasionally even preserving details of cell structure. *Petrified wood* is a fairly common example of permineralization.

Recrystallization is a type of preservation where the crystalline minerals forming an organism's hard parts fuse to form larger, more stable crystals. The original chemical composition can sometimes be preserved, but in other cases, unstable minerals, such as aragonite, recrystallize into a more stable, chemically identical form, such as calcite. The original chemical composition remains, and much of the original shape of the fossil is preserved, but the texture difference is obvious under the microscope and much of the fine detail of the structure is lost.

Replacement is a process in which an organism's original hard parts are dissolved by chemical action and replaced by another mineral, such as calcite, silica, iron, or pyrite. The result is a chemically different replica of the original fossil. The replacement process takes on the name of the secondary mineral, for example *silicification*, which is the most common form, entails replacement of the original mineralogy by silica (Figure 9.4); *pyritization* is the replacement of calcite or soft

Figure 9.4 Fossil from Glass Mountain in Texas, so named because the original material of many of the fossils has been replaced with silica. The fossils are extracted from the limestone by acid dissolution. The resulting silicifed specimens retain incredible detail and much fragile structure that would normally have been lost. Please see Color Plate section.

Figure 9.5 Pyritized fossil ammonite. Please see Color Plate section.
Source: Photograph by Randolph Femmer courtesy of the National Biological Information Infrastructure (NBII) office.

tissues with pyrite (like the sand dollars that you often see in stores) (Figure 9.5); *phosphatization* usually involves replacement of low-phosphate apatite with high-phosphate apatite; and *dolomitization* is usually the incorporation of magnesium into hard parts that were originally calcite-forming dolomite. Growth of a replacement mineral occurs at the expense of the original mineral components, destroying fine detail while preserving the original size and shape of the fossil.

Sometimes the original material dissolves away completely, leaving a void in the surrounding sediment and leading to the formation of *molds* and/or *casts*. Molds are surface impressions created in the sediment surrounding the original material. Think, for example, of leaving a footprint in the sand (Figure 9.6). Molds can be found around extant fossils or, if the original material dissolved from the rock matrix, are left behind as impressions called *external molds* (Figure 9.7). If the original fossil material becomes filled with sediment internally, an impression of the inside, called an *internal mold*, can also form (Figure 9.7). If a cavity in a fossil is completely filled in, it can form a nodular internal mold called a *steinkern*. A *composite mold* forms when internal and external mold surfaces are compacted together to produce impressions in the same layer of sediment.

Casts are formed when a mineral, sediment, or some other material fills in a mold and hardens to form a copy of the original structure. True casts are relatively rare in the fossil record, and internal molds (especially steinkerns) are often confused with them. A mold is a negative image of the original, while a cast is a positive image, or duplicate of the original. Since molds and casts occur as a result of the dissolution or destruction of the original material after the surrounding

Figure 9.6 Triassic predator footprint and an example of a fossil mold.

Figure 9.7 Block of limestone where the original brachiopod shells have dissolved, leaving internal and external molds.

matrix of sediment has hardened, only a limited amount of fine detail is usually preserved.

Carbonization occurs when organic material is preserved through rapid burial in an *anoxic* (very low or no oxygen) environment. The organics do not decay. Instead, the volatile elements, such as hydrogen, oxygen, and nitrogen are driven off, leaving behind organic-rich hard parts and soft parts to be preserved as thin, black films of organic carbon. The fossil loses its three-dimensional shape, but this process often will preserve the outline of soft tissues, hair, or feathers, and can reveal fine details that would have been destroyed by other forms of fossilization. Plants are also often preserved in this way (Figure 9.8). Some of the fantastic details of the 47-million-year-old fossil primate called Ida (incorrectly cited in the media as a missing link), such as the hair and *skin shadow* (an outline of an animal's soft tissue or flesh), are due to carbonization of the soft tissue around an animal's preserved bones. This combination gives an unprecedented amount of information about extinct organisms. Carbonized fossils are most commonly found on the bedding planes of sandstone and shale (Figure 9.9).

Different organisms have differing potential for fossilization. Organisms with hard parts are much more commonly preserved than organisms with no hard parts. The number and size of hard parts also affects preservation. Organisms with one or two large, robust shells are more likely to be preserved intact than organisms with lots of smaller, more delicate parts. The environment in which an organism lives also plays an important role in preservation. Organisms that live in environments where there is lots of chemical and physical weathering, and/or substantial erosion, such

Figure 9.8 Fossil plant material.

Figure 9.9 Carbonized fossil fish from the Green River Formation in southwestern Wyoming.

Figure 9.10 Impression of bark.

as mountains and deserts, are much less likely to be preserved than organisms that live in depositional environments, such as along the continental shelf. Plants are a special case. Different organs (leaves, stems, trunks, fruit, flowers, seeds, pollen, etc.) are only very rarely preserved together, so each part generally gets given its own species name (Figure 9.10).

Finally, do not forget about trace fossils, which include any preserved indicator of life and are typically biologically created sedimentary structures. This includes things like footprints and track ways, burrows (Figure 9.11), grazing marks, bite marks, nests, eggs (if the organism inside was not preserved as well), and coprolites (fossilized feces). Preservation of trace fossils works pretty much like it does with other sedimentary structures such as mudcracks and ripple marks. They must have

Figure 9.11 Trace fossil: fodinichnia. Burrows created by deposit feeders as they scoured the sediment for nutriment preserved in sandstone.

rapid burial and are preserved by lithification of the host sediments themselves, hopefully without significant distortion. Study of trace fossils (*ichnofossils*) is a specialized area of paleontology called *ichnology*. Because trace fossils are very rarely found directly in contact with the organisms responsible for their creation, ichnologists have developed their own separate methods of classification. Interested parties are directed to Seilacher (2007).

A Brief Introduction to the Classification of Fossils

In order to understand how fossils are named, it is important to understand a bit about *taxonomy* (the science of classification). Organisms are typically sorted into *taxonomic groups* or *taxa* (a named group of organisms) based on physical characteristics, such as body construction, with the intention of expressing how closely different organisms are related. *Linnaean taxonomy*, with which you might already be familiar, is a formal system for classifying and naming living things using a simple hierarchical structure. The major groups of organisms are subdivided into smaller and smaller groups on the basis of how closely the organisms resemble one another. The basic hierarchy, from broadest to narrowest grouping, is as follows: *kingdom, phylum, class, order, family, genus, species*. For example, humans are in the kingdom *Animalia* (animals), phylum *Chordata* (chordates or animals with nerve cords), class *Mammalia* (mammals), order *Primates*, family *Hominidae* (great apes and humans), genus *Homo* (humans), and species *Homo sapiens*.

Linnaeus originally recognized three kingdoms: plants (*Regnum vegetabile*), animals (*Regnum animale*), and minerals (*Regnum lapideum*). Since then, our understanding of living organisms has improved significantly. Scientists now recognize some much more subtle distinctions and have added a new taxonomic level: the *superkingdom* (or domain). There are two (or three) superkingdoms: *Prokaryota* (which is sometimes broken into Eubacteria and Archaea) and *Eukaryota*. Prokaryotes are the simplest single-celled organisms, lacking a discrete nucleus to house their genetic material. This group is called kingdom *Monera* in the "five kingdom" system. Archaea are the most primitive prokaryotes, and they are very poorly understood, which is why they are sometimes treated as a separate superkingdom. Recognizing most fossils from these groups requires specialized training and a scanning electron microscope (SEM).

Organisms whose genetic material (DNA) is encased in a nucleus are called *eukaryotes*. Within eukaryota, there are four kingdoms of organisms:

- **Protista**: single-celled or simple multicellular eukaryotes
- **Fungi (or Mycota)**: single-celled or multicellular eukaryotes that break down decaying organic matter
- **Plantae**: sexually reproducing multicellular organisms with chloroplasts (i.e. plants)
- **Animalia**: multicellular organisms that consume other organisms (i.e. animals).

Plant and fungi fossils are pretty rare and can be difficult to recognize, so they will not be considered here. Some Protista are important microfossils and will be introduced later in this chapter. For now, we will concentrate on the last group, Animalia, because the most common fossils are animal fossils.

The animal kingdom can be divided into two major branches, the *invertebrates* (animals without backbones) and the *vertebrates* (animals with backbones). There are more than 20 invertebrate phyla, but the chief ones that are preserved as fossils include: Porifera (sponges); Cnidaria formerly called Coelenterata (corals and jellyfish); Bryozoa (colonial moss animals); Brachiopoda ("lamp shells"); Superphylum Arthropoda (insects, crabs, shrimp, trilobites); Mollusca (clams, snails, octopi, ammonites); and Echinodermata (starfish, sand dollars, crinoids).

The vertebrates (including us) belong to phylum Chordata (referring to the nerve chord that extends down the center of the spine), subphylum Vertebrata (animals bone and/or cartilage supporting the nerve cord). Important classes in vertebrate paleontology (followed by their geologic timespan (see Table 4.1), **R** means recent) are: Superclass "Agnatha" (jawless fish) **Є-R**; Acanthodii (primitive jawed, spiny fish) **D-P**; Placodermi (archaic jawed fish) **S-P**; Chondrichthyes (cartilaginous fish) **D-R**; superclass Osteichthyes (bony fish) **S-R**; "Amphibia" (amphibians) **D-R**; "Reptilia" (reptiles) **Ᵽ-R**; Aves (birds) **Ŧ-R**; and Mammalia (mammals) **Ŧ-R**.

Paleontologists identify fossils by comparing them with known examples. If a paleontologist finds a species that has not been previously identified, they get to name the new species by publishing a paper about their find. The formal naming of organisms (both living and fossil) employs the *rule of priority*: whichever validly formed name for a species or genus was *published* first, even if only by days, is the name that must be used. If you have ever wondered what happened to *Brontosaurus*, this is the rule that explains its disappearance. In 1877, Othniel March published the name *Apatosaurus* for a juvenile and incomplete skeleton. Two years later when he described an adult and more complete skeleton, he gave it a new name: *Brontosaurus*. Later paleontologists determined that the two skeletons were really of the same genus and species, so, by the rule of priority, the name *Apatosaurus* is the proper name to use.

Each species has a *type specimen*, the originally identified and described organism to which all others are compared, which is *accessioned* (formally cataloged) in an appropriate institution, such as a museum. Thus, the *Apatosaurus* March named first is the type specimen against which all other skeletons are compared. Each genus has a *type species* and all other species are assigned to the genus based on their similarity to the type species. The rules beyond the genus level get rather complicated, especially since not everyone agrees which physical features are the most important for determining relationships between organisms.

In Linnaean taxonomy genera have one-word names (e.g. *Homo*, *Tyrannosaurus*) and the genus name is always capitalized and *italicized* (or <u>underlined</u>). Species have two-word names, the first part of which is the genus name (e.g. *Homo sapiens*, *Tyrannosaurus rex*). In species names the first part (the genus name) is always capitalized, the second part (called the *trivial nomen*) is always lower case, and the entire name is always *italicized* or <u>underlined</u>. Species names are properly abbreviated by using only the first letter of the genus name, followed by a period (not by a hyphen) as in *H. sapiens* and <u>T. rex</u>. All other taxon levels have one-word names, which are capitalized and presented in roman letters (i.e. they are not italicized or underlined) as in Dinosauria and Hominidae.

The Linnaean rules of classification call for all names to be unique and in Latin or Greek, or modified into a Latinized form. The names must also fit into the pre-existing nested hierarchy (species into genera, genera into families, and so forth). As you know, the primary unit of this system is the *species* (pl. species), but this is a tricky concept when you are working with fossils. The general biological definition of a species is "naturally occurring interbreeding populations that produce viable offspring." Since fossils don't do much breeding, this is not a particularly helpful distinction and paleontologists spent quite a bit of time hashing out when a specimen is really different enough to be considered a new species or genus. As you move up the taxonomic scale, things get increasingly more complicated. This is the reason that classification is sometimes awkward and why even the simplified list of common fossils in this chapter seems somewhat convoluted.

Over time, the Linnaean seven-fold hierarchical system has also turned out not to be nearly detailed enough. As more and more groups of organisms were identified, it became necessary to start creating additional subcategories. These include *tribe*, between family and genus, and the addition of both *division* and *cohort*, between class and order. Each category can also have prefixes added to create a higher grouping (super-), or lower (sub-, infra-) subdivision. So, in addition to the original categories, there are also such classifications as *superorder, suborder, infraorder, subgenus,* and *subspecies*, each arranged in nested ranks. This has led to a fairly unwieldy system when it comes to the classification of many fossils, especially the vertebrates. Also, of great inconvenience, the name most commonly applied to identifying a distinctive group can be from any of the levels of classification.

The reader is advised that the following information is simplified and introduces a basic classification scheme for some of the fossils that are most likely to be useful for a forensic worker. The most commonly applied group names are in bold text. As with other forms of classification, one standard reference should be chosen as a baseline for forensic identification purposes. The important thing for a forensic worker is to choose consistent descriptive terminology that can be shared with others to effectively communicate the identity of a fossil. Detailed classifications, if desirable, should be performed by a paleontologist.

The fossils that have the potential to be of the most importance in forensic geoscience are the most common invertebrate fossils and some important varieties of microfossils; so detailed information on only a few types of fossils will be included here. Identification and proper classification of vertebrate material generally requires a specialist. This is especially true when confronted with fragmentary material. Basically, if you are looking at bones (skull, ribs, pelvis, arms, legs, etc.), you are looking at a vertebrate. However, distinguishing between different types of vertebrates can sometimes be quite difficult.

While adult human skulls are highly distinctive, it can sometimes be tricky to differentiate other human and non-human mammal bones, especially when they are fragmentary, burned, or *comingled* (mixed up, also spelled *commingled*). According to Marks (1995) and Ubelaker (1999), the bones most commonly misidentified as human are from deer, dog, pig, cow, chicken, horse, and bear. Complete dog skeletons have been mistaken for infant human skeletons, and bear paws, sans fur and claws, have commonly been mistaken for human hands (Owsley and Mann, 1990). A study of Williams Bass's (2005) forensic caseload indicates that the proportion of non-human remains submitted for analysis rose to nearly 30% in the 1990s (Marks, 1995). This is mostly likely due to the combined effect of increased public interest in criminal investigations and the increasing incursion of population centers into formerly rural areas.

Differentiating between fossil material and recent material can also sometimes be problematic; generally speaking, the identification of fossil material calls for a vertebrate paleontologist, while the identification of more recent material would typically call for a forensic anthropologist, forensic archeologist, or a mammalian osteologist. More information on the identification of human bones can be found in Byers (2005), Bass (2005), and Ubelaker (1999). *Skulls and Bones* by Searfoss (1995) is a good basic guide to modern North American mammals and *Human and Nonhuman Bone Identification* by France (2008), though expensive, appears to be the best available resource for identifying the modern mammal bones most commonly confused with human remains. For general information on the discipline of vertebrate paleontology, see Benton (2004). Finally, if you happen to be interested in more information on dinosaur bone identification and classification see Martin (2006) and Lucas (2005).

Invertebrate Paleontology

Phylum Porifera

Phylum Porifera (Cambrian to Recent), or the sponges, are simple *colonial* organisms, which means that they always live in groups. They secrete a common skeleton (i.e. all of the organisms build a single structure) and filter food through pores in the walls of the structure. Sponge skeletons can be composed of spongin (an organic protein or the stuff that forms a natural bath sponge; Figure 9.12), silica (Figure 9.13), calcite, or aragonite. When sponges die, they usually disaggregate, scattering small *spicules* (the needle-like components that form the skeleton) across the seafloor. The fossil record for this group is rather poor, since whole specimens are rarely preserved. Sponge spicules are much more common and will be discussed in the micropaleontology section below.

Phylum Cnidaria

Phylum Cnidaria (formerly known as Coelenterata) includes corals, jellyfish, and sea anemones. This group includes both solitary and colonial organisms, all of which are characterized by *radial symmetry* (i.e. there is a regular arrangement of

Figure 9.12 Porifera: class Demospongea, formed of spongin.

Figure 9.13 Porifera: class Hexactinellida. The dried siliceous skeleton of the deep-sea glass sponge *Euplectella*.
Source: Photograph by Randolph Femmer courtesy of the National Biological Information Infrastructure (NBII) office.

body parts around a central axis such that a bisecting plane running from top to bottom through the center of the organism, introduced at any orientation, would produce two approximately equal halves; they have a top and bottom but no left and right) (Figure 9.14). Another defining characteristic are the stinging cells, called *nematocysts*, in their tentacles. Because many Cnidaria have only soft parts, they have a spotty fossil record (Figure 9.15). The following groups have a good fossil record (additional information is available on the companion website: www.wiley.com/go/bergslien/forensicgeoscience):

Class Anthozoa (Precambrian to Recent) includes soft and hard corals, sea pens, sea fans, and sea anemones. This group is exclusively marine and most of

Figure 9.14 Cnidaria: sea anemones displaying radial symmetry and tentacles with nematocysts.

Figure 9.15 Cnidaria: fossil Scyphozoa (*Rhizostomites*) from the Jurassic, Solnhofen Limestone, Germany (from Walcott, C. D. Fossil medusae. U.S.G.S, Mon 30, pl. 41, 1898).

them are *sessile* (i.e. live attached to the ocean floor). Anthozoans have the most diverse fossil record of the Cnidarians. The most important members of Anthozoa are the hard corals (subclass Zoantharia), all of which are very commonly found as fossils. Individual corals secrete calcium carbonate to create cups, called *corallites* (the skeleton of a single coral polyp) (Figure 9.16), which are surrounded by

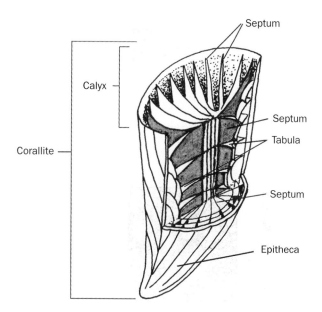

Figure 9.16 Basic morphology of solitary corals.
Source: Modified from McKinney, 1991.

skeletal walls called *theca*. The bowl-shaped depression where the animal resides is called the *calyx*. As the animal grows, it partitions off the older sections of the corallite with horizontal plates called *tabulae* (Figure 9.16). In some corals, vertical plates, called *septa*, also divide the corallite (Figure 9.16). Corals are classified by the arrangement of their septa, tabulate, and other skeletal features. There are three important groups of fossil corals:

Order Tabulata (the **tabulate** corals) (?Cambrian to Permian) were dominantly colonial and are characterized by their well-developed tabulae, with no or very small septa. There are four distinct skeletal types: *Favosites* (honey-comb corals) (Figure 9.17), *Halysites* (chain corals), *Syringopora* (organ-pipe corals), and *Aulpora* (small, trumpet-shaped corals).

Order Rugosa (the **rugose** or horn corals) (Ordovician to Permian) can be either solitary or colonial (Figure 9.18). They are distinguished from tabulates by their well-developed septa. Solitary Rugosa corals are shaped like wrinkled horns or cornucopias. Colonial Rugosa most often appear as masses, with the individual horns welded together into compact honeycombs, though irregular lumpy forms are also known.

Order Scleractinia (modern **hexacorals**) (Triassic to Recent) are the main reef builders in modern oceans. They can be either colonial or solitary and come in a wide range of forms (Figure 9.19). Scleractinia always possess well-developed septa, but never tabulae. The pattern of septa is basically six-rayed, and usually visibly differs from that of the Rugosa. If you have visited an aquarium, you are familiar with this group of corals. Usually, modern corals are fairly easy to distinguish from either of the ancient groups simply because they look younger (i.e. are whiter, less altered, and overall in better condition).

Figure 9.17 Photograph of a colonial tabulate coral (*Favosites*) showing the distinctive tabula and lack of septa.

Figure 9.18 A few examples of rugose corals showing their well-developed septa. The samples in the upper left and center show solitary corals (looking like horns or cornucopias), while the sample in the bottom right is a colonial rugose.

Figure 9.19 A few examples of modern scleractinian corals.

Phylum Bryozoa

Phylum Bryozoa (Ordovician to Recent) are colonial "moss" animals with calcite skeletons that superficially resemble corals (Figure 9.20). Individual bryozoa are tiny, only about 1 mm in size, so the perforations in their skeletons are much smaller than those found in corals. **Bryozoans** also do not have septa. This group is dominantly marine and sessile. Some common forms include bulbous, encrusting masses, thick branches, or radiating thin twigs. **Order Trepostomata** (Ordovician to Triassic), forms encrusting mounds or thick branches. **Order Fenestrata** (Ordovician to Permian) has delicate, mesh-like fans, including *Archimedes*, with hard parts that resemble a tubular screw (Figure 9.20e).

Phylum Brachiopoda

Brachiopoda (Cambrian to Recent) are large groups of solitary, marine organisms characterized by two shells (valves) that encase most of the animal. Though they are rare in modern oceans, **brachiopods**, or "lamp shells," are the most abundant of all Paleozoic invertebrate fossils and many are used as index fossils. They are typically small, ranging in size from 5 mm to over 8 cm, though a few, rare gigantic forms (up to almost 39 cm) have been found. They are distinguished by their shell symmetry. Most commonly, the top and bottom shells are quite different, but both exhibit bilateral (mirror) symmetry parallel to their length (i.e. left–right symmetry). Brachiopods are broken into two classes:

Class Inarticulata (Cambrian to Recent) are brachiopods with shells that lack teeth and sockets (Figure 9.21). Instead, the shells are held together by several sets of muscles. Though some have calcite shells, most have shells that are composed of a mixture of chitin and calcium phosphate, giving them a shiny, enamel-like luster. *Lingula*, mentioned previously, is an example of this group and has a shell that somewhat resembles an oval fingernail with ridges.

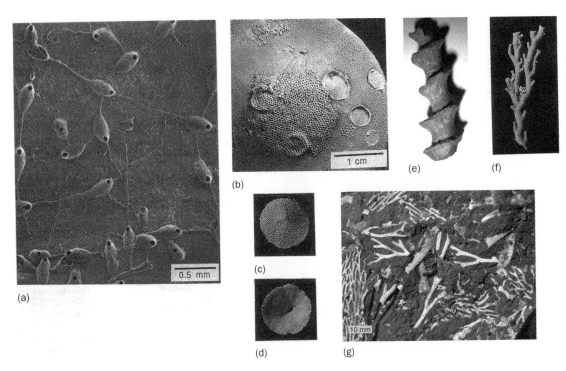

Figure 9.20 Basic bryozoan growth forms: (a) encrusting uniserial, (b) encrusting multiserial, (c and d) top and bottom of free-living conical form (*Cupuladria*), (e) erect colony of a tubular bryozoan (*Archimedes*), (f) erect colony (*Tabulipora*), (g) Ordovician bryozoans in an Estonian oil shale.
Source: (a–d and f) From McKinney, 1991; (g) photograph by Mark A. Wilson).

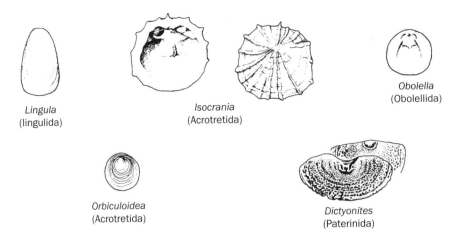

Figure 9.21 Brachiopoda, class Inarticulata. The shells lack teeth and sockets structure.
Source: McKinney, 1991.

Class Articulata (Cambrian to Recent) makes up some 95% of all known brachiopod genera. Articulates' shells are formed of calcite and have well-developed tooth and sockets hinges. Some brachiopods have an opening (called a *foramen*) in one of the shells (called the *pedicle valve*) through which the living animal would extend a fleshy stock (the pedicle) that it would use to attach to the substrate. A few of the most important groups are shown in Figures 9.22 and 9.23, with more details on the companion website.

Phylum Arthropoda

Phylum Arthropoda (relationships within this group are controversial and some define Arthropoda as a superphylum instead) (Precambrian to Recent).

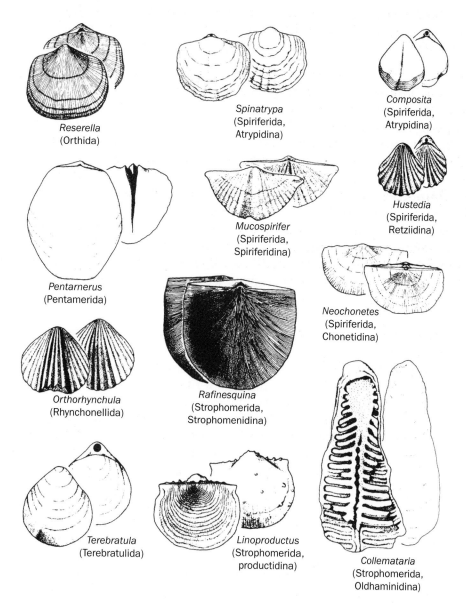

Figure 9.22 Brachiopoda, class Articulata.
Source: McKinney, 1991.

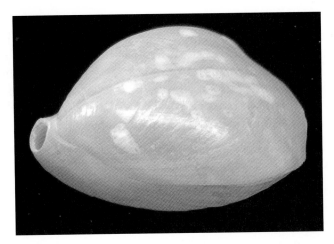

Figure 9.23 Brachiopoda, class Articulata, order Terebratulida. Recent brachiopod from the Philippines, displaying the distinctive shape that causes them to be called *lamp shells*. Please see Color Plate section.

Arthropods are the most diverse and numerous animals on the Earth today and probably have been since they first appeared. Over 80% of living animal species are arthropods; the group includes insects, spiders, crabs, shrimp, lobsters, trilobites, and eurypterids. Classification of this diverse group has always been controversial because their evolutionary history is poorly understood, but their uniting characteristics are that all arthropods have an exoskeleton, usually of chitin, and that they have segmented, bilaterally symmetrical bodies with paired appendages.

Subphylum Chelicerata (Cambrian to Recent) have bodies divided into two segments: the *prosoma*, which is a fusion of the head and thorax, and the *opisthosoma*. They do not have antenna and they are distinguished by *chelicerae*, specialized mouthparts modified out of their first pair of appendages. The second pair of appendages is also usually modified either into mouthparts or into pincers. These are followed by four pairs of walking legs, thus most Chelicerata, not just spiders, have eight legs.

Subclass Arachnida (Cambrian to Recent) includes spiders, scorpions, and ticks. They are rare as fossils.

Subclass Eurypterida (Ordovician to Permian) are also called "sea-scorpions." They range in size from 10 cm to more than two meters (Figure 9.24). The prosoma has six pairs of appendages, the first of which are fang-like chelicerae and the second of which are pincers. The next four pairs of appendages are legs, though the final pair is sometimes modified into paddles. The opisthosoma always has 12 segments followed by a telson, which could be a spike, a paddle, or a set of pincers.

Subclass Xiphosura (Cambrian to Recent) have a large prosoma and partially fused opisthosoma. The prosoma has six pairs of legs, all but the last of which have claws. There is one extant member of this group, the horseshoe crab (Figure 9.25).

Subphylum Crustacea (?Precambrian to Recent) are the most diverse marine arthropod group. This group includes crabs, lobsters, shrimp (Figure 9.26), crayfish, barnacles, ostracods, and a range of other shrimp-like organisms. Apart from ostracods and barnacles (Figure 9.27), this group has a poor fossil record.

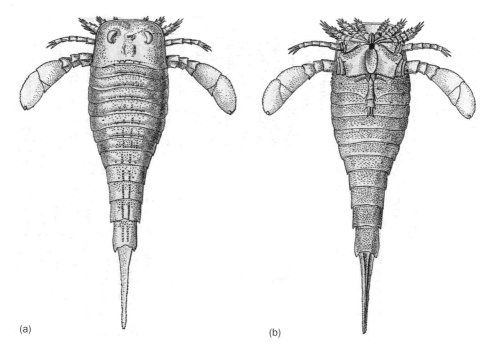

(a)   (b)

Figure 9.24 (a) Arthropoda, a Silurian Eurypterid (a) dorsal side, (b) ventral side, (b) *Eurypterus lacustris*.
Source: (a) After Clarke and Ruedemann, 1912, (b) plate 9 from Clarke and Ruedemann, 1912.

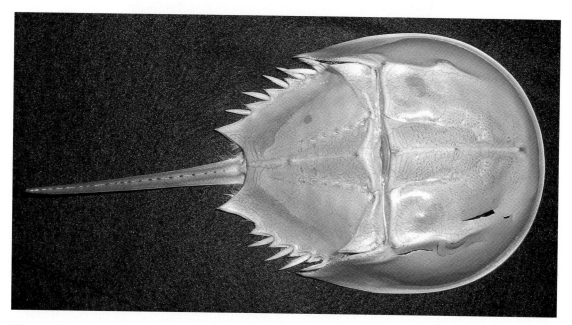

Figure 9.25 Arthropoda, Xiphosura. An example of a modern horseshoe crab.

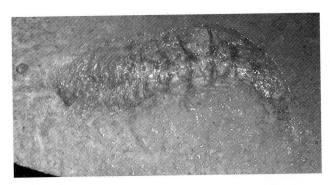

Figure 9.26 Arthropoda, Crustacea, Malacostraca. A shrimp fossil from the Solnhofen in Germany.

Figure 9.27 Arthropoda, Crustacea, Cirripedia. A group of *Balanus*, or acorn barnacles.

Ostracods, which resemble tiny clams, are dealt with in the section on micropaleontology below.

Subphylum Tracheata (Uniramia) (Cambrian to Recent) is the largest group of arthropods and includes the insects (**class Insecta** or **Hexapoda**; Devonian to Recent) centipedes and millipedes (both **class Myriapoda**; Carboniferous to Recent). Most members of Tracheata have a single pair of antennae. The body is divided into a cephalon with four to five pairs of appendages in front of the mouth. While the tracheates are most probably the most important animals on the planet, they have a very poor fossil record. Most fossil insects have been found in pieces of amber or as carbonized films.

Subphylum Trilobitomorpha (Cambrian to Permian) are one of the most commonly found fossils on Earth. **Trilobites** have a distinct head plate (called a *cephalon*), a segmented mid-section called the *thorax*, and a fused tail plate (the *pygidium*) (Figure 9.28). They were entirely marine and are commonly used as

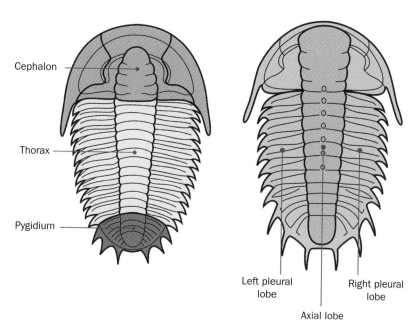

Cephalon

Thorax

Pygidium

Left pleural lobe

Right pleural lobe

Axial lobe

Figure 9.28 Diagram showing the major sections of a trilobite.
Source: With thanks to S. M. Gon III at www.trilobites.info.

index fossils. Some important groups are in Table 9.1 (available on the companion website) and are shown in Figures 9.29 through 9.31. For additional information, the websites http://www.trilobites.info and http://www.fossilmuseum.net are highly recommended.

Phylum Mollusca

Phylum Mollusca (Precambrian to Recent) is composed of an extremely diverse range of soft-bodied animals that appear to be radically different. The features that they have in common are a single, muscular "foot" used for locomotion and a cavity filled with ambient fluid (air or water) that is surrounded by a sheet of tissue called the *mantle*. They also lack segments and typically have an exterior calcite shell. **Mollusks** are the most common invertebrates in the marine realm. This group includes clams, snails, octopus, squid, nautilus, and ammonites.

Class Monoplacophora (Cambrian to Recent) includes the most primitive mollusks still extant. They have a single, cap-shaped shell and are the only mollusk with a segmented internal structure.

Class Gastropoda (?Precambrian to Recent) is by far the largest class of all the mollusks; gastropods are found in both marine and non-marine settings. This group includes snails, slugs, and whelks. Fossils from this group are most readily identified by their single, calcite, spiral-shells. The shells are usually hollow tubes that spiral down a central axis (Figure 9.32).

Class Cephalopoda (Cambrian to Recent) are the most modified from the primitive mollusk form. The single-foot is modified into a ring of tentacles around their mouths, and some forms have completely lost their calcite shell. This group includes the modern octopi, cuttlefish, and squids (**subclass Coleoidea**), which have a poor fossil record, as well as the following:

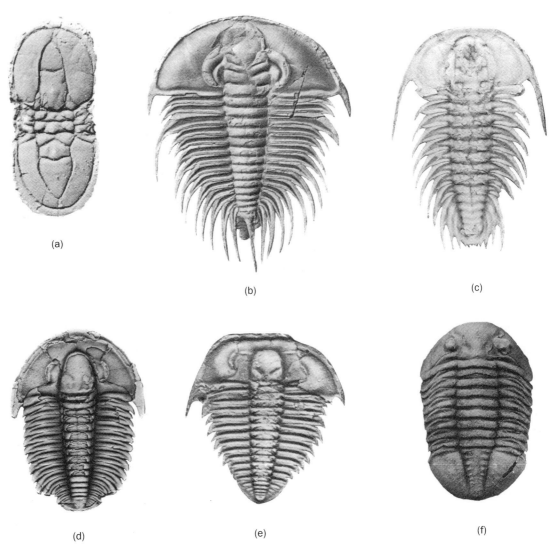

Figure 9.29 Arthropoda, Trilobitomorpha: (a) Agnostida *Ptychagnostus*, (b) Redlichiida *Olenellus*, (c) Corynexochida *Zacanthoides*, (d) Ptychopariida *Modocia*, (e) Ptychopariida *Olenus*, (f) Asaphida *Asaphus*.
Source: (a) From Robinson and Kaesler, 1985, (b) From Walcott, 1910, (c) Photograph by L. Gunther, from Robinson and Kaesler, 1985, (d) From Robinson and Kaesler, 1985. In: Boardman *et al.*, 1985, (e) Photograph by V. Jaanusson, from Robinson and Kaesler, 1985. In: Boardman *et al.*, 1985, (f) Photograph by M. H. Lawson, from Robinson and Kaesler, 1985. In: Boardman *et al.*, 1985.

Subclass Nautiloidea (Cambrian – Recent) are characterized by single shells that are straight, curved or *planispiral* (spiral around a point, like a coiled hose). Unlike gastropods, their shells are separated into chambers by calcite plates (septa). The living animal occupies only the largest, outermost chamber, having outgrown the previous chambers. The nautiloids have only one living member (*Nautilus*) but include a wide variety of important fossil forms. They are most commonly preserved as casts (Figure 9.33).

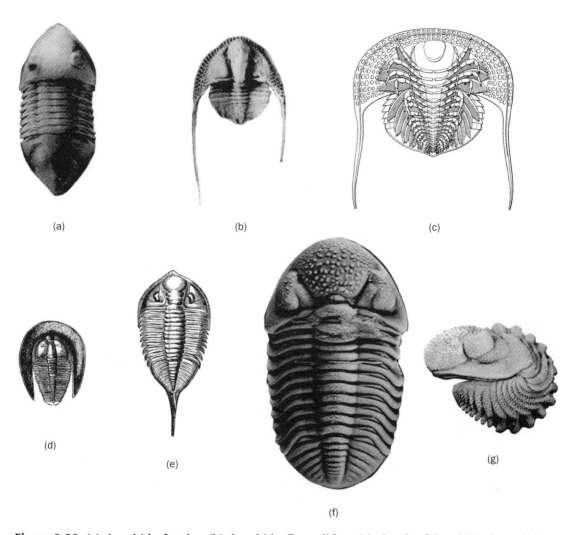

Figure 9.30 (a) Asaphida *Isotelus*, (b) Asaphida *Cryptolithus*, (c) sketch of Asaphida *Cryptolithus* with a reconstruction of the appendages, (d) sketch of Harpetida *Harpes*, (e) sketch of Phacopida *Dalmanites*, (f and g) Phacopida *Phacops* top view and enrolled.
Source: (a) Photograph by M. H. Lawson, from Robinson and Kaesler, 1985. In: Boardman *et al.*, 1985, (b) Courtesy of Smithsonian Institution, from Robinson and Kaesler, 1985. In: Boardman *et al.*, 1985, (c) From Raymond, 1920, (d and e) From Woods, 1909, (f and g) Photographs by Niles Eldredge, from Robinson and Kaesler, 1985. In: Boardman *et al.*, 1985.

Subclass Ammonoidea (Devonian to Cretaceous) are an extinct group of cephalopods that were once highly successful and very diverse. Like nautiloids, they can have straight or planispiral shells. However, while nautiloid shells have smooth surfaces with simple curved sutures (the points where the septa meet the exterior shell), ammonoid shells have simple zigzags to highly crenulated suture patterns and their exterior shells can be ornamented with bumps, knobs, and/or ridges (Figure 9.34). Ammonoid fossils are common, found as unaltered and altered shells, as well as casts. Because they changed so rapidly through geologic time, many serve as important index fossils.

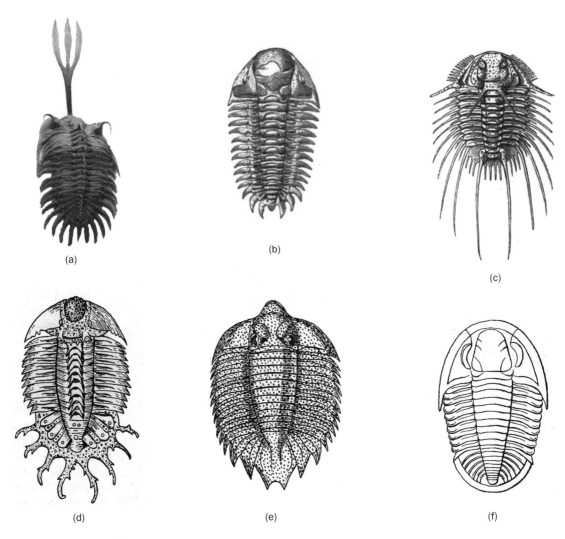

Figure 9.31 (a) Phacopida *Walliserops*, (b) Phacopida *Cheirurus*, (c) Lichida *Acidaspis*, (d) Lichida *Terataspis*, (e) Lichida *Lichas*, (f) Proetida *Proetus*.
Source: (a) Photograph by Kevin Walsh, (b and c) From Woods, 1909, (d) From Schuchert, 1924, (e) From Raymond, 1905, (f) From Grabau, 1899.

Class Bivalvia (formerly **Pelecypoda**) (?Precambrian to Recent) is the second-most-common group of mollusks and includes clams, oysters, scallops, and mussels (Figure 9.35). **Bivalves** are characterized by the hinged pair of calcareous shells that encase their bodies and are distinguished from brachiopods by their symmetry and lack of foramen. Bivalves are typically asymmetrical left–right, but have top and bottom shell symmetry. Because brachiopods have tooth and socket hinges, the shells are usually preserved closed and together. Bivalve shells have a ligament in the hinge area that acts like a spring, constantly pulling the valves apart. The living organism uses a pair of muscles to hold the shells shut, so when the organism dies, it automatically springs open. The shells usually become separated and most bivalve fossils are single, disarticulated shells.

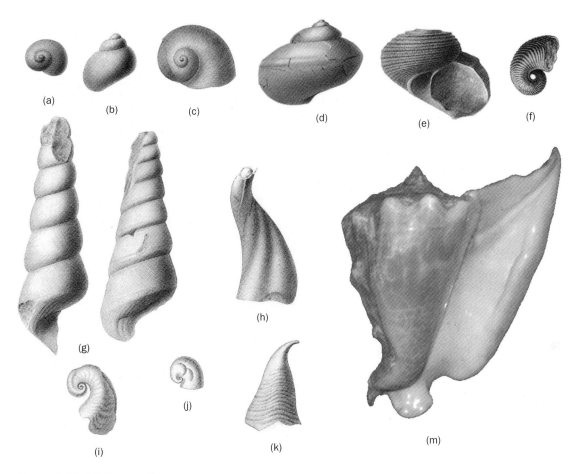

Figure 9.32 Mollusca, Gastropoda. (a and b) *Holopea*, (c) *Diaphorostoma* (now *Platyostoma*), (d) *Eotomaria*, (e) *Poleumita*, (f) *Tropidocyclus*, (g) two examples of *Coelidium*, (h–k) *Platyceras*, (m) *Strombus* (Conch) (not to scale).
Source: (a–k) From Clarke, 1908, (m) courtesy of NBII.

Use of Fossils to Help with a Question of Where . . .

In 1997, a reporter for a television station received an anonymous letter claiming that a woman, missing since December 6, 1996, had died in an accident and that her remains could be found somewhere in a creek in Gallatin County, Tennessee. The letter was followed by a second, both of which were turned over to the police. Apparently, whoever was mailing the letters was unsatisfied with the results because, in July 1997, the television station received a package containing a human mandible with an attached tag that had the missing woman's name written on it. The package had a Bowling Green, Kentucky postmark and contained a letter indicating that this was a way of bringing closure to the woman's family.

The jawbone was sent to a police lab, where it was positively identified as that of the missing woman based on her dental records. Investigators also obtained several hundred natural sediment grains from the mandible. The sediment contained grains of dolomitic siltstone, phosphatic siltstone, chert, and quartz, ranging in size from 0.06 to 4.5 mm in diameter. It also contained distinctive gastropod fossils.

The missing woman had last been seen in Clarksville, Tennessee some 70 miles away from Gallatin, so researchers examined stream sediment from both Clarksville and Gallatin. Stream sediments in Clarksville are derived from sedimentary layers that are entirely Mississippian in age and did not contain the gastropod fossil found associated with the mandible. However, investigators were able to link the evidentiary sediment to stream sediment in the Gallatin region, which was derived from Ordovician through Mississippian sedimentary rock formations, was composed of a similar geological material, and contained the noted gastropod fossils. Thus, the sediment and fossil material found on the mandible was distinctive enough to better define search areas for attempts to locate the rest of the woman's remains and enabled investigators to establish that the probable location of the jawbone was consistent with the location described in the two previously received anonymous letters.

In this case, the rest of the woman's remains were never located. The existence of the mandible did allow investigators to declare the woman dead and were used to establish that the victim's death was most probably a homicide. A few years later, the woman's husband was charged with first-degree murder. Investigators presented evidence that the husband was behind the anonymous mailing of the mandible and that he needed his wife declared dead so that he could collect the insurance money. For a variety of reasons, particularly centered on the lack of additional positively identified remains, the trial ended with a dismissal due to a lack of evidence.

(a)   (b)

(c)   (d)

Figure 9.33 Mollusca, Nautiloidea: (a) exterior and (b) interior of modern *Nautilus*, as well as casts of (c) straight, and (d) coiled fossil *Nautilus*.

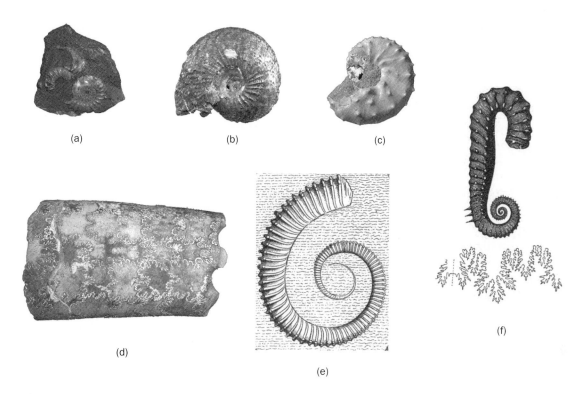

Figure 9.34 Mollusca, Ammonoidea: (a–c) coiled, (d) straight, and (e and f) exotic forms of ammonoids. Please see Color Plate section.
Source: (c) Photograph by Mark A. Wilson, (e and f) from von Zittel, 1900.

Figure 9.35 Mollusca, Bivalvia. A few of the myriad examples of bivalves.
Source: (a–e) Courtesy of the NNBII, (g) photograph by Kevin Walsh, (h) photograph by Mark A. Wilson.

Phylum Echinodermata

Phylum Echinodermata (Precambrian to Recent) includes starfish, sand dollars, sea urchins, crinoids, and blastoids. This group, commonly called **echinoderms**, is characterized by *pentameral* (five-sided) symmetry, and most have skeletons composed of small calcite plates. Higher-order radial symmetry (such as seven-fold or nine-fold) has evolved in some forms as a secondary modification.

Class Crinoidea (Cambrian to Recent) or sea lilies (Figure 9.36) are the most important fossil-forming group of echinoderms. They have a small "head" (*calyx*) with a variable number (five or more) of surrounding arms. Most fossil **crinoids** have a stem by which the calyx is attached to the seafloor. Their skeletons are composed of calcareous discs, and when they die the discs disarticulate, thus their fossils tend to look like scatterings of poker chips (Figure 9.37). They do have a five-part symmetry, but it is hard to see. Some crinoids are valuable index fossils.

Class Blastoidea (Silurian to Permian) a smaller group that is similar to the crinoids, **blastoids** have a "head" (called a *theca* instead of a calyx) that looks like an acorn with a five-point star superimposed on top (Figure 9.38). Once they disarticulate, blastoid stems look like poker chips, just like crinoid stems and arms. Without the distinctive theca intact on the top of the stem, it is virtually impossible to tell blastoid parts from crinoid parts.

Class Asteroidea (Ordovician to Recent) is composed of the starfish. They usually look just like five-pointed stars and have a poor fossil record (Figure 9.39).

Class Ophiuroidea (Ordovician to Recent) is composed of brittle stars. Their central body looks like that of a starfish but they have five long, whip-like arms (Figure 9.40). Ophiuroids also have a poor fossil record.

Figure 9.36 Echinodermata, Crinoidea. Photograph of a modern stalked crinoid (~5 in. across, ~13 cm) filter feeding on the caldera seafloor at West Rota volcano.
Source: Courtesy of NOAA.

Figure 9.37 Echinodermata, Crinoidea. A variety of examples of fossil crinoid pieces. Please see Color Plate section.

Figure 9.38 Echinodermata, Blastoidea. The distinctive acorn-shaped theca with the impression of a star.

Figure 9.39 Echinodermata, Asteroidea. Underside of modern starfish.

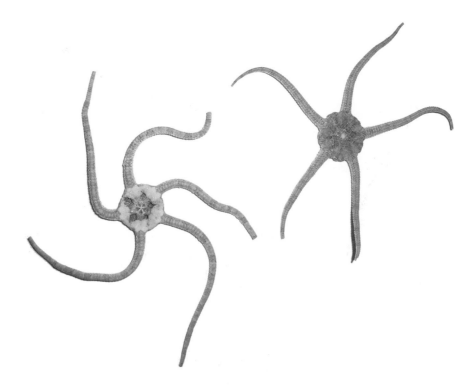

Figure 9.40 Echinodermata, Ophiuroidea.

Figure 9.41 Echinodermata, Echinoidea. A variety of examples of fossil and modern echinoids.

Class Echinoidea (Ordovician to Recent) includes sea urchins, sand dollars, heart urchins, and sea biscuits (Figure 9.41). **Echinoids** have biscuit-shaped, disc-shaped, or heart-shaped exoskeletons composed of multiple, tiny, interlocking plates of calcite. Most members of this group have spines made of single calcite crystals. Echinoids have been divided into two subgroups: *regular echinoids*, like sea urchins, which have nearly perfect pentameral symmetry and *irregular echinoids*, like sand dollars and heart urchins, which have secondarily altered symmetry. When an echinoderm dies, it will rapidly disarticulate into loose calcite plates; however, because most members of this group are burrowers, they are often preserved intact and so this group has a fairly good fossil record.

Phylum Hemichordata

Phylum Hemichordata, Subphylum Graptolithina (Cambrian to Pennsylvanian) was one of the greatest puzzles in paleontology. **Graptolite** fossils, which look like tiny hacksaw blades, are commonly found carbonized on Paleozoic black shale (Figure 9.42). They have been extremely important index fossils for centuries, but no one knew what they were. More recent finds of three-dimensional specimens show that they were tiny, colonial organisms that shared a central *stolon*, which apparently acted something like the nerve cord in chordates, allowing the individuals to communicate information. If you see something that looks like a small hacksaw blade, a bunch of connected hacksaw blades or even spirals with saw-teeth, you are probably looking at a graptolite (Figure 9.43).

This list included only very simple descriptions of the invertebrates most commonly preserved as fossils. For more information, consult an invertebrate paleontology text such as Clarkson, 1998.

(a)   (b)

(c)

(d)   (e)

Figure 9.42 (a, b, e) Graptolite fossils on black shale, (c and d) graptolite fossils in other sediments. Please see Color Plate section.
Source: (c and d) Photograph by Mark A. Wilson.

Micropaleontology

The term *micropaleontology* is used simply to describe the study of microscopic fossils. It includes a wide range of unrelated organisms, such as fossil pollen grains, animals, plankton, prokaryotic organisms, and even pieces of larger organisms (like conodonts). The only uniting factor is size. Microfossils can provide a variety of different types of information to the forensic scientist. Table 9.2 contains a summary of the most important types of microfossils. It should also be noted that classification schemes for most organisms are a bit complicated and argumentative. For one thing, the schemes used by biologists, which are based on structures of living organisms, can differ quite a bit from the schemes used by paleontologists, who typically only have mineralized hard parts by which to identify an organism. For much more detailed information about the types and distribution of such microfossils, and their living counterparts, Armstrong and Brasier (2005) is recommended.

Kingdom Protista includes many different kinds of plant-like (photosynthetic) and animal-like (consumers) single-celled organisms, as well as a few simple kinds of multicellular organisms. Organisms in this kingdom are all considered

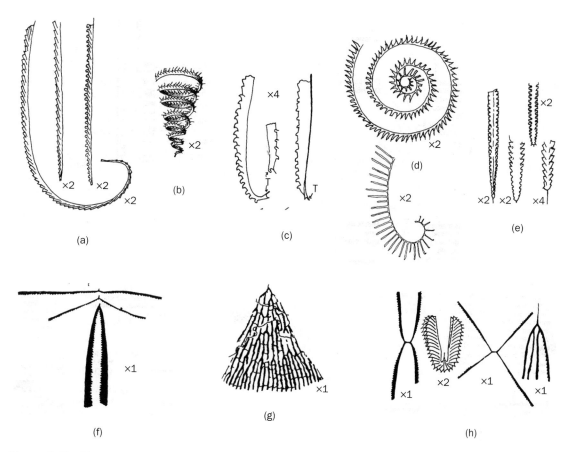

Figure 9.43 Sketches of some of the most stratigraphically useful graptolites:
(a–c) Monograptids, (d) *Monograptus spiralis*, (e) Diplograptids, (f) *Didymograptus*,
(g) Dendroid, (h) *Tetragraptus*.
Source: Churkin and Carter, 1972.

microfossils. Major groups of important Protistan microfossils are Foraminifera, Radiolarian, Diatoms, Coccoliths (calcarious nanoplankton), and Dinoflagellates.

Foraminifera

Order Foraminifera (Cambrian to Recent), the single-celled **foraminifera** are the most common, and geologically the most important, of the fossil protozoans. In the modern ocean, they comprise over 90% of deep-sea biomass. The foraminiferid cells have a flexible surface that both surrounds (ectoplasm) and is contained within (endoplasm) a shell, called a *test*. The tests are usually less than a millimeter across. The name translates to pore-bearing and refers to the numerous perforations (foramina) in the skeleton walls. Today, nearly all *forams* (nickname for the group) live in marine environments and are either bottom dwellers (*benthonic*) or float in the water column (*planktonic*). There are several beaches around the world where the sand is composed mostly or wholly of foraminifera tests. When foraminifera die, their shells accumulate on the seafloor, eventually compressing into limestone.

Forams can have single (*unilocular*) or multi-chambered (*multilocular*) tests (Figure 9.44). Multilocular tests are built by the addition of new chambers during

Table 9.2 The most important types of microfossils.

Fossil	Range	Shell Composition	Size	Morphology	Habitat
Foraminifera	Є to R / J to R	Calcite, aragonite, cemented grains, or organic	0.01–3.0 mm; some up to 100 mm, large ones are benthic	Multi-chambered shell, wide range of shapes	Marine, Benthic / Marine, Planktonic
Radiolaria	Є to R	Amorphous, opaline silica	0.03–2 mm	Symmetrical, lace-like globes or space ships	Marine only, Planktonic
Diatoms	K to R	Silica	0.02–1 mm	Circular, elongate or irregular pillboxes	Maine and Freshwater, Planktonic or Benthic
Coccolithophores	J to R	Calcite (Calcarious nannofossils)	0.002–0.02 mm (submicroscopic)	Layers of plates, rods, stars	Marine, Planktonic
Dinoflagellates	S to R	Organic (sporopollenin)	0.005–2 mm	Organic cysts	Marine and Freshwater, Mostly planktonic, some parasitic or symbiotic
Ostracods	Є to R	Calcite, come organic	0.5–3.0 mm +	Clam-like shell with dorsal hinge	Marine and Freshwater, mostly benthic
Conodont elements	Є to TR	Phosphate	0.1–10 mm	cone-like teeth, spikes, irregular knobby plates	Marine, free-swimming
Sponge spicules	Є to R	Calcite or silica		Needle like, pointed stars	Mostly marine, benthic

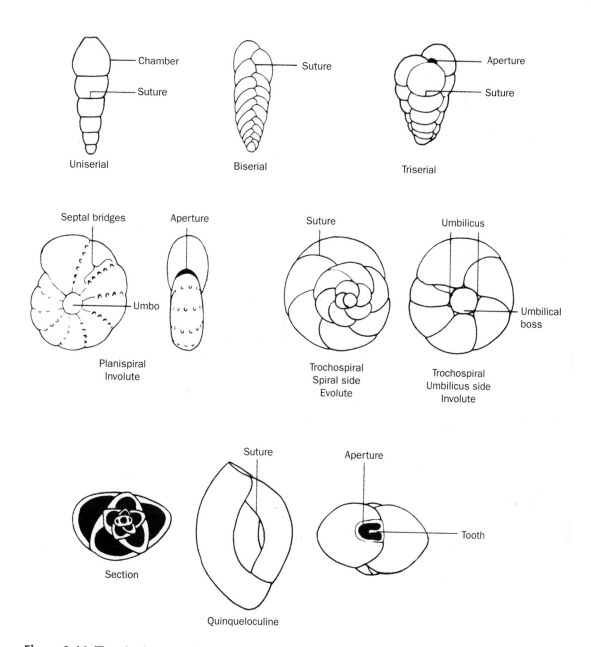

Figure 9.44 Terminology used to describe foraminifera tests.
Source: Buzas *et al.*, 1985. In: Boardman *et al.*, 1985.

life. The chambers are separated by partitions (septa; singular = septum), whose exterior expressions are termed *sutures*. Tests are composed either of calcite (CaCO$_3$) or are *agglutinated* (cemented foreign particles, e.g. sand or silt grains), with one exception. The calcite tests are usually white or light tan and translucent-looking, like porcelain. The classification of Foraminifera is based on (i) test microstructure, (ii) test symmetry, and (iii) aperture type (Figure 9.44). More information on foraminifera (and several other microfossils), plus numerous pictures, can be found online at Miracle: the microfossil image recovery and circulation for learning and education website (http://www.ucl.ac.uk/GeolSci/micropal/foram.html) and at the foraminifera gallery (http://www.foraminifera.eu/).

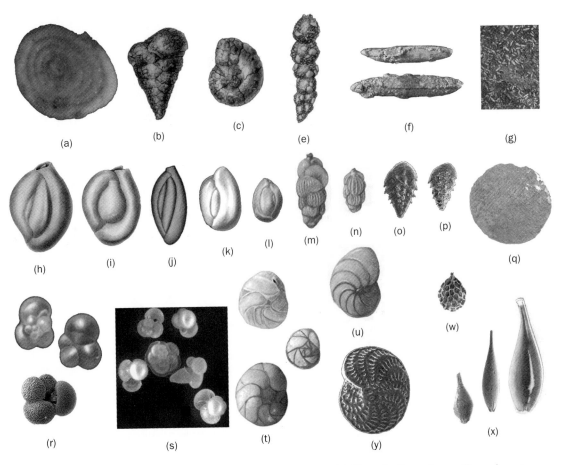

Figure 9.45 Foraminifera of paleontological importance: (a) Textulariina *Ammodiscus incertus*, (b–e) Textulariina (agglutinated tests), (f) Fusulinina, (g) Fusulinina fossils *in situ* looking like grains of rice, (h–l) Miliolina *Quinqueloculina*, (m and n) Rotaliina *Uvigerina*, (o) Rotaliina *Bolivina*, (p) Rotaliina *Brizalina*, (q) Rotaliina Nummulitaceadea, (r) Globigerina, (s) planktonic Globigerina in the Northern Gulf of Mexico, (t) Rotaliina *Cassidulina*, (u) Rotaliina *Valvulineria*, (v) Rotaliina *Elphidium*, (w) Lagenina, (x) Lagenina. (Images not to scale.) Please see Color Plate section.
Source: Images courtesy of the United States Geological Survey.

Particular groups of foraminifera of paleontological importance (Figure 9.45) are described in Table 9.3, which is available on the companion website.

Of special interest are the Nummulitaceadea ("nummulitids"), a superfamily within Rotaliina, which can be quite large (Figure 9.45q). The nummulitids existed during the Early Cenozoic, were especially common in the mid-Eocene, and are famous for their abundance in the limestone from which the Great Pyramids of Egypt were built (Figure 9.46). Their tests are planispiral and involute, coiling around the short axis.

Collection and Treatment

Living benthic foraminifera (along with other benthic microorganisms) can be collected by gathering samples of marine or estuarine seaweed or by collecting the top centimeter of mud from the intertidal area. To free the foraminifera from seaweed, the plants must be placed in water and shaken or whirled vigorously. The

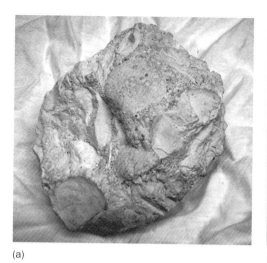
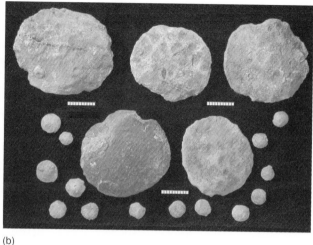

(a)   (b)

Figure 9.46 (a) Nummulitid limestone from Egypt, (b) Nummulitid foraminiferans from the Eocene near Al Ain, United Arab Emirates. Scale bars are 1 cm long and broken into mm segments.

Source: Photograph (b) by Mark A. Wilson.

plant/water or mud sample must then be washed through a 63 μm mesh No. 230 sieve. Do this over a large bucket and not into a standard sink. The residue in the sieve is then transferred into a Petri dish, or rinsed through filter paper and allowed to dry undisturbed, after which it can be examined under a microscope. Planktonic foraminifera can be captured directly but are sparse near-shore and are best collected in the open ocean using plankton nets or, if living specimens are desired, by scuba divers using glass jars. Deep-sea benthic forms are unlikely to be needed for forensic work, but the principle is the same. Dredge samples would be washed through a sieve and then the retained material examined under a microscope.

Foraminifera, as well as many other microfossils of interest, can also be collected from nearly any post-Triassic marine sediment, such as beach sand, and from many soft rocks. One common approach is to take loose sediment, or break fresh rock into 1–5 mm fragments, and place it in a beaker. Then fill the beaker with distilled water, add a couple of large spoonfuls of Na_2CO_3 (also called washing soda) or Calgon laundry detergent, mix well and place the beaker on a hotplate. Bring the mixture to a boil and then simmer, adding water as needed to keep the mix from boiling dry, until the rock shows no further signs of breaking down. The cooled, disaggregated mixture should be washed through a series of sieves using a gentle jet of water. In general, a 44 μm mesh sieve will retain small diatoms and dinoflagellates, while radiolarians, foraminifera, and conodonts will be retained by a 63 μm mesh sieve. Ostracods will predominantly be captured by a 250 μm mesh sieve. The final choice of sieves will depend on the material desired and the size of the clasts from which the organics must be separated. Note that strong acids should not be used, as these would destroy the foraminifera tests, along with all of the other organic and carbonate shell material.

The residue from each sieve should be placed into an evaporating dish. For the collection of foraminifera, radiolarians, conodonts, and ostracods, allow the residue to dry at low temperature on a hotplate or in a drying oven. The dried sample can then be analyzed under a microscope on a picking tray. The residue for samples that are to be examined for diatoms, or dinoflagellates, should be transferred into clean bottles with distilled water. Quick smear slides of such samples are created by adding Cellosize (a dispersant) to the distilled water, gently stirring and then transferring a small drop of sample to a cover slip. The cover slip is allowed to gently dry on a warm hotplate, undisturbed. The sample can then be examined directly under a microscope, or a permanent slide made. The mounting medium is placed on a clean slide and the cover slide placed on top (so that the residue lies on top of the medium). If too much material is recovered, pre-concentration of microfossils can be used to improve slide quality. A much more detailed discussion of collection, extraction, and sorting can be found in the Appendix of Armstrong and Brasier (2005).

Radiolarians

Subphylum Radiolaria (Cambrian to Recent) are single-celled marine plankton, found from the pole to the equator, which feed on other organisms. **Radiolarians** secrete symmetrical, amorphous, opaline silica shells that range in size from $30\,\mu m$ to 2 mm, typically around half the size of the foraminifera. They resemble nothing else so much as spiky, lace balls, and lace spaceships (Figure 9.47). Radiolarian skeletons are very transparent and glassy. Radiolarian shells are common contributors to the formation of the sedimentary rock *chert*. For more information on radiolarians visit the Radiolaria.org website (http://radiolaria.org/) or the Guide to Modern Radiolaria (http://gdcmp1.ucsd.edu/geol_coll/radlit/nm79titl.html). The two major fossil forming groups are:

Order Spumellaria (Cambrian to Recent): Radially symmetrical, often roughly spherical, ellipsoidal, or discoidal. May bear radial spines (Figure 9.47 a–d).

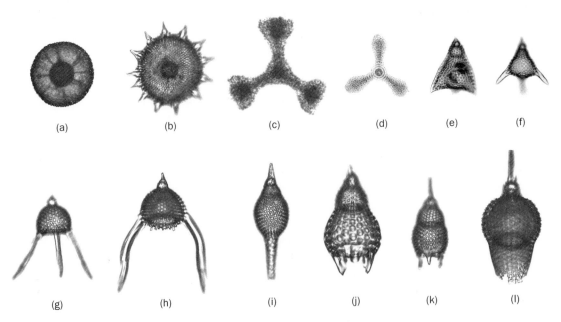

Figure 9.47 Radiolarians: (a–d) Spumellaria, (e–l) Nassellaria.
Source: Modified from NBII.

Order Nassellaria (Silurian to Recent): Radially symmetrical around
the long axis, forming roughly conical or pyramidal profile. This group
includes the radiolarians that look like spiky, lace spaceships
(Figure 9.47 e–l).

Diatoms

Infrakingdom (or subphylum) **Diatomea** – Round, Crawford, and Mann (1990)
employs **subkingdom Chromista, division Bacillariophyta;** this classification is
controversial – (Cretaceous to Recent): **Diatoms** are photosynthetic, planktonic,
golden-brown algae that can be solitary or colonial. They are found just about
anywhere that there is water and light, from the photic zone of the ocean to
freshwater ponds, damp soil, ditches, puddles, and even in polar ice. Most are
0.002–0.02 mm, though some get up to 1 mm, in diameter. Recent estimates
suggest that there are more than 200,000 different species of diatoms. Most species
have a preference for water of a specific salinity, which generally allows for the
distinction between freshwater and marine groupings, and some particular species
will be more commonly associated with ponds or streams than with lakes. Due to
their strong ecological preferences, diatom assemblages can change significantly
even over relatively short geographic distances.

Structurally, diatoms have a pillbox-shaped silica shell (*frustule*) consisting of two
overlapping *valves* that fit together in a nested fashion (like a Petri dish) (Figure
9.48). The larger valve is called the *epitheca* and the smaller valve the *hypotheca*. The
side-view of the frustule, where the two valves overlap, is called the *girdle*. In all
diatoms, a portion of the valve surface area (the top and bottom) is covered with
tiny pores, called *punctae*. The shape of the valve, the topography of the girdle view,
and the arrangement of punctae are the primary characteristics by which diatoms
are classified. Many species also display prominent septa (the partitions within the
valves). Another important characteristic of some diatoms is a smooth, punctae-free
groove or slit that runs through the center of each valve called a *raphe*. If the raphe
is elevated above the valve, it is called a *keel*. A central punctae-free area that does
not have an associated groove is called a *pseudoraphe*. For more information on
diatoms (and the controversy on classification) see the Diatom Home Page
(http://www.indiana.edu/~diatom/diatom.html), Diatoms of the United States
(http://westerndiatoms.colorado.edu), or the Diatom Collection website
(http://research.calacademy.org/izg/research/diatom).

Traditionally, there are two major divisions of diatoms, centric and
pennate, which are further divided into many different subgroups (a list of
these is available on the companion website), the initial taxonomic names are those
usually employed in paleontology, while the names listed [in brackets] are
commonly employed for living diatoms and are based on Round, Crawford and
Mann (1990):

Order Centrales [or **Class Coscinodiscophycidae**] (Jurassic to Recent): The
centric diatoms group is characterized by valves with radial symmetry around a
central point or points. Most are circular or triangular and none displays raphe
or pseudoraphe (see examples Figure 9.48a–d and Figure 9.49a–g).
Order Pennales (Bacillariales) (Eocene to Recent): this group of *pennate diatoms*
is characterized by bilateral symmetry (along a line). The valves tend to look like
rods (see Figure 9.48e and Figure 9.49h–n).

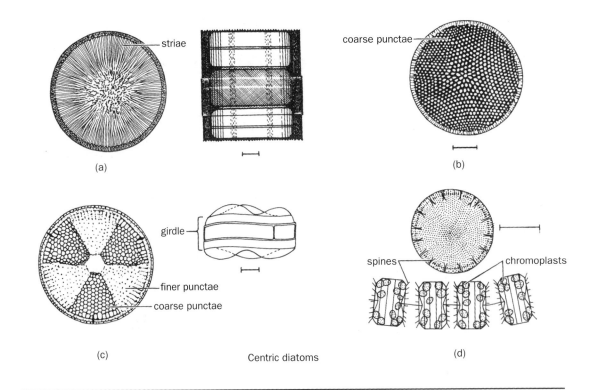

Centric diatoms

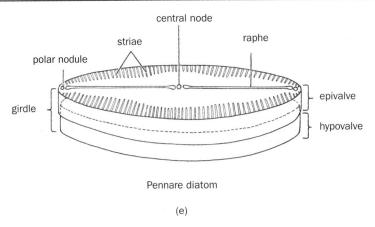

Pennare diatom

(e)

Figure 9.48 Centric diatoms: (a) Coscinodiscineae *Melosira* valve (left) and colony girdle view (right), (b) Coscinodiscineae *Coscinodiscus* valve view, (c) Coscinodiscineae *Actinoptychus* valve view, (d) Thalassiosirophycidae *Thalassiosira* valve view (above) and girdle view of colony (below), (e) Raphidineae *Pinnularia* oblique view. Scale bar equals 10 micrometers.
Source: Modified from Armstrong and Brasier, 2005; (a–d) after van der Werff and Huls, 1957–1963, (e) after Scagel *et al.*, 1965.

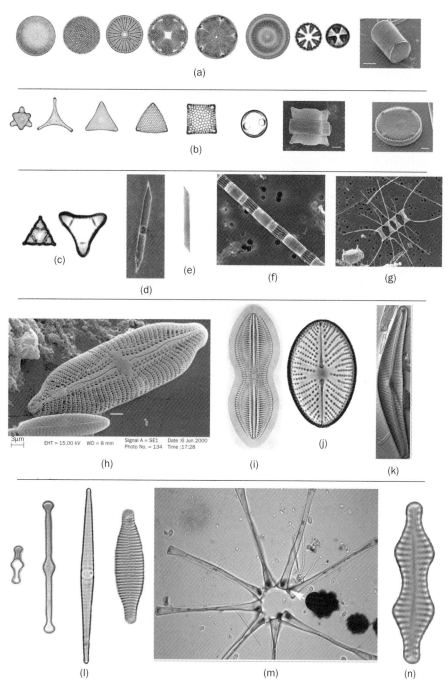

Figure 9.49 (a) Various examples of Centric Coscinodiscineae; (far right) frustule of *Melosira*, showing both valves and some girdle bands (size bar = 10 micrometers), (b) various examples of Centric Biddulphineae; (second from right) the whole shell of *Biddulphia* (size bar = 10 micrometers); (far right) a single valve of *Eupodiscus* (size bar = 20 micrometers), (c) Centric Lithodesmiophycidae, (d–e) Centric Rhizosoleniineae, (f) Centric Thalassiosirophycidae *Skeletonema*, (g) Centric Chaetocerotophycidae *Chaetoceros*, (h–k) Pennate Raphidineae, (l–n) Pennate Araphidineae.
Source: (a–c) Modified from NBII microphotography by Randolph Femmer, (a) and (b) SEM images at far right courtesy of Mary Ann Tiffany, San Diego University, (d, f–g) courtesy of Steve Morton, NOAA, Center for Coastal Environmental Health and Biomolecular Research, (h) courtesy of the United States Geological Survey, photograph by Sarah Spaulding, (j) courtesy of Diatoms of the United States image by Eduardo Morales, Ditmar Metzeltin, (k–n) courtesy of Diatoms of the United States, (n) photograph by Eduardo Morales.

For more detailed information on the identification and classification of living diatoms, see Round, Crawford and Mann (1990) and Hartley (1996) for information on living British diatoms. More details about fossil diatoms can be found in van Landingham's *Catalogue of Fossil and Recent Genera and Species of Diatoms and Their Synonyms* series. A simple and inexpensive introduction to genera classification of North American diatoms is given by Vinyard (1979).

Diatoms can be of great importance for forensic work in a variety of ways. For example, *Diatomaceous earth*, or *diatomite*, is a soft, gritty, white, or buff sedimentary rock formed from the compression of thick layers of silica diatom valves. Such deposits can often be characterized by species distribution, making them useful for geosourcing. In addition, diatomaceous earth is used for a range of commercial applications such as filtration (commonly in pool filters), as an absorbent (like kitty litter), as a fine abrasive, and as insulation in safes. Because the diatoms used for many of these commercial applications are mined from specific locations or are artificial mixtures of species, they can have a distinctive character. The species distribution in such mixtures can be used to establish a relationship between transferred material and a potential source.

Kidnapping and Diatoms

In the 1967, Kenneth Young, the 10-year-old son of the president of a Los Angeles savings and loan association, was kidnapped from his home and held for ransom. The ransom was paid, but the kidnapper escaped the trap set by law enforcement officers. Fortunately, the boy was released unharmed in Santa Monica. The car used to pick up the ransom was later found abandoned. It had been stolen just prior to the crime, and just after it had been through a commercial carwash during which both the exterior and interior had been cleaned. When the car was recovered, the FBI found a white footprint on the carpet of the backseat. The white power was vacuumed up and analyzed. It turned out to contain an odd mixture of freshwater and marine diatoms. The marine diatoms were mainly of the genus *Coscinodiscus*, commonly found in the Miocene Monterey Formation located along the coast of California, while the freshwater species were of genera *Melosira*, *Stephanodisus*, and *Cocconeis*, which are common in inland deposits in California, Oregon, Nevada, and elsewhere. Only an artificial mixture could contain this diverse group of diatoms.

Investigators theorized that, since pool filters commonly contain diatomaceous earth, perhaps the kidnapper was a pool maintenance man. Alternatively, perhaps the freshwater and marine diatoms were accidentally mixed in a warehouse or diatomite processing facility where both types were stored. Samples of the white footprint were sent to a specialist in the commercial production of diatomaceous earth. He reported that the Gerfco Company quarried marine diatomaceous earth from a deposit in the hills of southwestern Los Angeles County at the time of the kidnapping, and that periodically shipments of freshwater diatomaceous earth would be brought in for testing. One year, during a rainstorm, piles of freshwater and marine diatomaceous earth got mixed together and were deposited in a pond at the floor of the quarry. It turned out to be the same quarry where Kenneth had been held for a while.

For months, the FBI checked on persons known to work with or deal in diatomaceous earth. Then a strange series of events led them to a former Internal Revenue Service agent named Ronald Lee Miller who had used a government car on an inspection trip to that particular quarry a few days before the kidnapping. The mix of diatoms found in the government car (which had not been cleaned) matched those found in the white footprint and in the floor of the quarry. At trail, Ronald Lee Miller was linked to the kidnapping using a variety of lines of evidence, including the diatomaceous earth. He was sentenced to life.

Diatoms and Drowning

Another more complex forensic use of diatoms involves their application as a tool to determine whether a body was drowned and, if so, where. There has been significant controversy in the literature about the validity of this approach. One aspect of the contradictory results reported in the literature seems to stem from a misunderstanding of the biology of diatoms. Diatom populations constantly change through time as a result of changes in nutrient availability, water temperature, acidity, and other local factors. Thus, comparison samples need to be collected quickly in order to capture correct species distribution. Also, because diatoms are photosynthetic (sometimes called the *grass of the ocean*), there will be significantly fewer of them in winter. In spring, they tend to rapidly increase in number until they reach a peak, usually sometime in April or May in the northern hemisphere. Intense competition for resources then causes the population to crash so that in the summer months there may be few diatoms. In the late summer or early fall a second, smaller, population increase occurs as the diatoms rebound. Heading into late fall and then winter, the population collapses again. Thus, time of year plays a very important role in diatom availability. If comparison samples are taken too distantly in time, they could potentially bear little relevance to the forensic task. The same holds true if there has been a significant change in the water's chemistry or quality.

A second issue concerns the ubiquity of diatoms. Because they are so widespread in the environment and are used for a diverse range of commercial applications, people are commonly exposed to them. Many recent papers on the issue of diatoms and drowning site a study by Langer *et al.* (1971), reporting that cigars made with reconstituted tobacco, also known as tobacco sheets, contain significant percentages of inorganic material, including diatomaceous earth and often whole diatoms. The study also reported that smoke from these cigars contained fragments of respirable-size diatoms. This has been frequently reported as a potential reason to abandon the diatom method for establishing drowning. However, most of these papers fail to report that the conclusion of the 1971 study was that diatomaceous material was highly fragmented by the high temperatures involved in cigar smoking, thus significant quantities of whole diatoms could not be introduced into a body this way.

There have also been studies that have reported finding significant amounts of diatoms in fresh produce (Krstic *et al.*, 2002), seafood (Yen and Jayaprakash,

2007), and air transported from dusty desert regions of the Earth (Darwin, 1845; Delany et al., 1967; Krstic et al., 2002). Not to mention the significant possibilities of workplace exposures (manufacture of pool filters, etc.). Thus, it is not unusual for most people to have a few or even several diatoms incorporated into their organs and bones. However, when assessing drowning, it is not simply a question of the presence or absence of diatoms that is indicative. Instead, the question is relational, involving the correlation of diatom assemblages between the organs of a body and a body of water. The presence of the same mixture of species in the lungs, bone marrow, or other tissues or organs of a victim, and in a body of water, can clearly be used to establish a link to the probable location of immersion. Also, for a variety of reasons, such as the time of year, or dry drowning, many or even most drowning victims will not have detectable diatoms incorporated into their bodies, so the absence of diatoms does not exclude the possibility of drowning. With a better understanding of the limitations and valid parameters, diatom testing has the potential to provide vital information in a significant number of cases. More information on collection of biological samples for diatom analysis can be found in Pollanen (1998).

Place of Death

In May 2002, members of the California Highway Patrol found the body of a 15-month-old baby floating in Bear Creek. Just a few hours earlier, the police had received a 911 call concerning a teenage mother whose son had been kidnapped. The girl stated that she had been standing near a fountain in a park and that at around 9:45 p.m. a man grabbed the stroller with her sleeping son and ran off into the darkness. She gave chase but was unable to keep up. The body found in the creek was indeed that of her son.

Right away, investigators noticed that there was something odd going on. In a typical drowning, the victim struggles and then sinks. The body of the toddler was still floating. At autopsy, very little water was found in the stomach, again indicating that the victim did not struggle. Investigators used the term "a gentle drowning" to describe their findings, a concept that appeared to be completely at odds with the abduction being described by the tearful mother during her press conferences.

Investigators returned to the scene and collected water samples from both the fountain in the park and the creek where the body was found. The police sent these samples, along with the water collected from the baby's stomach, to the state crime laboratory, which was not able to perform an appropriate analysis. The lead investigator then sent out a plea for help, which came in the form of a University of Colorado at Boulder plant ecologist, Dr Jane Bock. She analyzed the water samples under a microscope and found that a vast number of diatoms were present. The water from the creek had distinctly different species of diatoms than the water from the fountain. And the water found in the baby's stomach contained diatoms from both the creek and the fountain. This implied that the child had been held underwater in the fountain before being thrown into the creek. When confronted with this evidence, the teenage mother confessed.

In a similar case from the United Kingdom, in April 1991, a two-year-old boy went missing from his home while his mother was distracted by a phone call. The mother started searching and was soon joined by a neighbor. Within a few minutes, the neighbor found the boy's body floating in a duck pond. At the time investigators attributed the death to accident. The mother told the police that her son must have somehow pushed a chair up to the door, climbed up to unlock it, and then left the house while she was on the phone. The boy then would have had to walk for some 200 yards, past ducks and geese, before reaching the pond.

In 1999, the drown child's mother and father divorced. The case was then reopened as a potential homicide when the child's grandparents and father contacted the police, 13 years after the fact, alleging that the child's mother had drowned him in the bathtub and then placed the body in the pond. The father had gained custody of the couple's other child, partially due to the mother's 1986 diagnosis of Munchausen Syndrome by Proxy, a condition that can lead mothers to inflict harm on their children. This fact had apparently not been disclosed at the time of the 1991 drowning.

As part of the criminal proceedings, investigators took 14 samples from around and inside of the pond to compare with the diatoms found in the child's lung tissue. They identified 37 different species of diatoms in the pond. The problem, however, was that the samples were collected 13 years after the drowning occurred and the investigators had no way of knowing what kinds of changes occurred during the time lapse. Examining the forensic samples, investigators found that there was significantly more species diversity in the pond water samples than in the lung tissue samples, a result that could be explained by a difference in season. However, the diatom assemblages in the lung tissue samples comprised many of the same species that were found in the pond. In this case, even with the significant time lapse, diatoms could be used to tentatively link the pond to the drowning and to exclude drowning in a bathtub.

On the basis of the scientific evidence, the timing of various events in the case, and the fact that she had been seen in dry clothing shortly after the boy disappeared, the mother was cleared of all charges. It also turned out that the diagnosis of Munchausen Syndrome by Proxy, which was used in order to bring the case to trial, had been made by a man subsequently discredited in several other court cases.

A detailed discussion on *collecting and extracting diatoms from field samples* is available at the companion website. A variety of different extraction methods have been published and are available online; however, it should be noted that most of them are *extremely* hazardous, relying as they do on strong acids or oxidizers, do not adequately describe the techniques involved, and sometimes will even destroy some or all of the diatoms in a sample. Whatever approach is used, it is important to note that prolonged exposure to strong acids can destroy diatoms, as can excess centrifugation. The fact that there is no universally accepted preparation procedure, and that many of the reported diatom/drowning studies in the literature employed significantly different methods, has contributed to considerable confusion as to the utility of using diatoms as indicators of drowning.

Coccolithophores

Phylum Haptophyta (Prymnesiophyta) (Early Jurassic to Recent) are marine, photosynthetic, calcareous nanoplankton. There are around 500 living species and many additional fossil species. The most common are the **coccolithophores** (Figure 9.50). Coccolithophores are tiny, golden-brown algae that form spherical balls generally ranging in size from 0.25 to 30+ μm armored by a layer of tiny calcareous plates, called *coccoliths*, that range in size from 2 to 15+ μm in diameter. These plates are so tiny that an SEM is necessary for visualization. Most commonly, they are disc- or button-shaped, but a wide diversity of forms have been found. When one of these organisms dies, the disarticulated plates rain down to the ocean floor, contributing to the formation of ocean sediments such as limestone. *Chalk* is limestone made almost exclusively of coccoliths. The White Cliffs of Dover are one of the most famous examples of such a deposit. The Cretaceous Period, when the material that forms the cliffs was deposited, means the age of chalk, from the Latin *creta*, meaning "chalk." More information on collection and identification of coccolithophores can be found in Winter and Siesser (2006).

Scanning Electron Microscope

SEMs are commonly employed in forensic laboratories for visualization and the analysis of ranges of samples (Figure 9.51). They have magnification ranges from 15× to 1,000,000× and up to a 300 times greater depth of field than an optical microscope, allowing clear three-dimensional views of extremely tiny objects. SEMs employ a small diameter primary electron beam, which is scanned across a sample's surface. The *incident electrons* strike the sample, generating a variety of electromagnetic signals. The signals that provide the greatest information are the *secondary electrons* emitted from the atoms occupying the top surface of the sample, which can be used to produce a readily interpretable high-resolution image of the surface, the *backscattered electrons*, which are "reflected" from atoms in the solid and can be used to show the distribution of different chemical phases in the sample, and finally *X-rays*, which as previously described can be used to determine

(a)   (b)

Figure 9.50 (a) SEM image of coccolithophore *Emiliania*, (b) Coccolithophore *Coccolithus pelagicus*.
Source: (a) From Buzas *et al.*, 1985, (b) photograph by Hannes Grobe/AWI.

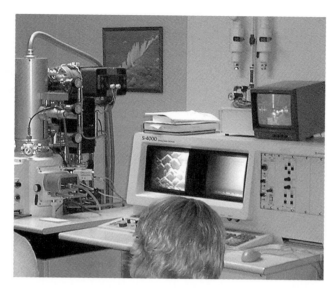

Figure 9.51 Scanning electron microscope in use at the Laboratory for Forensic Odontology Research, SUNY in Buffalo, New York.

elemental composition via energy dispersive X-ray spectroscopy (EDS or EDX) in a fashion similar to X-ray fluorescence (XRF).

Two disadvantages of most traditional SEMs are that they require a complete vacuum, so samples must be totally dry, and that the samples must be able to conduct electricity. For many samples, this requires the addition of a thin coating of gold or carbon. Beyond the coating step, however, the process is nondestructive and can provide a wealth of detailed information. Many laboratories have also acquired newer Environmental SEM systems that do not require a complete vacuum and can work under a wider range of conditions. SEM images of microfossils can provide a great deal of information and make the identification of some microorganisms/microfossils possible or at least much easier.

Dinoflagellates

Dinoflagellates (Silurian to Recent) are unicellular protists that possess two flagella and fall into **phylum Pyrrophyta**, or **division Pyrrhophyta** (literally *fire plants*), **class Dinophycaea**, depending on the reference used. They are commonly referred to as *algae* and are found in both marine and freshwater settings. Both photosynthetic and *heterotrophic* (eat other organisms) forms are known, with some species even falling into both categories. Approximately 90% are planktonic, though others live as symbionts or parasites (zooxanthellae in corals). Dinoflagellates are best known for their periodic population booms that result in "red tides," which may kill fish and shellfish.

Dinoflagellates lack mineralized hard parts and are instead formed of organic material (such as cellulose). They occur in a vast array of forms: some may resemble a top or a star, while others are covered with spines or ridged plates. All of them have a diagnostic indentation around their equator that holds the coiled flagella in life. The word "dinoflagellates" literally means "whirling whips." They potentially could be employed forensically in the same fashion as diatoms; however, their lack of a hard mineral shell makes recovery a bit more difficult. They also have a fairly sparse fossil record. More information on living dinoflagellates can be found in Taylor (1987) and at the All about Dinoflagellates (http://www. assurecontrols.com/info-dinoflagellates.htm) and Dinoflagellates (http://www. geo.ucalgary.ca/~macrae/palynology/dinoflagellates/dinoflagellates.html) websites.

Ostracods

Class Ostracoda (Cambrian to Recent), which falls into kingdom Animalia, phylum Arthropoda (or sometimes placed as a subclass of class Crustacea), is an extremely successful with group approximately 65,000 living and fossil species. **Ostracod** shells are very commonly preserved as fossils and, as a result, they have the most complete fossil record of any of the arthropods. Today, they are found in almost all aquatic environments including both fresh and marine waters, as well as in hot springs, caves, hypersaline environments, in some groundwater, and even in moist terrestrial habitats (Figure 9.52a). In fact, they can be found almost anywhere that is wet, even fleetingly so. Their great abundance and almost universal occurrence makes them potentially very useful for forensic science investigation.

Ostracods usually range in size from 0.1 to 2 mm, though forms up to 32 mm are known. Most are benthic (bottom dwelling), found crawling along or burrowing in the sediment at the bottom of a body of water, though the aptly named *Gigantocypris*, measuring in at 32 mm, and a variety of other species are *pelagic* (swim actively in the open water). Like other arthropods, ostracods have segmented bodies that display bilateral symmetry, with an appearance similar to that of a very small, laterally compressed shrimp. The paired body parts, however, are enclosed in a *dorsally* (back) hinged *carapace* (clam-like shell) consisting of two valves (shell halves) usually composed of low magnesium calcite.

The carapace of an ostracod is usually oval- or bean-shaped, often with one valve slightly larger than the other. The surface of the carapace can be smooth, have a variety of striations, and/or have significant surface ornamentation, usually referred to as *sculpture*. Distinctively, the ostracod carapace does not show growth lines, like those commonly found on bivalves and brachiopods. One of the most important characteristics used for classification is the shape of the dorsal hinge, including the presence or absence of tooth and socket structures. More information on ostracods, along with several pictures, can be found online at Miracle (http://www.ucl.ac.uk/ GeolSci/micropal/ostracod.html) and the Ostracod Research at Greenwich page (http://w3.gre.ac.uk/schools/nri/earth/ostracod/introduction.htm).

Distinction of fossil genera is based on the major features of the shells, while cataloging of extant ostracods is based on soft-body morphology. This has led to significant flux in Ostracoda classification. Based on Armstrong and Brasier (2005), there are three commonly recognized classes of true ostracods: **order Palaeocopida** (Ordovician to Triassic, ?Cenozoic; Figure 9.52g–k), **order Podocopida** (Ordovician to Recent; Figure 9.52d, e, o–r), **order Myodocopida** (Ordovician to Recent; Figure 9.52a, f, l–n), as well as two classes of putative ostracods, meaning their actual evolutionary relationship to modern Ostracoda is in question; **order Archaeocopida** (Cambrian to Early Ordovician); and **order Leperditicopida** (Ordovician to Devonian; Figure 9.52b, c). The putative Orders Bradoriida (Cambrian to Early Ordovician) and Eridostracoda (also extinct) have been excluded because of the likelihood that these groups do not belong in Crustacea.

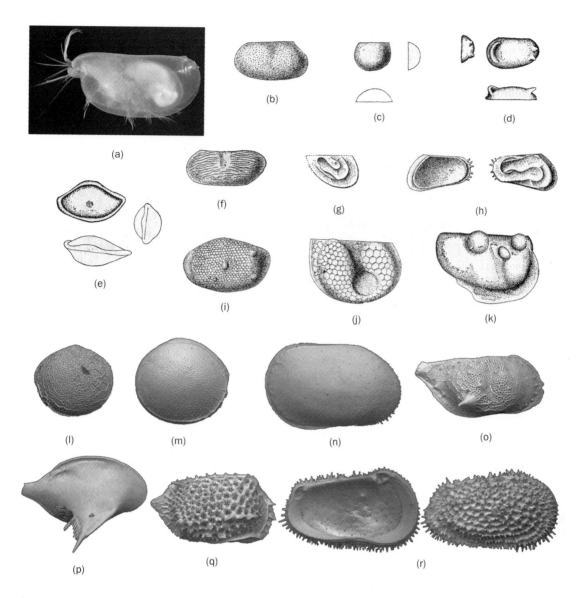

Figure 9.52 Ostracoda (a) A living ostracod, genus *Spelaeoecia* that is only from marine caves and occurs in Bermuda, the Bahamas, Cuba, Jamaica, and Yucatan, Mexico, (b) Leperditicopida *Isochilina*, (c) Leperditicopida *Leperditia*, (d) Podocopida *Moorea*, (e) Podocopida *Bairdia*, (f) Myodocopida *Entomis*, (g–h) Palaeocopida *Strepula*, (i) Palaeocopida *Primitiopsis*, (j) Palaeocopida *Primitia*, (k) Palaeocopida *Beyrichia*, (l and m) Myodocopida *Polycope*, (n) Myodocopida *Cytherella*, (o) Podocopida *Bythoceratina*, (p) Podocopida *Cytheropteron*, (q) Podocopida *Eucytherura*, (r) Podocopida *Henryhowella*. Please see Color Plate section. *Source:* (a) Image courtesy of Tom Iliffe, Bermuda: Search for Deep Water Caves 2009, http:// oceanexplorer.noaa.gov/explorations/09bermuda, (b–k) from Grabau, 1899, (l–r) from Yasuhara, Okahashi, and Cronin, 2009.

Figure 9.53 Microscopic spicules from a pachastrellid sponge.
Source: Courtesy of Islands in the Sea 2002, NOAA/OER, http://oceanexplorer.noaa.gov/explorations/.

Sponge Spicules

Sponges (phylum Porifera; Cambrian to Recent) are mostly marine organisms that live attached to the seafloor. When they die, they can disaggregate into smaller pieces called *sponge spicules* (Figure 9.53), which are the microscopic needle-like or multi-rayed skeletal elements secreted by sponge cells. Some groups of sponges secrete spicules of calcium carbonate, while other groups secrete spicules of silica or organic fibers (spongin). Sponge spicules are often recovered in marine sediments and are commonly mixed into beach sand. The composition and shape of the spicules can be used to identify species of sponges, but for simple microfossil work the spicules can simply be described by shape. A basic descriptive terminology for sponge spicules is shown in Figure 9.54, with additional information on the companion website.

Conodont Elements

Class Conodonta, of the phylum Chordata, subphylum Vertebrata, (?Precambrian to Late Triassic) are tiny, tooth-like, hard parts of an extinct organism. The name *conodont* means "cone-tooth" in reference to their shape. Conodonts are made of apatite (calcium fluorapatite), a phosphate mineral similar in composition to our teeth and bones (Figure 9.55). Most are 0.5–1.5 mm in size, though examples as large as 10 mm and as small as 0.1 mm have been found. They are extremely common in Paleozoic and Triassic marine sedimentary rock, especially organic-rich sandstone and limestone, and serve as the biostratigraphic index fossils of choice for the Paleozoic.

Most specialists now consider *conodont animals* to be chordates, and probably craniate as well, with bodies that would have resembled that of a modern lamprey or hagfish. The organism appears to have had a well-defined head, a notochord, and a distinct tail with fins. *Conodont elements*, the structures that are commonly simply called conodonts, would have ringed an anterior feeding apparatus.

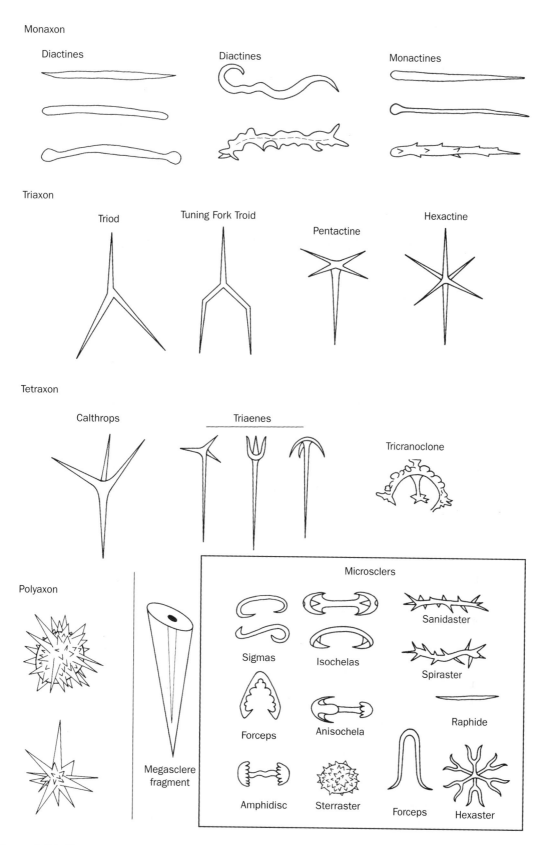

Figure 9.54 Terminology used to describe sponge spicules. Microscleres are shown in comparison to a megasclere fragment.
Source: Modified from Rigby, 1985.

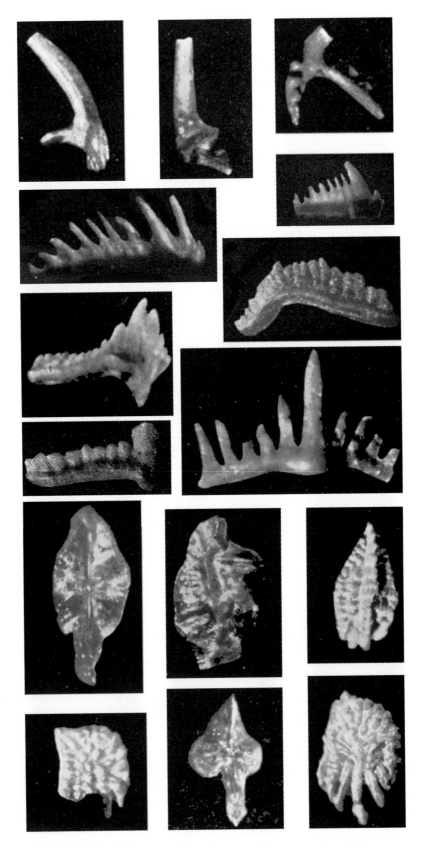

Figure 9.55 Some examples of Devonian conodonts from western New York.
Source: Bryant, 1921.

The main components of a conodont element are the *crown* (section that looks like one or more teeth) and the *basal body* (darker bit under the crown) (Figure 9.56a). The largest cone- or fang-shaped structure in the crown is called the *cusp*. For conodont elements composed of multiple cone-like structures, the largest is called the *main cusp*. The convex side of the cusp is denoted the *anterior* (also termed *rostral*, i.e. toward the nose), while the concave side faces toward the *posterior* (also called *caudal*, i.e. toward the tail). The smaller teeth-like, needle-like, or sawtooth-like structures on the crown that surround the main cusp are called *denticles*. The basal body of the conodont may have one or more projections, called *processes*, surrounding the main cusp. In addition to possible anterior and posterior processes, there can also be one or more *lateral processes* (processes that project to the "left" and to the "right").

Because conodont elements are parts of an extinct organism with no living counterparts, classification was originally based on morphology alone. However, now that conodont elements have been found *in situ* inside the fossilized bodies of conodont animals, the older classification schemes have largely been abandoned. Paleontologists are currently developing classification systems that clarify the evolution of conodont animals, each of which possesses multiple types of individual elements. Such classification schemes are not likely to be useful to a forensic worker confronted only with loose conodont elements. Therefore, a simplified list of descriptive morphological terminology is introduced here (Figure 9.56) with additional information provided on the companion website.

Fossil Pollen and Spores

Fossil pollen and *spores* are also considered microfossils. However, rather than including information about them here, readers are directed instead to the literature on the topic of *forensic palynology*, which is the forensic study of modern and fossil spores, pollen, and other microscopic plant remains. The following papers give a good introduction to the topic: Bryant and Jones (2006), Horrocks (2004), and Mildenhall (2006).

Stolen Dinosaurs

It is worth noting that not only are fossils useful tools in forensic investigation they can also be objects of crime. This is especially true for some special groups of fossils like those of mammals, and most especially dinosaurs. Dinosaurs are popular around the world, but dinosaur fossils are quite rare, especially those in good condition or nearly complete. This has resulted in a worldwide market for smuggled and/or stolen dinosaur bones, much to the detriment of our scientific understanding of them. Dinosaur bones can command high prices on the commercial market, such as the $8.36 million that the nearly complete *Tyrannosaurus rex* skeleton known as Sue fetched at a Sotheby's auction in 1997. The third most complete *T. rex* skeleton ever found, known as Sampson, went unsold at a Bonhams and Butterfields auction on October 3, 2009. The final high bid was $3.6 million, which is impressive but fell short of the minimum price set by its owner. A private buyer later purchased Samson and the skeleton has been placed on temporary displays at various locations. The Carnegie Museum of Natural History has a cast of the skull on permanent display. At the same auction an *Edmontosaurus annectens* (a type of duck-billed dinosaur) sold for $458,000 and in mid-2009 a giant *Triceratops* skull sold for $242,000. With legitimate pieces pulling in such large sums of money, it is no surprise that there has been rampant theft of bones from public lands and global smuggling of rare pieces.

In April 2000, scientists from the College of Eastern Utah Prehistoric Museum in Price discovered that looters had destroyed the *Stegosaurus* skeleton that they had painstakingly spent the past several years excavating. The looters broke the bones out of the ground using hammers, chisels, and large pry bars, destroying much of the skeleton in the process. Only a handful of complete *Stegosaurus* skeletons exist and only four partial skeletons have been found in Utah. The world's only known set of stegosaur footprints, located in the outback of Western Australia, were stolen in 1996. Thieves apparently used power tools to remove them. Two of the thieves were eventually caught and sentenced to two years in jail. One footprint was recovered in December 1998 but the rest remain missing. Apparently, it is becoming somewhat common for partly excavated skeletons to disappear overnight. In 1999, the National Park Service identified 721 documented incidents of fossil theft and/or vandalism between 1995 and 1998, and recent reports indicate such incidents are on the rise.

And it is not just fossils from remote locations that are disappearing. In 2003, a 110-million-year-old *Psittacosaurus sinensis* fossil, one of only 10 in the world, was stolen from a museum in southeastern Australia. The fossil of this parrot-beaked dinosaur was on loan from China as part of a display showing links between dinosaurs and modern birds. Thieves broke the security glass and only recovered part of the skeleton, leaving several bones in and around the museum.

A piece of *T. rex* jawbone that was stolen from the University of California at Berkeley in late 1994 was later found at the home of a European fossil collector. The FBI was able to recover the fossil after a colleague of Mark Goodwin, the museum's vertebrate paleontologist at UC Berkeley, visited a private museum in Wyoming and saw a replica of the jawbone. This meant that not only did someone have the stolen fossil but also that they were creating and selling replicas. Goodwin was able to find out the name of the dealer and passed that information, along with photographs of the missing fossil, on to the FBI. Several months passed, then the FBI called Goodwin and told him that they had recovered the jawbone. This is a rare happy ending, because most fossils that have been stolen are never seen again.

While dinosaur bones are probably the most popular and lucrative fossils that are stolen, all sorts of scientifically important pieces have been stolen from locations all around the world. While law enforcement agencies have become quite proactive when it comes to tracking and recovering stolen gems, jewelry, and artwork, there needs to be a more concerted effort to stem the tide of stolen fossils.

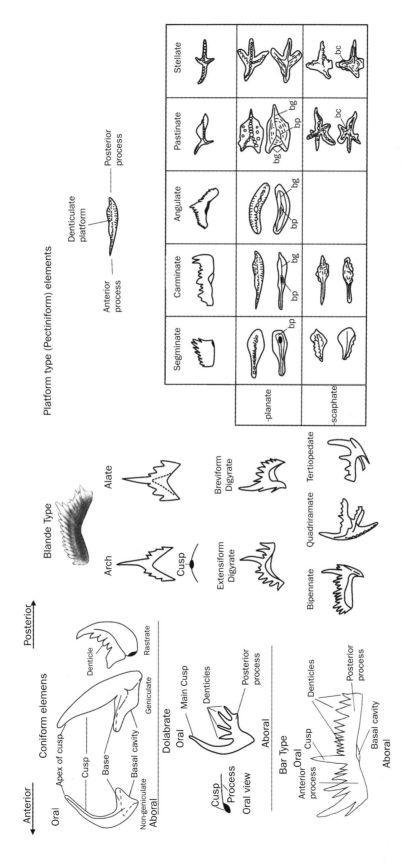

Figure 9.56 Descriptive morphological terminology for conodont elements. Abbreviations: bc, basal cavity; bg, basal groove; bp, basal pit. *Source:* Modified from Armstrong and Brasier, 2005, redrawn from Sweet, 1988; bar type element modified from McKinney, 1991; blade type element modified from Clark, 1985.

Is It Legal to Take This Fossil?

Most countries have strict fossil laws, but the United States historically has not, which is one of the reasons that fossil theft has been rampant here. However, on March 30, 2009, the Paleontological Resources Preservation Act (PRPA) was signed into law by President Obama (Public Law 111-11, Title VI, Subtitle D). The Act is intended to protect scientifically important fossils on public land, includes provisions for a permitting system to allow qualified researchers to collect fossils, and for casual collecting of "reasonable amounts of common invertebrate and plant resources for non-commercial private use." For most people, the new law will not have much of an affect. Any collecting that amateur paleontologists and rock hounds could legally do before the passage of PRPA is still legal now. The greatest changes from the previous state of affairs appear to be increased penalties for illegal collecting, structural changes in the governmental agencies responsible for oversight, and requirements for more cooperation between agencies.

The rules governing the removal of fossils vary according to who owns the land and the type of fossil. The following information is just a brief outline of the current state of affairs. If you intend to collect fossils, contact the appropriate local agencies in order to find out their policies. First off, if you own the land, any fossils you find are yours to do with as you please. Any type of fossil may be collected and sold if it is collected with the permission of a private landowner. Public lands, however, are held in trust by federal, state, or local governments, each of which can and do establish their own policies. Individual states have the power to grant permits for state lands, and the rules vary widely. Some states allow the free collection of certain types of fossils, while others have stricter rules prohibiting most collection. Most states do prohibit collection in parks. To confuse matters, locally held government land, such as county recreation areas, can have vastly different rules regarding fossil collection than state lands do. Interested parties are advised to contact local agencies to find out what policies apply.

Then you have the rules for federally held land. Five federal agencies control virtually all of the publicly owned land in the United States, and each has its own regulations. One thing that all federal lands have in common is that the commercial collection of any type of fossil is not allowed, except for petrified wood, which has been designated a mineral, and is therefore managed under the Mineral Materials Act (43 CFR 3622).

On Bureau of Land Management and US Forest Service land, the policy generally states that "reasonable amounts" of common invertebrates (such as ammonites, trilobites, and brachiopods), plants, and petrified wood may be collected for personal use, but not for sale, barter, or trade. For petrified wood, up to 25 lb plus one rock per day but no more than 250 lb a year is considered reasonable, but there does not appear to be a clear definition of what constitutes a "reasonable amount" of fossils. No mechanized equipment or explosives can be used and you must not create a significant ground disturbance. No vertebrate fossils or rare invertebrate fossils may be collected without a permit. Permits are only granted to professional paleontologists who agree to preserve their finds in a public museum, college, or university to ensure their future availability to researchers and the public. Finally, collection is strictly prohibited in specially classified areas such as designated Wilderness Study areas or Areas of Critical Environmental Concern.

On National Park Service and Bureau of Reclamation land, permits are required for the removal of *any* material, including fossils. Technically, it is illegal even to remove dirt or leaves from National Parks. As before, permits are only granted to professional paleontologists who agree to preserve their finds in a public museum, college, or university. If you find fossils in a National Park, you are requested to note their location and report them to a park ranger. Similarly, on US Fish and Wildlife Service lands, special permits are required for the removal of any fossilized material.

One common dodge used by fossil thieves is to claim that they found their fossils on private land. By comparing the rock matrix and sedimentological structures of the material surrounding the bones, it is sometimes possible to determine whether a fossil was really acquired on private land or stolen from public land. This can work if the thieves are found close to the site of acquisition, but there are limitations to this tactic. Obviously, such a determination could take significant time and most park rangers do not have this type of training. Also, once a fossil has been transported away from a site, it can be very difficult to establish from where it might have originated.

Rare-earth Elements

A new approach for determining the original location of fossil bones that shows promise is the analysis of *rare-earth elements* (REEs). REEs range from atomic number 57 (lanthanum) to 71 (lutetium), all of which possess similar chemical characteristics and occur primarily in the trivalent (3+) oxidation state (Figure 2.3). Living organisms do not incorporate REEs into their structures and fresh bone contains very low concentrations of REEs (<20 ppm). After death, however, organic remains are transformed into fossils in concert with the surrounding sediment as it lithifies into sedimentary rock. Vertebrate bones are composed of very fine-grained, poorly crystalline hydroxyapatite, and during the process of fossilization the hydroxyapatite recrystallizes and grows. Because REEs are similar in size to calcium, they are able to substitute for it in the hydroxyapatite. REEs are exchanged between bone and surrounding sediments, so the fossils should, theoretically, end up sharing the REE signature of the surrounding rock.

Geologists are interested in this information because REE signatures can be used to learn about the paleoenvironmental (ancient environmental) conditions (Metzger, Terry, and Grandstaff, 2004). Environments with different chemistries (pH, redox, salinity, etc.) will cause certain REEs to become either concentrated or depleted in sedimentary layers. And because environmental conditions change over time, the bones found in successive sedimentary layers may have very different REE signatures. Thus, REE signatures are an obvious tool for potentially linking stolen vertebrate fossils with their points of origin. Researchers currently use inductively coupled plasma–mass spectrometry (ICP–MS) for this analysis, which is a destructive analytical method, and there is only a limited amount of information available concerning REE signatures for different bone beds, but this tactic shows great promise.

Summary

The goal of this chapter was to provide enough information that readers become familiar with the broad scope of fossil (and some associated living) organisms, and the manners in which they can be potentially useful as forensic tools. Descriptions of the basic morphology of many organisms, and simplified classification schemes, have been provided in order to enable the recognition of common fossils.

References

Armstrong, H. A. and Brasier, M. D. (2005) *Microfossils*. Blackwell Publishing, Malden, MA.

Bass, W. M. (2005) *Human Osteology: A laboratory and field manual*, 5th edn. Missouri Archeological Society, Springfield, MO.

Benton, M. J. (2004) *Vertebrate Palaeontology*, 3rd edn. Wiley-Blackwell, Malden, MA.

Bryant, W. L. (1921) The Genesee conodonts with descriptions of new species. *Bulletin of the Buffalo Society of Natural Sciences* **13**(2).

Bryant, V. M. and Jones, G. D. (2006) Forensic palynology: Current status of a rarely used technique in the United States of America. *Forensic Science International* **163**: 183–97.

Buzas, M. A., Douglass, R. C., and Smith, C. C. (1985) Kingdom Protista. In: R. S. Boardman, A. H. Cheethan, and A. J. Rowell (eds), *Fossil Invertebrates*. Blackwell Scientific, Oxford.

Byers, S. N. (2005) *Introduction to Forensic Anthropology: A textbook*, 2nd edn. Pearson Education, Boston.

Churkin, D. L. and Carter, C. (1972) Graptolite identification chart for field determination of geologic age. *United States Geological Survey Oil and Gas Investigations Chart OC* **66**.

Clarke, J. M. (1908) Early Devonic History of New York and Eastern North America. *New York State Museum Memoir* **9**.

Clarke, J. M. and Ruedemann, R. (1912) The Eurypterida of New York. *New York State Museum Memoir* **14**.

Clarkson, E. N. K. (1998) *Invertebrate Paleontology and Evolution*, 4th edn. Blackwell Science, Malden, MA.

Darwin, C. (1845) An account of the fine dust which often falls on vessels in the Atlantic Ocean. *Proceedings of the Geological Society* **2**: 26–30.

Delany, A., Delany, A., Parkin, D. *et al.* (1967) Airborne dust collected at Barbados. *Geochimica et Cosmochimica Acta* **31**: 885–909.

France, D. L. (2008) *Human and Nonhuman Bone Identification: A color atlas*. CRC Press, Boca Raton, FL.

Geological Society of America (2009) Walker, J. D., and Geissman, J. W., compilers, Geologic Time Scale: Geological Society of America, doi: 10.1130/2009.CTS004R2C. Available online at http://www.geosociety.org/science/timescale/.

Grabau, A. W. (1899) Geology and Palaeontology of Eighteen Mile Creek and the Lake Shore Sections of Erie County, New York. Published by the Buffalo Society of Natural Sciences

Gupta, V. J. (1989) The case of the peripatetic fossils: Part 2. *Nature* **341**: 11–12.

Gupta, V. J. (1990a) A response to the co-authors. *Nature* **343**: 307–8.

Gupta, V. J. (1990b) Indian palaentology under a cloud: Discussion. response by V. J. Gupta and replies by S. B. Bhatia, J. A. Talent, and U. K. Bassi. *Journal of the Geological Society of India* **35**: 649–64.

Gupta, V. J., Rhodes, F. H., and Austin, R. L. (1967) Devonian conodonts from Kashmir. *Nature* **216**: 468–9.

Hartley, B. (1996) *An Atlas of British Diatoms*. Biopress Ltd, Bristol.

Horrocks, M. (2004) Sub-sampling and preparing forensic samples for pollen analysis. *Journal of Forensic Science* **49**(5): 1–4.

Jayaraman, K. S. (1994) Fossil inquiry finds Indian geologist guilty of plagiarism. *Nature* **369**: 698.

Krstic, S., Duma, A., Janevska, B., Levkov, Z. *et al.* (2002) Diatoms in forensic expertise of drowning: A Macedonian experience. *Forensic Science International* **127**: 198–203.

van Landingham, S. L. (1967–1971, 1975, 1978) *Catalogue of Fossil and Recent Genera and Species of Diatoms and their Synonyms: Parts I–VIII*. J. Cramer Verlang, Germany.

Langer, A., Mackler, A., Rubin, I. *et al.* (1971) Inorganic particles in cigars and cigar smoke. *Science* **174**: 585–7.

Lewin, R. (1989) The case of the "misplaced" fossils. *Science* **244**: 277–9.

Lucas, S. G. (2005) *Dinosaurs: The textbook*, 5th edn. McGraw-Hill, New York.

Marks, M. K. (1995) William M. Bass and the development of forensic anthropology in Tennessee. *Journal of Forensic Sciences* **40**(5): 741–50.

Martin, A. J. (2006) *Introduction to the Study of Dinosaurs*. Blackwell Science, Malden, MA.

McBirney, A. R. (2004) *Faulty Geology: Frauds, Hoaxes and Delusions*. Bostok Press, Eugene, OR.

McKinney, F. (1991) *Exercises in Invertebrate Paleontology*. Blackwell Scientific, Oxford.

Metzger, C. A., Terry Jr., D. O., and Grandstaff, D. E. (2004) Effect of paleosol formation on rare earth elemental signatures in fossil bone. *Geology* **32**(6): 497–500.

Mildenhall, D. C. (2006) *Hypericum* pollen determines the presence of burglars at the scene of a crime: An example of forensic palynology. *Forensic Science International* **163**: 231–35.

Oliwenstein, L. (1990) Fossil fraud. *Discover* **11**(1): 43.

Owsley, D. W. and R. W. Mann (1990) Medicolegal case involving a bear paw. *Journal of the American Podiatric Medical Association* **80**(11): 623–5.

Pollanen, M. S. (1998) *Forensic Diatomology and Drowning*. Elsevier Science: Amsterdam, The Netherlands. [Warning, this book contains graphic images of drowning victims and bodies in various stages of dissection.]

Raymond, P. E. (1905) The Trilobites of the Chazy Limestone. *Annals of the Carnegie Museum* **3**(2).

Raymond, P. E. (1920) *The Appendages, Anatomy, and Relationships of Trilobites*. Connecticut Academy of Arts and Sciences.

Rigby, J. K. (1985) Phylum Porifera. In: R. S. Boardman, A. H. Cheethan, and A. J. Rowell (eds), *Fossil Invertebrates*. Blackwell Scientific, Oxford.

Robinson, R. A. and Kaesler, R. L. (1985) Phylum Arthropoda. In: R. S. Boardman, A. H. Cheethan, and A. J. Rowell (eds), *Fossil Invertebrates*. Blackwell Scientific, Oxford.

Round, F. E., Crawford, R. M., and Mann, D. G. (1990) *The Diatoms: Biology and morphology of the genera*. Cambridge University Press.

Scagel, R. F. R. J., Bandoni, G. E., Rouse, W. E. *et al.* (1965) *An Evolutionary Survey of the Plant Kingdom*. Blackie, London.

Searfoss, G. (1995) *Skulls and Bones: A guide to the skeletal structures and behavior of North American mammals*. Stackpole Books, Mechanicsburg, PA.

Schuchert, C. (1924) *Textbook of Geology: Part II. Historical Geology*. John Wiley & Sons Ltd, Chichester.

Seilacher, A. (2007) *Trace Fossil Analysis*. Springer, Berlin/Heidelberg/New York.

Spaulding, S. A., Lubinski, D. J., and Potapova, M. (2010) Diatoms of the United States, http://westerndiatoms.colorado.edu, accessed 24.11.11.

Sweet, W. C. (1988) *The Conodonta: Morphology, taxonomy, paleoecology and environmental history of a long extinct animal phylum*. Oxford Monographs in Geology and Geophysics 10. Oxford University Press, New York.

Talent, J. A. (1989a) The case of the peripatetic fossils. *Nature* **338**: 613–615.

Talent, J. A. (1990) The case of the peripatetic fossils: Part 5. *Nature* **343**: 405–6.

Talent, J. A. (1995) Chaos with conodonts and other fossil biota: V. J. Gupta's career in academic fraud: Liographies and a short biography. *Courier Forschungsintstitut Senckenberg* **182**: 523–51.

Taylor, F. J. R. (ed.) (1987) *The Biology of Dinoflagellates*. Botanical Monographs No. 21. Blackwell Scientific, Oxford.

Ubelaker, D. H. (1999) *Human Skeletal Remains: Excavation, analysis, interpretation, 3rd edn. Manuals on Archeology: Vol. 2*. Taraxacum, Washington.

Vinyard, W. C. (1979) *Diatoms of North America*. Mad River Press, Inc., Eureka, CA.

Walcott, C. D. (1910) Smithsonian Miscellaneous Collections, 1910.

van der Werff, A. and Huls, H. (1957–1963) *Diatomeeeriflora van Nederland* (in seven parts). Published privately.

Winter, A. and Siesser, W. G. (eds) (2006) *Coccolithophores*. Cambridge University Press.

Woods, H. (1909) Trilobites. In: S. F. Harmer and A. E. Shipley (eds), *The Cambridge Natural History*. Macmillan, London.

Yasuhara, M., Okahashi, H., and Cronin, T. M. (2009) Taxonomy of the Quaternary deep-sea ostracods from the western North Atlantic Ocean. *Palaeontology* **52**(4): 879–931, http://digitalcommons.unl.edu/usgssta pub/242, accessed 12.11.11.

Yen, L. W. and Jayaprakash, P. T. (2007) Prevalence of diatom frustules in non-vegetarian foodstuffs and its implications in interpreting identification of diatom frustules in drowning cases. *Forensic Science International* **170**: 1–7.

von Zittel, K. A. (1900) *Textbook of Palaeontology: Vol. 1*, (trans. C. R. Eastman). Macmillan, London.

Chapter 10
Geology and People: Forensic Anthropology and Forensic Archeology

The term *anthropology* means the study of humans and encompasses cultural, linguistic, and sociological aspects as well as physical structure and human biology. Forensic anthropology is basically the application of the methods of physical anthropology to legal questions, such as the identification of, and detection of trauma on, human remains. There are many sub-disciplines in the field of anthropology, but the one most applicable here is *human osteology*, the study of the human skeleton. The term *archeology* means the study of ancient things. Forensic archeology is the application of archeological theory and methods to crime scene excavation and the proper recovery of human remains. As you can tell, there is some overlap in these fields. In the United Kingdom they are treated as two separate scientific disciplines, while in the United States archeology is often treated as a sub-discipline of anthropology. Whatever the categorization, both fields bring important tools to the table and have some important areas of overlap with forensic geoscience.

Forensic anthropologists typically assist with the identification of human remains. By examining bones, they can provide information such as sex, approximate age at death, ancestry, and general physical appearance, such as height and weight. They are also trained to identify trauma on skeletal remains and to distinguish between *pre-mortem* (or ante-mortem, which means before death), *peri-mortem* (at or around the time of death), and *postmortem* (after death) modifications. Anthropologists use information about weathering, decomposition, and insect activity to help determine time since death. Some are also trained in the techniques of facial reconstruction using structural features from skulls. For more information on forensic anthropology, see Bass (2005), Byers (2005), and Ubelaker (1999).

An Introduction to Forensic Geoscience, First Edition. Elisa Bergslien.
© 2012 Elisa Bergslien. Published 2012 by Blackwell Publishing Ltd.

Forensic archeologists are usually involved with ground searches, scene surveys, and site mapping. They use their skills with geophysical search techniques to help locate human remains and are specially trained in excavation techniques, the proper recovery and preservation of remains, and artifact collection. They are proficient at site documentation, creating detailed photographic and written records. Their skills are also employed to help work out the history of a crime scene and to determine what important changes may have occurred in the interval between the crime and the investigation of the scene. This is important for the improved recovery of dispersed skeletons and associated materials. Forensic archeologists use their investigative techniques for the reconstruction of crimes and the interpretation of events at a scene. Many forensic archeologists also have special training in field sample collection of soil, pollen, botanical samples, and insects (entomology). Some recommended resources on forensic archeology are Dupras et al. (2006), Hunter and Cox (2005), and Hunter, Roberts, and Martin (1996).

Locating Ground Disturbances

The fields of archeology, anthropology, and geology intersect most clearly in the area of forensic field investigation, such as the search for clandestine graves or mass burials. Anyone who has ever dug a hole in the ground knows that when you try to fill in that hole one of two things will happen: the soil will either have compacted so that when you put it back into the hole you have too little to fill it, or the soil will have bulked out so that you will have soil left over after you have filled the hole. Both of these results in observable topographic features that can make it possible for trained personnel to spot the location of the burial, such as a depression in the ground (called a *primary burial depression*), an area of mounded earth over the burial site, or areas where excess soil from the hole has been piled up or distributed (Figure 10.1a).

If you are sneaky, you may decide to try to level out your hole by either adding or removing dirt and making sure that the ground is approximately flat. This works temporarily to level the soil out, but over time the soil will start to regain its original texture, once again resulting in an observable surface feature, usually in the form of a depression. Compacting excess soil will also only work in the short term. In the long term the location of the ground disturbance usually will become obvious due to the lack of plant growth in the over-compacted soil (Figure 10.1b, e). Secondary burial depressions can also form when the chest cavity of a body collapses, and in any case, the disturbed soil will have different characteristics than the surrounding undisturbed soil.

Digging a hole mixes the soil column, affecting the physical properties of the soil and potentially affecting plant-growth long term. Loose soil and the addition of nutrients, like a body, generally tend to promote plant growth, causing the vegetation over the gravesite to grow taller and for there to be greater species diversity over the grave (Figure 10.1c). On the other hand, if the body was wrapped in plastic or covered in rocks, this tends to suppress vegetation so that the spot over the grave is barren or has stressed vegetation (Figure 10.1d). Whatever the tactic used, it is almost impossible for the original soil to be replaced without causing significant disturbance and without creating indicators of the location of the hole. However, the strength and longevity of these indicators is dependent on a number of factors, such as soil type and climate. Generally speaking, the longer a grave is left undisturbed, the harder it will become to locate.

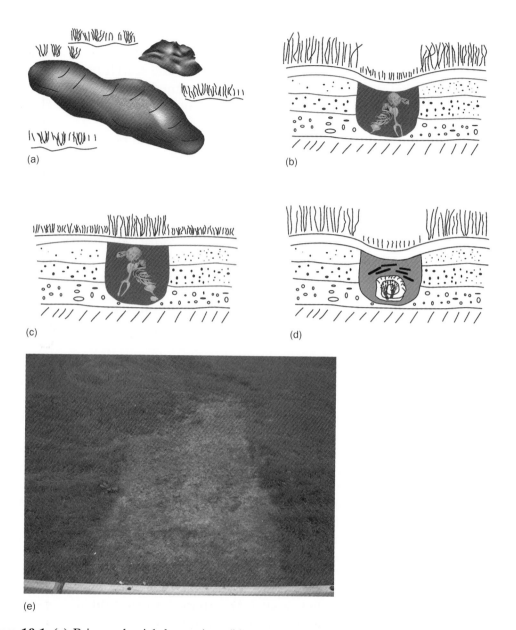

Figure 10.1 (a) Primary burial depression, (b) over-compacted soil reducing plant growth, (c) loose soil and nutrients from burial increasing plant growth over site, (d) lack of plant growth over burials sites where the body is either wrapped in impermeable material or covered with slabs of rock, (e) photograph of burial site with stressed vegetation.

The biggest problem with locating clandestine graves is knowing where, exactly, to look.

Search

A *search* is the application and management of systematic procedures, combined with appropriate equipment, to locate a specific target or targets (Anon 2006; Donnelly, 2009). A forensic search cannot be a random exploration but must be systematic and well designed in order to achieve a designated goal. There are a variety of different forms of search. In the most basic sense, searches can either be searches *of*, such as the search of a crime scene, or searches *for*, such as the search for a victim or weapon. Whatever the case, the goals of the search need to be clearly defined so that the search can be properly designed and equipped from the start. Harrison and Donnelly (2008, 2009) have summarized some of the different types of searches as follows:

In a *search and rescue* the goal is to locate a person or some people who are lost. The typical approach is to make some assumptions about the maximum distances that they might have traveled and in which direction(s), in order to set a search perimeter. These searches are generally large scale and would not usually require the specialized assistance of a forensic geologist.

Scenario-based searches are developed from information gathered by investigators that they use to develop a working hypothesis. This includes data such as behavioral profiles (typical activities), geographical profiling (many offenders work a particular territory or prefer a specific type of terrain), "last sighting" reports, and records of movement. The search would be fact based and generally have very specific objectives. Geologists can often be of use in such searches because they can provide information about terrain types and regional or local topographic features that can help narrow search parameters.

Feature-based searches are generally developed purely from the study of geographical features such as distance from crime scene, distance from local roads and/or houses, and area lines-of-sight. In general, though certainly not always, murderers do not want to drive far with a body in the car, they do not want to carry the body far to dispose of it, and they definitely do not want to be seen. Perpetrators will generally try to dispose of bodies out of the line-of-site of houses and roads, and most of the time, though not always, perpetrators will carry bodies downslope to dispose of them rather than upslope. Plus, bodies are heavy and awkward to transport, so most of the time they will be found near a point of access. Geologic factors can play an important role in these kinds of searches. The thickness and texture of regional sediment, location of terrain markers, and regional topography should all be considered when mounting a search under these conditions. The general diggability of the soil and depth to water should also be considered. It is harder to hide a body in soil that is too shallow, too hard to dig in, or in places where the water table is really close to the surface so that any holes dug will tend to collapse. Chapter 5 of Ruffell and McKinley (2008) has a good discussion of the relationship between geography and crime.

Intelligence-led searches are developed from specific information gathered by police investigators, for example from a confession or tips received in the course of an investigation. Technically speaking, all searches need to be based on some kind of intelligence, but in this case the search area would be defined based on reliable intelligence developed during the investigation of a crime. A geomorphologist would be well suited to assist in such searches, determining the changes that have occurred in an area since an alleged burial took place.

The search type and background are used to determine a logical place to start the search and allow a comprehensive search plan to be developed. Usually, the first step is to collect and analyze as much information about the search area as possible through the use of topographic, geologic, and geographic maps, aerial photographs, and historical information. This information is vital to developing a viable search plan and to determining the boundaries of the search. Older maps will often have information that has become lost or hidden with the passage of time, such as the location of mineshafts, abandoned dirt roads, building foundations, or garbage dumps.

Aerial and ground level photographs taken from different directions and at different times of day can also prove quite useful in looking for topographic features in a landscape. This is especially true when the sun is low in the sky, casting long shadows along the ground, an effect that can sometimes be locally mimicked using the high beams on a car. Aerial photographs taken during a period of drought are also often very useful in locating variations in soil conditions, making disturbed earth stand out prominently. Donnelly (2009, 2010) advocates using all of this information to develop a conceptual geologic model for the search plan that narrows the area down to defined units and determines which methods and equipment will be employed in the search. Conventional and innovative geologic techniques and methods began to be applied to police searches to help locate burials in the United Kingdom in the mid-1990s (Donnelly 2002, 2003, 2011).

The production of a *Red–Amber–Green (RAG) map* for the search area may assist with the prioritization of search areas (Figure 10.2). A RAG map is a color-coded

Key to Symbols

Green ▨ Not diggable without mechanical scraping or use of mechanical excavators

Amber ☐ Diggable with a spade or shovel with moderate ease

Red ▦ Diggable with ease with hand held tools

Figure 10.2 Red–Amber–Green diggability survey map based on the estimated engineering properties of the superficial deposits. Red indicates areas of high diggability, which should be investigated first, and green indicates areas of low diggability, which are highly unlikely to be of interest. Classification was verified by *in situ* field observations and the performance of diggability tests.

Source: Donnelly and Harrison, 2011.

map that delineates the boundaries of each unit in the search area and identifies high/medium/low-priority (usually designated red/yellow/green respectively) search areas (Donnelly and Harrison, 2011). Effective communication is vital to the successful employment of RAG maps, ensuring that everyone is interpreting the maps correctly (Donnelly and Harrison, 2011). For most police and military forces the color red is used to indicate the highest priority search areas, while for many field geologists and environmental scientists red tends to be used to indicate dangerous areas (falling rock zones) or hot zones (i.e. do not enter without appropriate personnel protective gear).

Geologists, more specifically *geomorphologists* (scientists who study landforms and the Earth's surface processes), can potentially be very helpful during this phase of search design because of their expertise in differentiating natural topographic features from anthropogenic ones. Because different surface processes result in different thicknesses and textures of accumulated sediment, they may be able to point out areas where it could be possible to conceal a body, or hide materials, in a particular landscape. Their insight can prove significant in prioritizing search areas. Geomorphologists are also adept at understanding how landscapes change over time. For example, *fluvial systems* (rivers and streams) are often employed for the disposal of murder victims, weapons, and other items. Such systems evolve through specific processes over time and geomorphologists would best be able to determine potential points of origin for things found downstream, or to help locate where objects might have ended up after being dumped. Ruffell and McKinley (2008) is strongly recommended for a detailed discussion of the role of geomorphology in forensic investigation.

Usually the next step pre-search is *field observation* and/or a *reconnaissance visit* (a walk-over survey). Searchers visually survey an area looking for physical evidence, like bones or artifacts, or for surface features such as mounds of earth, depressions, soil scars, seeps, color changes in the soil, and variations in vegetation. Different techniques are used for searches depending on the ultimate goal of the search. Initial overviews and landscape surveys are usually best handled by a small number of trained investigators, but investigation of large areas may sometimes still call for significant manpower. The pre-search observations are used to *ground truth* (check the validity of remotely collected/developed data with real physical features) the conceptual search model and verify the accuracy of the search maps that will be employed. It can also be used to field test any equipment that searchers intend to use, such as geophysical instruments. An idealized search strategy and methodology, which may be conducted in three distinct phases, known as the pre-search, search, and post-search phases, is proposed by Donnelly (2009, 2010).

The actual search would usually begin with securing the search area and the establishment of a search grid, search lines, and referenced benchmarks. A systematic, non-invasive visual search would then typically begin. Searchers are assigned to a particular search unit(s) and briefed on what to look for and the type of search pattern to use. The most common search patterns are line (or strip) searches, grid searches, and spiral (or circle) searches (Figure 10.3).

To perform a *line search* (Figure 10.3a), searchers literally start out in a straight line and are positioned closely enough that their fields of view overlap. A line leader determines pacing and everyone moves at roughly the same rate while scanning the ground. No one should get significantly out of pace with the rest of the line, nor should the line "flex" so that portions of the ground are not covered. Each member

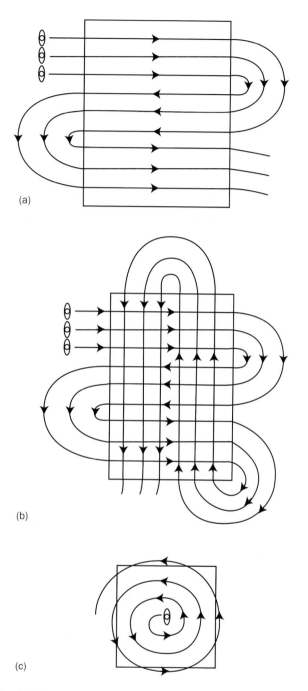

(a)

(b)

(c)

Figure 10.3 Patterns for (a) line (or strip) searches, (b) grid searches, and (c) spiral (or circle) searches.
Source: Darrel Kassahn and Elisa Bergslien.

of the line usually carries flags or some other kind of marker that they can place when they spot anything potentially significant. Marked areas are then inspected in more detail after the line has passed. *Grid searches* (Figure 10.3b) work basically the same way that line searches do except that upon completion of a line the search team will search the same area again working perpendicular to the original direction. This way the ground is viewed from at least two different perspectives. Grid searches are often employed in rough or steeply sloped terrain.

Spiral searches (Figure 10.3c) are best suited to small areas and/or when there are only a small number of searchers. There are two basic ways that a spiral search can be conducted. In the first case, the search starts at a central point and works outward in increasingly wide concentric circles. For example, if bones or some other piece of evidence is found in a central location, it can be helpful to work outward from that point to find additional bones, or other items. Some authorities also suggest that the most efficient one-person searches are conducted by starting at an outer perimeter and moving inward in decreasing concentric circles to ensure that the entire area is effectively covered. Clearly, the search methodology employed will be determined by terrain, available personnel, and a variety of other factors.

Geophysical Tools

It is usually after the initial field searches have been conducted, and areas of interest have been identified, that a geophysicist or expert in the use of geophysical tools would be brought in to help. Geophysical survey tools can be quite useful for forensic investigation because they are non-invasive and can help narrow down or prioritize areas for further, intrusive exploration. They are also much faster to employ than an excavation. One area of significant overlap between the worlds of forensic archeology and forensic geoscience lies in the use of geophysical tools.

Geophysicists (people who study the physics of the Earth) use a variety of techniques to measure variations in the Earth's properties, such as changes in magnetic field strength or the local gravitational field. When a hole is excavated, and a body or item secreted in the ground, whatever has been introduced into the soil, and perhaps even the fill material itself, will have different geophysical properties than the surrounding soil. The goal of a forensic geophysical investigation is to locate subsurface areas of such differences, called *anomalies*. Geophysical tools are not magic wands, and each of them is subject to some limitations, but they can be extremely helpful in determining where it would be most profitable to perform intrusive investigation, such as excavation.

Magnetometry

If you place a clear cover over a strong bar magnet and then sprinkle it with iron filings, the filings will align themselves around the magnet in a distinctive pattern (Figure 10.4a). That curved pattern outlines the *magnetic lines of force* created by the magnetic field of the magnet. There are a couple of distinct components to the arrangement. First, the iron filings, which become very small magnets themselves, are aligned parallel to the bar's center and arch around at increasingly steep angles to converge around each end of the magnet. This is known as *inclination*, or the angle of the magnetic field above or below horizontal, (Figure 10.4b). Second, the spacing of the lines is indicative of the *intensity* of the magnetic field. The closer together the lines get, the stronger the field gets. The areas of convergence are known as *magnetic poles*. Because the magnet has two areas of convergence, a north (+) pole and a south (−) pole, it is called a *dipole*.

The Earth's magnetic field is generated by the flow of its molten iron outer core, but it can be roughly modeled as a simple bar magnet type dipole. The Earth is

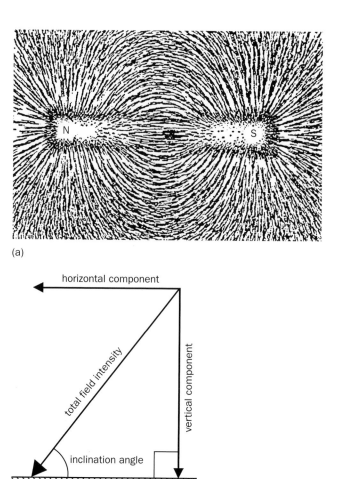

Figure 10.4 (a) Magnetic lines of force created by the magnetic field of the magnet, (b) inclination of the magnetic field.

surrounded by magnetic lines of force that form a complex three-dimensional shape, the poles of which are offset from the axial poles of the Earth by approximately 11° (Figure 10.5). The strength of the field changes both with latitude, getting stronger toward the poles, and temporally, slowly reflecting changes in the thermal currents of the Earth's outer core (known as *secular variation*). Magnetic variations in the near surface of the Earth can be measured using *magnetometry*. More information and charts of the Earth's magnetic field are available from the United States Geological Survey's National Magnetism Program (http://geomag.usgs.gov/) and from the British Geological Survey (http://www.geomag.bgs.ac.uk/earthmag.html).

For forensic work, the important thing to know about the Earth's magnetic field is that regionally it provides a steady background against which much smaller changes can be measured. The overall intensity of the Earth's magnetic field is quite small, ranging from approximately 70,000 nT (nanotesla; the tesla is the SI unit of measurement for magnetic field strength) at the magnetic poles and

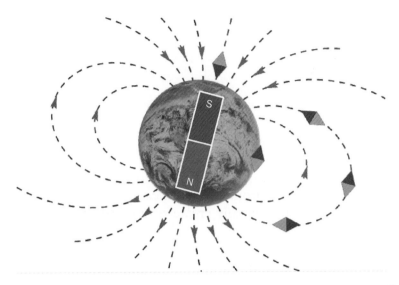

Figure 10.5 The Earth's magnetic field. The Earth's north magnetic pole is actually a south pole, attracting the north poles on compass needles.

dropping down to around 30,000 nT at the magnetic equator. For comparison, the magnetic field strength of a typical refrigerator magnet is ~10 mT (i.e. 1×10^7 nT or 10,000 times stronger than the Earth's field). In the United States, the total intensity ranges from around 58,000 nT near the Canadian border to around 46,000 nT in southern Florida. Field strength in the United Kingdom varies from around 50,000 nT in the north to 48,000 nT in the south. Against this background, it is possible to measure very small variations in the magnetic field (usually in the order of 1 nT) using instruments called *magnetometers*.

There are basically two phenomena that are of interest for forensic investigation. The first results from materials that react to the Earth's magnetic field becoming magnetized themselves. Such materials create secondary magnetic fields of their own and will show up as localized anomalies within an otherwise uniform background. The ease with which a material can become magnetized is referred to as *magnetic susceptibility*; the higher the susceptibility, the more easily the material becomes magnetized. The second has to do with how magnetically susceptible sediments become orientated or disarranged in the Earth's magnetic field.

If the Earth were composed of a uniform material, like a bar magnet, then the lines of force around the planet would be uniformly parallel, like the pattern the iron filings make around a bar magnet. However, because the Earth is composed of a variety of different materials, which have differing magnetic susceptibilities, the lines of force around the Earth get distorted. These distortions are called anomalies. Anomalies can be both negative (a weakening of the local field) and positive (a strengthening of the local field).

Materials are divided into classifications based on their magnetic susceptibility. *Diamagnetic materials* have low-value negative susceptibilities. This means that magnetization develops in the opposite direction to the applied field and that the resulting strength of the field is very small (10^{-6} SI). Mineral examples include quartz, calcite, and feldspar. *Paramagnetic materials* have low-positive susceptibilities, i.e. weak magnetization develops in the same direction as the applied field ($10^{-4} - 10^{-5}$ SI). Paramagnetic minerals include olivine, pyroxene, amphibole, and many clay minerals. Neither diamagnetic nor paramagnetic materials create anomalies of significance; instead, they form part of the background signature of an area.

Materials that react strongly to magnetic fields are *ferromagnetic*. *True ferromagnetic* materials have high magnetic susceptibilities, strong magnetic properties, and can become permanently magnetized if exposed to a strong enough field. Examples include pure metals like iron, nickel, and cobalt, none of which occurs naturally on Earth, though they are found in meteorites. However, anthropogenic objects made of iron, steel, and cast iron are common and fall into this category. Such materials generally create strong measurable anomalies (Figure 10.6). Also important are *ferrimagnetic materials*, which have relatively high susceptibility and strong magnetic properties. The atomic magnetic moments of these materials are unequal, resulting in net spontaneous magnetism. Minerals such as magnetite (i.e. lodestone), maghemite, titanomagnetite, ilmenite, and pyrrhotite fall into this category and play an important role in the second phenomenon mentioned concerning the orientation of sediment particles. Plus, there is a final classification, *antiferromagnetism*, resulting from a situation in which the atomic

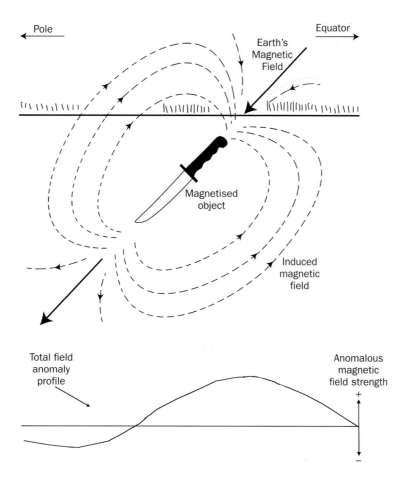

Figure 10.6 Strong detectable anomaly caused by burial of ferromagnetic object.
Source: Darrel Kassahn and Elisa Bergslien.

properties of a material cause the magnetic moment to cancel out (i.e. to be zero). As a result, these materials have weak, positive susceptibilities due to the random motion of electrons. Some iron oxides, like some variations of hematite and goethite, fall into this category. These materials also do not create significant anomalies.

The second phenomenon of interest for forensic investigation is caused by the preferential orientation of magnetic minerals in rocks and soil. This is classically demonstrated in geology by the example of molten igneous rock rising along mid-ocean ridges. The iron-rich minerals forming in the molten mass become aligned to the Earth's magnetic field. Once the lava solidifies, that magnetic alignment is permanently frozen in the rock. This type of *thermoremanent magnetism* has helped geologists to determine the placement of the Earth's continents at various points in the geologic past.

Layers of sediment can also have a magnetic signature, though much weaker than that found in iron-rich igneous rocks. As ferrimagnetic mineral clasts settle through a layer of fluid, or are carried by the wind, they will also become aligned to the Earth's magnetic field. The resulting bands of sediment have *natural remanent magnetism* (NRM). In addition, there are biogenic and diagenetic processes by which ferrimagnetic minerals form and that result in *secondary remanent magnetism* (SRM) in sediment layers. SRM is especially prominent in topsoil and weathered sediment profiles.

Topsoil (O and A horizons) formed in temperate, subtropical, tropical, Mediterranean, and steppe climate zones typically shows significantly higher magnetic susceptibility than the underlying subsoil or bedrock does, sometimes by an order of magnitude. The exact reasons for this enhancement are unclear and are currently being debated in the literature, but in general as long as the soil's parent material contained iron and the soil is well drained the resulting topsoil can reasonably be expected to show magnetic enhancement. Well-developed (mature) soil is also more likely to have higher magnetic susceptibility than the topsoil of a poorly developed (immature) soil. Plus, topsoil that receives significant amounts of atmospheric pollution, such as from the burning of fossil fuels, or industrial or manufacturing processes, also typically demonstrates enhanced magnetic susceptibility. Such enhancement is often seen in urban soil and downwind from industrial zones, and has frequently been used to map and monitor pollution on regional scales. Another factor that can cause distinctive enhancement of soil magnetic susceptibility is fire. Fires that reach sufficient temperatures (600–900 °C), such as forest fires, fireplaces, and under ancient kilns, will result in the formation of fine-grained magnetic minerals in soils that have iron available. This also makes features like fired bricks and ancient coal pits stand out in surveys as positive anomalies.

In contrast, constant high-moisture levels in soils will decrease susceptibility, even in soils with available iron. Water-logged (gley) soils that are under reducing conditions have been shown to contain considerably reduced levels of ferrimagnetic minerals and therefore to have very low magnetic susceptibilities. This process is quite rapid, with laboratory and field tests showing that after 1–10 years of substantially water-logged reducing conditions, ferrimagnetic minerals can be completely destroyed. However, it is important to note that short-term moisture levels do not significantly affect magnetic susceptibility. It is possible to employ magnetometry on soils that are wet, as water is diamagnetic and will not affect measurements. The effect being noted here is due to long-term seasonal conditions and suggests that soils which spend a significant part of the year, each year, water-logged, as in bogs or wetlands, should not be expected to show enhanced susceptibility. For a more detailed discussion of the magnetic enhancement of soil, see Blundell *et al.* (2009).

Enhanced magnetic susceptibility in soil is of interest because it becomes possible to locate disturbances in the soil profile if such conditions exist. Areas where the topsoil has been altered by mixing with the subsoil or by reorientation can contrast strongly with surrounding areas of undisturbed soil, showing up in magnetometry surveys as distinctive anomalies. In areas where the topsoil has developed a high magnetic susceptibility, the act of digging a hole will reorient the soil particles so that they are no longer aligned with the Earth's magnetic field, creating a negative anomaly. Mixing of less susceptible subsoil with the topsoil as a hole is filled in will also cause a negative anomaly. However, it is important to remember that this only works in soils and sediments that contain sufficient ferrimagnetic minerals. Glacial tills, gravel, course-sand deposits (where the clasts are too large to become oriented), and soils that lack sufficient ferrimagnetic minerals will not display distinctive anomalies in areas of disturbance.

Where soil conditions are right, this method is excellent for delineating large-scale earthworks, and for this reason magnetometry surveys have long been a staple in archeological investigations. Magnetometry has seen much less use for forensic investigation. Part of this probably stems from a poor understanding of the constraints of the method. Buck (2003) is an oft-cited forensic study in which magnetometry was employed at three sites with poor results. However, two of the sites were clearly not good contenders for the technique, due to the presence of large quantities of ferrous material (metal flower holders at each of the graves in one case and "large quantities of metal trash" and power lines in the second case). The third case mentioned (the CRM trench) has so little background information provided it is hard to assess.

A magnetometer was used, along with ground-penetrating radar (GPR), on a recently refilled trench that was 2.5 m deep by 1.5 m wide and long enough that a 10 × 10 m grid could be established over "part of the trench." The author refers to the test as "an ideal scenario," points out there were no nearby sources of interference, and indicates surprise that neither method employed was able to pick out the location of the trench. But no information is given about soil type or condition. Without knowing about soil type (was it clay-rich loam or gravelly sand?) and condition (was it water-logged or immature and lacking well developed topsoil?), it is hard to know whether either method could have reasonably been expected to work. The other issue with this study concerns the configuration of the magnetometer employed. France *et al.* (1992), report that they did detect anomalies over areas of excavation, but again no details about soil type or character are included. In future, it is to be hoped that there will be more controlled and detailed analysis of the utility of this method and that it will be performed under appropriate conditions.

Human bodies in and of themselves have very low magnetic susceptibility and cannot be directly detected with magnetometry. However, many associated objects, such as knives, barrels, and belt buckles, are ferromagnetic and result in strong detectable anomalies, regardless of soil conditions (Figure 10.7a). If no such objects were buried in the area of interest, variations in the magnetic susceptibility of the soil can be used to locate disturbances. It has also been suggested that mass burials, with large numbers of bodies decaying, may result in locally anaerobic conditions and the reduction of ferrimagnetic minerals in and around the gravesite.

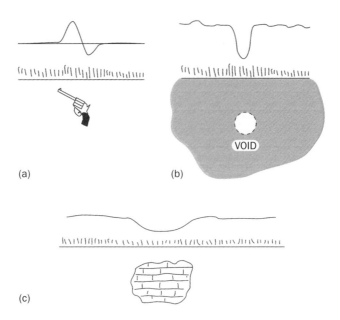

(a)

(b)

(c)

Figure 10.7 Magnetic anomalies created by (a) a buried ferromagnesian object, (b) a void space, and (c) a block of limestone.
Source: Darrel Kassahn and Elisa Bergslien.

The upshot could be a detectable negative anomaly, but this has not yet been tested. Voids in the Earth, such as tunnels and caves, can also stand out in surveys as distinctive negative anomalies (Figure 10.7b). Plus, negative anomalies can be created by the placement of low-susceptibility foundation materials like limestone or quartz sandstone blocks into soil with high susceptibility (Figure 10.7c).

Also of interest for forensic work, magnetometer surveys can be carried out over bodies of water. Davenport (2001) reports a case in which an informant provided police with the particulars of a murder where the victim had been placed into the truck of a vehicle that was then pushed into the Missouri River. Based on the make, year, and model of the vehicle, geophysicists calculated the most likely size of the magnetic anomaly to be expected due to the engine block. Then, using an aluminum boat (aluminum is paramagnetic), a marine magnetometer was towed along the river in conjunction with a GPS to provide positioning. Five anomalies of roughly the correct size were located. A dive team was given a prioritized map showing the location of each of the anomalies. They located a vehicle at the first anomaly and bought the license plate to the surface. It turned out to be the plate from the vehicle of interest and when the vehicle was recovered a body was found in the trunk. This case resulted in a first-degree homicide conviction. Marine magnetometers are also routinely used to locate murder weapons like guns and knives that have been dumped into bodies of water.

There are a variety of different types of magnetometers (flux-gate magnetometers, proton-precession magnetometers, alkali-vapor magnetometers), each of which has pros and cons. Though many authors recommend one instrument or another, the ultimate choice of instrument type should be made by a person familiar with the technology and with reference to the characteristics of the particular site to be investigated. For more detailed information on the physics behind magnetometry and its application, see Burger, Sheehan, and Jones (2006).

Figure 10.8 Scintrex SM4G Cesium magnetometer showing vertical gradiometer configuration. *Source:* Photo courtesy English Heritage.

Regardless of the instrument type chosen, a *vertical gradiometer configuration* should be employed (Figure 10.8). This necessitates the use of two sensors separated by a fixed distance, usually between 0.5 and 1 m, in a single vertical tube. The lower sensor should generally be close to the ground (between 0.3 and 0.5 m), though in some cases higher elevations would be used depending on the depth of the feature(s) of interest (Davenport, 2001; Silliman, Farnsworth, and Lightfoot, 2000). The closer the sensor is to the ground, the more attuned the instrument is to small-scale variations and near surface anomalies (<1 m depth). If the features of interest are deeper, the lower sensor should be raised. The upper sensor delineates broad-scale features and is basically used as a filter for corrections. The adjusted difference between the two sensors is optimized to identify near-surface anomalies while screening out diurnal variations in the Earth's magnetic field and any large-scale geologic anomalies. This technique also helps to screen out *cultural noise* (i.e. building, bridges, and other metal structures). The best instruments will allow the user to adjust sensor base height and separation. Optimally, an area of interest will be surveyed at least twice with the lower sensor at different heights. According to Cheetham (2005), another of the reasons that Buck (2003) had unsatisfactory results with the magnetometer is that a single sensor alkali-vapor instrument was used which allowed natural and cultural variations to mask near-surface features.

It should be noted that geological surveys are almost always going to use sensors configured to measure deeper, large-scale anomalies, as in the search for ore deposits or to delineate geologic formations. For ground-level geological work, the lower sensor will be raised significantly above ground surface, 2 m or more, and the vertical separation between sensors, if a vertical gradiometer configuration is used at all, will be at least 1 m. The standard sensor arrangement employed for geophysical surveys should not be employed for forensic investigation, as subtle near-surface anomalies will not be detected. In fact, geological surveys are most efficiently carried out using aircraft (these surveys are known as *aeromagnetics*). Prior to conducting a shallow-surface investigation, it can sometimes be worthwhile to see whether aeromagnetic maps are available for your search area.

For forensic investigation, the area to be explored should be broken into square or rectangular blocks and then methodically surveyed by conducting a series of equally spaced parallel traverses. The separation between traverses should generally be around 1 m, though to identify smaller features, like post-holes, the traverses should be at most 0.5 m apart (Figure 10.9). Measurements need to be taken at

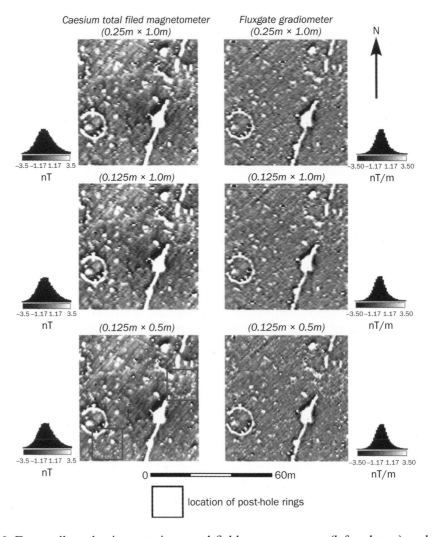

Figure 10.9 Data collected using a cesium total field magnetometer (left column) and a flux-gate magnetometer (right column) at different sampling intervals, illustrating the effect of increasing sampling density.
Source: Courtesy English Heritage.

regular intervals along each traverse, no more than 0.25 m apart, or the instrument set for continuous reading. Working with a grid is essential to acquiring usable data. Simply sweeping a magnetometer over an area and observing the result (i.e. using the instrument like a metal detector) will only be of profit if the target being sought is ferrous and likely to create a very large anomaly. This approach will not be successful for detecting weak or subtle anomalies and is, in general, strongly discouraged.

Magnetometry will not work effectively in areas with high levels of magnetic "noise." Electrical power lines, pipes, metal fences, steel beams, and the like will all mask the subtle magnetic anomalies usually sought in forensic investigations. Using a vertical gradiometer configuration will help screen noise out, but when such sources are within ~20–30 m of the instrument there may be too much noise for a survey to be of use. The person holding the instrument also needs to be stripped of potential contaminants. Things like belt buckles, keys, metal zippers, coins, metal eyeglasses, pens and clipboards will interfere with readings. It is also worth noting that each individual person will have a slightly different effect on the instrument as well, so the same operator should be responsible during a grid survey. Local geologic features can also have a pronounced influence. Rock units with strong remanent magnetism or ore bodies will profoundly affect regional magnetic field strength.

If a vertical gradiometer configuration is not employed, survey data must be corrected to compensate for short-term variations in the magnetic field (called a *diurnal change*). Over the course of a day, due to variations in solar wind and solar weather, magnetic field intensity can fluctuate by tens of nanotesla. The most effective way of screening out such fluctuations is to return to a base station every hour or so and correct for any drift. A more accurate alternative is to have a continuously reading magnetometer set up at a base station. The field readings are adjusted by subtracting the base readings from the raw data.

It is important to realize that anomaly size and target size are not related. A small ferromagnetic object, like a powerful magnet, can create an anomaly that is the same size or larger than that created by a large ferrimagnetic object, say the engine block of a car. Also, while some anomalies can be detected in the field and investigated immediately, post processing of data is usually required to detect subtle anomalies. You might remember that the inclination of the magnetic field of the Earth changes with position. When searching for ferromagnetic objects, like a weapon, the anomaly profile sought is actually a point of inflection, where the magnetic field drops and then peaks (Figure 10.10). This profile is created because of the inclination of the Earth's magnetic field. In most of North America, Europe, and Australia, the inclination of the magnetic field is steep enough that the magnetic high is nearly centered over the magnetic object. When working nearer the equator, the magnetic low will be centered over the object.

During the initial grid survey, if an anomaly is detected, the point of maximum (or minimum) intensity or the point of maximum inflection (the transition from local maximum to local minimum) should be marked and the initial survey completed. Additional surveys should then be conducted over the marked area(s). If the object creating the anomaly is small, say a gun or a knife, chances are that it will be offset from the primary survey line. The general approach to isolating the position of the anomaly is to narrow the grid spacing and perform a second survey orientated perpendicular to the initial survey. Repeat as needed, alternating orientation, to narrow down the location of maximum, or minimum, intensity. A

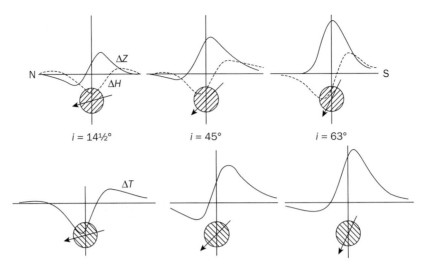

Figure 10.10 Vertical (ΔZ), horizontal (ΔH), and total field (ΔT) magnetic anomaly profiles over a uniformly magnetized sphere in regions of different geomagnetic inclination, with declination of the magnetic field at zero.
Source: Sharma, 1997, modified from Haalck, 1953.

similar approach may also be warranted after post processing of survey data to narrow down the location of more subtle anomalies.

A related instrumental approach that may also be of interest is a topsoil magnetic susceptibility survey. Magnetic susceptibility meters are placed directly against the ground and generally determine the properties of the soil to a depth of 2–10 cm. This approach is commonly used for delineating the impact of air pollution in urban or industrialized areas and for locating burn areas.

Electrical Resistivity (ER)

Another geophysical technique that shows promise for forensic investigation is *electrical resistivity* (ER). Before this will make sense, we need to introduce some terminology. Voltage is a measurement of potential, kind of like gravitational potential. With gravitational potential, the higher an object is above the ground, the more energy it will have when it is released and hits the ground (within limits). *Voltage* (V) is technically the potential of a charge (like an electron) to move through space. Since that is a rather odd concept, it is probably easier to think of voltage as a difference between two points. The greater the difference between the two points, the more current (within limits) will flow between those two points. This difference is measured in *volts* (also V). Another analogy that is often used to describe voltage is that it is the pressure pushing the charge along. The greater the difference in "pressure," the more current can be pushed along a wire.

Current (I) is the number of charges that are flowing along a wire between two points. Current is measured in *amps* (A). The third fundamental concept used to describe electricity is *resistance* (R): how much the material that the current is flowing through resists that flow. It works kind of like friction. Resistance is measured in *ohms* (Ω). If you picture voltage as a large, high-pressure water tank, with the water as the current, then resistance is the size of the pipe coming out of the tank. The smaller the pipe (i.e. the greater the resistance), the less water (i.e. current) will flow. Voltage, current, and resistance are related by *Ohm's law.*

$$R = V/I \text{ (or } \Omega = V/A)$$ (10.1)

Different materials have inherently differing resistance to the flow of current, a property that is usually referred to as *resistivity* (ρ) and reported in ohm-meters (Ωm).

The goal of using ER is to measure the electrical resistance of materials in the subsurface. As you might guess, rock and soil are generally quite poor at conducting electricity (i.e. have high resistivity). But electricity flows well through water that contains dissolved salts, like surface water and groundwater. This means that resistance of the ground to the flow of electricity depends predominantly on water content, thus an ER survey is really predominantly measuring variations in saturation (Table 10.1). The range of resistivities found in the Earth's surface materials is huge, from lows of 10^{-5} to highs of $10^{16}\Omega$m (Table 10.2). Wet soils have much lower resistance than dry soils do, and bedrock will have very high resistance. Clay can hold much more water than sand, so clay will typically have lower resistance than sand. In areas where the soil has been disturbed, the porosity of the fill will often be lower (or sometimes higher) than the surrounding material. This in turn means the disturbed soil will retain more (or less) moisture, creating a detectable disturbance.

To measure electrical resistance, a current must be induced by applying a voltage to the ground. This is accomplished by inserting two electrodes (*current electrodes*) into the earth and applying steady direct current (DC, like a battery) or preferably low-frequency alternating current (AC). In a wire, current has a restricted path but when you apply a voltage difference from the ground's current the current generated spreads out, only coalescing at the entry and exit points where the probes have been inserted into the ground. In a homogeneous material, the current induced in the ground will be distributed horizontally and vertically in a predictable pattern that is dependent on the separation of the probes

Table 10.1 Moisture content versus resistivity.

Moisture % by weight	Typical Resistivity (Ωm)		
	Sandy loam	Topsoil	Clean quartz sand
0	>10,000,000	>10,000,000	–
2.5	1500	2500	3,000,000
5	430	1650	50,000
10	185	530	2100
15	105	190	630
20	63	120	290
30	42	64	–
Saturated	30	–	–

Source: Modified from Gill, 2008.

Table 10.2 Typical ranges of electrical resistivities of Earth materials.

Material	Typical Resistivity (Ωm)
Pyrite ores	0.001–100
Magnetite	0.01–1000
Graphite ore	0.01–10
Salt/sea water	0.1–2
Graphic Schist	0.1–500
Wet to moist clay	1–100
Weathered or fractured bedrock	1–low 100s
Fresh water	1–300
Shale and Sandstone	5–1000
Wet to moist silt and sand	10–1000
Glacial tills	50–5000
Dry sand and gravel	100–10,000
Limestone, Dolostone	1000–100,000
Unweathered basalt	50–100,000
Unweathered granite	500–100,000
Ice	10,000–100,000

Source: After Palacky, 1987; Milsom, 2003.

(Figure 10.11). Deviations from the predicted pattern indicate the presence of heterogeneities (Figure 10.12).

You can measure the pattern of the current flowing through the ground by inserting two more electrodes (*potential electrodes*; P) into the ground and using a resistivity meter. Probes must be used in pairs to counteract the influences of poor probe-to-ground contact, called *contact resistance*. If you inserted only one set of probes, the effects of contact resistance would dominate the measurement, masking everything of interest. Instead, two current probes and two potential probes are used and the potential difference measured. If there is significant contact resistance, the current drops, but the resistivity, which is a ratio, stays the same. If the probes are inserted into a completely homogeneous material, the resultant measurement is the true resistivity of the material. In this case, it does not matter how the probes are configured or where they are positioned: the same value is always measured. In reality, the subsurface contains heterogeneities with differing resistivities. Depending on how the probes are arranged and where they are located, the measured resistivity is different. Because different combinations of heterogeneities can create the same measurements, the resultant data collected are actually a measurement of *apparent resistivity*.

The whole point of an ER survey is to measure distortions (anomalies) in the current flow pattern. With careful analysis, the variations in apparent resistivity can be interpreted as particular types of phenomena, such as low-resistivity ore bodies or zones of high saturation in sediments. The key to the success of ER lies in the

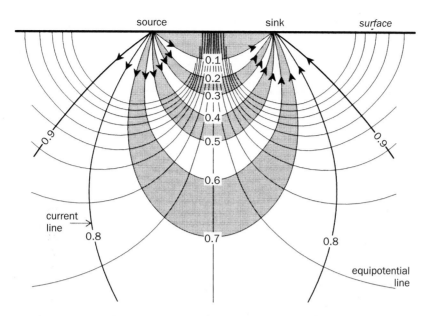

Figure 10.11 Cross-section of idealized current flow between a source and a sink in homogeneous material. The numbers on the current lines indicate the fraction of current flowing above the line.
Source: Lowrie, 1997, after Robinson and Çoruth, 1988, based on Van Nostrand and Cook, 1966.

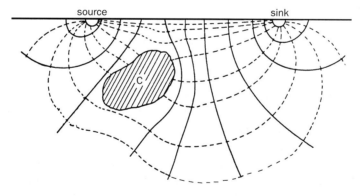

Figure 10.12 Effect of a conductive object (C) on the pattern of current flow (broken lines) and equipotentials (solid lines) between two electrodes.
Source: Modified from Sharma, 1997.

appropriate arrangement and separation of the electrodes, known as the *array configuration*. This in turn determines the scale of the anomalies detected (resolution) and the depth of the features revealed. There are actually a variety of different arrays adapted for different purposes.

The most common configuration for geophysical surveys is the *Wenner array*, four electrodes evenly spaced a distance (*a*) apart with the two current electrodes (C) on the outside and the two potential electrodes (P) on the inside (Figure 10.13a).

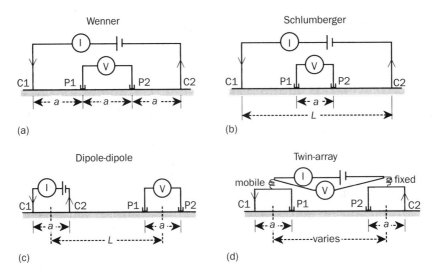

Figure 10.13 Commonly used electrode array configurations. C1 and C2 denote locations of current electrodes, while P1 and P2 denote the locations of the potential electrodes: (a) Wenner array, (b) Schlumberger array, (c) dipole–dipole array, and (d) twin-array configuration with mobile- and fixed-electrode pairs connected by a cable.
Source: (a–c) Modified from Lowrie, 1997.

As electrode separation increases, a greater percentage of the current will penetrate deeper into the subsurface. Theoretically, the depth of penetration by a Wenner array is approximately equal to the probe separation distance divided by five. This makes the Wenner array useful for vertical profiling of the subsurface. By keeping the midpoint of the survey fixed and increasing the electrode separation (a) in a stepwise fashion, the apparent resistivity measured represents a deeper and deeper slice of the subsurface. Other common configurations are the Schlumberger (Figure 10.13b), pole–dipole and dipole–dipole (Figure 10.13c), each of which has advantages and disadvantages.

The standard array used for most archeological exploration, and most likely to be appropriate for forensic investigation, is called the *twin-array configuration* (Figure 10.13d). Strangely, this array appears not to have been used much in the United States and is not even mentioned in most common geophysics textbooks. Here, one current electrode and one potential electrode separated by a short fixed distance and attached to a frame (called the *mobile pair*) are moved along the survey grid while the other current and electrode pair (*fixed pair*) are located at a remote distance away (Figure 10.14). A length of low-resistivity cable connects the twin probe sets.

With this configuration, the separation distance between the electrodes of the mobile pair determines the depth of penetration. The wider the separation, the deeper the anomalies detected, but at the cost of resolution. For most archeological applications, a probe separation of 0.5 m is seen as the best compromise. At this separation, features down to around 1 m beneath the surface should be detected. If the feature of interest is deeper, a wider probe separation should be used, but at a

Figure 10.14 Electrical resistance survey of an archeological site using a twin-probe array. Note the yellow trailing cable connecting the survey probes to the remote probes. *Source:* Photograph courtesy of Tapatio.

separation of much greater than 1 m resolution drops significantly, making it difficult to impossible to isolate smaller features.

The twin-array configuration works because as long as the fixed set of electrodes are positioned far enough away from the mobile set, the measurements appear to be made at the same points relative to their separation. A quick rule of thumb is that the separation between the mobile electrodes and the fixed electrodes should be at least 30 times the space between the probes on the frame. Thus, if a 0.5 m separation is used, the fixed probes must be, at a minimum, 15 m away from the area to be surveyed. If a 1 m separation is used for the mobile probe set, the fixed electrode set must be 30 m away. For the detection of graves or other smaller features, a high sampling density is also needed. Most authorities recommend using a twin probe with 0.5 m separation and a 0.5 × 0.5 m sampling grid, though under some circumstances a different array may be preferred.

ER has been a bit out of favor in the United States due to misunderstandings of its application and because some have called it labor- and time-intensive (e.g. Buck, 2003). The time and labor involved are directly related to the array chosen. In the past the electrodes had to be placed individually by hand. Data acquisition time has been improved dramatically by mounting the electrodes on a bar or mobile cart so that they can be quickly moved into position (Figure 10.15) and by the use of the twin-probe array, which requires only two of the four probes be moved (Figure 10.14). Mobile cart or mobile fixed-frame arrangements of Wenner, twin-probe,

(a)   (b)

Figure 10.15 Geoscan RM15 earth resistance meter (a) with a multi-electrode array and (b) mounted on an MSP40 square array cart with a flux-gate magnetometer also attached. *Source:* (a) Photograph courtesy of Roger Walker, Geoscan Research Ltd, (b) Courtesy of English Heritage.

and other configurations are available. Once a survey grid is in place, measurements are quickly acquired by simply walking the grid and pushing the probes into the ground in the appropriate locations. Resistance measurements are only obtained when all four probes are in the ground, as air has infinite resistance.

There are some significant advantages and disadvantages to using an ER survey. ER surveying will not work in waterlogged, frozen, or very hard soils, or in soils that are very dry, but other than that it has few restrictions on soil or geology. Helpfully, they are not very sensitive to "cultural noise," like metal structures and electrical interference, which makes ER a good choice for urban investigations. Plus, ER instruments are less expensive than other types of geophysical survey equipment and often cited as having the lowest training requirements. It is also worth noting that multi-channel systems with 64 or more electrodes are also available that can be used for high-resolution 2-D and 3-D imaging of the subsurface. This approach, called *electrical resistivity tomography* (ERT), is more sophisticated and requires a significantly greater investment in time and labor but can yield very detailed images of the subsurface.

Pigs in the Ground and an Application of ER

Jervis, Pringle, and Tuckwell (2009) conducted an assessment of how well an ER survey could detect graves using pigs buried at a controlled research site on a campus. Pigs are commonly used as human analogues in forensic research because they are approximately the same size and weight as a person. Plus, they have generally the same metabolism as a human, a similar fat-to-muscle ratio, and are hairless. The soil at the research site was considered a typical urban soil: "red brown, slightly clayey, slightly gravelly sand with occasional brick fragments" and sandstone bedrock was located at a depth of around 2.6 m (Nicholls Colton Geotechnical, 2005). Average rainfall in the area was 75 mm and the average temperature was 9 °C.

The three "graves" under consideration in the project were: a grave with a "naked" pig, an empty grave, and a grave with a pig wrapped in a semi-permeable, woven, polyethylene tarpaulin. Each grave was approximately 1.5 m long, 0.75 m wide, and 0.6 m deep. They were refilled by returning the excavated soil, tramping it down until the pit was filled to ground level, and then replacing the turf on top. Excess soil was disposed of off-site. Resistivity data were collected using a twin-probe array 11 days before the pigs were buried and every 28 days after burial for one year. The mobile electrodes were mounted on a frame separated by 0.5 m and the fixed electrodes were placed approximately 1 m apart and located ~17 m from the survey area. A 0.25 × 0.25 m survey grid was used.

During the entire course of the survey the data showed a distinctive low-resistivity anomaly at the location of the unwrapped pig burial, clearly indicating the position of the grave. The empty gravesite did not produce an anomaly of any sort, at any point during the study. The results for the wrapped pig were much more complex. Initially, the wrapped pig showed a small, high-resistivity anomaly (at 28 and 140 days after burial) that transitioned into a small, predominantly low-resistivity anomaly by the 192-day survey. In the 252-day post-burial survey, the region of the wrapped pig burial was dominated by a low-resistivity anomaly offset from the actual burial site. By the 364-day survey, there was once again a small positive anomaly located at the center of the burial site. This clearly indicates that the presence of a wrapping or cover can have a profound effect (Figure 10.16).

The researchers believe that the low-resistivity anomaly associated with the naked pig grave was caused by the release of decomposition fluids. Changes in resistivity can be caused by both changes in moisture content and by changes in groundwater chemistry. To determine which of these factors was dominant, researchers monitored both. Changes in the moisture content of the soil were monitored by taking soil cores from a second empty grave and a control point in an undisturbed area. Volumetric water content was determined by comparing the mass of the moist core with its dry weight and assuming a standard density for the solid particles (as discussed in Chapter 7). The researchers monitored groundwater conductivity using a *lysimeter*, which is a long, thin plastic pipe with a porous ceramic tip from which water samples can be drawn. The samples were pulled to the surface using a hand pump, and the conductivity of each sample determined using a multi-parameter meter.

They found that the porosity of the gravesites was higher than that of the undisturbed control area, but that the average moisture levels were roughly the same. This suggests that the graves used for the ER study would also have approximately equal moisture content. However, the conductivity of the groundwater in the monitored empty grave and the control point remained relatively constant (near zero), while the conductivity of the groundwater in the second naked pig grave rose substantially, to nearly 3000 mS/m by the end of the project. These results strongly suggest that the observed anomalies are related to changes in groundwater chemistry due to the decomposition of the cadavers.

This idea is supported by many studies that have shown the presence of a cadaver results in a pulse of nutrients that can be detectable in the soil even after several years. During the decomposition process, electrolytes, volatile fatty acids, and the byproducts of bacterial action are released into the surrounding soil. This increases the amount of cations and anions in the groundwater, significantly changing the pore water chemistry and increasing conductivity (σ), which is the reciprocal of resistivity.

$$\sigma = 1/\rho \tag{10.2}$$

Other studies have also demonstrated the presence of resistivity anomalies over test gravesites, though the occurrence and persistence of the anomalies vary (Bray, 1996 as cited in Cheetham, 2005; Jervis, Pringle, and Tuckwell, 2009). Also pertinent, Jervis *et al.* (Jervis, Pringle, and Tuckwell, 2009; Jervis *et al.*, 2009) found that the increased porosity of the soil caused by the construction of a gravesite was insufficient to create a detectable anomaly. This suggests that ER might not be as useful for the detection of graves once decomposition is complete and the remains have been skeletonized. However, their research was conducted in an urban, synthetic soil. Contrasts between natural, undisturbed soil and graves might be more pronounced. ER also works better in periods when moisture levels are higher or lower than normal and soil temperatures high, rather than when moisture levels are consistent. Under other circumstances, ER has been used to detect disturbed earth and areas with higher or lower porosity than the surrounding soil. Environmental conditions and soil type clearly play a role that needs further investigation.

Electromagnetic Induction (EMI)

Electromagnetic induction (EMI) is related to ER, in that it measures the electrical properties of material, and to magnetometry, as the flow of the induced current generates a detectable magnetic field. The difference is that while magnetometry measures the magnetic field and ER measures resistance to current flow, EMI measures induced *conductivity*, or the ability of a material to conduct electricity. Conductivity, which is measured in milliSiemens per meter (mS/m), is the inverse of resistivity, so the two techniques should generally produce similar data. A technical discussion of EMI is available on the companion website (plus Figure 10.17 and Tables 10.3 and 10.4; wiley.com/go/bergslien/forensicgeoscience).

There are a variety of types of EMI systems. The simplest mobile EMI systems, called *slingrams*, can be carried by one person (Figure 10.18). The entire instrument is a long rod with the Tx on one end and the Rx on the other. It is held horizontally (or vertically) to the ground and carried by a sling hung on the shoulder. Technically, this is considered a mobile transmitter/receiver method. Ground conductivity meters are related systems, with two large coils separated by a length of wire. These systems require two people and are used for a deeper profiling of the subsurface. There are other systems, predominantly for lateral profiling and

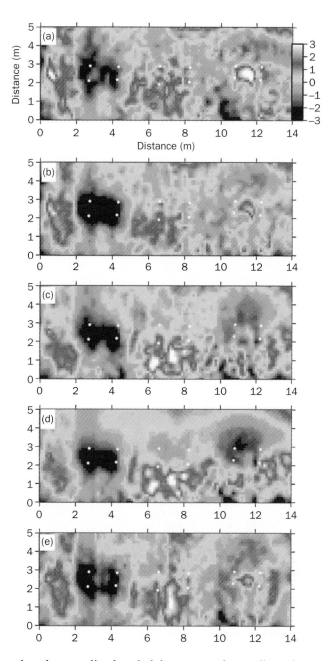

Figure 10.16 Processed and normalized resistivity survey data collected: (a) 28 days after burial, (b) 140 days after burial, (c) 192 days after burial, (d) 252 days after burial, and (e) 364 days after burial. The corners of each grave are marked with white circles. The color scale is the same for all datasets.
Source: Jervis, Pringle, and Tuckwell, 2009.

depth sounding that use a fixed Tx coil, while the Rx coil is moved around (Figure 10.19).

Just like with magnetometers, EMI systems are sensitive to the presence of conductive metallic material, like fences, pipes, and steel buildings, and are affected by the presence of power lines. The operator must remove metal items like belt buckles, keys, and coins to reduce noise. However, a major advantage of EMI is that, unlike magnetometry, it can be used to detect nonferrous metallic objects, like

Figure 10.18 The EM31-MK2 Ground Conductivity Meter.
Source: Photograph courtesy of Genoics, Ltd.

Figure 10.19 When the coils are not mounted in a rigid structure, the distance between them can be changed to provide increasing depths of investigation with increasing separation between the coils. The EM34 is being used in the horizontal dipole mode.
Source: Photograph courtesy of Genoics, Ltd.

lead casing and copper shells. Some other significant advantages of EMI surveys are that one person working alone can handle an instrument, surveys can be conducted very rapidly, and they do not need ground contact, though near-ground contact helps to reduce noise. When working in rough, vegetated terrain, on hard-packed soil or paved ground, this is actually quite an advantage over ER, which requires the probes to be inserted into the soil.

While, technically, both ER and EMI surveys are measuring the electrical properties of the soil, there are cases when both might be useful. EMI strongly detects magnetic objects, while ER does not, which means that ER may detect anomalies that are masked in EMI surveys. It is also generally quite easy to change the depth of an ER survey by changing probe separation distance. Most EMI equipment is sold at a fixed probe separation that cannot be altered by the user. All this said, EMI has seen very limited use in either forensic or archeological contexts, and more information is needed to really understand the strengths and weaknesses of EMI in a forensic setting. For a much more detailed discussion of EMI, see Sharma (1997) or Witten (2006), though the reader is advised that some of the equations in the latter are in error.

Specialized EMI: Metal Detectors

Metal detectors are actually a type of EMI instrument that is designed specifically to detect conductive metal objects in the very near surface (Figure 10.20). The head size of the detector determines the depth of penetration and the size of the object that can be detected (Figure 10.21). For small objects like coins, metal detectors can achieve a maximum depth of penetration between 20 and 40 cm, depending on the instrument type and head size. Larger heads (>24 cm) will penetrate deeper into the soil, but will not be able to detect small items, like shell casings, while small heads (<12 cm) can, but have very limited depth of penetration.

Figure 10.20 A systematic metal detector survey of an area divided into a 10 m grid.
Source: Photograph courtesy English Heritage.

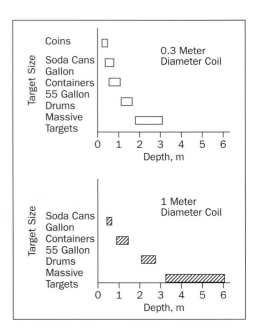

Figure 10.21 Approximate metal detector detection depths for various targets with two coil sizes. *Source:* Benson, Glaccum, and Noel, 1983.

Dupras *et al.* (2006) state that for general-purpose use heads measuring 18–23 cm (7–9 in.) represent a good compromise between scanning depth and the minimum target size. Larger targets can be located at greater depths. There are also metal detectors known as two-box systems, which have separate transmitting and receiving coils, designed to achieve significantly greater depth of penetration, theoretically up to around 6 m (20 ft).

Metal detectors must be tuned to local soil conditions before use and work best in areas with relatively dry, compact soil. The operator must hold the instrument horizontally, as close to the ground as possible, and must move the coil in a sweeping motion back and forth. Metal detectors produce an audio signal that gets louder as the coil moves over a conductive target. To pinpoint the location of a target, it should be approached from multiple directions. Once again, the operator must not be wearing metal, and cultural noise can mask the objects of interest. While metal detectors have several disadvantages, they are relatively inexpensive, easy to use, can be used to quickly scan a search area, and will work through concrete, asphalt, and thick foliage. Killam (1990) has an excellent discussion on the forensic use of metal detectors.

Ground-penetrating Radar (GPR)

Ground-penetrating radar (GPR) differs significantly from all of the methods previously discussed. Magnetometry, ER, and EMI all report on bulk properties. The data represent a summation value for the measured property that combines the information for the feature or features of interest plus all of the surrounding material. This means that there can be multiple interpretations of the source of each anomaly. Remember that a small, highly conductive item can create the same kind of magnetic anomaly that a larger, lower-conductivity item produces. That is also why we refer to *apparent* resistivity or conductivity. Sometimes, the features of interest contrast so strongly with the surrounding media that the resulting anomalies are easy to spot. At other times, a significant post-processing of data is needed to reveal their location because the differences recorded are so subtle. GPR is distinctive because it is a true wave method, where the instrument interacts directly with the subsurface, revealing the shape and location features of interest, usually pretty much in real time.

The word *radar* originally came from the acronym RADAR, which stood for *radio detection and ranging*. It refers to the use of electromagnetic waves to determine the range, altitude, direction, and/or speed of a variety of objects such as aircraft, ships, and weather formations. The idea behind GPR is basically the same, except it is used to reveal subsurface features. GPR broadcasts radio or microwave frequencies (roughly 10–1000 MHz, also called the VHF and UHF bands) into the subsurface and uses the reflected signals to create a pattern that reveals layers/features in the subsurface that have differing physical properties.

Similar to EMI, a GPR system basically consists of a transmitting antenna, a receiving antenna, and a data-logger/display unit, but because a different range of frequencies is used, the behavior of the waves in the subsurface is also different (Figure 10.22). The transmitting antenna (Tx) is a point source that emits rays of a selected frequency in all directions. The receiving antenna (Rx), which is located very close to the Tx but shielded from the outward pulse, records the incoming signal as a function of time for a user-specified duration. As long as no signal is being received, the recorder registers the equivalent of a zero (i.e. no deflection). When a reflected wave arrives, the recorder registers a deflection, returning to a zero once the reflected wave passes.

If you remember from Chapter 6, when electromagnetic energy (visible light or a radio wave) encounters an *interface*, the energy will be broken up so that part is refracted (i.e. enters a second medium), part is absorbed, and part is reflected (Figure 6.6). The wave energy that is refracted will travel at a new speed and angle inside of the new medium, while the wave energy that is reflected bounces back

Figure 10.22 Annotated photograph of a Sensors and Software Pulse Ekko 1000 GPR system. *Source:* Photograph courtesy English Heritage.

toward its source at the same wave speed and at 2θ (i.e. at the same angle as the incident ray but the opposite side of a line normal to the boundary) (Figure 6.6). Because the separation between the Tx and Rx for a GPR is so small in comparison to the distance that the electromagnetic wave travels before being reflected, the angle of incidence is treated as approximately zero. This means that you can envision the incident and detected waves as being parallel to each other.

In principle, there is a very simple relationship between the ray's travel time and distance to an interface. When a wave moves through homogeneous material to encounter a horizontal interface, the depth of the interface (d) is equal to the velocity of the wave (v) multiplied by the travel time (t) divided by two.

$$d = (vt)/2 \qquad\qquad (10.3)$$

You have to divide by two because the wave has traveled from Tx to the interface and then back to the Rx (i.e. twice the distance to the interface). The only problem is that in practice you do not know the exact velocity of the wave.

Wave speed is related to the physical properties of the media through which it travels. Some materials transmit VHF and UHF frequency waves better than others, a property known as *dielectric conduction*. What GPR actually detects are variations in the dielectric conduction of subsurface materials. This is partly expressed by a material's *dielectric constant* (ε_r), a measurement of how much energy is required to propagate a wave through a material in comparison to a standard material, in this case air. It is a dimensionless value that ranges from 1 to 81 for common materials (Table 10.5). The velocity (v) at which a radio wave (or microwave) travels through a material is roughly determined by that material's dielectric constant. The lower the dielectric constant, the faster the wave can travel and vice versa (Table 10.5). As long as the conductivity of the media is less than 100 mS/m (using frequencies between 10 and 1000 MHz), wave speed can be calculated by:

$$v = c/(\varepsilon_r)^{1/2}, \qquad\qquad (10.4)$$

where c is the speed of an electromagnetic wave in free space (also known as the speed of light) or 3×10^8 m/s. Because most geological materials are heterogeneous mixtures, the actual dielectric constant will not be known. In the field, one way to help determine the actual relationship between depth and time is to bury a metal object at a known depth and then run the GPR over the area. This information helps establish a reference that can be used to convert time into depths for the survey area.

Another important physical property that affects the behavior of the wave is conductivity. As the wave travels it loses energy, i.e. attenuates, eventually dissipating entirely. Part of this attenuation (α) is accounted for by the dielectric constant, but the more conductive a material is the more energy is transformed by its interaction with the media. The dielectric constant of most earth materials does not vary over much more than a factor of ten, but conductivity can vary over several orders of magnitude, thus it turns out that conductivity is actually the most important controlling variable affecting *skin depth*, which roughly translates to signal depth (see the companion website for a more detailed discussion).

Skin depth decreases with increasing conductivity. If you compare freshwater with saltwater, the dielectric constant is roughly the same, but because seawater is

Table 10.5 Typical dielectric constant (ε_r), conductivities, wave propagation velocities (v), and resultant signal attenuation (α).

Material	ε_r	v (m/ns)	Average conductivity (mS/m)	Signal attenuation (α) (dB/m)
Air	1	0.30	0	0
Sandstone (dry to moist)	2–10	0.21–0.1	0.001–10	2–20
Asphalt (dry to wet)	2–12	0.21–0.09	1–100	2–20
Ice	3–4.3	0.17–0.14	0.01–0.5	0.01
Dry clay/silt	3–6	0.17–0.12	1–1000	1–100
Permafrost	4–6	0.15–0.12	0.002	0.1–5
Dry salt	4–7	0.15–0.11	0.01–1	0.01–1
Limestone	4–8	0.15–0.11	0.5–2	0.5–20
Granite	4–9 typ.	0.15–0.1	0.01–1	0.01–5
Dry sand/gravel	4–10	0.15–0.1	0.01	0.01–1
Concrete (dry to wet)	4–20	0.15–0.07	1–100	2–25
Cement (dry/wet)	6–11	0.12–0.09	0.01–10	0.01–5
Wet clay/silt	7–40	0.11–0.05	50–1000	20–640
Wet sand/gravel	10–30	0.09–0.06	0.01–1	0.03–5
Fresh water	81	0.03	0.05	0.01–0.1
Saline/Saltwater	81	0.01*	3000	100–1000

*Wave velocity in seawater is different from that of freshwater because the conductivity is so high is cannot be considered a low-loss material.
Source: Modified from Sharma, 1997; Davis and Annan, 1989.

so much more conductive, the wave energy is quickly attenuated, resulting in a significantly reduced skin depth. Skin depth also decreases based on the number of interfaces encountered. At each interface the wave is broken into reflected, refracted, and absorbed components. The more times this happens, the greater the energy loss from the wave. However, the skin depth relationship does not equate to depth of penetration for GPR, because the frequency of the signal also plays a role.

GPR systems actually work over a range of frequencies, denoted *bandwidth*, and are characterized by their *center frequency* (f_c), which is defined as the difference between the highest operating frequency and the lowest operating frequency divided by two. Remember that frequency (*f*) and wavelength (λ) are inversely proportional ($\lambda = v/f$, where v is equal to the wave velocity which is equal to c in free space); so the higher the frequency, the shorter the wave, and vice versa. GPR systems with a higher center frequency (400–1000 MHz) have shorter wavelengths and can therefore resolve smaller features, i.e. have higher resolution, but have a lower depth of penetration (Figure 10.23). Low-frequency systems (10–50 MHz)

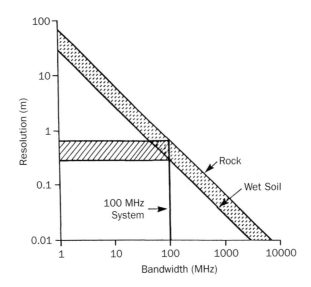

Figure 10.23 The relationship between instrument bandwidth and resolution. *Source:* After Davis and Annan, 1989.

have longer wavelengths, which blur out smaller features, reducing resolution, but penetrate deeper into the subsurface. As pointed out with XRD, the size of the waveform is roughly equivalent to the size of the feature that can be observed, so big waves "see" big things, like geologic structures, and smaller waves "see" smaller things, like pipes or graves. The idea is to select a center frequency that is the best compromise between the necessary resolution versus depth of penetration for the feature(s) of interest given the physical properties of the subsurface.

Many forensic books and articles cite "recommended" GPR antenna frequencies, but there really is no such thing. Selection has to be made specifically to optimize for the conditions at the site being investigated. The ability to use multiple antennae is optimal, clearly providing the most detail about a site, but this means using different GPR systems or a system with exchangeable antennae. One cannot simply change the frequency of a system like tuning a radio, because antenna size and center frequency are related. High-frequency GPR antennae are smaller. For example, there are small handheld units that are used to examine concrete for voids or defects. The lower the frequency, the larger the antenna needed to generate the longer wavelengths (Figure 10.24). These systems sometime require two people or even a car to move them.

Roughly speaking, GPR typically has penetration of about 20 m, increasing to 50 m under ideal conditions and decreasing to less than 2 m in high-conductivity materials. A couple of rough rules of thumb are $d_{max} < (30/\alpha)$ and $d_{max} < (35/\sigma)$. Relative conductivities from ER or EMI surveys can be very helpful for determining whether and where GPR surveys may be useful. For forensic work, attaining an ideal understanding of the maximum theoretical depth of penetration is not that important anyhow. Instead, it is usually more important to achieve a high enough resolution to detect the features of interest. Clandestine graves, for example, tend to be in the order of 0.4–0.6 m deep and roughly rectangular in plan-view (1.5 × 0.5 m or so) (Pringle *et al.*, 2008). Thus, center frequency

Figure 10.24 A 200 MHz antenna made by GSSI (Geophysical Survey Systems, Inc) in use. *Source:* Photograph courtesy of Kevin Williams.

selection should be based on achieving a high enough resolution to detect the smallest dimension of the feature of interest. A discussion of how vertical and horizontal resolution is determined (including Figures 10.25 and 10.26 and Tables 10.6 and 10.7) is on the companion website.

There are a variety of ways that GPR surveys can be refined, such as changing antenna orientation and whether the GPR antenna is *monostatic* (which means that a single device both transmits and receives) or *bistatic* (which means that the transmitter and receiver are separate units) and which will affect survey parameters and data clarity, but are outside the scope of this introduction. For more details about GPR, interested parties are directed to Davis and Annan (1989) and Conyers (2004) as good starting points.

In the past there has been both too much optimism and too much pessimism about GPR. It is important to remember that no geophysical method is a magic wand. The main point is that, like the other geophysical methods discussed, the utility of GPR is highly site-specific. Given the physical properties of common geologic media, you can see that GPR will not work well in the presence of high surface conductivity, like through concentrations of wet clay minerals or bodies of saltwater, because the signal attenuates very quickly. Instead, GPR works best through low-conductivity, low-dielectric-constant materials like ice, snow, dry sandy soil, and dry concrete. It can also work reasonably well through low-conductivity, clay-sized particles, like rock flour, which is clay-sized but not composed of high-conductivity clay minerals. GPR also works better when there are fewer clear interfaces, which means in areas overlain by till or construction debris there can be so much "noise" that the profile can be very difficult to interpret. A GPR survey will only be successful if the surface and subsurface conditions are favorable.

There are a few other important points to consider. To work properly, the GPR needs to be in good contact with the ground, something called *coupling*. Elevating the antenna by more than a very few centimeters creates a new interface, that between the air and the ground. Clearly there is a huge contrast in properties at this interface, so there will be a large reflected signal right at ground level. A large proportion of the energy will be lost (potentially 50% or more), radically reducing the amount of subsurface information that can be gathered. This requirement can clearly create issues in areas with rough terrain. On the other hand, GPR works very well over flat, paved or covered surfaces, such as asphalt parking lots, porches, and in basements.

Establishing the correct temporal and spatial sampling intervals is also important (see companion website). While it is possible to conduct preliminary grid-less GPR surveys to get basic information about a site, it is much better practice to establish a survey grid and collect multiple sections just as you would with any other geophysical technique. Ideally, survey lines should run perpendicular to the long dimension of the feature of interest, but since knowing the actual orientation of the features of interest is somewhat rare in forensic investigation, it is preferable to run two sets of survey lines oriented perpendicular to each other (i.e. an *orthogonal survey*). *Line spacing*, also known as *profile spacing*, is the distance between parallel survey lines and is determined by the data density required to locate a target. For the best resolution, or to create a three-dimensional view of the subsurface, line spacing should also be approximately equal to the vertical resolution. Sometimes, due to time constraints, that is not possible and wider line spacing must be used. Very roughly speaking, the spacing should not be more than the size of the smallest dimension of the feature of interest. For comparison, most forensic surveys have employed a spacing of 0.5 m.

Once the basic parameters of a study have been determined, it is finally time to collect some data. Basic operation of the instrument is quite simple. First, operators need to ensure that their cells phones are turned off and that they are away from radio frequency interference from cell towers and transmission lines. The GPR is then moved along the ground, usually at a slow, constant walking pace. The transmitting antenna will emit a burst of energy at a predetermined interval (step size) and at the same time the instrument will start recording the energy being received by the receiving antenna (Figure 10.27a). The received signal is plotted vertically as a function of time with increasing time downward (Figure 10.27b). A single line is called a *trace*. When no energy is being received, the line is simply a straight vertical line. When energy is received, there is a "blip" that is recorded either as a line as it is deflected either to the left or to right or as a darker or lighter shade of gray in a grayscale display. A single trace cannot actually give you much information. Instead, as the GPR unit is moved along the ground, a series of traces (one from each station) are plotted in parallel, but shifted to the right (Figure 10.27b). The resulting pattern is called a *radargram*.

The radargram created by a single horizontal interface is simply a horizontal line of blips (Figure 10.27; see A, the reflection of the air-soil interface). As the signal encounters more interfaces, there are more blips on each trace and the geometry of the interface affects the shape of the pattern of blips across the radargram (Figure 10.27; see B and C). Simple interfaces such as horizontal or tilted stratigraphic layers are fairly easy to interpret, but features such as troughs will create bowtie like patterns that are a bit trickier to understand. The kinds of anomalies most commonly associated with graves create a discrete hyperbola (Figure 10.28). As you can see, depending on the subsurface topography and the number of reflectors, radargrams can be quite complicated (Figure 10.29). Deciphering them requires practice, but when working under the correct conditions, the amount of

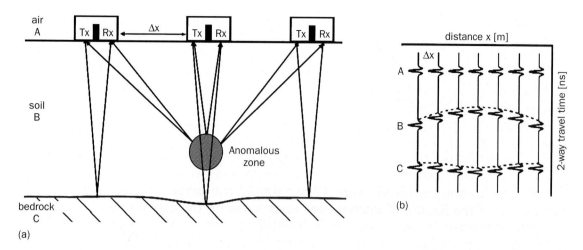

(a)

(b)

Figure 10.27 (a) Conceptual illustration of the GPR unit being used to profile the subsurface, (b) idealized version of the resultant radargram generated over a localized anomaly. *Source:* Adapted from Annan and Cosway, 1992.

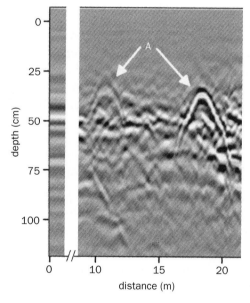

Figure 10.28 Example of the parabolic shape (A) created by a discrete buried object. *Source:* Courtesy of Kevin Williams.

information that can be quickly gathered is amazing because each radargram basically represents a two-dimensional slice of the subsurface.

One of the major advantages is that in many cases it is possible to use digitally produced radargrams in real time to identify subsurface anomalies, a mode called *scanning*. However, post processing of a detailed area survey will reveal significantly more information about the subsurface and potentially locate more anomalies of interest. There are many ways of using the data for a better visualization, from creating a series of time slices to generating virtual, false-perspective 3-D

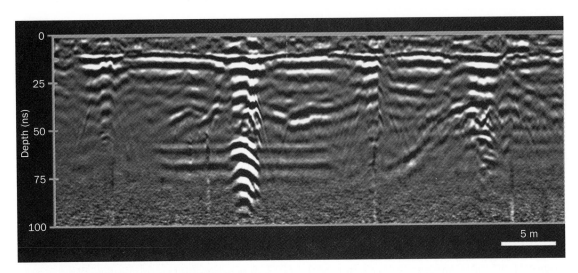

Figure 10.29 Example of an actual radargram.
Source: Courtesy of Kevin Williams.

representations. Due to its high vertical and horizontal resolution, GPR is probably the most commonly used geophysical tool for forensic investigations. Its application as a serious forensic investigation tool in the United States appears to have started with project PIG (pigs in ground) by a group now more formally known as NecroSearch International (France *et al.*, 1992). A nice discussion of the history of NecroSearch International can be found in Jackson (2002). There are also several projects and practitioners using GPR for a variety of forensic investigations, such as searching for a sunken Jet Ski after an accident in a freshwater lake (Ruffell, 2006), isolating the location of the graves of 1918 Spanish flu victims in the permafrost of Norway (Davis *et al.*, 2000), and performing time-lapse studies at Keele University (Pringle *et al.*, 2008). There have been a significant number of papers published since the turn of this century on a variety of potential uses for GPR.

Helpful Negative Results

After decades of violence, in the 1990s the Irish Republican Army (IRA) called a series of ceasefires that eventually led to the 1997 All Party Peace Talks, the 1998 Good Friday Agreement, and to the current peace in Northern Ireland. The complex history of the IRA, PIRA, and OIRA is completely beyond the scope of this text and interested parties are directed to English (2004). The point is that starting in the late 1990s, members of the IRA expressed their willingness to begin to disclose the locations of people who had been murdered and their bodies hidden during the Troubles, those called the "Disappeared."

As part of the campaign to locate the bodies of the missing, Ruffell (2005) used GPR to evaluate a possible burial site. The site was located in a grave at a church graveyard that had been identified by police intelligence. The site was located only a few hundred meters from a Nationalist housing estate, so conventional investigative approaches would have proven unwise as the residents would be highly likely to misinterpret the actions of investigators. Instead, a rapid, non-invasive, non-threatening technique was needed for investigation, so GPR was chosen.

The site of the investigation was a grave plot delineated by a double headstone 2 m high and 1.5 m wide with iron railings on the other three sides of the plot. The gravesite had been cleared of vegetation as part of a larger clearance in the graveyard and the headstones indicated the presence of three burials in the plot. One plot contained two known burials, one from the 1870s and one from the early 1970s. The other plot contained a burial from the 1860s. It was the later plot that intelligence suggested might also be the location of the victim. The known 1970s burial was made in a wooden coffin at a roughly known depth and could therefore be used as an internal reference for relating velocity to approximate depth in the local soil. This eliminated the need for burying a reference object or conducting velocity surveys (another way to relate velocity to depth), saving valuable time.

The topsoil at the site was thought to extend down 20–30 cm, with mixed subsoil to 1 m underlain by glacial till to a depth of 2 m beneath which Paleogene basalts and dykes underlay the whole area. Data were collected after a prolonged (two-month) dry period, so even though the soil was apparently clay-rich, surface high conductivity was not an issue. The survey was conducted with the lines perpendicular to the long dimension of the plot. Each line started outside the area of the plot, progressed across the known and suspected burial sites, and then continued again outside the plot.

Dr Ruffell had prior experience in the general area from conducting a geophysical study of a nearby quarry the previous month. Based on this, three antennae were used at the site: 100 MHz, 200 MHz, and 400 MHz. The results from the 100 MHz transverses were interpreted in the field, while the 200 MHz and 400 MHz data were post-processed, also in the field. They found that the 400 MHz data provided a good definition of the shallow features and the 100 MHz data provided information about regional stratigraphy. The 200 MHz data were reportedly essentially the same as the 400 MHz data.

The data from the 400 MHz survey showed what is possibly the most interesting finding of the paper, that the anomaly detected over the known 1970s grave was not the hyperbolic curve structure so commonly reported in the pig burial studies but a synclinal structure most likely the result of grave collapse. The syncline was broader near the headstone, continued for a distance perpendicular to the headstone, and the reflections showing disturbed soil extend approximately 2 m deep. Grave soil is often distinctive because it is imported soil or soil that is a mixture of different layers. Interestingly, the ringing over the burial area was different from that in the undisturbed areas. *Ringing* (also called *multiples*) is a common phenomenon that typically appears as a series of horizontal bands near the top of each profile and is generally caused by sympathetic resonance in the antennae, i.e. system noise. It frequently occurs when there are strong reflectors in an area, such as the metal fence around the gravesite. Ruffell (2005) suggests that the ringing over the disturbed soil was reduced because the less cohesive soil was able to absorb more energy.

The 100 MHz data also showed a disturbance at the location of the 1970s grave and post processing clarified the locations of the glacial till layer, the infill of the grave area, and even the existence of a platform that the gravedigger probably made in order to excavate the grave to a sufficient depth. None of these types of features was apparent in the suspected burial area. The soil in the suspected area showed no evidence of reworking. Based on an on-site evaluation of the data collected, investigators were able to conclude that there was no burial in the suspected site. This conclusion was later confirmed when the victim being sought was subsequently discovered elsewhere.

There are a couple of important points made in this study. First is that, as stated previously, there is no such thing as a one-size-fits-all, preferred antenna frequency for all forensic site evaluation. Some authors suggest that higher-frequency antenna would be preferred given the size of the target. However, using a well-designed survey plan, the lower-frequency antennae provided enough information for a satisfactory evaluation of the site. It is unclear whether use of higher-frequency antennae would have added anything of import to the survey. The other important point is that even though the results were "negative" (i.e. the victim was not located at the site), the geophysical investigation was not a failure. Too many reports speak of the results of a geophysical investigation as unsuccessful if the victims or objects sought are not found. That wholly misses the point of most forensic geophysical exploration. Geophysical tools allow for the rapid assessment of an area and thus the better utilization of resources. In this light, the site investigation was perfectly successful. Investigators did not waste time or resources attempting an excavation in an area where such activity would have been viewed with suspicion, if not outright hostility, especially given that the results were negative.

Here? Really?

Nobes (2000) provides an interesting perspective on the use of geophysical tools in the search for a missing woman. She had disappeared over 11 years earlier under suspicious circumstances, and even though the police had a suspect, nothing could progress without a break in the case. Under some kind of difficulty, her killer confessed in 1994 and agreed to lead investigators to the gravesite in Woodhill State Forest, north of Auckland, New Zealand. Due to normal forestry operations, several of the features of the site had changed in the intervening time, most prominently the orientation of the roads. This made locating the burial site quite difficult, so geophysical tools were brought in to help identify the most likely location of the body.

The soil in the area was a well-drained, medium-grained sand with occasional lenses of silt or clay. Conditions had been dry for a prolonged period. Both EMI and GPR were chosen for use at the site. Because it could be used very rapidly, the initial site survey was conducted with EMI. Given that both positive- (usually from metal in clothing and jewelry) and negative-conductivity anomalies could be potentially expected from a grave, researchers looked for any reasonably sized anomaly that contrasted with the surroundings. It is also logical to measure conductivity prior to employing GPR, since the latter method is so severely affected by high-conductivity materials.

Investigators designed a survey plan that included a primary search area, an alternate search area, and extensions outside of the primary search area for an improved understanding of the site's background. The primary search area contained several large stumps that complicated the survey. Several small anomalies were located, most of which turned out to be discarded saw blades. Other anomalies in the primary search area were investigated with GPR and identified as tree stumps and roots. There was one significant anomaly in the alternate search area, which turned out to be a large ceramic item. There was also a clear isolated anomaly in the extended search area that had originally been surveyed merely for additional geophysical background. It was in association with this last anomaly that the body was located. Given the time and budget constraints of the investigation, no GPR data were collected over the actual burial site.

Given the complexity of the site, which contained subsurface clay and silt lenses, the extensively reworked ground due to logging, and large numbers of tree stumps and roots, it is unsurprising that EMI turned up several anomalous features. However, most of them were too small to reasonably be a gravesite. Nobes found only eight clear anomalies considered primary targets and 30 smaller anomalies in all. This included the site where the body was eventually uncovered. The investigators in charge of the case were apparently very impatient, however, and began excavations before the geophysical survey was complete and before the data could be analyzed. Allowing geophysicists to complete their work would have substantially reduced the number of sites excavated. The other point worthy of note is that the body was found not in any of the "most likely" areas identified by the killer, but instead in the expanded survey area. Based on such experiences, Nobes (2000) recommends that searches always be expanded beyond the expected location of the body or object sought.

Search and Post-search Operations

During the active phase of a search, there may be multiple deployments of one or more geophysical instruments. All of the anomalies identified would be marked, but there would be no invasive exploration at this point. Instead, the whole search area would be investigated using a range of non-invasive techniques, ideally to isolate the most promising anomalies for further study. The search area would also be mapped, photographed, and possibly even videotaped. Information regarding the type and location of each anomaly and other detailed information would be collected.

The next phase of the search would generally involve invasive techniques, such as probing the area using soil augers or hand probes. Each identified anomaly would be investigated and potentially excavated by qualified personnel. The goal is to use the least intrusive method possible to gather the maximum amount of information.

Geoscientists would not typically be actively involved at this point, though they should be on hand to relate the geophysical information gathered to the physical conditions revealed by excavation. Instead, depending on the circumstances, a forensic anthropologist, forensic archeologist, or crime scene investigator would typically direct any excavations in order to ensure the appropriate recovery of evidence.

At the completion of the activities at the site, there should be a post-search *debriefing*, often called a *tailgate meeting* at environmental sites. Each member of the search team would be involved to discuss whether the search objectives were met, hash over any issues or problems that were encountered, and to discuss the lessons learned at the site. All of the information gathered from the search would be used to produce a final report documenting the results.

Elemental and Mineralogical Analysis of Human Bone

Human bones and teeth are mineralized tissues composed of a calcium phosphate that is similar in composition and structure to the mineral group apatite. The organically produced form of apatite is usually referred to as *bioapatite* and most closely resembles the geologically occurring mineral hydroxylapatite, which has the idealized formula $Ca_{10}(PO_4)_6(OH)_2$. The hexagonal crystalline structure of apatitic minerals is extremely flexible and allows a wide range of *substitutions*, where ions of different elements replace some portion of the idealized elemental structure. Geologically occurring apatite can incorporate trace amounts of half of the elements in the periodic table into its structure.

Bioapatite also contains many elements and molecular species other than calcium and phosphate, but from a much more limited subset than geologic apatite. In living organisms, fluorine (F^-) most commonly substitutes for the hydroxyl ion (OH^-). Modern use of fluoridated toothpaste and water alters tooth enamel to fluorapatite, which is more resistant to acid. This can also occur postmortem in bioapatite that is exposed to fluoridated water. The ions reported in the literature to substitute for calcium (Ca^{2+}) in bioapatite include Na^+, K^+, Fe^{2+}, Zn^{2+}, Sr^{2+}, Mg^{2+}, Cd^{2+} Ba^{2+}, Mn^{2+}, and Pb^{2+}, and the ions reported to substitute for phosphorous (P^{5+}) include As^{5+}, V^{5+}, Si^{4+}, S^{6+}, and Sb^{5+}. Many of these elements are essential nutrients that are stored in the skeleton, while others are toxic substances that will not accumulate to high levels (the host would die first). The types and amounts of trace elements that are available to be incorporated into the mineralized tissues of any given individual are dependent on two main factors.

The first is local geology, which governs the trace element loads found in the local water supply, in the soils in which local crops are grown, and in the dust particulates in the air that you breathe. The trace element signature of the location where an individual has spent the majority of their life will be reflected in their bodily trace element load. There had been some idea that the food distribution systems that now exist in the United States and other industrialized nations would act to homogenize trace element loads in the various nations, but more recent work shows that is probably not the case. Contact with local soil and water still ensures that local geology plays a significant role in determining the trace element loads of individuals living in industrialized nations and is arguably the primary factor for individuals living in pre-industrialized nations or in isolated rural areas.

Diet does, however, play an important role in determining bodily trace element loads. For example, studies in animals have demonstrated that Sr:Ca ratios generally decrease as one moves from bedrock → soil → plants → herbivores → carnivores. Since humans are typically omnivores, their Sr:Ca ratio should lie somewhere between that of herbivores and carnivores, though marine and freshwater shellfish, and marine fish, have very high levels of strontium, thus a diet rich in seafood would elevate Sr levels. Based on this, vegans should have higher strontium levels than someone who eats a significant amount of red meat. Fishermen and others who eat large quantities of seafood should also have very high strontium levels. Other research has found a significant positive correlation between the levels of Ni, Co, Mn, Cr, Mg, Al, Ag, and Ca in bone and seafood consumption, a negative correlation between Zn and frequency of alcohol consumption, and a positive correlation between Cu and fruit consumption. This suggests that significant variations in diet from a regional norm, such as macrobiotics or veganism, or high levels of seafood or meat consumption, should have a discernible impact on an individual's trace element load.

The other major source of trace elements is the anthropogenic load in the local environment. In general, people who live in urban areas will have higher loads of heavy metals in their bodies than people who live in rural areas do. This is also true of people who live in close proximity to a factory, power plant, or mine. Obviously, people who work in one of these industries will also commonly have a significant body burden of associated trace elements. Interestingly, metals carried in the body in the form of bullets, and presumably other metal fragments as well, also increase the bodily loads.

The total trace element load of any one individual is the summation of a variety of exposures over the course of their life. The rate at which the bodily trace element loads change is a point of contention, though most authors suggest bone-remodeling rates of 7–10 years. However, for people who have had a high level of exposure to bone-seeking elements, the elemental half-life can be significantly longer. Trace elements can be released from bone into the bloodstream only to be reincorporated into bone. The best understood example of this is with lead, which is commonly recycled through the bloodstream back into bone, giving it a half-life of between 15 and 30 years. Similar behavior is believed to exist with other trace elements, though the rates of exchange are thought to be significantly different. This means that someone who has recently moved into an area from a significantly different environment could have a notably different body burden of trace elements than the locals that will take years or even decades to equilibrate with the local environment.

All of this suggests both that there could be a rather wide range of trace element signatures for human bone and that there might be some utility in using trace element loads in a forensic context, such as to help figure out where a person might have come from or to help differentiate commingled remains. Unfortunately, at the current time, reliable and well-sourced data on the minor and trace element composition of human bone are quite scarce. With no reliable baseline for comparison and relatively little understanding of how trace element loads vary in populations, relying on such analysis for the identification of human bone is fraught with uncertainty. Someday, with the development of a sufficient reference database, trace element analyses may have several interesting applications.

What is in That Urn?

In some ways, forensic geoscience is simply a special perspective that is employed to address questions of trace evidence analysis. It intersects several different fields of study, such as biology, chemistry, and archeology, and you never know what background information might be most pertinent. For example, when presented with a container of whitish-brown powder, the forensic examiner is not also given a label that says, "This material is geological in nature." Thus, the materials that fall under the broad category of "geologic" evidence can often be quite unexpected, such as diatomaceous earth, kitty litter, glass particles, building materials, or the unexpected contents of an urn.

In early 2002, people across the United States watched in shock as the story of the Tri-State Crematory unfolded before them on national television. It started in February, when a woman walking her dog found a skull. Authorities were called in, and within hours, they found three dozen bodies scattered around the site, some simply stacked next to tools in a storage shed. At the time, the Walker County coroner, Dewayne Wilson, remarked, "The worst horror movie you've ever seen – imagine that 10 times worse. That is what I am dealing with." Things went downhill from there.

The more investigators explored the site, the more bodies they discovered. Some had hospital toe-tags still attached, while others were buried in shallow pits or stuffed into vaults. Some bodies were mummified, while others were badly decomposed. Given the unprecedented situation, a portable morgue was set up on-site and a federal disaster team called in to help. Workers began the laborious process of trying to sort out the commingled remains and identify the bodies. All told, 339 bodies were eventually recovered, setting off a vast exercise in forensic anthropology. Only about 200 were ever identified.

Apparently, when Ray Brent Marsh took over the facility from his father in the mid-1990s he just stopped performing the contracted cremations and instead started dumping bodies unceremoniously around his property. The only explanation he ever offered was that the crematorium had stopped working, but when tested the oven was found to be in working order. To add to the confusion, most bodies received prior to a certain date were actually cremated, though not all, and after that date, most of the bodies were dumped, but some of them appear to have been sent to other facilities for proper cremation. Suddenly, thousands of families were uncertain as to the contents of the urns in their possession.

Many families turned to the local authorities for help and a significant number of urns were sent in for an analysis of the contents. Initially, the question of how to identify the contents of the Tri-State urns presented difficulties simply due to the enormity of the problem, as most forensic laboratories do not have large trace evidence analysis sections. Working with the tools that were available, laboratories found that instead of the cremated remains of loved ones, many of the urns turned out to contain cement dust, silica, rock, burnt wood chips, or other debris. While the sheer scale of the problem presented by Tri-State is unusual, questions concerning the contents of urns are actually quite common. Trace evidence workers are often confronted with strange jobs, such as the need to identify random materials that may or may not be geological in nature.

To determine the nature of the materials in an urn, a detailed trace element analysis is not necessary. The bulk elemental composition of the material in an urn typically allows workers to clearly differentiate between powderized cremated bone and typical filler materials. Calcium and phosphorus will always dominate bioapatite and the Ca:P ratio should fall in the range roughly of 1.61 to 2.02 for modern samples, with archeological samples ranging up to 2.58 (Ubelaker *et al.*, 2002). Samples with bulk elemental compositions that are dominated by other elements, or lack either Ca or P, will not be bone. For example, plaster of Paris and drywall compounds contain significant amounts of sulfur and are almost totally lacking in phosphorus. Most other building materials, like concrete, grout, and mortar, also lack significant phosphorous. In addition, building materials will generally contain significant levels of other elements that are typically excluded from bioapatite or found only at very low concentrations. Grout, mortar, and quick-setting concrete all contained moderate to high concentrations of titanium. Concrete and geologic apatite will contain high levels of silicon, aluminum, and iron. Wood ash and charcoal ash will normally contain moderate to high levels of strontium and iron. The materials that Mr March used to fill the Tri-State urns were mostly building debris and wood, which should be clearly differentiable from cremated bone.

In investigations of this type, powder X-ray diffraction (XRD) can also be used to differentiate bioapatite from other materials. As demonstrated in Chapter 7, XRD is used to identify crystalline materials, including differentiating components of mixtures. The elevated temperatures achieved under burning, or cremation, cause the recrystallization of bioapatite, which actually simplifies its analysis. The resultant material is distinguishable from geologic apatite and can be clearly differentiated from the materials reportedly used as fill-in the urns from Tri-State because they would have a completely a different mineral content (Bergslien, Bush, and Bush, 2008).

Ray Brent March was ultimately charged by the State of Georgia with 787 separate criminal counts, including theft by deception (i.e. being paid to perform a service that he did not perform), abuse of a corpse, and burial service fraud, among others, and faced a prison sentence of potentially several thousand years. The case was unique and would have presented many difficulties at trial, but the potential legal circus was aborted in 2005 when Marsh pled guilty in agreement with the terms of a plea deal. He was sentenced to 12 years in prison and 75 years of probation. Just to keep things interesting, in February 2008, defense attorneys submitted to the state parole board the novel defense that Mr Marsh was suffering from mercury toxicity due to exposure by a faulty ventilation system while cremating bodies with mercury dental amalgams and thus was not wholly responsible for his actions.

Summary

This has been just the briefest introduction to the fields of forensic anthropology, forensic archeology, and some of the geophysical tools that have the most potential to be useful in forensic investigation. It was meant to give the reader an appreciation for the kinds of information that can be gathered and the conditions under which these tools can be used. This background should help the reader to be better informed about when such tools would be applicable and what it really takes to use them in the field. It is important to remember that geophysical tools are not magic wands. They will only work under appropriate conditions and when used correctly by a trained operator. The information that can be gathered in the correct circumstances, however, can prove invaluable.

Further Reading

Dupras, T. A., Schultz, J. J., Wheeler, S. M., and Williams, L. J. (2006) *Forensic Recovery of Human Remains: Archaeological approaches.* CRC Press, Boca Raton, FL.

Hunter, J. and Cox, M. (eds) (2005) *Forensic Archaeology: Advances in theory and practice.* Routledge, New York.

Hunter, J., Roberts, C., and Martin, A. (1996) *Studies in Crime: An introduction to forensic archaeology.* B. T. Batsford, London.

Killam, E. W. (1990) *The Detection of Human Remains.* Charles C Thomas, Springfield, IL.

References

Annan, A. P. and Cosway, S. W. (1992) Ground penetrating radar survey design. *Proceedings of the Symposium on the Application of Geophysics to Engineering and Environmental Problems*, April 26–29, 1992, Oakbrook, IL.

Anon (2006) Practice advice on search management and procedures. In: M. Harrison, C. Hedges, and C. Sims (eds), National Policing Improvement Agency (NPIA), London

Bass, W. M. (2005) *Human Osteology: A laboratory and field manual*, 5th edn. Missouri Archeological Society, Springfield, MO.

Benson, R. C., Glaccum, R., and Noel, M. (1983) *Geophysical Techniques for Sensing Buried Wastes and Waste Migrations.* US Environmental Protection Agency, Washington, DC, Contract No. 68-03-3050, National Water Well Association, Worthington, OH.

Bergslien, E. T., Bush, M., and Bush, P. J. (2008) Identification of remains using x-ray diffraction spectroscopy and a comparison to trace elemental analysis. *Forensic Science International* **175**: 218–226.

Blundell, A., Dearing, J. A., Boyle, J. F., and Hannam, J. A. (2009) Controlling factors for the spatial variability of soil magnetic susceptibility across England and Wales. *Earth Science Reviews* **95**: 158–88.

Buck, S. C. (2003) Searching for graves using geophysical technology: Field tests with ground penetrating radar, magnetometry, and electrical resistivity. *Journal of Forensic Sciences* **48**(1): 5–11.

Burger, H. R., Sheehan, A. F., and Jones, C. H. (2006) *Introduction to Applied Geophysics: Exploring the shallow subsurface.* W. W. Norton & Co, New York.

Byers, S. N. (2005) *Introduction to Forensic Anthropology: A textbook*, 2nd edn. Pearson Education, Boston.

Cheetham, P. (2005) Forensic geophysical survey. In: J. Hunter and M. Cox (eds), *Forensic Archaeology: Advances in theory and practice*. Routledge, New York, pp. 62–95.

Conyers, L. B. (2004) *Ground-penetrating Radar for Archaeology*. AltaMira Press, Rowman & Little Publishers, Lanham, MD.

Davenport, G. C. (2001) Remote sensing applications in forensic investigations. *Historical Archaeology* **35**(1): 87–100.

Davis, J. L., and Annan, A. P. (1989) Ground penetrating radar for high-resolution, mapping of soil and rock stratigraphy. *Geophysical Prospecting* **37**: 531–51.

Davis, J. L., Heginbottom, J. A., Annan, A. P. *et al.* (2000) Ground penetrating radar surveys to locate 1918 Spanish flu victims in permafrost. *Journal of Forensic Sciences* **45**(1): 68–76.

Donnelly, L. J. (2002) Finding the silent witness. *Geoscientist* **12**(5): 16–17.

Donnelly, L. J. (2003) The applications of forensic geology to help the police solve crimes. *European Geologist* **December**: 8–12.

Donnelly, L. (2009) The geological search for a homicide grave. *The Investigator* **July/August**: 42–9.

Donnelly, L. J. (2010) The role of geoforensics in policing and law enforcement, http://facstaff.buffalostate.edu/bergslet/ForensicGeology/Papers/Donnelly2010.pdf, accessed 19.11.11.

Donnelly, L. J. (2011) The renaissance in forensic geology. *Teaching Earth Science* **36**(1): 46–51.

Donnelly, L. and Harrison, M. (2011) *Geomorphological and geoforensic interpretation of maps, aerial imagery, conditions of diggability and the colour coded RAG prioritisation system in searches for criminal burials*. Presented at the 3rd International Workshop on Criminal and Environmental Soil Forensics, 2–4 November (2010), Long Beach, California. Geological Society of London [publication pending].

Dupras, T. A., Schultz, J. J., Wheeler, S. M., and Williams, L. J. (2006) *Forensic Recovery of Human Remains: Archaeological approaches*. CRC Press, Boca Raton, FL.

English, R. (2004) *Armed Struggle: The history of the IRA*. Oxford University Press, New York.

France, D. L., Griffin, T. J., Swanburg, J. G. *et al.* (1992) A multidisciplinary approach to the detection of clandestine graves. *Journal of Forensic Sciences* **37**(6): 1445–58.

Gill, P. (2008) *Electrical Power Equipment Maintenance and Testing*, 2nd edn. CRC Press, Boca Raton, FL.

Harrison, M. and Donnelly, L. J. (2008) Buried homicide victims: Applied geoforensics in search to locate strategies. *Journal of Homicide and Major Incident Investigations*. Produced on behalf of the Association of Chief Police Officers (ACPO) Homicide Working Group, by the National Policing Improvement Agency (NPIA).

Harrison, M. and Donnelly, L. J. (2009) Locating concealed homicide victims: Developing the role of geoforensics. In: K. Pye and D. J. Croft (eds), *Forensic Geoscience: Principles, techniques and applications*. Geological Society of London, Special Publications 232, pp. 197–219.

Hunter, J. and Cox, M. (eds) (2005) *Forensic Archaeology: Advances in theory and practice*. Routledge, New York.

Hunter, J., Roberts, C., and Martin, A. (1996) *Studies in Crime: An introduction to forensic archaeology*. B. T. Batsford Ltd, London.

Jackson, S. (2002) *No Stone Unturned: The true story of the world's premier forensic investigators*. Pinnacle Books, New York.

Jervis, J. R., Pringle, J. K., Cassella, J. P., and Tuckwell, G. W. (2009) Using soil and groundwater data to understand resistivity surveys over a simulated clandestine grave. In: K. Ritz, L. Dawson, and D. Miller (eds), *Criminal and Environmental Soil Forensics*. Springer Publishing, Dordrecht, The Netherlands, pp. 271–84.

Jervis, J. R., Pringle, J. K., and Tuckwell, G. W. (2009) Time-lapse resistivity surveys over simulated clandestine graves. *Forensic Science International* **192**: 2–13.

Killam, E. W. (1990) *The Detection of Human Remains*. Charles C Thomas, Springfield, IL.

Lowrie, W. (1997) *Fundamentals of Geophysics*, Cambridge University Press: Cambridge.

Milsom, J. (2003) *Field Geophysics*, 3rd edn. John Wiley & Sons Ltd, Chichester.

Nicholls Colton Geotechnical (2005) A ground investigation for the Moser Centre Building, University of Keele, Staffordshire, engineering report. Nicholls Colton & Partners Ltd.

Nobes, D. C. (2000) The search for "Yvonne": A case example of the delineation of a grave using near-surface geophysical methods. *Journal of Forensic Sciences* **45**(3): 715–721.

Palacky, G. J. (1987) Clay mapping using electromagnetic methods. *First Break* **5**: 295–306.

Pringle, J. K., Jervis, J., Cassella, J. P., and Cassidy, N. J. (2008) Time-lapse geophysical investigations over a simulated urban grave. *Journal of Forensic Science* **53**(6): 1405–16.

Ruffell, A. (2005) Searching for the I.R.A. Disappeared: Ground-penetrating radar investigation of a churchyard burial site, Northern Ireland. *Journal of Forensic Sciences* **50**: 1430–1435.

Ruffell, A. (2006) Under-water scene investigation using ground penetrating radar (GPR) in the search for a sunken jet ski, Northern Ireland. *Science and Justice* **46**(4): 221–30.

Ruffell, A. and McKinley, J. (2008) *Geoforensics*. John Wiley & Sons Ltd, Chichester.

Sharma, P. V. (1997) *Environmental and Engineering Geophysics*. Cambridge University Press: Cambridge.

Silliman, S. W., Farnsworth, P., and Lightfoot, K. G. (2000) Magnetometer prospecting in historical archeology: Evaluating surveying options at a 19th century rancho site in California. *Historical Archaeology* **34**(2): 89–109.

Ubelaker, D. H. (1999) *Human Skeletal Remains: Excavation, analysis, interpretation*, 3rd edn. Manuals on Archeology: Vol. **2**. Taraxacum, Washington.

Ubelaker, D. H., Ward, D. C., Braz, V. S., and Stewart, J. (2002) The use of SEM/EDX analysis to distinguish dental and osseus tissue from other materials. *Journal of Forensic Sciences* **47**(5): 940–943.

Witten, A. J. (2006) *Handbook of Geophysics and Archaeology*. Equinox Publishing, London. [This is an excellent, very readable introduction to geophysics that, unfortunately, has some significant formula errors in the text.]

Chapter 11
Environmental Forensics: Tracking Pollution to Its Source

Environmental crimes characteristically involve willful violations of federal environmental law. Not only must an environmental law be broken, the person or people responsible must be deliberately violating the law. This means that the situation cannot be the product of an accident or mistake, though gross negligence is considered a deliberate violation. Detailed knowledge of the specific statutes or regulations involved, however, is not required. For the violation to be considered "willful," it is sufficient for the violator to be aware that their conduct is illegitimate or probably unlawful. One of the best resources for information on environmental crimes in the United States is the Environmental Protection Agency's (EPA) website (http://www.epa.gov/oecaerth/criminal/investigations/environmentalcrime .html). A brief introduction to federal environmental laws and regulations in United States is available on the companion website (www.wiley.com/go/bergslien/ forensicgeoscience).

In the most basic sense, environmental forensics usually involves tracking pollutants to their sources. This can sometimes be quite a complex undertaking. While some pollutants are exclusively synthetic, like polychlorinated biphenyl (PCB) and trichloroethene (TCE), others, like arsenic and lead, can occur both naturally or be released into the environment as a result of human actions (i.e. they are *anthropogenic*). Many areas of the United States have long histories of industrial use that left behind extraordinary amounts of waste products, while other areas have hot spots of contamination due to naturally occurring geologic deposits. Sorting out new spills from old buried materials from natural deposits can be a daunting task. Determining the ultimate point of origin for a particular pollutant involves excluding alternate potential sources in order to satisfactorily pinpoint the cause or causes of a particular problem.

An Introduction to Forensic Geoscience, First Edition. Elisa Bergslien.
© 2012 Elisa Bergslien. Published 2012 by Blackwell Publishing Ltd.

Sometimes, the source might be an illegal discharge pipe, or a pit full of rusting barrels. These types of sources are called a *point source*, because the pollution can be tracked back to a specific point. Other times, pollutants come from a broad area, such as oil sprayed over dirt roads to keep down dust, or the fertilizers/pesticides used on farmland. These types of sources are called *non-point source pollution* and are generally much harder to pin down or control. Environmental pollutants can travel through the air and the water, they can contaminant soil, and they can kill wildlife. Sometimes, all these things occur at once. Environmental forensics is a huge subject that involves disparate disciplines, such as civil engineering and toxicology. This chapter is just a brief introduction to a few of the basic earth science systems concepts most likely to be encountered in a forensic context.

Firing the Public Imagination

Just before noon on June 22, 1969, the Cuyahoga River just southeast of downtown Cleveland, Ohio caught fire. Floating on the water was an oil slick composed of "volatile petroleum derivatives" and other flammable debris that had been discharged into the river, probably from one or more of the surrounding industries, though no one knows its ultimate source. Cuyahoga is a Native American word for "crooked river," and debris very commonly built up in the sharp bend of the slow-moving river. Authorities believed that sparks from a passing train probably ignited the oil, resulting in the implausible sight of a river roaring with flame. Reportedly, the fire reached a height of five stories. A fireboat, which regularly patrolled the river, and three fire battalions managed to put the blaze out in just over 20 minutes. The burning oil slick damaged two railroad trestles that passed over the river, noticeably warping the tracks of one of them. In total, an estimated $50,000 worth of damage was done.

This was not actually the first time that the Cuyahoga River had burned, nor was it the worst. There had been at least ten fires on the river; the earliest was probably in 1868. The largest fire occurred in 1952 and caused well over a million dollars in damage to boats and riverside property. Cleveland had become so used to river fires that in the major local papers the 1969 fire only merited a couple of small front-page pictures of the aftermath and no full-length stories. The fire was actually almost out before photographers had even arrived on the scene. According to the chief of the Cleveland Fire Department, William E. Barry, "it was strictly a run of the mill fire." In fact, Cleveland had long since become heartily sick of the condition of their waterways. So much so that the previous year voters had overwhelmingly passed a $100-million bond issue to allocate funds to improve existing sewage facilities and build 25 miles of trunk-line sewers plus a modern sewage treatment plant. At the time of the 1969 fire, Cleveland had already begun to clean things up.

In fact, these kinds of fires were almost routine throughout the industrial northeast, but for some reason the 1969 Cuyahoga fire caught the public imagination. The August 1, 1969, issue of *Time* magazine included an article with a rousing indictment of the state of US waterways:

Among the worst . . . is the 80-mile-long Cuyahoga . . . No Visible Life. Some river! Chocolate-brown, oily, bubbling with subsurface gases, it oozes rather than flows. "Anyone who falls into the Cuyahoga does not drown," Cleveland's citizens joke grimly. "He decays." . . . It is also – literally – a fire hazard.

The article also included a dramatic and rather misleading photo not of the 1969 fire but of the substantially worse 1952 blaze. This coverage was followed up by an article entitled "Sad, Soiled Waters: The Cuyahoga River and Lake Erie" in the December 1970 issue of *National Geographic*. The "river that burns" became one of the rallying points of the environmental movement, and its story, in a rather mangled form, became one of the catalysts that led to the formation of the EPA in 1970 and the passage of the Federal Water Pollution Control Act in 1972. Many of the environmental laws in the United States have a similar history with some key event that captures the public imagination and leads to the passage of laws to prevent similar future occurrences.

Water: Our Most Precious Natural Resource

As observed by Arthur C. Clark, our Earth could more properly be called Ocean, as water covers approximately 70% of the planet's surface. The oceans store ~97.25% of the Earth's total volume of water. Solar radiation striking the surface of bodies of water will give a small portion of the water molecules enough energy to move into vapor phase, i.e. *evaporate* (1 on Figure 11.1) and rise into the atmosphere, leaving behind the dissolved salts that make ocean water undrinkable for terrestrial organisms. At any given point, the atmosphere holds just 0.001% of the total water on the planet, but this seemingly tiny amount circulates over the globe (4) and constantly replenishes the world's freshwater supplies. Water vapor rises in the atmosphere, where it cools and *condenses* (3) to form clouds. When conditions are right, the water is released in the form of *precipitation* (5), as rain, snow, sleet, or hail. Most of this precipitation falls directly back onto the oceans, but a significant fraction falls onto the landmasses, where some of it flows and accumulates on the surface in the form of streams and lakes, called *runoff* (6), while another portion of the water sinks or *infiltrates* (7) into the ground, recharging subsurface *groundwater* (8) supplies. Eventually, water from the landmasses runs downhill and returns to the oceans. This process, called the *hydrologic cycle*, is constantly taking place (Figure 11.1).

The traditional hydrologic cycle consists of water moving from the oceans into the atmosphere, falling onto (or sinking into) the land, and then finally returning to the oceans. However, there are other paths. For example, precipitation can fall directly over the ocean, bypassing land altogether. If the rain falls into an area of thick vegetation, some or all of it might be *intercepted* by the plants so that the water never reaches the ground. The moisture can also return directly back to the atmosphere from plants, a process called *transpiration* (2 on Figure 11.1). Water can also become trapped in the upper layers of the soil, bound to organic matter, or used in biological activity. Soil moisture and the biosphere account for ~0.00504% of the water on Earth.

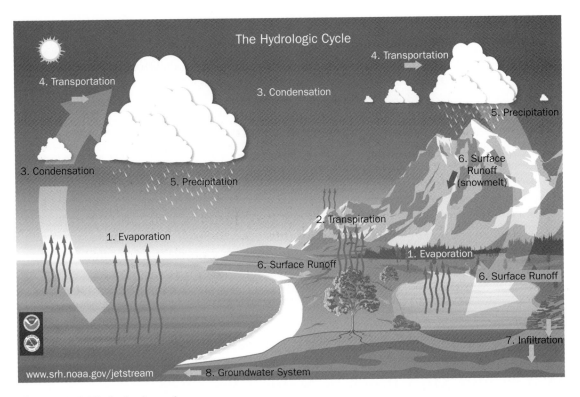

Figure 11.1 Hydrologic cycle.
Source: Modified from a National Weather Service/National Oceanic and Atmospheric Administration graphic.

Only ~2.75% of the water on Earth is freshwater, and just under 75% of that (or ~2.05% of the world's water) is locked up in ice caps and glaciers. This means that a meager 0.7% of the world's water supply is flowing freshwater. The vast majority of the liquid freshwater, just over 97% of it (i.e. 0.68% of the world's water), is stored underground. All of the rivers, streams, and lakes on Earth hold just 0.0101% of the total volume of water on the planet. This has two major implications. First, it should be clear that the usable freshwater supply on Earth is limited. Saltwater is undrinkable and cannot be used to grow crops. It takes vast amounts of energy to *desalinate* water from the ocean (i.e. remove dissolved salts to make the water drinkable), and even more energy to move any appreciable amount. Second, this means that groundwater is a vital drinking water resource. Given these considerations, it is extremely important to protect our freshwater supplies from contamination.

Surface Water

The study of the movement of surface water is called *hydrology*. In general, the flow of water on land is controlled by surface topography. As the old saying goes, water flows downhill. When precipitation hits the ground, it flows down gradient and collects at points of locally low elevation, such as streams, lakes, reservoirs (natural or artificial impoundment used to store drinking water), and wetlands. *Stream* is the

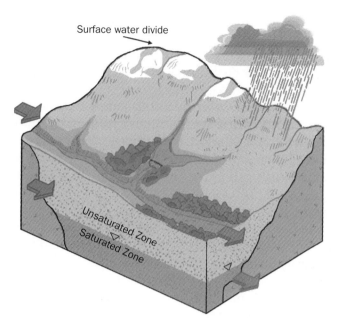

Figure 11.2 An example of a watershed divide. Precipitation that falls in front of the divide becomes part of the watershed surrounding the river. Precipitation that falls behind the divide flows backward to a watershed out of sight.

technical term for any channelized flow, regardless of its size. The words *river* and *creek* are colloquial terms whose use is locally determined. Though streams make up a very, very small fraction of the water on Earth, they are responsible for the vast majority of sediment transport on land. They are also extremely important water resources, supplying water for municipalities, irrigation, power generation, manufacturing, and industry. As you continue to move down gradient, small streams feed into ponds, lakes, and larger streams. Eventually, the water will coalesce into large streams that outlet into the ocean.

A *watershed* (also called a *drainage basin* or *catchment*) is the total land area from which the water drains to a common point, like a stream, wetland, or lake (Figure 11.2). Watersheds can be quite small or extremely large, depending on what common point you choose to examine. The watershed of a local creek could be only tens of square kilometers, while the Amazon River basin is in the order of 6.16 million square kilometers (2.38 million square miles). Each watershed is delineated by its surrounding points of highest elevation, like ridges or hills, which are designated *surface water divides*. To manually delineate a watershed, use a topographic map. First, identify the lowest point on the map for the water feature of interest and mark it with an "O." This is the *watershed outlet*. Now identify the *tributaries* (streams that feed into the water body of interest). Next, draw a line to the right of the outlet and ensure that the line is always at a right angle to the contour lines it crosses. All of the tributaries should be to the left of this line. Continue the line until just before it hits a contour line that forms part of a series with trending lower elevation. Repeat this process to the left of the outlet, ensuring that tributaries are to the right of the line.

For small watersheds, or if you get lucky with a large watershed, the lines that you have drawn from the right and the left of the watershed outlet will end near each other and it will be a simple matter to connect them along the zone of highest elevation, thereby enclosing all of the tributaries and outlining the watershed. For more complex watersheds, these lines will not end near to each other and you will need to identify a series of *breakpoints* (points of highest elevation surrounding the feature of interest and all of its tributaries). This requires a careful examination of contour lines, while envisioning the direction that water would flow in the landscape. Mark breakpoints with a series of "x" marks. Once there are breakpoints surrounding the water feature and all of its tributaries, draw a line to connect them, ensuring that it is perpendicular to every contour line it crosses. The final product delineates the watershed.

In the United States, watersheds are the basic unit of water resource management. The total inflows and outflows of water are monitored to ensure that there is enough to supply local needs, and flood control measures are now adapted for individual watersheds. A watershed approach is also used for addressing issues of water quality planning and monitoring for pollutants. If you live in the United States, you can "Surf Your Watershed" on the EPA's website (http://cfpub.epa.gov/surf/locate/index.html).

Clean Water Act

The Clean Water Act (CWA) 33 USC sections 1251–1387 was established in 1977 as an amendment to the Federal Water Pollution Control Act of 1972, which set the basic structure for regulating discharges of pollutants to the waters of the United States. The act underwent major amendment in 1987. The CWA gave the EPA the authority to set effluent standards (i.e. how much and what can be released) on an industry basis, reaffirmed the requirements to set water quality standards for all contaminants in surface waters, and required the EPA to prepare a list of priority pollutants to control. This law also provided grants to municipalities to build municipal wastewater treatment plants.

In addition, the CWA established the National Pollutant Discharge Elimination System (NPDES), which provides for the issuance of permits for discharges to surface water bodies. The CWA makes it unlawful for any person to discharge any pollutant from a point source into navigable waters unless a permit is obtained, though there are exclusions for household septic tanks. Notice that there are two important caveats: it only applies to navigable waters and to a discharge of point source pollution. In most cases, the EPA uses a watershed-based NPDES permitting approach that is supposed to address the impact of the total pollutant load to a hydrologically defined area, rather than on a permit-by-permit basis. The goal is to ensure that the water quality of the entire watershed is maintained.

One rather odd thing about water law in the United States is that surface water and groundwater are treated as separate things. The federal government has control over most surface water issues, while the state and local municipalities have control over groundwater issues. For an introduction to the CWA, visit http://www.epa.gov/r5water/cwa.htm; for the entire text of the law, visit http://www4.law.cornell.edu/uscode/33/ch26.html.

Mom, This Water Tastes Funny

For decades, Woburn, Massachusetts obtained its drinking water from six wells (named A–F) drilled into an aquifer that lay beneath Horn Pond. When this drinking water supply became inadequate for the growing population, city officials decided to drill two new wells, well G located near the Aberjona River in 1964, and well H located about 153 meters (500 feet) north of well G and even closer to the river in 1968. Almost as soon as the wells went online, residents of East Woburn complained that the water smelled and tasted bad. Repeated routine testing showed that the water was unpleasant, but safe to drink. Then in May 1979, a new round of testing was conducted because officials discovered that someone had dumped nearly 200 barrels of industrial waste in the vicinity of the wells. The new tests showed that the contents of the barrels had not contaminated the wells, but that both wells G and H were contaminated with several chlorinated organic compounds, including TCE and tetrachloroethene (PCE), both of which are commonly used industrial solvents. Tests for these particular contaminants had not been specifically conducted in the past. The wells were closed on May 22, 1979.

The discovery of the contaminants led to studies showing that Woburn's childhood leukemia rate was significantly higher than the national average (i.e. it was a *cancer cluster*) and that most of the cases were among the families that had received the bulk of their household water supply from wells G and H. Six families, later increasing to eight, filed lawsuits against three major corporations, alleging that their improper handling of industrial solvents had contaminated the local groundwater supply, causing leukemia, as well as several other illnesses, including cardiac arrhythmia and neurological disorders. Seven children of the plaintiffs had contracted leukemia and five of them died. The spouse of another plaintiff died of acute myelocytic leukemia. Plus, there were additional deaths in families that did not enter the lawsuit.

There were several industries that could have been responsible for the contamination, but the lawsuit focused on three. W. R. Grace and Company operated the Cryovac Division manufacturing plant, located approximately 732 meters (2400 feet) from the wells. The company used solvents to clean equipment. The Beatrice Food Company purchased property in the area that had formally been the site of a tannery. As part of a subsequent business deal, Beatrice agreed to retain liability on any environmental issues after reselling the land back to the Riley family. Dozens of degenerating 55-gallon drums of waste were found on sections of the property. Finally, UniFirst Corporation operated an industrial dry-cleaning business that used PCE about 610 meters (2000) feet from the wells. Tests on UniFirst property revealed high levels of PCE in both the soil and the underlying groundwater.

Due to the complexity of the issues involved, the actual trial did not start until nearly four years after the suit was originally filed. There was a significant amount of drama involving a variety of pretrial motions and a prolonged *discovery period* (the time during which parties have the right to collect information from each other and from third parties). As part of discovery, a variety of environmental tests were conducted. These tests established that the water in the shallow wells behind the Cryovac plant contained >8000 μg/L of

TCE, while water in the wells installed on the land Beatrice Foods retained liability for had concentrations of >400,000 μg/L of TCE. During all of the pre-trial maneuvering, UniFirst agreed to pay a $1.05-million settlement under the condition that the company did not admit to any responsibility for contaminating the wells. Unusually, the settlement also required that the money be used to finance the case against Grace and Beatrice.

The court case had been assigned to Judge Walter J. Skinner, who decided to break the trial into three separate phases. The first phase would deal exclusively with the movement of the contaminants into the wells. Only if one or both of the defendants were found guilty of contaminating the wells would the trial move to the second phase, which would require that causality be established between the contaminants and leukemia and/or the other illnesses alleged in the lawsuit. If a causal relationship were demonstrated, then the third phase would set penalties and damages.

The first phase of the trial finally began on March 10, 1986, and lasted for 78 days. It was almost entirely technical and mostly involved the testimony of expert witnesses. In the adversarial trial system of the United States, experts are in essence hired by the prosecution or defense, which unfortunately means that some experts view themselves, either consciously or unconsciously, as working for one side or the other. This can bias the view of the unwary expert. There are many ways that otherwise well-intentioned scientists can find themselves out on a limb. Lawyers, with either side of a case, can tempt experts to speak to topics that are outside their area of expertise, which can sometimes lead to a witness's entire testimony being thrown out or disregarded by the jury.

There are also scientists who appear, for whatever reason, not to quite understand the gravity of courtroom testimony. One frequently cited example of this is found in the book *A Civil Action* by Jonathan Harr (1996), which tells the story of the Woburn lawsuit through the perspective of Jan Schlichtmann, the lead attorney for the plaintiffs. A significant portion of the complex case involved scientific issues pertaining to *hydrogeology* (study of the occurrence and movement of groundwater) and *toxicology* (the study of toxic substances and their effects on living organisms). One of the major obstacles for the plaintiffs was that their lawsuit was filed before the EPA or United States Geological Survey (USGS) had completed their investigations. The area had been listed as a Superfund Site in 1983.

In order to address the complex issue of how the contaminants could have moved from the spill or dump sites on corporate-owned lands into the groundwater table and under a river finally to arrive at municipal drinking water wells, Schlichtmann introduced a renowned specialist, Princeton Professor George Pinder. Unfortunately, Dr Pinder appears not to have done sufficient preparation for his trial appearance and his testimony was demolished during cross-examination. Harr's book describes several difficulties with Dr Pinder's testimony and one fatal error that is sometimes referred to as Dr Pinder's *shower epiphany*. Never tell a jury that the answer to a complex question just occurred to you while you were taking a shower that morning. They will not be impressed.

In simple terms, Dr Pinder testified that although the contaminated groundwater flowing into the municipal wells moved from the corporate properties and underneath the Aberjona River, the Aberjona River itself contributed no water to the wells. The defense lawyers were able to demonstrate this was wrong and the USGS later determined that between 40 and 50% of the water to the wells came from the river. In Dr Pinder's defense, it is unclear if the story told in Harr's book is exactly what transpired. Whatever the case, the errors in Pinder's testimony came off as arrogant conjecture rather than honest mistakes and fundamentally undermined the plaintiffs' case. This story is used as a cautionary tale in hydrogeology classes throughout the United States.

After nine days of deliberation, the jury found that there was insufficient evidence against Beatrice Foods, while W. R. Grace was found liable for contaminating the wells. However, the convoluted and very poorly worded *jury interrogatories* (a written list of specific yes/no questions that the judge instructs the jury to answer) led to significant confusion and the jury could not agree as to when negligent behavior began. The first phase of the trial ended with the case against Beatrice Foods being dismissed. Theoretically, the trial would move into phase two against Grace but, as a consequence of the confusing findings by the jury, the judge suggested that he was likely to grant a motion for a new trial due to the inconsistency between the earliest date that the jury found Grace could be held negligent and the dates when some of the plaintiffs' children were diagnosed with leukemia. Such a move would cause a retrial for the first phase of the proceedings.

Hours before the second phase of the trial was scheduled to begin, W. R. Grace settled with the families for $8 million, with Grace maintaining that it was not responsible for the contamination. Most of the money went toward legal fees and trial expenses. Plus five other families being represented by Schlichtmann who were suing in another court, where the trial had not begun, were to end their suit and share in the settlement. After years of turmoil, each family received less than $300,000 from the settlement. At the same time, Judge Skinner ruled in favor of an appeal filed by Grace for a mistrial, vacating the jury's verdict, and ordering that the first phase of the trial against Grace be held over again. This in effect ended the trial and the families involved never even got a chance to speak on the witness stand.

In 1987, the USGS released a report entitled "Area of Influence and Zone of Contribution to Superfund Site Wells G & H, Woburn Massachusetts," which clarified the regional groundwater flow in the area and showed that contamination had in fact reached well G and H from the Beatrice Foods property. There were additional appeals and other legal maneuvers, but for the families that had lost children the process was ultimately disappointing. In July 1991, the EPA finalized a $69.5-million settlement with five responsible parties for the cleanup of wells G and H of the Superfund Site. The parties found responsible were W. R. Grace, UniFirst Corporation, New England Plastics, Olympia Nominee Trust, and Wildwood Conservation Corporation, the new owners of the now defunct Beatrice Foods property. The site is still on the National Priorities List. For additional background on this extraordinary and ultimately unsatisfactory case, Dan Kennedy, a reporter who followed the case, has an excellent website at http://www.dankennedy.net/woburn-files, as does reporter Charles Ryan (http://www.northshoreonline.com/woburn/). Plus, the EPA maintains an archive of all of its activities at the site (look up "Wells G & H" at http://www.epa.gov/region01/superfund).

CERCLA and SARA

In 1980, Congress passed the Comprehensive Environmental Response, Compensation, and Liability Act (CERCLA) to deal with sites like Love Canal and Woburn's Wells G and H. Before CERCLA, the federal government had virtually no ability to play an active role in the response to the discovery of hazardous waste sites or accidental spills. CERCLA provided the EPA with both the authority to take action and the necessary funding to initiate a response, such as the remediation or removal of contaminants. Congress established a $1.6-billion fund to implement the massive cleanup program that was to last for five years, after which the program was to be self-funding. The intention was to use the funding to identify hazardous waste sites and clean them up, then establish liability before recovering the costs from the identified potentially responsible parties. Unfortunately, CERCLA set some unrealistic goals and deadlines in terms of the technology available, and significantly underestimated the number of toxic waste sites that would qualify, so that at the end of the initial five-year period only six sites had been cleaned up (http://www.epa.gov/superfund/policy/cercla.htm).

CERCLA was amended with the Superfund Amendments and Reauthorization Act (SARA) in 1986. Today, the term *superfund* refers to both laws and the cleanup program thus mandated. SARA was essentially a complete overhaul of CERCLA. Congress increased the level of funding to $8.5 billion for the cleanup of abandoned sites and added an additional $500-million fund specifically for dealing with leaking petroleum tanks. SARA also established the Right-to-Know provisions, which require industries to plan for emergencies and inform the public of hazardous substances being used. Additional information on CERCLA and SARA can be found at the EPA website (http://www.epa.gov/superfund/).

Groundwater

Groundwater, or subsurface water, is simply any water that exists below the Earth's surface. The study of the movement and contamination of groundwater is called hydrogeology. Groundwater and surface water are inseparably linked, no matter what the legal framework suggests. When one is affected, so is the other. The difficulty with groundwater issues is two-fold. One is that groundwater is a "hidden" resource: out of sight, out of mind. It is generally fairly easy to determine whether surface water is polluted (though there are new classes of contaminants like pharmaceuticals which are difficult to track), and if the problem is bad enough it can be readily observed via the color/condition of the water and by the lack of biodiversity. Groundwater pollution is hidden deep underground and completely out of sight unless you happen to have a well that taps into a polluted source. Once you locate polluted groundwater, it can be quite difficult to track down where contaminants are entering. The other issue is temporal. Surface water moves quickly, but groundwater moves very, very slowly. Most groundwater moves less than ten meters per year (10 m/yr). It can take years for a problem to even be detected. With effort, surface water can be remediated relatively quickly, but the inaccessibility and slow movement of groundwater makes cleanup quite difficult. For these reasons, many environmental forensic cases involve groundwater.

As you saw in the hydrologic cycle (Figure 11.1), when precipitation falls onto the ground, some of it will flow along the surface as runoff, while another portion of it will sink into the ground, replenishing the groundwater supply. The Earth beneath our feet is composed of a complex series of rock and sediment layers with differing compositions and properties. Deep underground, the Earth's surface is formed of dense, crystalline bedrock, like granite or gabbro, usually buried under layers of sedimentary rock, like sandstone, shale, and limestone. Whatever the underlying bedrock is composed of, the nearer you come to the surface, the more that geological material is subjected to weathering, which physically and chemically breaks the rock down into fragments. These fragments can be transported or form the basis of an *in situ* soil. Most landmasses are draped in layers of loose sediment deposited by the actions of wind, ice, and/or water. All of these fragments of weathered rock and layers of loose sediment contain vast amounts of *void space* (or pore space), the empty areas that lie between individual grains. It is through these void spaces that water, under the force of gravity, sinks into the Earth (Figure 11.3).

Water will travel through the soil and loose sediment layers down to the underlying bedrock. Some types of bedrock, like sandstone, also contain lots of void spaces, which water flows through or accumulates in, while other types of bedrock, like unfractured shale and crystalline granite, are almost completely lacking in voids. All types of bedrock, however, can potentially become fractured, creating additional pathways for water to flow. Some types of bedrock, such as limestone, can even be dissolved by contact with water, creating caves and large

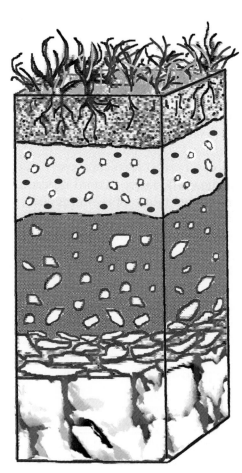

Figure 11.3 Vertical cross-section from the surface to bedrock. The relative amount and size of void space decreases with increasing depth.

cavities, though this is generally rare. Underground streams that flow through caves large enough to row a boat in (à la *Phantom of the Opera*) are an extremely rare and highly localized phenomenon.

If you were to cut a vertical cross-section of a landmass on Earth, at the top you would generally find a layer of soil grading into fragments of the underlying sedimentary bedrock (Figure 11.3). This would be underlain by a series of sedimentary rock layers, eventually reaching a dense bedrock layer formed of tight, unfractured crystalline rock. In some of these layers, water will move relatively freely, while other layers will impede flow. When the downward migration of water is blocked, it tends to migrate laterally (horizontally) inside of layers where the water can flow more freely, often ultimately seeping into surface bodies of water like streams and lakes, or escaping directly into the oceans.

The term used to describe the amount of void space in a material is *porosity*, which is technically defined as the volume of void space (Vv) in a sample divided by the total volume (Vt) of the sample and ranges from 0% (solid rock) to 100% (pure void).

$$n = (V_v/V_t) \times 100\% \tag{11.1}$$

In general, porosity depends on the shape of the grains (the more spherical the grains, the greater the void space, while the more angular the grains, the less the void space), the sorting of the grains (the more uniform the grains, the more the void space), and the efficiency of grain packing (loose, chaotic assortments have more void space than neatly aligned, well-nested arrangements) (Table 11.1). However, porosity encompasses all voids, even those that are isolated and through which no flow would be possible. To truly describe flow, another descriptive property is needed.

Permeability is used to describe how much water can flow through a material. Rocks, sediments, and soils vary greatly (over 13 orders of magnitude) with respect to their ability to transmit fluid. Part of this is related to the interconnectivity of the void spaces. If you can look closely enough at a block of sandstone or a layer of sediment, you can see that not all void spaces are connected (Figure 11.4). Some spaces are totally isolated, while others lead only to dead ends. Each of these dead ends will actually impede flow, even though they technically count as void space. Thus, high porosity does not always equate to high permeability. This is especially obvious with rocks like vesicular basalt. Gas bubbles create numerous little spaces in the basalt, but none of them is connected, which means that while the porosity of vesicular basalt can be quite high permeability is extremely low.

The other difference has to do with interactions between the medium and the fluid. As with porosity, grain shape, sorting, and packing all help to determine the degree of permeability. The more irregular the grains and the tighter the packing, the lower the porosity and the lower the permeability. However, one significant factor is different. Because porosity is a purely geometric description, the size of the grains does not really matter because you are looking at a ratio. Beach balls, ping-pong balls, and round sand grains all have the same porosity as long as they are stacked the same way. With permeability, size does matter.

Remember that water is a highly polar, asymmetrical molecule with two hydrogen atoms bonded to one atom of oxygen. The side of the molecule with the hydrogen atoms has a slight positive charge, while the other side of the molecule has a slight negative charge. The result is a somewhat boomerang-shaped molecule

Table 11.1 Porosity and hydraulic conductivities of common geology materials.

Sediment/Rock	Porosity (n) %	Hydraulic Conductivity (K) cm/sec	Effective Porosity (n_e) %
Gravel, unconsolidated	25–40	10^{-1}–10^{+2}	13–26
Sand, fine, or silty, unconsolidated	20–50	10^{-5}–10^{-3}	10–28
Sand, coarse, med., unconsolidated	20–50	10^{-3}–10^{-1}	20–35
Silt, unconsolidated	35–50	10^{-7}–10^{-3}	3–19
Mud/clay, unconsolidated	40–70	10^{-9}–10^{-6}	0–5
Basalt	1–50	10^{-9}–10^{-6} unfractured 10^{-4}–10^{+3} fractured	
Karst limestone	5–50	10^{-4}–10^{+4}	5–40
Sandstone	5–35	10^{-8}–10^{-3}	0.5–10
Limestone/dolostone, unfractured	0.1–25	10^{-7}–10^{-3}	0.1–5
Shale, fractured	30–50	10^{-7}–10^{-4}	
Shale, unfractured	1–10	10^{-11}–10^{-7}	0.5–5
Unfractured crystalline rock	0.01–1	10^{-12}–10^{-8}	0.0005
Fractured crystalline rock	1–10	10^{-8}–10^{-4}	0.00005–0.01

Source: After Sanders, 1998.

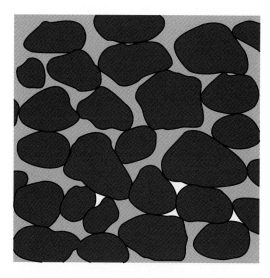

Figure 11.4 Pore space in a layer of unconsolidated sediment. The gray areas are connected, while the white areas represent dead-end void spaces.

that acts somewhat like a tiny magnet. The opposite charges on water molecules are strongly attracted to each other (*hydrogen bonding*), giving water all sorts of interesting properties (like high surface tension and a high boiling point). This also means that water molecules are attracted to other charged surfaces, which is why your windshield gets covered in water droplets when it rains. Geological materials, which contain unsatisfied/broken bonds, have charged surfaces that will attract a thin layer of water molecules. These *adsorbed* (attached to the surface) water molecules will in turn attract more water molecules, creating a thin layer of immobile, bound water.

The thickness of the adsorbed layer is determined by the balance between the strength of the attraction versus the force of gravity. The upshot is that this layer of immobile water molecules can completely block small voids, but the larger the void, the smaller the affect. Larger grains create larger void spaces, so that the adsorbed water will only occupy a small portion of the void and the media will have a higher permeability. Coarse, unconsolidated sediments like clean sands and gravels are highly permeable. In fine-grained materials with smaller void spaces, like very fine sand and silts, the adsorbed water will block a larger portion of the void space so these media tend to impede flow. Clays tend to have highly charged surfaces so that water will become strongly bonded. Clay actually has very high porosity but most of the water is bonded so that clay layers generally have quite low permeability.

The subsurface is vertically broken into two major groundwater zones: the *vadose zone* (or zone of aeration), which is characterized by void space that contains a mixture of air and water, and the *zone of saturation* (or the phreatic zone), in which the void space is completely filled with water (Figure 11.5). The vadose zone is further subdivided into the soil moisture belt, the intermediate zone, and the

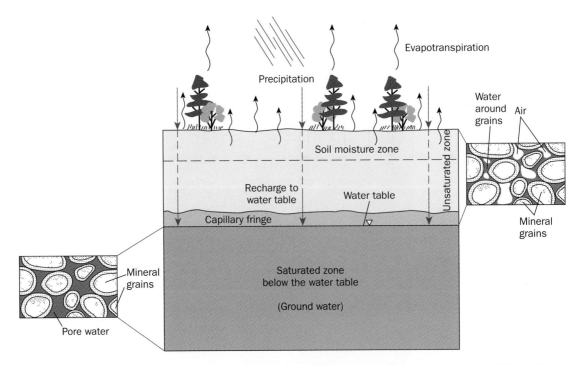

Figure 11.5 Cross-section showing the relative locations of the vadose zone, water table, and saturated zone.
Source: Modified from a United States Geological Survey graphic.

capillary fringe. When water first percolates into the ground, some portion of it will become bound to the clay-sized particles and decaying organic matter found in O and A soil horizons (described in Chapter 7). Another portion of this water will be pulled in by plant roots or otherwise used by the biosphere. The zone where all of this occurs is referred to as the *soil moisture belt*. The water that continues to flow downward next passes into the *intermediate zone*, which contains little to no organic matter, and is generally characterized by larger void spaces. A small amount of the water will be attracted to the surfaces of the geologic media in the intermediate zone but most of it will travel unimpeded downward under the influence of gravity until it reaches the capillary fringe.

The *capillary fringe* is a section of the vadose zone that is almost completely filled with water that is unable to flow. In this area of the vadose zone, the water is bound within the pores of the geologic media by *capillary forces*, an expression of hydrogen bonding. The attraction of the water to geologic media actually overcomes gravitational pull such that the water will "climb up" from the underlying saturated zone. You can see this effect in glass tubes or even plastic drinking straws. Water will climb up a thin glass tube because of the strong hydrogen bonding between the water molecules and the unsatisfied oxygen atoms, and terminal hydrogen atoms, that form the surface of the glass (which is basically SiO_2, i.e. quartz). The intermolecular attraction is balanced against gravity, which will pull the liquid back down. The narrower the diameter of the tube (or the smaller the void spaces) the higher upward the water will travel because the thinner the column of water is, the less it weighs (Figure 11.6). The larger the tube (or the bigger the voids), the thicker and the heavier the column of water will be, limiting the height of the rise. Fine-grained soil with small, well-connected voids is akin to bundles of very small glass tubes. In clays, the capillary fringe can be as much as

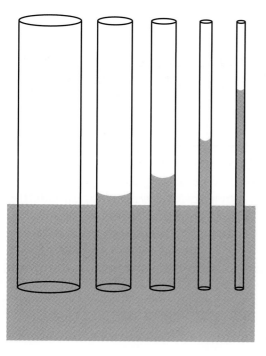

Figure 11.6 Capillary rise in a series of increasingly smaller diameter tubes.

10 m thick and in silts the rise is often as high as 1–3 m. On the other hand, capillary rise in coarse sands and gravels usually varies from near zero to a few centimeters.

The boundary between the vadose zone, at the base of the capillary fringe, and the zone of saturation is called the *water table*. Technically, the water table is the elevation at which the pressure of the overlying water column is equal to the atmospheric pressure. The term is rather unfortunate in that it implies a fixed, horizontal border. In actuality, the water table is usually a suppressed reflection of the surface topography, standing somewhat higher under hills and lower in valleys. Plus, its position fluctuates. Under wetter conditions, the water table rises, while at the height of summer, or during a drought, the water table will fall. There are even small daily variations caused by changes in barometric (air) pressure, as weather fronts move through. Where the water table intersects (or lies above) the ground surface, springs, lakes, swamps, or rivers are present.

Strata in the saturated zone are characterized as being either *aquitards*, which block or retard water flow, or *aquifers*, through which water flows easily. Strictly speaking, an aquifer is a zone with high porosity and high permeability that is saturated. Aquifers can be found in almost any area, though they are most commonly associated with layers of sandstone or sand. Highly fractured shale or bedrock can also make a very good aquifer.

Aquifers are further subdivided depending on whether they have a direct connection to the atmosphere. *Unconfined aquifers* are in direct contact with the atmosphere and the vadose zone. The water table forms the upper boundary of an unconfined aquifer. A *confined aquifer* is confined between two aquitards that act to block the flow of water from the overlying units and prevent water from migrating further downward. In this case, the water is not in direct contact with the atmosphere and, technically, there is no water table. The water in a confined aquifer is pressurized so that if you were to drill a well the water would rise above the base of the confining layer to a level called the *potentiometric surface*, or theoretically the level the water would be at if the aquifer were unconfined.

A key characteristic that determines whether a unit is considered an aquifer is if it will yield "useful quantities" of water to intersecting wells. There is not a hard-and-fast definition for what qualifies as a useful quantity of water. It mostly depends on local conditions and the local climate. In dry areas, some subsurface reservoirs of water are called aquifers that would not merit any interest in areas where surface water is plentiful. Whatever the local interpretation, it is the degree of permeability that separates aquifers from aquitards.

Permeability, however, cannot be determined directly. Instead, it is typically measured indirectly using a relationship called *Darcy's Law* and reported in terms of a property called *hydraulic conductivity*. Darcy's Law is named after a French engineer, Henry Darcy, who was responsible for designing a water filtration system for Dijon, France. In order to improve the efficiency of his design, he made the first systematic study of the movement of water through a porous medium. With the publication of his report, in 1856, the quantitative science of groundwater hydrology – hydrogeology – was born.

Hydraulic conductivity (K) is a characteristic of both the medium and the fluid, in this case water. In addition to hydrogen bonding potential, the viscosity of a fluid will also profoundly affect flow rate. Think of the difference between flowing honey and flowing water. If you are interested in the flow of oil or brine, a slightly different relationship is used. According to Darcy's Law, hydraulic conductivity can be defined as:

$$K = (Q/A) \times (\Delta l/\Delta h), \tag{11.2}$$

where Q is the flow rate expressed as a volume over time, A is the cross-sectional area of the zone of flow, $\Delta l = (l_a - l_b)$, the linear distance between two points of measurement, and $\Delta h = (h_a - h_b)$, the change in the elevation of the water table between the two measurement points or vertical drop (technically called the *change in head*). Hydraulic conductivity is reported in distance over time, typically centimeters per second, but should not be confused with velocity. This value does not describe how fast the water is moving. Hydraulic conductivities for common geologic media are in Table 11.1.

To determine approximately how fast groundwater is moving, the following relationship is used:

$$v = (K/n_e)(\Delta h/\Delta l), \tag{11.3}$$

where v is average groundwater velocity, K is hydraulic conductivity, n_e is the effective porosity, and, as before, Δh is the change in head and Δl is the distance traveled. *Effective porosity* (n_e) is somewhat akin to permeability and is defined as the percentage volume of a geological material that consists of interconnected void space. Effective porosity will always be less than or at most equal to the total porosity of a medium.

The expression ($\Delta h/\Delta l$) is known as the *hydraulic gradient* and is a way of conveying the "steepness" of the water table or the potentiometric surface. If you take a ball and roll it down a slope, the steeper the incline, the faster the ball will roll. Hydraulic gradient basically works the same way: the greater the difference between two points, the faster the water will move between them. The major difference is that hydraulic gradient includes pressure differences as well as elevation differences. The good news is that the *hydraulic head* (h), a parameter that expresses the energy of the groundwater due to both elevation (i.e. potential energy) and pressure, can typically be measured quite easily. You just need to determine the elevation to which water rises in a well or in a *piezometer*, which is essentially a pipe with a very short *screened* (i.e. open) *interval* that allows water to enter.

For surface bodies of water, the elevation of the water surface equals its hydraulic head. For groundwater sources, you need to use a *water level meter*, which is essentially a long tape measure with a weighted conductivity sensor on one end and a buzzer and/or light on the other. Water level meters are used to measure the distance from the top of the well casing down to the top of the water column in the well. The sensor is slowly lowered into the well, and when the sensor contacts water a circuit is completed, activating the buzzer and/or light. The depth to the water is then measured off the tape.

By itself, a depth-to-water measurement is not particularly useful. This information needs to be translated into an elevation, which is then equivalent to the hydraulic head at that point in the groundwater system. To convert a depth-to-water into a hydraulic head, the elevation of the top of the well casing (often abbreviated TOC) must be known. Usually, this information is on the drill log from the installation of the well, or a surveyor can determine the elevation. The depth-to-water is subtracted from the elevation of the top of the well to obtain the water level elevation. For example, if the elevation of the well casing is 315 m above mean sea level and the depth to water is 7 m, then the elevation of the water table is 308 m.

To determine groundwater flow, hydraulic head measurements from two (or more) wells are needed. Picture two wells, "A" and "B," that are located a short distance (Δl) from each other. The elevation of well A is 338 m and the elevation of well B is 347 m. To determine which way the water is flowing (from A to B or from B to A), you need to determine the hydraulic head for each well. Using a water level meter, you find that the water level in well A is 3 m BTOC (below top of casing) and that the water level in well B is 12 m BTOC. Subtracting the depth-to-water from the elevation of the well casing, you find that the elevation of the water table is 335 m in well A and 332 m in well B. Water flows downhill or, more technically speaking, from areas of high hydraulic head to areas of low hydraulic head. This means that water is flowing from well A to well B. It is important to recognize that groundwater flow direction does not necessarily have anything to do with local surface topography, though regional topography does matter. One of the most important jobs for a hydrogeologist is determining the direction of regional groundwater flow. This knowledge is critical for understanding the migration of contaminants in the subsurface, for determining risks to drinking water supplies, and for planning remediation efforts.

Groundwater flow is broadly horizontal, but it can be vertical as well. Wells "A" and "B" can be positioned at the same physical location at the surface, but tap into different depths in the subsurface. Such a construction is called a *piezometer nest*, though they usually have three or more piezometers installed side by side. By determining the difference in hydraulic head (Δh) between the discrete elevations at which the screened intervals are positioned (Δl), you can find the vertical gradient. This can be especially useful in areas where there is more than one aquifer. If an upper aquifer is contaminated and a lower aquifer is a municipal drinking water supply, it is important to understand the dynamics between the two aquifers.

Information from two wells only tells you the water flow direction between those two points. For a more detailed understanding of regional groundwater flow, hydraulic head data must be collected from as many sources as possible in order to construct a *water level map*, for data from unconfined aquifers, or a *potentiometric surface map*, if you are using data from a confined aquifer. These maps are similar to topographic maps, but show hydraulic potential and regional flow patterns instead. There is a detailed exercise, called A Criminal Case of LUST, available on the companion website that describes how to construct a water level (potentiometric surface) map and determine the hypothetical source of a contaminant (includes Figures 11.7 and 11.8 and Tables 11.2 and 11.3). For a more detailed introduction to hydrogeology, see Fetter (2000).

Contaminant Hydrogeology

When something is leaked into groundwater, the bulk of a dissolved contaminant will migrate in the direction of groundwater flow. By looking at a map showing regional flow, it is possible to narrow down the likely source of contamination. However, to further narrow down the possible location of a spill, it is often necessary to determine the concentrations and locations of the contaminants in the subsurface. As dissolved materials, or *solutes*, move in the subsurface, they spread out, forming a subsurface *plume* (zone of contamination), the shape of which is determined by three fundamental processes. The first is *advection*, which is simply the solute being carried along by the flowing water. You can roughly determine the speed at which the bulk of the contaminants will be transported, via advection, by calculating the average linear groundwater velocity.

Next is *dispersion*, which is caused by mechanical mixing of the water mostly due to variations in groundwater velocity. This might make more sense if you imagine covering the end of a garden hose with your thumb, narrowing the opening and causing the water to squirt out at high velocity. If you move your thumb, the water pours out the wider opening more slowly. As the groundwater flows, some of it will move through narrow void spaces, so that it has a higher-than-average velocity, while some of it will move through large voids, resulting in a lower-than-average velocity. This variation in velocities causes the plume to spread out in the direction of flow, known as *longitudinal dispersion*. While the majority of dispersion is longitudinal, causing the plume over time to get longer and longer, some *lateral* (sideways) *dispersion* also occurs because as the water flows it will be divided and reunited following the tortuous pathways around grains. This causes the plume to get "fatter" and more diffuse the further it flows.

The final process is *chemical diffusion*, the movement of solutes from areas of high concentration to areas of low concentration at a rate that is unrelated to groundwater flow velocity. Diffusion occurs even when water is still. Imagine putting a few drops of food coloring into a cup of still water and just leaving it alone. Over time, the dye will spread evenly throughout the water as a result of diffusion. Notably, this means that, due to chemical diffusion, some portion of the contaminant plume is actually moving faster than the highest groundwater velocity of the plume. Diffusion occurs in all directions away from the plume, even backward, causing the plume to widen and lengthen as well. The effects of mechanical dispersion and chemical diffusion can be impossible to distinguish, so they are often combined into a single factor called *hydrodynamic dispersion*.

The combined effects of advection and hydrodynamic dispersion result in an elongate subsurface plume that is narrower and more concentrated near the source of the leak or spill and that gets wider and less concentrated in the direction of flow (Figure 11.9). With an active leak, the concentration of the contaminant will be greatest close to or beneath the source and will become less concentrated as the plume migrates and is diluted with clean water. Spills or temporary leaks create disconnected plumes or *slugs* that can be much harder to link back to their source. Slugs get larger and more diffuse the further they travel.

To delineate the location of a subsurface plume, you plot the measured concentrations for a single contaminant at each of the wells on a water level map. All of the values used must have been measured at roughly the same time. Next, pick an appropriate contour interval, usually a whole number value, though sometimes logarithms are used when the concentrations range over many orders of magnitude. Each contour line is drawn so that it encircles values greater than its own (Figure 11.10). The result should be a series of concentric shapes that outline the shape of the plume and indicate the locations of the *hot spot(s)* or area(s) with the highest level(s) of contamination. The outer boundary of the plume is sometimes designated with a dashed line that is designated "non-detect" (N/D), "less than level of detection" (<LOD), or "below-detection-level" (BDL). Figure 11.11, showing the benzene plume for the on-line Harpers Corners exercise, is available at the companion website.

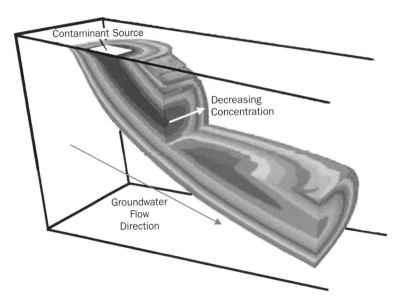

Figure 11.9 Combined effects of advection and hydrodynamic dispersion result in an elongated subsurface plume.

This was just a simplified introduction to contaminant hydrogeology. There are many complicating factors. For example, different classes of contaminants behave differently in the environment, groundwater flow patterns can vary seasonally, and multiple leaks can occur in the same area. In order to sort it all out, one would typically call in a *contaminant hydrogeologist* (someone who specializes in the study of how contaminants move in the subsurface).

Safe Drinking Water Act

The Safe Drinking Water Act (SDWA; which applies to all public wells and to private wells with more than 25 users) originally passed in 1974 to ensure that public water supplies were maintained to a high quality. It was amended 1986 to require that the EPA set national primary drinking water standards, including *maximum contaminant levels* (MCLs), which are legally enforceable limits, and *maximum contaminant level goals* (MCLGs), which are not legally enforceable but instead are aspirational levels that would best prevent potential health problems but are, for technological or other reasons, not easily achieved. The SDWA also sets secondary standards, which are also non-enforceable guidelines based on aesthetics, such as taste, odor, and color, and on cosmetic effects. The SDWA focuses on all waters actually or potentially designated for drinking use, whether from aboveground or underground sources, so it overlaps somewhat with the CWA. For more information on the SDWA, visit the EPA website (http://www.epa.gov/safewater/sdwa/sdwa.html).

Water-quality Measurements

Water sampling is conducted for a wide variety of reasons, from determining the chemical state of natural waters and understanding how different reservoirs interact with surrounding geological materials, gases, and other fluids to determining the relationships between watersheds and measuring the concentrations and movement

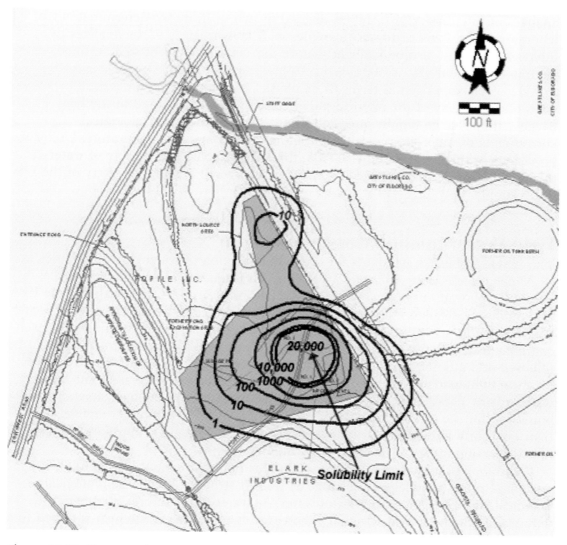

Figure 11.10 Contour plot delineating the location of a subsurface plume of pentachlorophenol (PCP).
Source: Courtesy of the US Environmental Protection Agency.

of contaminants in water supplies. There are two major types of water sampling: groundwater sampling and surface water sampling. Groundwater sampling is significantly more complex and requires more equipment, but many of the basic parameters measured are the same.

First, before any environmental sampling occurs, a collection protocol must be established. This protocol includes information about how many and what types of samples will be collected, what equipment will be used for collection and describes the sampling process in detail. Information such as the types of sample containers to be used (glass or what type of plastic, for example), any pre-treatment needed, and how to decontaminate the sampling equipment between collections is also included. For environmental work, maintaining chain-of-custody and QA/QC (quality assurance and quality control) procedures are just as important as they are in police laboratories. As part of the QA/QC measures, not only are environmental samples collected in the field but also each protocol will specify the number and type of reference samples, *blanks* (samples that should contain no trace of the analytes of interest), *splits* (samples divided in two or more for determinations of laboratory precision), and *spikes* (samples with known levels of the analytes of interest) that must also be included before the samples are sent to a certified laboratory for analysis.

Water is characterized in terms of its physical, chemical, and biological components. Many of the physical parameters must be measured in the field because they change rapidly after collection. The chemical and biological constituents of water are typically determined in an analytical laboratory. Under the provisions of the CWA and the SDWA, the EPA established a variety of water quality parameters that are used to monitor the overall health and safety of the US water supply.

Field Water-quality Measurements

The USGS publishes standards for the field collection of water quality samples in its *National Field Manual for the Collection of Water Quality Data*, which is available online (http://water.usgs.gov/owq/FieldManual/index.html). On its website, the EPA also publishes detailed information on how samples are to be collected, stored, and analyzed (http://www.epa.gov/safewater/methods/analyticalmethods.html). For each component of a water analysis, the EPA has recommended procedures that must be followed and a timeframe by which all analysis must be completed. While analysis for most contaminants must be conducted in a laboratory, there are some basic water quality parameters that can, or must, be measured in the field that can give you a feel for the general "health" of a body of water. For a more complete discussion of water quality issues, see Perry and Vanderklein (1996), and for a very nice discussion of *field hydrogeology*, see Sanders (1998).

Temperature is known as a *controlling variable*, one that once it begins to change will cause other parameters to change as well. It strongly affects the biological, chemical, and physical properties of water. As water temperatures go down, the amount of dissolved oxygen actually increases. If too much waste heat is added to a surface body of water from power generation, industrial, or manufacturing processes, it can severely affect an aquatic ecosystem. Fish, insects, zooplankton, phytoplankton, and other aquatic species all have a preferred temperature range. If the temperature deviates too much, the number of species in a body of water will decrease until there is none left. Water temperatures affect organisms' metabolic rates, as well as their ability to absorb nutrients and their sensitivity to waterborne diseases. The reproductive cycles of many aquatic organisms and their development from juvenile to adult forms are controlled by temperature cues as well. Many types of activities, like cutting down trees along a stream, and thereby reducing shade, laying down an asphalt parking lot, or constructing a dam, can significantly change the temperature of a stream.

Temperature variations in a lake or stream can also be used to track a warm plume back to its discharge point, to delineate recharge and discharge areas in a watershed, and to differentiate multiple inputs into a water system. For such an important water quality parameter, temperature can be measured quite easily using a thermometer. For all measurements, a calibrated field thermometer or probe should be used. For surface water measurements, lower the thermometer from a bridge, or boat, into the center of the body of water at a point that is far away from the shoreline. The thermometer, or tip of the probe, should be immersed in the water and out of direct sunlight. You can also wade out, if it is safe (wearing boots or waders, rubber gloves and goggles if there is a chance that the water is contaminated), and set the thermometer into the stream or pond if the water is only elbow deep. If it is deeper, hold the thermometer tip in the water until the temperature reading remains constant for two minutes and record the temperature in °C. Be sure to hold only the top of the thermometer in order to decrease the effect of body heat changing the thermometer reading. You should also measure the air temperature at all points where water quality measurements are taken.

Groundwater measurements are best taken by lowering a downhole temperature probe into a well that has been *purged*. This means the stagnant water that has been sitting in the well has been pumped or bailed out and that the well has started to refill with water from the surrounding aquifer. If a probe is not available, reasonably accurate temperature measurements can be taken by quickly pulling up a water sample, pouring it into a thermos, and waiting for a couple of minutes for the thermos to equilibrate with the water temperature. Dump the water out, get a new sample from the well, and pour it into the thermos. Insert a thermometer, wait a minute, and then take a reading.

A primary factor indicative of the chemistry of a natural water system is *pH*. It also has a profound impact on biological activity and is an important factor in the overall health of a body of water. Pure water (H_2O) is both a weak acid and a weak base. This means that hydrogen ions (H^+) can be readily transferred from one water molecule to another, resulting in the formation of one hydroxide ion (OH^-) and one hydronium ion (H_3O^+) per pair of water molecules. In pure water at 25 °C, the concentrations of these ions is equal at 1.0×10^{-7} moles per liter (M).

$$[H_3O^+] = [OH^-] = 1.0 \times 10^{-7} \text{ M [at 25 °C]} \qquad (11.4)$$

When an *acid* (molecules that lose H^+), like hydrochloric acid (HCl), is added to water, the concentration of the H_3O^+ ion increases, because the HCl will *dissociate* (break apart) into H^+ and Cl^- ions in aqueous solution (aq). The free hydrogens will then combine to make more hydronium.

$$HCl_{(aq)} + H_2O \Leftrightarrow H_3O^+_{(aq)} + Cl^-_{(aq)} \qquad (11.5)$$

At the same time, the OH^- ion concentration decreases because the H_3O^+ ions produced in this reaction will neutralize some of the OH^- ions, forming more water molecules.

$$H_3O^+_{(aq)} + OH^-_{(aq)} \Leftrightarrow 2 H_2O \qquad (11.6)$$

The same thing would occur if a *base* (things that lose OH^-) were added to the water. Some of the free hydroxyl ions would react with the hydronium to form water. These reactions occur to drive the system back to an equilibrium state. At equilibrium, the product of the concentrations of the H_3O^+ and OH^- ions is constant, no matter how much acid or base is added to water. In pure water at 25 °C, the product of the concentration of these ions is always equal to 1.0×10^{-14}.

$$[H_3O^+][OH^-] = 1.0 \times 10^{-14} \qquad (11.7)$$

This value is an example of an *equilibrium constant* (K_{eq}): a numeric value that expresses the constant ratio of *activities* (effective concentrations) of the chemical components in a reaction that has reached equilibrium. Equilibrium constants are used to understand a wide variety of different natural processes.

In 1909, the Danish biochemist S. P. L. Sørensen suggested that, because the range of actual concentrations of H_3O^+ and OH^- ions in aqueous solutions varied over so many orders of magnitude, concentration should be reported on a more manageable logarithmic scale, which he named the *pH scale*. Because the H_3O^+ ion concentration in water is almost always smaller than 1, the log of these concentrations would be a negative number, so as to avoid having to constantly work with negative numbers, Sørensen defined pH as the negative of the log of the H_3O^+ ion concentration. Many textbooks substitute a hydrogen ion (H^+) for hydronium in this relationship, but naked protons do not really exist in a solution. The hydrogen ion form of the equation is chemical shorthand.

$$pH = -\log_{10}[H_3O^+] = -\log_{10}[H^+] \tag{11.8}$$

Applying the pH equation compresses the natural range of H_3O^+ ion concentrations from 1 to 10^{-14} into a pH scale that goes from 0 to 14, which is much easier to handle. If you remember, the concentration of the H_3O^+ ion in pure water at $25\,°C$ is $1.0 \times 10^{-7}\,M$. This translates into a pH for pure water of 7.

$$pH = -\log_{10}[H_3O^+] = -\log_{10}[1.0 \times 10^{-7}] = 7 \tag{11.9}$$

If there are more H_3O^+ ions, then there is more hydrogen ion (H^+) activity, the pH will be less than 7 and the solution is called *acidic*. If there are more OH^- ions in water, the pH will be greater than 7 and the solution is called *basic* or *alkaline*.

By itself, water forms only a very small number of H_3O^+ and OH^- ions but when other acids or bases are added, the concentrations of the H_3O^+ and OH^- ions in the water changes. There are many naturally occurring acids, such as carbonic acid (H_2CO_3), which forms wherever carbon dioxide (CO_2) is in contact with water, and sulfuric acid, which is found wherever there is volcanic activity. Due to the presence of CO_2 in the atmosphere, the typical pH of natural rainwater is around 5.66. However, many human activities add carbon dioxide, sulfur, nitrogen oxides, and other compounds to the atmosphere that acidify rainwater. Rain with a pH lower than 5.66 is considered *acid rain*. The more a surface body of water deviates from the 6–8 pH range, the more an aquatic ecosystem is affected. Typical pH values for a variety of environments are in Table 11.4.

As you might already have guessed, pH is temperature-dependent, and changes in temperature alter the point of equilibrium. At $60\,°C$, the neutral point is 6.51, while at $10\,°C$ it is 7.26. This means that pH testing should be done immediately after a water sample is collected. As with the temperature test, the water samples used for pH testing should be collected away from the shore and below the surface, and groundwater testing should be performed with a downhole probe. This also means that pH measurements taken in the field must be corrected for temperature, but fortunately most pH probes do this automatically.

A typical procedure for taking a measurement is to insert a calibrated pH probe into a water sample so that the electrodes are completely submerged. Gently stir the probe to make sure that there are no air bubbles trapped. Continue to stir the probe for two minutes and then read the display. Keep stirring for another two minutes and read the display again. The two measurements should not vary by more than 10%. If the difference is greater than 10%, keep stirring the probe and

Table 11.4 Some typical values for pH and EC.

Type of water	Typical pH ranges	Typical electrical conductivity (µS/cm)
Rainwater	4–7	2–100
Freshwater streams and lakes	6.5–8.5	50–1500
Groundwater	6–8.5	50–50,000
Oilfield brines	Near neutral	up to 500,000
Ocean water	7.8–8.4	42,000–53,000
Acid mine drainage	<5.5	~>500
Wetlands and bogs	6.5–9	200–2000

Source: Modified from Sanders, 1998.

taking readings at two-minute intervals until two sequential readings are within 10% of each other, and then record the final reading.

Electrical conductivity (specific electrical conductance) (EC) is a measure of the ability of water to conduct electricity. In general, the higher the concentration of dissolved salts in the water, the easier it is for electricity to pass through the water. Conductivity is reported in microsiemens per centimeter (µS/cm). EC is temperature-dependent, so most meters automatically perform a correction to report in (µS/cm) at 25 °C. Pure water has a very low EC, less than 0.1 µS/cm, and water roughly >500 µS/cm is generally considered unfit for drinking. The measurement procedure requires an EC probe and is essentially the same as for pH. Typical values for EC are shown in Table 11.4.

Turbidity relates to the amount of particulate matter that is suspended in water. High turbidity makes it difficult to treat water for microbial contaminants, and therefore is regulated in drinking water supplies by the EPA. Turbidity is measured using an instrument called a *portable turbidimeter*. The turbidimeter measures the light transmittance of a sample in NTUs (nephelometric turbidity units, a standard measure).

These are just a few of the different types of measurements of water quality that are taken in the field. Others include hardness, Eh (oxidation-reduction potential), and DO (dissolved oxygen). All of these measurements are used as indicators for monitoring the health of aquatic systems and for determining whether a particular body of water has been compromised or can be used as a drinking, agricultural, or industrial water supply.

Water Contamination

Under the provisions of the CWA, the SDWA, and other environmental laws, the EPA has established water quality criteria not only for physical parameters like pH and turbidity but also MCLs, legally enforceable limits, for a range of biological and chemical pollutants as well. The criteria vary, depending on the designated uses of the reservoir of water in question. Drinking water supplies and potential drinking water supplies have the strictest standard, while water bodies designated for recreational, agricultural, or industrial use have less stringent standards. More information about US water quality standards and criteria can be found on the EPA Office of Water website (http://www.epa.gov/ow).

The European Union's water quality standards can be found on the European Commission Environment website (http://ec.europa.eu/environment/water/). Additional European water quality information can be found at the WISE (Water Information System for Europe) website (http://water.europa.eu/en/welcome).

A wide variety of different types of contaminants can be found in water, each of which will behave differently in the environment and have different requirements for sampling and analysis. A few of the more important classes of water pollutants are as follows.

Dissolved inorganics are substances that are of mineral origin and do not typically contain carbon. Calcite, dolomite, and other mineral forms of carbon are considered inorganic. Typical inorganic constituents of water include dissolved ions of calcium, sodium, magnesium, and other elements, as well as trace metals, such as arsenic, mercury, and strontium. Because these substances are dissolved in the water, they will migrate in the direction of flow. Most inorganics have both natural and anthropogenic sources. Some, like calcium, are necessary nutrients that only become objectionable at high levels of concentration, while others, like mercury, can render water unfit for consumption at very low levels.

Nutrients at first glance might not seem like a class of contaminants, but in this case too much of a good thing can cause significant damage to aquatic ecosystems. In fact, non-point source runoff of nutrients is one of the top causes of water quality degradation. The addition of too much dissolved nitrogen and phosphorous, the two nutrients of primary concern, causes the algae (and/or other microorganisms) in the water to become highly productive, resulting in an *algal bloom*. The rapid proliferation of algae uses up most of the available oxygen in the water, creating dead zones, where most other aquatic life cannot survive. Water bodies in this state, with less than two parts per million oxygen, are technically referred to as *hypoxic*. Excess nutrients come from a variety of sources, such as agricultural and residential overuse of fertilizer, agricultural runoff containing animal waste, overflows from septic systems, storm-sewer overflows, and discharges from water treatment plants.

Organic compounds are naturally occurring or synthetic substances that contain carbon, excluding allotropes of carbon (diamond, graphite, etc.), carbonates, and carbon oxides. Some are naturally occurring, such as petroleum, which is a mixture of organic hydrocarbon compounds, but most of the organic compounds of interest as water pollutants are synthetic. Organic compounds can be further subdivided into a few important categories:

- *NAPLs* (non-aqueous phase liquids) are organic compounds that have very low solubility in water, literally mixing "like oil and water." They are of interest as water pollutants because, though they have low solubility, the concentration of the dissolved phases are still many times higher than the MCLs set by the EPA. For example, TCE, a commonly used solvent, has a solubility in water of 1100 parts per million but the MCL is 0.005 parts per million. NAPLs will collect and migrate in the subsurface as a separate liquid phase, creating long-term groundwater pollution sources that can last for centuries. After a spill or leak, NAPL will flow under the force of gravity, migrating downward into the vadose zone where some of it will collect in the void spaces. The disconnected globules will partially volatilize into the soil air so that some of the spill will migrate to the surface as a gas. The remainder of the liquid spill will continue moving down until it hits the capillary fringe. At this point its behavior depends on the density of the liquid. LNAPLs are *lighter* than water (<1 g/cm³) and will float on the water table, generally moving in the direction of the steepest hydrologic gradient (Figure 11.12). Common LNAPLs are gasoline and fuel oil. DNAPLs are *denser* than water (>1 g/cm³) and will sink through the water column to settle on an underlying low-permeability layer. DNAPL contamination source zones can be extremely difficult to locate because the bulk fluid will follow the subsurface topography of low-permeability layers, which is usually unknown and has nothing to do with the groundwater flow direction (Figure 11.13).

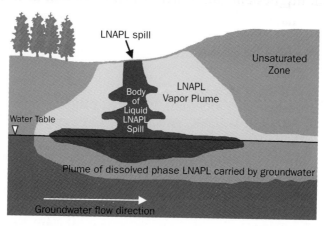

Figure 11.12 Behavior and distribution of a light non-aqueous phase liquid (LNAPL) contaminant in the subsurface.

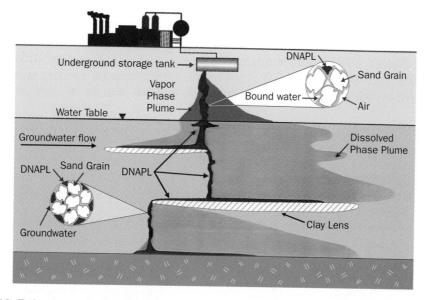

Figure 11.13 Behavior and distribution of a dense non-aqueous phase liquid (DNAPL) contaminant in a granular aquifer system showing the development of several fingers due to heterogeneities in the sedimentary layers. DNAPLs will follow bedding and bedrock topography.

- *Pesticides* are organic compounds designed to be very toxic to insects, animals, and plants. Though they are intended for use against specific targets, they typically are toxic to all forms of life at low concentrations.
- *Volatile organic compounds* (VOCs) are organic compounds that technically are defined as having a boiling point less than or equal to 100 °C and/or a vapor pressure greater than 1 mmHg at 25 °C, but more simply put are substances that evaporate readily at 20 °C. This includes a wide range of common organic compounds such as the components of gasoline, most industrial solvents, and dry-cleaning fluids. The EPA uses the term VOC slightly differently, however. It defines a VOC as "any organic compound that participates in atmospheric photochemical reactions (react in sunlight) except those designated as having negligible photochemical reactivity" (http://www.epa.gov/iaq/voc2.html). Basically, what this means is that the EPA is predominantly interested in chemicals that volatilize into the air and then react to form ozone or smog. Several volatile organic compounds are considered toxic water pollutants, but these do not fit the EPA definition of a VOC.

Pathogens are microorganisms, including bacteria, protozoa, parasites, and viruses, that cause disease in humans, animals, and plants. They can become introduced to bodies of water from agricultural runoff containing animal waste, overflows from septic systems, storm-sewer overflows, and accidental discharges from water treatment plants. Because testing for individual pathogens would be extremely costly and difficult, water is typically monitored for the presence of *indicator organisms* (organisms whose presence would suggest that the water is contaminated) instead. For example, the presence of more than one coliform bacterium per 100 mL of water suggests that the water is contaminated with human feces, and would not be safe to drink.

Radionuclides are radioactive compounds that can be naturally occurring, such as radon and radium-226, or are the byproducts of nuclear weapons testing or other nuclear activities, like tritium and strontium-90.

A list of monitored drinking water contaminants and their MCL can be found on the EPA website at (http://www.epa.gov/safewater/contaminants/index.html).

The San Mateo Mystery Spill

It was January 2002 and once again mystery oil slicks were appearing off the coast of California, and birds, completely covered in oil, were washing ashore and being found floating helplessly in the ocean. It had almost turned into an annual winter event, probably triggered by the large storms that took place each year in the late fall and winter. The powerful storm currents were thought to be pushing the pollution hiding in the ocean toward the shore. This year, things were weirder than usual, though. The first oiled seabirds were found in late November, followed by a new group that started appearing further south in early December. Then, very strangely, in the middle of December, searchers found struggling pelicans coated with vegetable oil.

By January 20th, globs of oil had been washing up on beaches from Point Reyes to Monterey for eight weeks but there was no obvious oil slick in the ocean. Instead, the oil floated in transient patches as a rainbow-colored sheen, while noxious tar blobs washed up on shore. And birds kept dying. Between late November, 2001, and January 20, 2002, 1254 oil-coated birds — mostly common murres but also Pacific loons, Xantus's murrelets, western grebes and even a few brown pelicans – had been found. Most of them, over 700, died. Worse, scientists estimated that for every bird found 10–15 birds got oiled and died at sea. That meant a more realistic estimate of bird deaths was probably somewhere between 12,500 and 19,000.

Signs pointed to at least three different sources, and the various spills were sampled and analyzed to determine their oil "fingerprint." Crude petroleum is a complex mixture of thousands of *organic hydrocarbons* (compounds formed of carbon and hydrogen), including *alkanes* (chains formed with single carbon–carbon bonds), *cycloalkanes* (three-dimensional structures formed with single carbon–carbon bonds), and *aromatic hydrocarbons* (compounds that contain one or more benzene rings). The proportion of different hydrocarbons varies by source, as does the percentage of sulfur and other impurities. However, in its raw form, crude oil does not have many uses. It must be refined, or separated into fractions based on boiling point. For example, *gasoline* is composed of alkanes from 6 to 10 carbons long (C_6H_{14} to $C_{10}H_{22}$), with boiling points between 70 and 180 °C, and *kerosene* is composed of $C_{11}H_{24}$ and $C_{12}H_{26}$, which boil at 180–230 °C. The manufacturing processes used to refine oil into a final product also leave distinctive signs, making it theoretically possible to connect spilled oil, both raw and refined, back to its source.

For example, in 1998 the *Command*, which had a crack in its tank, spilled a small amount of fuel in Half Moon Bay while refueling. The Coast Guard ordered the ship to undergo repairs, but instead the ship took off for the southbound shipping lanes, leaking 4000 gallons of fuel as it went. The Coast Guard chased the ship all the way down the coast, intercepting it near Guatemala. The leaked fuel contaminated 15 miles of beach in San Mateo County alone, and oil fingerprinting was able to clearly demonstrate that the source of the fuel was the vessel *Command*.

There were several potential sources for the oil for the 2001–2002 mystery oil spill along the California coast. More than 1000 ships a day use the nearby shipping lanes and it has been estimated that nearly 70% of the oil pollution in the ocean comes from the routine legal operation of tankers and freighters. Oil is often spilled during loading, unloading, and refueling. Ships often leak fuel from their tanks or their cargo holds. International law also allows ships to pump diluted oily water overboard when 200 miles from land. The strong winter storm currents push these oil slicks toward shore each winter and they can sometimes coat everything in their path with tar-like goo. Passing freighters and tankers clean out oily bilge and fouled fuel tanks and these ships are known to sometimes illegally dump the wastewater without proper dilution and/or too close to shore. Plus, there are several sunken ships that leak oil from their tanks and holds. Offshore drilling operations occasionally have blowouts, and spills of raw crude can even come from naturally occurring oil seeps or leaking oil pipelines.

Given the myriad possibilities, researchers began to collect and test as many spill and source samples as possible, employing a technique known as *chemical fingerprinting*. The basic idea of chemical fingerprinting is that mixtures of compounds, such as crude oil, industrial solvents, and gasoline, have distinctive characteristics. These can be special additives, the ratios of specific compounds in a mixture, their content of different types of biological markers, their stable isotope ratios, and/or specific types of degradation products generated as the original product breaks down in the environment. Each load of oil, batch of a product, or collection of wastes will be unique to some degree, so this approach allows researchers to link spills and disconnected slugs of contamination back to specific sources.

Researchers investigating the San Mateo mystery spill found that the dominant oil in the 2001–2002 season was not a highly refined product, nor was it Alaskan crude, though it was pretty close to other types of crude oil. There were no buried pipelines in the area and the oil did not match the natural seeps from the Monterey Formation, so they could not be the source. Instead, the tar balls' composition led researchers to believe that it could be a variety of *bunker fuel* (the thick, sludge-like fuel used by ships) mixed with a kerosene-like diluent (i.e. a thinning agent). Researchers had a breakthrough when they found that the majority of the samples collected turned out to have the same chemical fingerprint as tar balls sampled during the 1997–1998 mystery spill, which killed more than 2000 birds around Point Reyes. In fact, a variety of area mystery oil spills, dating as far back as 1992, appeared to have come from the same source.

The fact that most of the tar balls had the same fingerprint as found in past events strongly suggested the source was a sunken vessel, since the composition of oils in working ships changes substantially over time. Each time a ship takes on a new load or refuels, the composition of the oil in the tanks is altered, sometimes making it difficult to catch polluting ships in time to chemically match them with a spill. Since the composition had not changed over such a substantial period, the most probable source was a sunken ship, but which one? One candidate in the area was the *Puerto Rican*, which carried several different types of oil, and exploded, burned, and sank in 1984. Other possibilities were the *Sierra*, a wooden-hulled lumber ship with two diesel engines that sank in 1923, or the *Henry Bergh*, a troop carrier that ran aground in the 1940s.

Attempts to match the chemical fingerprints of the mystery oil spill with known wrecks, like the *Puerto Rican*, were all negative until investigators got a new clue. On December 8, a flight team spotted an oil slick approximately seven miles long, southeast of the Farallon Islands. The slick dissipated down to three miles by the next day and disappeared shortly afterward. The oil slick was also visible in satellite images. Investigators from the Coast Guard, the California Office of Spill Prevention and Response, the National Oceanic and Atmospheric Administration, and the State Lands Commission met and compiled records of shipwrecks, coming up with a total of some 1500 known sunken ships off the Californian coast. They had the onerous task of wading through all of them to try to find the culprit. In late January or early February, researchers came across records of the *SS Jacob Luckenbach*.

The *Luckenbach* was a 4615-ton freighter hauling automotive parts and supplies to Korea during the Korean War. On July 14, 1953, the *Luckenbach* left port in San Francisco, headed out into the fog and, oddly enough, collided with its sister ship, the *Hawaiian Pilot*, which was headed inbound with a load of sugar, molasses, and pineapple. According to a Coast Guard investigation, the accident was caused by negligence. Both ships actually were aware of each other, but neither took evasive action so the two ships collided. No one was killed and within 30 minutes the *Luckenbach*, loaded with an estimated 457,000 gallons of oil in its fuel tank, sank not far from the Farallon Islands in the general area where the oil slicks had been spotted in December 2001.

Investigators lowered a remote-controlled submersible with a video camera down to the wreckage and found that the ship had split into three sections. They also collected samples of oil from the wreck and sent it to two different laboratories for analysis. Both confirmed that the oil was a match for the mystery oil spills that had been affecting California's coast for years. Due to negligence over 50 years ago, tens of thousands of seabirds had been killed, miles of coastline polluted, and over $6 million dollars spent just to prevent future leaks. For more information on this topic, *Oil Spill Environmental Forensics*, edited by Wang and Stout (2007), is recommended.

As this chapter was being written, the topic of oil spills suddenly took on even greater poignancy. On April 20, 2010, a large explosion occurred on the *Deepwater Horizon*, a deep-ocean drilling rig owned by Transocean, but under lease to BP. The rig was located in the Gulf of Mexico approximately 40 miles offshore of Louisiana. Eleven workers were killed in the explosion. On April 22, the fire on the rig was finally extinguished as the rig crumpled and sank beneath the water. In the process, the well on the ocean floor, located over 1.5 km (almost a mile) below sea level, was extensively damaged and started leaking a substantial quantity of oil into the waters of the Gulf. The leak lasted pretty much unabated for approximately 86 days until July 15, when a tight-fitting cap was put into place. Estimates for the total amount of oil spilled range widely and are the subject of much controversy because some of the fines BP and others will have to pay are based on the total amount of oil spilled. However, it is pretty clear that this has become the largest oil spill in US history (though not the largest spill in the world).

In mid-May 2010, several tar balls washed up on shore in Florida, but tests demonstrated that they were not from the *Deepwater Horizon*. Tar balls that have washed up since have been identified as coming from the *Deepwater Horizon*. To date, sections of shoreline in all of the Gulf states have been affected. Differentiating the impact of the *Deepwater Horizon* spill from other spills that have occurred in the Gulf will clearly be an ongoing issue in environmental forensics. Given that some portion of the oil has probably reached the Gulf Loop Current, which flows around the coast of Florida and into the Atlantic Ocean, the likelihood is that scientists will be using chemical fingerprinting methods for years to come in order to determine whether tar balls or other byproducts that wash up along coastlines have the *Deepwater Horizon* as their ultimate source.

Analytical Techniques for Chemical Fingerprinting

Gas chromatography is one of the most commonly employed analytical techniques used for the identification of organic compounds and is a mainstay in chemical fingerprinting. The basic theory behind chromatography was described previously in Chapter 9. The difference here is that gas chromatography uses gas as the mobile phase rather than a liquid. In this case, a very small volume of a sample liquid or gas is injected into the *gas chromatograph* (GC) either manually using a special syringe or via an autosampler. Inside the heated injection chamber, the sample is vaporized and mixed with an inert carrier gas, such as nitrogen, helium, hydrogen, or argon. The gas transports the sample through a long *column*, which acts as the stationary phase and allows separation of the sample into its components. The different molecular components of the sample will partition between the mobile phase (carrier gas) and the solid phase (the column) based on their size and affinity for the stationary phase.

Different types of columns are used, depending on the compounds of interest. The result is that smaller molecules, like light alcohols, will move through a column quickly, while large compounds, like PCBs, will do so much more slowly. The separated components of the sample are eventually swept by the carrier gas out of the column and into a detector, which generates an electrical signal proportional to the amount of a compound present. The detector response is plotted as a series of peaks versus time to create a chromatogram (Figure 11.14). Using software, the time position of the peak can often serve to identify the compounds present in the sample, while the height of the peak can be translated into a quantitative amount.

There are a variety of different detectors for GCs, each of which takes advantage of a unique characteristic of a molecule and uses that characteristic to generate a measurable electrical signal. One of the most common is a *flame ionization detector* (FID). As the sample exits the column, it is mixed with hydrogen and air before flowing out of a stainless steel jet. The mixture is ignited at the tip of the jet and

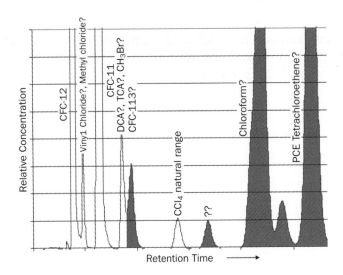

Figure 11.14 Gas chromatogram showing an example of the separation and peak shape of a variety of common groundwater contaminants.
Source: Courtesy of the United States Geological Survey.

the organic compounds burning in the flame are ionized. The ions are attracted to a metal electrode that is located near the flame, thereby generating an electric current. The current is amplified into the millivolt range and this signal used to create the chromatogram. FID is sensitive to almost all molecules that contain hydrocarbons, such as the components of gasoline, other VOCs, and PCBs, and can detect compounds in the low-part-per-billion to high-part-per-trillion range. The downside to this process is that the FID destroys the sample. Other types of detectors include the *photoionization detector* (PID), which consists of a special ultraviolet lamp mounted on a low-volume flow-through cell, the *electron capture detector* (ECD), and the *electrolytic conductivity detector* (ELCD), also known as a Hall detector, which is a halogen-specific detector that operates on electrolytical conductivity principles. Some types of detectors are nondestructive and can be used in series with FID in order to get more detailed information about a sample.

GCs can also be incorporated with mass spectrometers (GC/MS). A *mass spectrometer* separates charged atoms and molecules based on their masses and their behavior in electrical and/or magnetic fields. The gas effluent from the GC is carried directly into the MS and bombarded by a stream of electrons, which ionize the constituents of the sample. The positively charged ions will have the same molecular weight as the complete molecules, but can be accelerated by a high-voltage field and directed toward a *collimator*, a series of spaced slit plates, that shapes and directs the beam toward a mass analyzer. The ion beam enters an electromagnetic field that deflects the ions into specific circular pathways. The lighter ions are deflected less than the heavier ones, so they follow a shorter pathway. The sorted ion beams are directed to re-converge as they leave the magnetic field and the positively charged ions are then focused toward a mass analyzer that differentiates the ions according to their mass-to-charge ratios.

This means that a mass spectrometer will not only identify specific compounds but also differentiate between the isotopes of a given element, allowing for a comparison of isotopic ratios to be made. The output from the detector is displayed as a *mass spectrum*, which usually takes the form of a vertical bar graph that relates the intensity of the mass peaks to their mass-to-charge ratio. The height of each peak is related to the relative abundance of the identified isotope. Mass spectra are compared to a database for identification based on the general fragmentation pattern and the peak ratios. The combined output of a GC/MS includes the chromatogram from the GC and the mass spectra (Figure 11.15) for each chromatographic peak, making GC/MS a very powerful analytical tool for the positive identification of compounds and for the quantification of the different organic components of a mixture.

Isotopes in the Environment

Isotopes are atoms of the same element that have different atomic masses because they have different numbers of neutrons. Most elements have multiple isotopes, some of which are stable and some of which are radioactive. Radioactive isotopes spontaneously *decay* or breakdown over time to form new elements until they reach a stable state. Both types of isotopes have utility as environmental indicators and have a variety of potential forensic uses.

Stable light (atomic mass <40) isotopes are of interest because they can fractionate in the environment. *Fractionation* is the partitioning of isotopes between phases or products during physical, chemical, or biological processes. With heavy

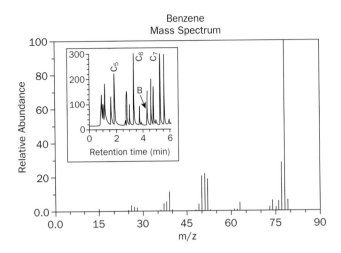

Figure 11.15 Combined output of a GC/MS includes the chromatogram from the GC (inset) and the mass spectra for each chromatographic peak, in this case the mass spectra for the benzene peak (labeled "B") in the chromatogram is shown.
Source: Mass spectra courtesy of the NIST Chemistry WebBook (http://webbook.nist.gov/chemistry).

isotopes (atomic mass >40), the percentage mass difference between isotopes of the same element is not significant, so they will behave identically in the environment. However, the lighter the isotope, the greater the percentage mass difference becomes, leading to slight differences in chemical behavior. Light isotopes vibrate at higher frequencies and as a result are less strongly bonded to other atoms than their heavier isotopic counterparts are. Natural processes in the environment exploit the variations in bonding strength, resulting in measurable differences in isotopic concentrations in different reservoirs.

For example, there are two stable isotopes of hydrogen (^{1}H and ^{2}H), and three stable isotopes of oxygen (^{16}O, ^{17}O, and ^{18}O), though one (^{17}O) is so rare that its presence is generally ignored. This means that there are six common possible isotopic permutations of water – ^{1}H$_2$^{16}O, ^{1}H^{2}H^{16}O, ^{2}H$_2$^{16}O, ^{1}H$_2$^{18}O, ^{1}H^{2}H^{18}O, and ^{2}H$_2$^{18}O – that range in molecular mass from 18 to 22. It takes more energy to move a heavier molecule into gas phase than a lighter molecule, so when water evaporates the molecules with more ^{1}H and ^{16}O will tend to move into the atmosphere more readily, leading the vapor phase to become enriched in the lighter isotopes. Thus, the water vapor in the atmosphere over the ocean will be isotopically lighter than the ocean water from which it evaporated.

Some of this oceanic water vapor will move over landmass, cooling and condensing to form clouds. In this processes, the heavier isotopes are more likely to condense into the water phase, so the precipitation that falls from these clouds will contain more ^{2}H and ^{18}O, i.e. be enriched with heavier isotopes. The precipitation that falls from these clouds in coastal areas will be enriched with heavier isotopes, leading the clouds themselves to become even more enriched with lighter isotopes and increasingly depleted in heavy isotopes. As the moist air mass moves inland, the isotopic signature of the precipitation gets lighter and lighter as the heavier isotopes get rained out.

The partitioning of isotopes between substances or phases is described by its *isotopic fractionation factor* (α), defined as:

$$\alpha = (R_A)/(R_B), \tag{11.10}$$

where R_A is the ratio of the heavy isotope to the light isotope in phase (or molecule) A, and R_B is the ratio of the heavy isotope to the light isotope in phase (or molecule) B. The fractionation factor for any given pairing of isotopes is constant at a given temperature. The higher the temperature, the more energy is available, so the less fractionation will occur. The lower the temperature, the greater will be the degree of fractionation. Since temperature varies roughly with latitude, less fractionation occurs near the equator while more occurs toward the poles. This means the water that rains out in equatorial coastal areas will have isotopic ratios closer to that of ocean water, while the nearer you move to the poles, the lighter the isotopic signature of the water vapor will become.

The combined dynamics of fractionation and temperature variations around the globe means that different reservoirs of water on the surface of the Earth will have different stable isotopic signatures. Contact with geologic media can also leave a distinctive component to a reservoir's isotopic signature. Environmental researchers use the stable isotopic signatures of water bodies to answer a wide variety of questions. For example, researchers can determine whether the water feeding a wetland, lake, or stream is mainly from precipitation or due to infiltration from the underlying groundwater system. They can determine whether the water in a stream is coming from high altitude snowmelt off an inland mountain or from sources nearer the coast. Studies of isotopic signatures can even be used to establish how much mixing from different sources occurs.

Because the isotopic variations measured are very small, they are usually reported in *delta notation*, as parts per thousand (‰) versus a relevant standard, calculated as follows:

$$\delta = [(R_A - R_{standard})/R_{standard}] \times 1000, \tag{11.11}$$

where R_A is the heavy to light isotopic ratio of sample A and $R_{standard}$ is the heavy to light isotopic ratio of a relevant standard. For example, for oxygen, R is the ratio of ^{18}O to ^{16}O. There are several different standards used depending on the element(s) of interest. For hydrogen and oxygen, SMOW (standard mean ocean water), and now V-SMOW (Vienna standard mean ocean water), are the standards most commonly used. The delta value of oxygen would be determined using:

$$\delta^{18}O \ (‰) = \{[(^{18}O/^{16}O)_{sample} - (^{18}O/^{16}O)_{SMOW}]/(^{18}O/^{16}O)_{SMOW}\} \times 1000 \tag{11.12}$$

Positive $\delta^{18}O$ values indicate that the sample is enriched in ^{18}O relative to the standard, while negative values indicate that the sample is depleted in ^{18}O relative to the standard. For example, if a sample yielded an $\delta^{18}O$ value of -10 ‰, it would mean that the water was depleted in ^{18}O by 10 parts per thousand relative to the standard. Similar standards and relationships have been established for other stable isotopes, each of which has applications to different types of geologic and environmental investigations. The other stable isotope ratios most commonly used for environmental work are $^7Li/^6Li$, $^{11}B/^{10}B$, $^{13}C/^{12}C$, $^{15}N/^{14}N$, and $^{34}S/^{32}S$.

$\delta^{18}O$ and δD (where D means deuterium, another name for ^2H) ratios have been used extensively in studies of paleoclimate and for archeological investigations of populations. It turns out that the $\delta^{18}O$ values found in bioapatite are related to the $\delta^{18}O$ of local bodies of water, specifically local sources of drinking water. Archeologists have been using oxygen, hydrogen, and other light isotopes for tracking migration patterns in ancient populations. Generally speaking, the light isotopic signature of an organism can be used to ascertain the relative latitude, in comparison to specific watersheds, at which an organism lived and to elicit information about its diet. Theoretically, this information can be used to study modern populations, though the advent of bottled water, treated tap water, and imported foodstuffs has complicated the relationship in complex and currently poorly understood ways.

Radioactive Isotopes

Radioactive isotopes are most commonly used for determining the relative or absolute ages of materials. The rate at which radioactive materials decay is constant over time, technically an example of first-order kinetics, and can be expressed as:

$$N = N_o e^{-\lambda t},$$
(11.13)

where N is the number of radioactive *parent atoms* (i.e. atoms in the original state) that remain at time t, N_o is the initial number of parent atoms that were present at time t = 0, e is the exponential function, and λ is the radioactive decay constant of the isotope. It is simple enough to rearrange Equation 11.13 to solve t, but there is still a difficulty. We do not know directly how many parent isotopes were in our sample when it formed. Therefore, we need to amend this relationship slightly to include only terms that we can measure directly.

Each time a radioactive parent isotope decays, it produces a *daughter product*. Daughter products can be either stable or radioactive and are termed *radiogenic isotopes*. In the simplest case, at time t = 0, there would be no daughter product and over time, as the parent isotope decays, daughter product is produced, i.e. $D = N_o - N$. Substituting this relationship into 11.13 creates the expression:

$$D = N(e^{-\lambda t} - 1),$$
(11.14)

where both D and N can be measured using a mass spectrometer. Equation 11.14 can also be rearranged to solve explicitly for time:

$$t = 1/\lambda \ln(D/N + 1)$$
(11.15)

Both expressions can also be corrected for more complex situations, such as when there is some amount of daughter product incorporated during the formation process.

Many of the minerals that form on Earth incorporate enough radioactive material that *radiometric dating methods*, i.e. analysis of the ratios of specific isotopes, can be used to determine their age. Applying radiometric dating methods to the minerals that form igneous rocks will yield the age of the rock from the point that it solidified. For metamorphic rocks, radiometric dating methods will determine the time since metamorphism ceased. With sedimentary rocks, things are a bit more complicated. The ages determined for clastic sedimentary materials would be the dates at which the individual clasts formed, not the date of formation for the sedimentary rock. There are a few rare minerals that form during some sedimentary processes, like glauconite, which can be dated using radiometric techniques, but more often than not, sedimentary rocks cannot be given absolute ages. This is why the relative dating methods discussed in Chapter 10 are still so important.

For environmental and forensic work, the actual age of the rock is not of primary interest. Instead, we can take advantage of how the ratios of isotopes in different bodies of rock vary in order to track materials back to their original source zones. This works because different bodies of rock often have significantly different isotopic and radiogenic compositions based on their age, original chemical composition, and subsequent history. It is easy to see how a chunk of rock broken out of a geologic layer will have the same radiogenic signature as its parent rock. This also holds true, with certain caveats, of the sediments and soils formed from the parent rock. In fact, it holds true for much more complex processes as well.

If you remember from Chapter 10, humans, other animals, and plants all acquire a trace elemental signature from contact with their local geologic and anthropogenic environment. This elemental signature includes acquisition of the isotopic ratios of materials from their local environment. Remember that while light isotopes fractionate in the environment heavy isotopes, which includes virtually all isotopes used for radiometric dating except carbon-14, do not. So the specific radiogenic isotopic signature of a body of rock can be transferred into the environment and even into a person. Radiogenic isotopes that do not undergo fractionation, like lead, strontium, and neodymium, are commonly used as tracers. Theses isotopes accumulate in bones, teeth, and other tissues, offering the possibility of providing a link between an organism and the area where that organism lived.

Stable and radiogenic isotopes can be used in a variety of different ways to trace pollutants through the environment, to study the behavior of different reservoirs over time, and to link materials and organisms to their points of origin. For a more detailed discussion of the theory and application of stable and radioactive isotopes, see Faure and Mensing (2004). For a more detailed discussion of how isotopes are used in hydrogeology see Clark and Fritz (1997). Finally, for a discussion of how isotopes can be applied forensically, see Pye (2004).

Ötzi the Iceman

On September 19, 1991, two hikers spotted a body lying in the melting ice of a glacier in the Ötztal Alps located between Italy and Austria. Authorities initially believed the body to be the modern corpse of a hiker who had gone missing on the mountain years ago, but after a difficult excavation it started to become clear that the discovery was something else entirely. After their removal and transport to the Institute of Forensic Medicine in Innsbruck, the body and associated artifacts were examined by archeologists. They tentatively announced that the ~46-year-old man had lived at least 4000 years ago, making the discovery an international sensation. Radiocarbon dating was later used to determine that Ötzi, as the body was dubbed, had lived sometime between 3350 and 3100 BCE during the Neolithic-Copper Age of Europe, over 5100 years ago. This made Ötzi the oldest natural human mummy that has ever been discovered in Europe and one of the oldest mummies in the world.

To help establish Ötzi's point of origin and the extent of his travels, researchers preformed stable (O and C) and radiogenic (Sr and Pb) isotopic studies of his tooth enamel, which archives childhood elemental exposures, and his dentine, cortical, and trabecular bone, which reflect adult elemental exposures (Müller et al., 2003). Stable oxygen isotope ratios were used to deduce information about the watershed source of Ötzi's drinking water. Researchers took water samples from several rivers in the valleys surrounding the discovery site. Assuming that weather patterns had not changed drastically, their isotopic signatures could be used to help establish where Ötzi had lived.

The rivers north and south of the discovery site showed distinctive differences. The northern watershed areas receive precipitation from air masses that have traveled a great distance from the cool Atlantic Ocean, while the watersheds to the south receive precipitation that comes from the nearby warm Mediterranean Sea. This means that the water in the south is less depleted in $\delta^{18}O$ than the water in the north. Plus, due to snowmelt, there is an overall reduction in $\delta^{18}O$ values across the region toward the west. After correcting for metabolic fractionation, researchers found that the $\delta^{18}O$ of the Iceman's enamel was closest to that found in the southeast of the research area (Müller et al., 2003). Based on this, they suggested he had spent his childhood in the south, probably somewhere near the confluence of the Eisack and Rienz rivers. Ötzi's cortical and trabecular bone samples showed slightly higher levels of $\delta^{18}O$ depletion, indicating that he had migrated during his adulthood to another area, though still in the south, such as the Etsch valley.

To complement the stable isotopic study, researchers used an analysis of radiogenic isotope ratios to link Ötzi to specific geologic environments. The potential areas of habitation included four lithological units that could be distinguished isotopically: Mesozoic limestone, Permian rhyolite, Eocene basalts, and a heterogeneous group of phyllites and gneisses. The enamel from Ötzi's teeth was characterized by high $^{87}Sr/^{86}Sr$ ratios consistent with the composition of the phyllites and gneisses, suggesting that he spent his childhood in the Schnals valley or the Ulten, or Adige/Etsch valleys (Müller et al., 2003). These results were consistent with the results of the $\delta^{18}O$ study, and taken together allowed researchers to suggest that the Iceman had spent his childhood somewhere in the Eisack valley, though they listed other possible sites as well. Ötzi's bones showed lower $^{87}Sr/^{86}Sr$ ratios, indicating that he had migrated to another region during the last one or two decades of his life, which is again consistent with the results of the $\delta^{18}O$ study. The researchers had more difficulty in narrowing down where the Iceman had lived during adulthood, possibly because he did not live in a permanent settlement but instead migrated seasonally.

Seeking Justice for Adam

On the afternoon of September 21, 2001, a commuter walking south across Tower Bridge in London spotted a small body floating in the Thames and immediately called the police from his cell phone. Police constables recovered the body about a mile downriver approximately 20 minutes later. The facts of this case are unfortunately quite graphic and will be avoided here, but the child was very clearly not the victim of an accident. At a postmortem conducted the following day, a doctor established that the body was that of a well-nourished boy and estimated that he was 5 to 7 years old. The investigating officers thought that they would soon be able to link the child with a missing person's report, but months passed with no luck and, further, they determined that the circumstances of the child's murder were unique in the United Kingdom.

By January 2002, investigators believed that they had exhausted what conventional forensic techniques could tell them. The child, whom investigators had named Adam, was still unidentified and a £50,000 reward, the highest that had ever been offered at that time, remained unclaimed. Following a special meeting of forensic scientists and senior detectives, investigators made the decision to start employing more cutting-edge techniques in order to determine where Adam had been born. A variety of forensic specialists volunteered their services.

Much interesting work was done using different DNA investigation techniques, but it turns out that an environmental geologist played a vital role as well. Dr Ken Pye, a professor of environmental geology from Royal Holloway, University of London, and head of a private consulting firm, performed an isotopic trace element analysis on samples of Adam's bone and tissue. Based on strontium and neodymium isotopic ratios, Dr Pye was able to determine that Adam had spent all of his life, other than the last few weeks or so, on or close to exposures of Precambrian rock in West Africa. He was able to narrow Adam's home down to three specific regions. First, was an area that stretched from north of Accra in Ghana, east through Togo and Benin, and into Nigeria, where the rock formed the Yoruba Plateau. The second area was the Jos Plateau in Nigeria and the third was an area that bordered Cameroon on the eastern side of Nigeria. At the time, the geologic and soil map data that were available on Nigeria were of an inadequate quality to further isolate Adam's point of origin.

Dr Nick Branch, also from Royal Holloway, performed a palynological study on Adam's remains. The pollen assemblage he found had largely been produced by the combination of grasses and cultivated cereals native to northern and northwestern Europe that would have been growing in early September. Layla Renshaw, senior lecturer at Kingston University, performed a separate isotopic analysis of skin and fat samples in order to determine how much time Adam had spent in the United Kingdom. She was able to conclude that Adam had undergone a significant dietary change approximately four weeks prior to his death. These facts strongly suggested that Adam had been killed in the London area and that he had been brought there only a few weeks before his murder.

In order to make more progress on the investigation, a team from the United Kingdom traveled to Nigeria in November 2002. Because there was insufficient reference material available, they had decided to collect soil, rock, and bone samples themselves. The investigators hoped to be able to narrow the search area down enough to be able to locate the child's family. For three weeks, they traveled over 17,000 km and collected more than 150 samples, plotting the exact location of each collection point using GPS systems. The samples were then analyzed by Dr Pye, who was able to conclude that Adam had come from somewhere in a small area of southwest Nigeria, south of Ibadan, north of Lagos, and bordered in the east by Benin City. Based on information gathered, Benin City was thought to probably be Adam's home, though he could have come from anywhere in this small region.

Dr Branch also discovered that though Adam's stomach was empty, indicating a period of fasting, there was an odd collection of foreign matter in the child's lower intestine. Dr Pye analyzed the material, also taking SEM images, and found that it included angular fragments of calcium phosphate (possibly ground fragments of bone), weathered quartz grains, pellets of clay that contained metallic particles, chiefly of gold, carbon particles with traces of heavy metals, and remnants of a highly poisonous bean indigenous to the Calabar region of West Africa. Collectively, the material was tropical in origin and had been feed to the child some time before his death.

In an attempt to locate Adam's family, an investigative team visited the region of Nigeria identified by Dr Pye in February 2003. For two weeks, they visited villages, towns, and cities, plus they offered a reward to encourage anyone with information to come forward. They heard about other children going missing, but there were no official records to view and investigators received strangely little information during the trip. Adam was buried in an unmarked grave in 2007, but, while Adam's case remains unresolved, the techniques pioneered to help police investigate his murder have since been used to solve other cases. The ability to so narrowly pinpoint where an unidentified child, who could have come from literally anywhere in the world, had lived is a remarkable feat and demonstrates the amazing potential of forensic investigation. Meanwhile, the investigators working on this case still hope to someday put a name to this small child. If you are interested in more information on Adam's case, see O'Reilly (2007).

Summary

This chapter gave just the briefest introduction to a few aspects of the field of environmental forensics, concentrating on water, water contamination, and the use of isotopes. Other major areas of endeavor include investigations of the sources of air pollution and the contamination of soils. The techniques described here have both a direct and indirect application to a variety of forensic endeavors. For a more detailed introduction to the field of environmental forensics, see Sullivan, Agardy, and Traub (2001). For an introduction to environmental law in the United States, see Kubasek and Silverman (2011). A discussion of the legal framework of environmental forensics for the United Kingdom can be found in Hester and Harrison (2008).

References

Clark, I. D. and Fritz, P. (1997) *Environmental Isotopes in Hydrogeology*. CRC Press, Boca Raton, FL.

Faure, G. and Mensing, T. M. (2004) *Isotopes: Principles and applications*, 3rd edn. John Wiley & Sons Ltd, New York.

Fetter, C. W. (2000) *Applied Hydrogeology*, 4th edn. Prentice Hall, Upper Saddle River, NJ.

Harr, J. (1996) *A Civil Action*. Vintage Books, New York.

Hester, R. E. and Harrison R. M. (eds) (2008) *Environmental Forensics: Issues in environmental science and technology*. Royal Society of Chemistry, Cambridge.

Kubasek, N. K. and Silverman, G. S. (2011) *Environmental Law*. Prentice Hall, Upper Saddle River, NJ.

Müller, W., Fricke, H., Halliday, A. N. *et al.* (2003) Origin and migration of the Alpine Iceman. *Science* **302**(October 31): 862–6.

O'Reilly, W. (2007) The "Adam" Case, London. In: T. Thompson and S. Black (eds), *Forensic Human Identification: An introduction*. CRC Press, Boca Raton, FL.

Perry, J. and Vanderklein, E. L. (1996) *Water Quality: Management of a natural resource*. Blackwell Science, Malden, MA.

Pye, K. (2004) Isotope and trace element analysis of human teeth and bones for forensic purposes. In: K. Pye and D. J. Croft (eds), *Forensic Geoscience: Principles, techniques and applications*. Geological Society, London (Special Publications) **232**: 215–236.

Sanders, L. (1998) *A Manual of Field Hydrogeology*. Prentice Hall, Upper Saddle River, NJ.

Sullivan, P. J., Agardy, F. J., and Traub, R. K. (2001) *Practical Environmental Forensics: Process and case histories*. John Wiley & Sons Ltd, New York.

Wang, Z. and Stout, S. A. (eds) (2007) *Oil Spill Environmental Forensics: Fingerprinting and source identification*. Elsevier/Academic Press, Burlington, MA.

Index

Page numbers in **bold** refer to information in figures.
Page numbers in *italics* refer to information in tables.
color plate(s) indicates relevant figure is also located in the color plates section of the book